# Recent Titles in This Series

# Exploiting Symmetry in Applied and Numerical Analysis

# Lectures in
# APPLIED
# MATHEMATICS

## Volume 29

# Exploiting Symmetry
# in Applied
# and Numerical Analysis

1992 AMS-SIAM Summer Seminar
in Applied Mathematics
July 26–August 1, 1992
Colorado State University

Eugene L. Allgower
Kurt Georg
Rick Miranda
Editors

**American Mathematical Society**
Providence, Rhode Island

The Proceedings of the 1992 Summer Seminar on Exploiting Symmetry in Applied and Numerical Analysis were prepared by the American Mathematical Society with support from the following sources: the National Science Foundation through NSF Grant DMS-9121007 and the U.S. Department of Energy through Department of Energy Grant DE-FG02-92ER25131.

1991 *Mathematics Subject Classification.* Primary 20Cxx,35Qxx;
Secondary 58F05,58F14,58G35,65H10.

**Library of Congress Cataloging-in-Publication Data**
Summer Seminar on Applied Mathematics (22nd : 1992 : Colorado State University)
    Exploiting symmetry in applied and numerical analysis : 1992 AMS-SIAM Summer Seminar in Applied Mathematics, July 26–August 1, 1992, Colorado State University / Eugene L. Allgower, Kurt Georg, Rick Miranda, editors.
        p.   cm. — (Lectures in applied mathematics ; v. 29)
    Includes bibliographical references.
    ISBN 0-8218-1134-7
    1. Numerical analysis–Congresses. 2. Symmetry groups–Congresses. I. Allgower, E. L. (Eugene L.) II. Georg, Kurt. III. Miranda, Rick, 1953- . IV. Title. V. Series.
QA297.S846   1992
515′.35—dc20
                                                                                        93-30491
                                                                                        CIP

10 9 8 7 6 5 4 3 2 1     98 97 96 95 94 93

# Contents

# Foreword

Symmetry arises and plays an important role in nearly all branches of science. This is particularly true in theoretical physics, where representation theory has long been a primary tool. Thus it is remarkable that symmetry aspects have only recently been exploited in some branches of mathematics. In applied analysis, the study of symmetry is much more widespread, and in the more traditional area of classical differential equations the understanding of symmetry groups as a central tool has been realized since the last century. The relatively newer area of bifurcation theory has systematically incorporated aspects of symmetry. But the numerical study of bifurcation has rarely been concerned with symmetry. Other areas of numerical analysis usually have considered aspects of symmetry only on an ad hoc basis. However, there is now a growing collection of researchers in numerical analysis who are currently attempting to use symmetry groups and representation theory as a fundamental tool in their work. The time therefore seemed ripe for bringing these varied communities together to exchange ideas and move the field of numerical analysis forward in the aspect of incorporating symmetry and group theoretical concepts systematically into numerical methods.

With this aim in mind, the twenty-second AMS-SIAM Summer Seminar in Applied Mathematics was held July 26–August 1, 1992, at Colorado State University. The seminar was sponsored by the American Mathematical Society, the Society for Industrial and Applied Mathematics, and the Department of Mathematics at Colorado State University. It was supported by grants from the National Science Foundation and the Department of Energy. The conference provided a wide-ranging survey of the exploitation of symmetry in applied and numerical analysis. It had both an entry-level summer school component intended for young researchers and a frontier-level research aspect. About 100 scientists representing numerous countries, universities, and laboratories took part in the conference.

A further purpose of the seminar was to stimulate interaction between aspects of Applied Mathematics (e.g., PDE's, integral equations, bifurcation), Numerical Mathematics (e.g., numerical linear algebra, boundary and finite element methods), Pure Mathematics (e.g., representation theory of groups), and Classical Physics (e.g., Taylor and Bénard problems).

The papers in this volume are mostly more formal versions of some of the lectures which were presented at the conference. However, they have all undergone a standard refereeing process. The papers are in final form and will not be published elsewhere.

It gives us pleasure to acknowledge the help and support of several people who contributed to the success of the conference. We are grateful to our chairman Robert Gaines for his support and encouragement. Ms. Donna Salter of the AMS was of immense help as Conference Coordinator and in her work in the planning and organizing stages. In the preparation of this volume we received excellent help from Ms. Donna Harmon of the AMS. We also wish to express our thanks to the external members of the Organizing Committee for their expertise:

- M. Golubitsky (University of Houston),
- K. Kirchgässner (University of Stuttgart, Germany),
- P. Olver (University of Minnesota).

Finally, we thank the conference participants for their contributions toward its success.

Ft. Collins, June 1993

E. L. Allgower
K. Georg
R. Miranda

Lectures in Applied Mathematics
Volume **29**, 1993

# Hidden Symmetries of Nonlinear Ordinary Differential Equations

B. ABRAHAM-SHRAUNER*
AND
P. G. L. LEACH**

ABSTRACT. Hidden symmetries of ordinary differential equations of type I and type II are analyzed for three equations of physical interest: the modified Painlevé-Ince equation, the Pinney equation and semiconductor transport equations.

## 1. Introduction

The analytical solution of nonlinear, ordinary differential equations (NLODEs) was put on a systematic footing by Sophus Lie[1] and is discussed in many monographs.[2-10] Many ad hoc approaches known before his work were shown to be feasible because of the underlying symmetries of the NLODEs. The symmetries of the ODEs were identified by the condition that the ODEs be invariant under Lie point (continuous) groups. The ODEs invariant under a one-parameter group were shown to be reducible in order by one; those invariant under a two-parameter group were shown to be reducible by two orders if for nonabelian two-parameter groups the group invariants of the normal subgroup were chosen to reduce the original ODE first by one order. Investigations of the invariance of ODEs under multi-parameter groups by Lie showed that the NLODEs invariant under Lie point groups could be reduced to quadratures if the associated Lie algebra was solvable and had dimension equal to or greater than the order of the ODE.[8,10]

Two systematic methods were developed by Lie, one of which is called the classical Lie method, in the use of symmetries to solve ordinary differential equations of both linear and nonlinear varieties. The classical Lie method determines the Lie point groups symmetries of a set of ODEs by an invariance condition. The method also applies to PDEs but we discuss only ODEs here. From the resultant group variable transformations to ODEs invariant under translations in one variable can be

---

1991 Mathematics Subject Classification. Primary 34A34, 22E70.

Supported by the Southwestern Bell Foundation Grant.

This paper is in final form and no version of it will be submitted for publication elsewhere.

found. Such transformed equations can be reduced in order by one if the new variables are the differential invariants of the group. The order of the ODE is reduced by one for each group used. The inverse method, also developed by Sophus Lie, starts with a group as represented by the group generator and calculates the most general form of the NLODE. Tables of the the general forms NLODEs invariant under common Lie groups have been compiled.[3,6]

Neither of these methods finds all the symmetries. In recent years nontrivial applications of contact symmetries, also developed by Sophus Lie, have occurred in the transformation of nonlinear PDEs to linear PDEs.[9,11] Generalized (Lie-Bäcklund) symmetries have been introduced to describe the symmetries of evolution NLPDEs which support solitons and used to find additional solutions.[8,12] Contact and generalized symmetries depend on variable transformations which are functions of the first derivatives and all orders of derivatives respectively. However, there are variable transformations that depend on integrals of the variables. Potential symmetries are examples where the NLPDEs must be put in conservation form.[9,13] Alternatively, inverse differential operators have been developed to treat these nonlocal transformations in a direct fashion but assumptions are made about the form of the inverse operators.[14,15] Nonlocal group generators have been studied briefly.[8] These nonlocal symmetries are evidence that the direct methods for finding Lie point symmetries, contact symmetries or generalized symmetries may be missing some symmetries.

All the present direct methods for calculating nonlocal symmetries have limitations. The introduction of inverse operators is restrictive since the nonlocal variables are chosen to be integrals of the dependent variables integrated over one of the independent variables. From an analysis of nonlocal symmetries of NLODEs we have found that the nonlocal variables may be integrated over an intricate combination of independent and dependent variables. The potential method requires that the PDEs be put in conservation form which is not always possible and make the special assumption that the integration constant is zero. The Direct Method depends on the proper choice of ansatz for the solution. The nonclassical method can be difficult to apply as the right additional symmetry relation must be appended to the original PDEs. However, one may apply the inverse method as has been done for Lie point symmetries and determine general forms of NLDEs. This approach has the advantage that one does not need an a priori assumption about the form of the differential equations or the inverse differential operators. It is well suited to the exploration of the different forms of NLDEs with hidden symmetries, the determination of the types of variable transformations between different reduced NLDEs and the possible forms of the nonlocal group generators. The inverse method for the hidden symmetries of which two types have been identified so far[16,17] possesses some unique aspects not found in the inverse method for Lie point symmetries. For type I hidden symmetries first-order ODEs were investigated first. They were chosen because there is no direct method for finding the symmetries of first-order ODEs that is easily applicable yet the equations for many practical problems reduce to first-order ODEs. The direct method is not useful since the determining equation for first-order ODEs, although a linear PDE, is so complicated that solving it has yielded few solutions. Instead ad hoc guesses and the inverse method for Lie point transformations have been applied to find the symmetries. For type II hidden symmetries the hidden symmetries of second-order ODEs were investigated first by choosing a ubiquitous first-order ODE which was then increased in order by a suitable variable transformation. As the first-order ODE is constructed to be invariant under a one-parameter group, it can be reduced by the direct method to quadratures. The type I hidden symmetries of ordinary differential equations are the symmetries of normal subgroups that vanish when a higher-order ODE is reduced in order by one by the non-normal subgroup differential invariants. The resultant reduced, lower-order ODE possesses the hidden symmetry and is invariant under an extended nonlocal group generator. No general direct method of finding these nonlocal group generators from a given ODE has yet been devised. The type I hidden symmetry was noted in an example by Olver[8] and has been investigated recently. A systematic study of the type I hidden symmetries of first-order ODEs associated with nonabelian, two-parameter projective subgroups has been reported.[16] The solved form, that is the highest-order derivative appears to the first power, of the general second-order ODE invariant under the two-parameter group was first found for each subgroup. These second-order ODEs were then reduced to first-order by the variables which are the differential invariants of the non-normal subgroup. These first-order ODEs lost the point symmetry of the normal subgroup. Alternatively, the second-order ODEs were reduced to separable first-order ODEs by the normal subgroup differential invariants. These first-order ODEs can be reduced to quadratures by the direct method as they are invariant under a known one-parameter group. A nonlocal variable transformation between the variables of these two first-order ODEs enabled a solution to be given for the ODE with the hidden symmetry. The inverse method is more indirect than that used for Lie point symmetries in that the general forms of the second-order ODEs invariant under nonabelian two-parameter groups were first calculated and then reductions made to the first-order ODEs which had the hidden symmetries. Applications of type I hidden symmetries to the Vlasov characteristic equation and a reaction-diffusion equation have been made.[16]

Type II hidden symmetries are those which appear when the order of an ODE is reduced by one and the reduced ODE is determined to be invariant under a new group. An example of these hidden symmetries was given by Olver and they have been studied for energy conserving equations since these equations are so widespread in science and engineering.[17] In this hidden symmetry the procedure for determining general forms of the energy-conserving equations with hidden symmetries was to choose the first-order ODE invariant under a group and then to increase the order of the ODE by using the differential invariants of the projective subgroups. The group under which the first-order ODE was invariant was chosen such that it was not a symmetry of the resultant second-order ODE. The direct classical Lie method was applied to the the second-order ODE to ascertain that the symmetry of the first-order ODE was lost as a point symmetry in the transformation to the second-order ODE. Again an inverse method was used since the hidden symmetry of the second-order ODE can not be determined directly. To avoid hidden symmetries one reduces by invariants of normal subgroups.

The two inverse methods for determining the hidden symmetries of ODEs differ from each other and from the inverse method for point symmetries. For the type I hidden symmetry the general form of an ODE of some order that is invariant under a multi-parameter group is calculated. The ODE is then reduced in order by variables which are the differential invariants of a non-normal subgroup; the resultant ODE has one or more hidden symmetries corresponding to the lost symmetries of the associated normal subgroups. On the other hand for the type II hidden symmetries the general form of an ODE invariant under one or more groups is calculated and then the order of that ODE increased where at least one of the symmetries of the lower order ODE is lost.   For the point symmetries the inverse method is simpler; one calculates the ODE in any given order for a group by means of the differential invariants.

The application of the inverse methods to finding hidden symmetries of three ODEs is presented here.   In section 2 the Painlevé-Ince equation is discussed.    In section 3 the Pinney equation is discussed. In section 4 a semiconductor transport equation is discussed. The conclusions are presented in section 5.   Since we deal with specific differential equations, we use a modified inverse method. The modified procedure is to increase the order of the ODE once and then to reduce the higher-order ODE, which for our three cases is a third-order ODE, by the differential invariants of the various groups under which the third-order ODE is invariant.   The delicacy in the procedure is in the choice of the group by which differential invariants one increases the order of the original ODE.   In the choice one is guided by the tables compiled with projective subgroups of type I and type II hidden symmetries.[16,17]   No attempt was made to increase the order of the three second-order ODEs to higher orders than three but in principle there is no reason not to try increasing the order of the ODEs more.

## 2. Modified Painlevé-Ince Equation

We begin with the modified Painlevé-Ince equation to illustrate the technique for particular NLODEs.[18]   The NLODE where ' denotes differentiation with respect to $x$ is

$$(1) \qquad\qquad y'' + \sigma yy' + \beta y^3 = 0.$$

The modified Painlevé-Ince equation arose in the solution of the modified Emden equation.   The original Emden equation modeled the thermodynamic behavior of a spherical cloud of gas molecules interacting by gravitational attraction.[19]   For $\beta \neq 1/9$ Eq.(1) is invariant under a two-parameter group and for $\beta = 1/9$ Eq. (1) is invariant an eight-parameter group.    This was determined by Leach and collaborators.[20,21]  Consequently, even for $\beta \neq 1/9$ this equation can be reduced to quadratures. However, the solutions reported in the dictionary edited by Kamke[22] or found by Leach[20] involve several integrations.

The technique here is to transform Eq.(1) to a third-order ODE which can then be reduced to another second-order ODE of simpler form.    As has been already stated there is at present no systematic method for choosing the group for increasing the order of Eq.(1) here.   We can look at the groups under which Eq.(1) is invariant. We find that

$$(2) \qquad\qquad U_{1a} = \frac{\partial}{\partial x} \; , \qquad U_{2a} = x\frac{\partial}{\partial x} - y\frac{\partial}{\partial y} \; .$$

The scaling transformation represented by $U_{2a}$ suggests a Riccati transformation. The Riccati transformation, which increases the order of an ODE, uses the differential invariants of the scaling transformation in the dependent variables.[16]  Calculations with  different scaling symmetries of third-order  ODEs show that if one of  two scaling symmetries is used to reduce  the order of the  third-order ODE, the  reduced second-order ODE will retain a scaling symmetry in the new variables.   Furthermore scaling groups are common non-normal subgroups. If we let

(3)
$$y = \frac{\Gamma u_z}{u} \ , \ \ x = z \ ,$$

we find

(4)
$$u^2 u_{zzz} + M u u_z u_{zz} + N u_z{}^3 = 0$$

for $M = \sigma\Gamma - 3$ and $N = 2 + \beta\Gamma^2 - \sigma\Gamma$.

Eq.(4) is invariant under three groups represented by the group generators

(5)
$$U_{1b} = \frac{\partial}{\partial z} \ , \ \ U_{2b} = z\frac{\partial}{\partial z} \ , \ \ U_{3b} = u\frac{\partial}{\partial u} \ .$$

The differential invariants of all three groups lead to a general form for the third-order ODE

(6)
$$F\left( \frac{u^2 u_{zzz}}{u_z{}^3} \ , \ \frac{u u_{zz}}{u_z{}^2} \right) = 0 \ .$$

Comparing Eqs.(4) and (6), we see that Eq.(4) is linear in the invariants together with an additive constant. Reduction of Eq.(4) by the differential invariants of $U_{1b}$ leads to a linear ODE where the transformation is

(7)
$$\bar{y} = \left( \frac{\Gamma u_z}{u} \right)^2, \ \ \bar{x} = \Gamma \ln u$$

The resulting second-order ODE is

(8)
$$\frac{d^2\bar{y}}{d\bar{x}^2} + \frac{\sigma d\bar{y}}{d\bar{x}} + 2\beta\bar{y} = 0$$

which has well-known solutions for RLC circuits or for damped harmonic oscillators among others. The third-order Eq.(4) has a type II hidden symmetry since Eq.(8) reduced from it is invariant under an eight-parameter group, an increase over the two-parameter group expected.

Two other reductions of order of Eq.(4) are possible. Reducing Eq.(4) by the differential invariants of $U_{2b}$ results in an ODE with a type I hidden symmetry. Besides the symmetry represented by the group generator $U_{2b}$ the symmetry represented by $U_{1b}$ has been lost. If we reduce Eq .(4) by the differential invariants of $U_{3b}$, we recover the original ODE, Eq.(1).

The advantage in transforming to Eq.(8) over reducing the original Eq.(1) to quadratures is that the parametric solutions are simpler.[18] The transformation from Eq.(1) to Eq.(8) was originally guessed by experience with the projective subgroups. The direct variable transformation between the two Eqs.(1) and (8) is nonlocal and is

(9)
$$\bar{y} = y^2, \quad \bar{x} = \int y \, dx$$

## 3. Pinney Equation

The Pinney equation has wide applications in accelerator physics as well as in the solutions of the one-dimensional Vlasov-Maxwell equations.[23,24] It is

(10)
$$y'' + \omega^2(x)y = \frac{M}{y^3}$$

for M a constant and the prime ' denotes differentiation with respect to $x$. It is the first integral of the following equation as has been shown by Leach.[25] He derived the differential equations and evaluated their Lie point symmetries for which the hidden symmetries are here analyzed. The equation is

(11)
$$u''' + 4\omega^2 u' + 4\omega\omega' u = 0$$

where $u = y^2$ and again the prime ' denotes differentiation with respect to $x$. The Pinney equation is invariant under a three-parameter group. On the other hand the third-order ODE, Eq.(11), is invariant under a seven-parameter group which is the maximal dimension of a group under which a third-order ODE is invariant.

The group structure of the third-order ODE, Eq.(11), is of interest since it possesses a type II hidden symmetry and a second-order ODE reduced from it has a type I hidden symmetry. The group generators under which Eq.(11) is invariant are

(12)
$$U_{1c} = u \frac{\partial}{\partial u}, \quad U_{2cj} = f_j(x)\frac{\partial}{\partial u}, \quad U_{3cj} = f_j(x)\frac{\partial}{\partial x} + u f_j'(x)\frac{\partial}{\partial u}$$

for $j = 1, 2, 3$ and $f_j(x)$ are linearly independent solutions of Eq.(11). If the third-order ODE (11) is reduced in order by the differential invariants associated with $U_{1c}$, then the second-order ODE has lost the three point symmetries associated with $U_{2cj}$. We note that

(13)
$$[U_{2cj}, U_{1c}] = U_{2cj}$$

where the groups associated with $U_{2cj}$ are the normal subgroups and that associated with $U_{1c}$ is the non-normal subgroup. Reducing Eq.(11) by the non-normal subgroup variables of $U_{1c}$ loses the symmetries of the normal subgroups of $U_{2cj}$. The inherited point group associated with the second-order ODE reduced from Eq.(11) by the differential invariants of $U_{1c}$ is a three-parameter group represented by $U_{3cj}$. This is expected since the group of $U_{1c}$ is used in the reduction and the loss of the three symmetries of $U_{2cj}$ leave three of the symmetries of Eq.(11). However, when the reduction is made the actual group under which the reduced ODE is invariant is an eight-parameter group! The variable transformation in the reduction of order is

(14)
$$v = \frac{u_z}{u}, \quad z = x .$$

The reduced equation is

(15)
$$v_{zz} + 3vv_z + v^3 + 4\omega^2 v + 4\omega\omega_z = 0.$$

Three of the group generators are of the form of the first extension of $U_{3cj}$ in the $(z,v)$ variables but the function $f_j(x)$ is now replaced by $p_j(z)$. They are

(16)
$$U^e_{3cj} = p_j(z)\frac{\partial}{\partial z} + (up_j'(z) - p_j''(z))\frac{\partial}{\partial u}$$

with $j = 1, 2, 3$. The equation satisfied by $p_j(z)$ has a more general form than that for $f_j(x)$. The five other group generators are associated with groups that are not symmetries of the third-order ODE (11).

The type I hidden symmetries associated with Eq.(15) are the three symmetries of the normal subgroup of $U_{2cj}$. The type II hidden symmetries associated with Eq.(11) are the new symmetries found for Eq.(15). Not only are five new groups found but the three groups which have a group generator of the form of $U_{3cj}$ are more general. The type II hidden symmetry is of a sort that not only adds new symmetries but modifies those already present. This last property was not been seen before in ODEs with a simpler group structure.

## 4. Semiconductor Transport Equation

The semiconductor transport equations for a steady-state, (no time dependence) one-dimensional system with a constant collision frequency are the differential form of Gauss's law, the momentum conservation equation and the vanishing of the divergence of the current density.[26] These can be combined and dimensionless coordinates defined to give single second-order ODE with ' denoting differentiation with respect to $x$

(17)
$$2\bar{v}^2\bar{v}'' + 2\bar{v}\bar{v}' + 2\beta\bar{v}\bar{v}' + \bar{v} - 1 = 0 .$$

where $-v = v/v_E$ for $v_E = J/q_0n_0$, the parameter $\beta = 1/\sqrt{2}\,\omega_p\tau$ is not the same $\beta$ as in section 2 and $x = X/x_0$ for $x_0 = J/(\sqrt{2}\,q_0n_0\,\omega_p)$. Here $v$ is the carrier speed, J is the current density, $q_0n_0$ is the background charge density, $\omega_p$ is the carrier plasma frequency and $\tau$ is the collision time (inverse collision frequency) and X is the position of the carrier with respect to the emitting electrode.

Eq.(17) is invariant under translations in $x$. To discuss the nonlocal transformation we use the differential invariants, $w_z = dw/dz$ and $z = x$. For ease of calculation we let

(18)
$$\bar{v} = \frac{1}{w_z} .$$

With this transformation Eq.(17) becomes a nonlinear third-order ODE.

$$(19) \qquad 2w_z w_{zzz} - 6 w_{zz}{}^2 + 2 \beta w_z{}^2 w_{zz} - w_z{}^4 + w_z{}^5 = 0 .$$

It is invariant under translations in $z$ and in $w$. The group generators are

$$(20) \qquad U_{1d} = \frac{\partial}{\partial z} , \qquad U_{2d} = \frac{\partial}{\partial w} .$$

The differential invariants of $U_{2d}$ were used to increase the order of Eq.(17) to the third-order ODE. On the other hand the differential invariants of $U_{1d}$ can be used as the variables to reduce the second-order ODE to a another second-order ODE. We let the new variables be

$$(21) \qquad \bar{v} = 1/w_z \text{ and } T = w/\sqrt{2} .$$

The likely procedure would be to use $w_z$ as the new dependent variable and then transform the results in the second-order ODE. Here we do two transformations of the variables in one step. We find

$$(22) \qquad \frac{d^2 \bar{v}}{dT^2} + \sqrt{2\beta} \frac{d\bar{v}}{dT} + \bar{v} = 1$$

with $\quad T = \omega_p \int_0^x \frac{dX'}{\bar{v}} .$

The new second-order ODE, Eq.(22), is invariant under an eight-parameter group since it is a linear ODE; Eq.(19) is invariant under a two-parameter group. Hence, the third-order ODE, Eq.(19), has a type II hidden symmetry. Interestingly, Eq.(22) is of the same form as Eq.(8). Both for the modified Painlevé-Ince equation and the set of semiconductor transport equations a nonlocal transformation to the well-known, linear harmonic oscillator equation has been shown.

## 5. Conclusion

Hidden symmetries of ordinary differential equations originating from three second-order ODEs of considerable physical and mathematical interest have been demonstrated. This has been done by increasing the order of the original ODEs and then reducing the order of the third-order ODEs by the differential invariants of the various subgroups. Both type I and type II hidden symmetries have been identified. The type I hidden symmetries occur when the order of the ODE is reduced by the non-normal subgroup variables and the resultant lower-order ODE loses a symmetry. The type II hidden symmetries occur when the order of an ODE is reduced and new symmetries not present in the higher order ODE occur. The occurrence of these hidden symmetries is associated with nonlocal transformations between ODEs of the same order. The transformations are not local since the order of the associated Lie algebra differs in the two ODEs whereas a point transformation preserves the order of the Lie algebra.

## REFERENCES

1. Sophus Lie and F. Engel, *Theorie der Transformationgruppen*, B. G. Tuebner, Leipzig, Vol.I, 1888, Vol. II, 1890, Vol. III, 1893, reprinted Chelsea, New York, 1970.

2. R. Hermann, *Lie Groups: History, Frontiers and Applications*, Vol I: Sophus Lie's 1880 Transformation Group Paper, trans. M. Ackerman, 1975, Vol. III: Sophus Lie's 1884 differential Invariant Paper, 1976, Math Sci. Press, Boston.

3. A. Cohen, *An Introduction to the Lie theory of One-Parameter Groups with Applications to the Solution of Differential Equations*, D. C. Heath, New York, 1911.

4. J. M. Page, *Ordinary Differential Equations with an Introduction to Lie's theory of the Group of One Parameter*, McMillan, London, 1897.

5. G. W. Bluman and J. D. Cole, *Similarity Methods for Differential Equations, Springer*, New York, 1974.

6. L. V. Ovsiannikov, *Group analysis of differential Equations*, trans. ed. by W. F.Ames, Academic, New York, 1982.

7. W. F. Ames, *Nonlinear Partial differential Equations in Engineering*, Vol. I, 1965, Vol. II, 1977, Academic Press, New York.

8. P. J. Olver, Applications of Lie Groups to Differential Equations, Springer-Verlag, New York, 1986.

9. G. W. Bluman and S. Kumei, *Symmetries and Differential Equations*, Springer, Berlin, 1989.

10. H. Stephani, Differential equations, Cambridge University Press, Cambridge, 1989.

11. G. W. Bluman and S. Kumei, "Symmetry-based algorithms to relate partial differential equations, I. Local symmetries," Euro. Jnl of Applied Mathematics 1 (1989), 189-216.

12. R. L. Anderson, S. Kumei and C. E. Wulfman, "Generalization of the concept of invariance of differential equations. Results of applications to some Schrodinger equations," Phys. Rev. Lett. 28 (1972) 988.

13. G. W. Bluman and S. Kumei, "Symmetry-based algorithms to relate partial differential equations:II. Linearization by local symmetries, Euro. Jnl. of Applied Mathematics 1 (1989), 217-223.

14. A. M. Vinogradov and I. S. Krasil'shchik, "On the theory of nonlocal symmetries of nonlinear partial differential equations," Sov. Math. Dokl. 29 (1984), 337-341.

15. I. S. Krasil'shchik and A. M. Vinogradov, "Nonlocal symmetries and the theory of coverings: an addendum to A. M. Vinogradov's 'Local symmetries and conservation laws'," Acta Applic. Math. 2 (1984), 79- 95.

16. B. Abraham-Shrauner and A. Guo, "Hidden symmetries associated with the Projective group of nonlinear first-order ordinary differential equations," J. Phys. A: Math. Gen. 25 (1992) 5597-5608.

17. A. Guo and B. Abraham-Shrauner, "Hidden symmetries of energy conserving differential equations," IMA J. of Appl. Math. (submitted for publication).

18. B. Abraham-Shrauner, "Hidden symmetries and linearization of the modified Painlevé-Ince equation," J. Math. Phys. (submitted for publication).

19. H. T. Davis, *Introduction to Nonlinear differential and Integral Equations*, Dover, New York, 1962.

20. P. G. L. Leach, "First integrals for the modified Emden equation $^{TM}$ $q + \alpha(t)\dot{q} + q^n$," J. Math. Phys. **26** (1985), 2510- 2514.

21. P. G. L. Leach, M. R. Feix, and S. Bouquet, "Analysis and solution of a nonlinear second-order differential equation through rescaling and through a dynamical point of view," J. Math. Phys. **29** (1988), 2563-2569.

22. E. Kamke, *Differentialgleichungen Losungsmethoden und Losungen,* Chelsea, New York, 1971.

23. B. Abraham-Shrauner, "Lie transformation group solutions of the nonlinear one-dimensional Vlasov equation," J. Math. Phys. **26** (1985), 1425-1435.

24. B. Abraham-Shrauner and O. A. Anderson, "Space-charge effects in warm ion sheet beams in the Vlasov- Maxwell approximation," Phys. Fluids B **2** (1990), 2217-2225.

25. P. G. L. Leach (unpublished work).

26. B. Abraham-Shrauner, W. Weeks, IV and R. N. Zitter, "Current-voltage curves for a spatially periodic diodes," J. Appl. Phys. (in press).

*DEPARTMENT OF ELECTRICAL ENGINEERING, WASHINGTON UNIVERSITY, ST. LOUIS, MO 63130.
E-*mail* address: bas@wuee1.wustl.edu
**DEPARTMENT OF MATHEMATICS AND APPLIED MATHEMATICS
UNIVERSITY OF NATAL
KING GEORGE V AVENUE
DURBAN, 4001
REPUBLIC OF SOUTH AFRICA
E-*mail* address: LEACH@ph.und.ac.za

Lectures in Applied Mathematics
Volume **29**, 1993

# Liapunov-Schmidt Reduction for a Bifurcation Problem with Periodic Boundary Conditions on a Square Domain

E. ALLGOWER, P. ASHWIN, K. BÖHMER, AND Z. MEI

ABSTRACT. We consider a model bifurcation problem of a reaction-diffusion PDE with doubly periodic boundary conditions on a square domain and $Z_2$ symmetry of the reaction term. Due to extra symmetries of the low order terms of the bifurcation equations, the truncated bifurcation equations are not determined at third order (there are three and four dimensional tori of solutions although the equations are only equivariant under a two dimensional group). However, including the fifth order terms for the $(1, 2) : (2, 1)$ mode at bifurcation gives a determined system for which we find all bifurcating steady solutions.

## 1. Introduction

We study the branching of steady solutions from bifurcations of the following model problem motivated by reaction-diffusion systems:

$$(1) \qquad G(u, \mu) := \begin{cases} \Delta u + \mu(u + f(u)) = 0 \text{ on } \Omega := R^2 \\ \text{with } u(x, y) = u(x + 2\pi, y) = u(x, y + 2\pi) \end{cases}$$

where $f : R \to R$ is a smooth and odd function and

$$Df(0) = 0, \ D^3 f(0) \neq 0.$$

Evidently, $G$ maps $X := C^{2,s}(\Omega)$ into $Y := C^{0,s}(\Omega)$ (with $s \in (0,1)$) and $\{(0, \mu); \ \mu \in R\}$ is a trivial solution. Due to the periodic boundary conditions the problem (1) is equivariant under a two dimensional symmetry group $Z_2 \times D_4 \times_s T^2$

1991 *Mathematics Subject Classification.* 58G28,35B32.
Partially supported by the National Science Foundation under grant number DMS-9104058
Supported by the Royal Society European Exchange Program, UK.
Supported by Deutsche Forschungsgemeinschaft, FRG.
This paper is in final form and no version of it will be submitted for publication elsewhere.

(where $\times_s$ means a semi-product of two groups: the action of $D_4$ does not commute with that of $T^2$) and its bifurcation scenarios reflect this. However, if we perform Liapunov-Schmidt reduction of (1) to the kernel of $G'$, we find there may be extra symmetries in the $k$-jets (i.e. the truncated Taylor expansions including up to $k$th order terms) of the bifurcation equations. In our case, the bifurcation equations truncated to third order have much more symmetry than the original problem. Thus, the bifurcation equations cannot be determined at third order (cf. [9], [1]). As a consequence one has to consider the higher order terms in the Taylor expansion of bifurcation equations in order to give a complete bifurcation scenario of (1).

Crawford et al. [6, 7] and Gomes and Stewart [10, 11] use singularity theory to find normal forms for mode interactions in doubly-periodic problems on squares and rectangles. As in the results of Armbruster and Dangelmayr [8] for mode interactions on 1-D domains, higher order terms (dependent on mode number) break any extra symmetry in the low order terms and determine the branching.

In this paper we investigate this degeneracy of the *nonlinear* terms for our model example, firstly by using an iterative Liapunov-Schmidt method to find analytically the 3-jets of bifurcation equations of (1) at its bifurcation points. This puts the original equation (1) into the normal forms of Crawford, Gomes et al.. We find up to two dimensional degeneracies in the solution of the 3−jet, even if the linearisation of the equation at the bifurcation point has a *generic* 4 or 8 dimensional kernel. Secondly, we look in detail at the $(1,2):(2,1)$ mode bifurcation in the example (1). We find that the 5−jet determines the bifurcation, and we use the degenerate solutions of the 3−jet to find the full bifurcation behaviour of (1) at this bifurcation.

## 2. Basic properties

Linearising the equation (1) at $(0, \mu_0)$ we get

$$(2) \qquad G'(0, \mu_0)(u, \lambda) = \partial_u G(0, \mu_0) u = \Delta u + \mu_0 u = 0$$

which has eigenfunctions composed of the real and imaginary parts of

$$\exp i(\pm kx \pm ly)$$

with eigenvalue $\mu = k^2 + l^2$. We assume that $(0, \mu_0)$ is a bifurcation point of (1) and the linearisation (2) has a kernel of dimension $8M + 4N$, $N = 0$ or 1; i.e. not necessarily the $D_4$ generic cases $M = 1$, $N = 0$ which have irreducible representation on the null space. For example, $M = N = 1$ for $\mu_0 = 50\pi^2 = (7^2 + 1^2)\pi^2 = (5^2 + 5^2)\pi^2$. This means that

$$\mu_0 = k_i^2 + l_i^2 = m^2 + m^2$$

with $\{k_i \neq l_i, \ k_i \cdot l_i \neq 0 \ : \ i = 1 \ldots M\}$ are sets of distinct critical mode numbers. The case $k \cdot l = 0$, $|k| + |l| > 0$ corresponds also to $N = 1$. We parameterise a point in the (real) kernel using $z \in C^{4M+2N}$ in the following way:

$$(3) \qquad u(z)(x,y) := \sum_{i=1}^{M} \sum_{j=1}^{4} z_{j,i} \psi_{j,i} + \sum_{i=M+1}^{M+N} \sum_{j=1}^{2} z_{j,i} \psi_{j,i} + \text{c.c.}$$

where '+ c.c.' means we add the complex conjugates to give real quantities, and we define the eigenfunctions to be the real and imaginary parts of

$$(4) \qquad \left. \begin{aligned} \psi_{1,j} &:= \exp i(k_j x + l_j y) \\ \psi_{2,j} &:= \exp i(k_j x - l_j y) \\ \psi_{3,j} &:= \exp i(l_j x + k_j y) \\ \psi_{4,j} &:= \exp i(l_j x - k_j y) \end{aligned} \right\} \text{ for } j = 1 \cdots M$$

$$\left. \begin{aligned} \psi_{1,j} &:= \exp i(m_j x + m_j y) \\ \psi_{2,j} &:= \exp i(m_j x - m_j y) \\ \text{or} \\ \psi_{1,1} &:= \exp(ik_j x) \\ \psi_{2,1} &:= \exp(ik_j y) \end{aligned} \right\} \begin{aligned} &\text{for } j = M+1, \ \mu_0 = 2m_j^2 \text{ or } \mu_0 = k_j^2, \\ &\text{hence } N = 1. \end{aligned}$$

We shall call the corresponding 8 or 4 dimensional system obtained by fixing a value of $j$ and looking at this subspace of the null space, a *block*. Note that $4M + 2N$ is not bounded above; see Crawford et al. [7] and Ashwin [2] for discussion of how this comes from the fact that the equation is the restriction of a Euclidean equivariant problem to a square.

**2.1. Equivariance properties.** The equation (1) inherits several equivariance properties from the symmetries of the square and translations in the $x$ and $y$ directions which preserve the doubly periodic boundary conditions. Firstly, $D_4$ is generated by the following two symmetries acting on $C^{0,s}(\Omega)$:

$$(5) \qquad \begin{aligned} \kappa_1 &: u(x,y) \mapsto u(x,-y) \\ \kappa_2 &: u(x,y) \mapsto u(y,x) \end{aligned}$$

which leave the domain, boundary conditions and equation invariant. The translations

$$(6) \qquad \rho_{a,b} u(x,y) \mapsto u(x+a, y+b) \text{ with } a, \ b \in R \ (\text{mod} 2\pi)$$

give a $T^2$ symmetry. Finally, replacing $u$ by $-u$ gives a $Z_2$ symmetry

$$(7) \qquad \kappa_3 : u(x,y) \mapsto -u(x,y).$$

This means that the full problem has symmetry

$$(8) \qquad \Gamma := Z_2 \times D_4 \times_s T^2$$

generated by the transformations (5)-(7). These groups induce the following actions on blocks $i$ ($C^4$ or $C^2$) spanned by the $\psi_{j,i}$ in (4). We define $(z_j, k, l) := (z_{j,i}, k_i, l_i)$, and in the case that $i$ is in $1 \ldots M$,

$$(9) \quad \kappa_1 : (z_1, z_2, z_3, z_4) \;\mapsto\; (z_2, z_1, z_4, z_3)$$

$$(10) \quad \kappa_2 : (z_1, z_2, z_3, z_4) \;\mapsto\; (z_3, \bar{z}_4, z_1, \bar{z}_2)$$

$$(11) \, \rho_{a,b} : (z_1, z_2, z_3, z_4) \;\mapsto\; (z_1 e^{i(ka+lb)}, z_2 e^{i(ka-lb)}, z_3 e^{i(la+kb)}, z_4 e^{i(la-kb)})$$

$$(12) \quad \kappa_3 : (z_1, z_2, z_3, z_4) \;\mapsto\; (-z_1, -z_2, -z_3, -z_4),$$

otherwise for $N = 1$ and $i = M + 1$, we use $(z_j, m) := (z_{j,i}, m_i)$ for $\mu_0 = 2\mu_i^2$,

$$\kappa_1 : (z_1, z_2) \;\mapsto\; (z_2, z_1)$$

$$\kappa_2 : (z_1, z_2) \;\mapsto\; (z_1, \bar{z}_2)$$

$$\rho_{a,b} : (z_1, z_2) \;\mapsto\; (z_1 e^{im(a+b)}, z_2 e^{im(a-b)})$$

$$\kappa_3 : (z_1, z_2) \;\mapsto\; (-z_1, -z_2),$$

and $(z_j, k) := (z_{j,i}, k_i)$ for $\mu_0 = k^2$,

$$\kappa_1 : (z_1, z_2) \;\mapsto\; (z_1, \bar{z}_2)$$

$$\kappa_2 : (z_1, z_2) \;\mapsto\; (z_2, z_1)$$

$$\rho_{a,b} : (z_1, z_2) \;\mapsto\; (z_1 e^{ika}, z_2 e^{-ikb})$$

$$\kappa_3 : (z_1, z_2) \;\mapsto\; (-z_1, -z_2).$$

**2.2. Scaling properties.** Let us define an epimorphism $\beta_n : \Gamma \mapsto \Gamma$ for any integer $n \in N$ by

$$\beta_n(\kappa_i) = \kappa_i, \; i = 1, 2, 3, \quad \beta_n(\rho_{a,b}) = \rho_{na,nb}.$$

The kernel of $\beta_n$ is the finite cyclic group:

$$\ker(\beta_n) = Z_n := \{\rho_{2\pi i/n, 2\pi i/n}, \; i = 1, \ldots, n\}.$$

For the fixed point spaces

$$Y^{Z_n} = \{u \in Y \mid u(x - \frac{2\pi}{n}, y - \frac{2\pi}{n}) = u(x, y)\}, \qquad X^{Z_n} = X \cap Y^{Z_n},$$

if we define a mapping $h_n : u \in Y \mapsto h_n u \in Y^{Z_n}$ by

$$h_n u(x, y) = u(nx, ny) \qquad \text{for all } u \in Y,$$

then $h_n$ maps $X$ into $X^{Z_n}$ and

$$h_n \rho_{a,b} = \rho_{na,nb} h_n, \qquad h_n \kappa_i = \kappa_i h_n, \quad i = 1, 2, 3.$$

Moreover, the following scaling laws hold:

$$n^2 h_n G(u, \mu) = G(h_n u, n^2 \mu) \qquad \text{for all } (u, \mu) \in X \times R, \; n \in N.$$

Hence, solutions of the equation (1) possess the scaling property, i.e., $(u(t), \mu(t))$ is a solution branch passing through $(0, \mu_0)$ if and only if $(h_n u(t), n^2 \mu(t))$ is a solution branch passing through $(0, n^2 \mu_0)$, see e.g., [5] and [12].

The scaling laws imply that all solution branches at $(0, \mu_0)$ can be shifted to those at $(0, n^2 \mu_0)$ and vice versa, if at these two bifurcation points the problem (1) has same number of solution branches. Thus the set of bifurcation points can be divided into "equivalent" classes. More precisely, for any $\mu_0 = k^2 + l^2$ in which $k, l$ have no common factors, let $S_{\mu_0}$ denote the set of points $(0, n^2 \mu_0)$, $n \in N$ where the number of solutions of (1) is same as at $(0, \mu_0)$. Then solution branches of (1) at all points in $S_{\mu_0}$ can be generated from those at a single point $(0, \mu_0)$. Hence, solution branches at different points in $S_{\mu_0}$ are treated as "equivalent". We consider $(0, \mu_0) \in S_{\mu_0}$ with the smallest $\mu_0$, i.e., the case $\mu_0 = k^2 + l^2$ and $k, l$ have no common factor. Thereafter, the bifurcation of (1) is relevant only for the mode $(k, l) = $ (odd, even) or (even,odd), denoted as (odd,even):(even,odd).

**2.3. Liapunov-Schmidt reduction.** The self-adjoint nature of the problem (2) means that we can split the space $C^{0,s}(\Omega)$ into a direct sum of the kernel and image of $\partial_u G = \partial_u G(0, \mu_0)$. We define the orthogonal projector $P$ on $L^2(\Omega)$ in the following way:

$$P(v) := v - \sum_{i=1}^{8M+4N} \langle \psi_i, v \rangle \psi_i$$

which restricts to give

$$P : C^{0,s}(\Omega) \to C^{0,s}(\Omega)$$

so that $\text{Im}(P) = \text{Im}(\partial_u G)$ and $\text{Ker}(P) = \text{Ker}(\partial_u G)$ where the inner product is defined to be

$$\langle a(x,y), b(x,y) \rangle := \frac{1}{4\pi^2} \int_{x=0}^{2\pi} \int_{y=0}^{2\pi} \overline{a(x,y)}\, b(x,y)\, dy\, dx.$$

In fact, the standard regularity theory ensures that the projection $P$ is also well defined under the Hölder norm, see e.g. Healey-Kielhöfer [13]. We use a slight modification of the iterative Liapunov-Schmidt method detailed in [2, 3] and [4], i.e. we define

$$\mu := \mu_0 + \lambda \quad \text{and} \quad w_k : C^{4M+2N} \times R \cong \text{Ker}(\partial_u G) \times R \to \text{Im}(\partial_u G)$$

by

$$w_1(z) = 0, \ z = \sum z_{i,j} \psi_{i,j},$$
$$w_{2k+1}(z) := -(\partial_u G)^{-1} P\, j_{2k+1} G(u(z) + w_{2k-1}(z), \mu_0 + \lambda),$$

where $j_k(v(z))$ is the $k$-jet of $v(z)$ and $u(z)$ as in (3). Note that $PG(u, \mu) = \mu P f(u)$ means that in this case $w_{2k+1}$ is the solution of

$$\partial_u G w = -\mu P j_{2k+1} f(u(z) + w_{2k-1}(z)).$$

The truncations of the bifurcation equations are defined by

$$B_{2k+1} : \mathrm{Ker}(\partial_u G) \times R \to \mathrm{Ker}(\partial_u G)$$
$$B_{2k+1}(z) := (I - P)\, j_{2k+1} G(u(z) + w_{2k-1}(z), \mu_0 + \lambda)) = 0$$

and are also correct to $(2k + 1)$st order.

## 3. The third order bifurcation equations

Using the iterative Liapunov-Schmidt method we obtain that the first truncated bifurcation equations have the form, see (3) for $u(z)$,

$$
B_3(z) \quad := \quad (I - P) j_3(\lambda u + \mu_0 f(u))
$$

$$
= \quad \left\{
\begin{array}{ll}
\begin{pmatrix}
\langle \psi_{1,i}, \lambda u(z) + au^3 \rangle \\
\langle \psi_{2,i}, \lambda u(z) + au^3 \rangle \\
\langle \psi_{3,i}, \lambda u(z) + au^3 \rangle \\
\langle \psi_{4,i}, \lambda u(z) + au^3 \rangle
\end{pmatrix}
& \text{for } i = 1 \cdots M \\[4ex]
\begin{pmatrix}
\langle \psi_{1,i}, \lambda u(z) + au^3 \rangle \\
\langle \psi_{2,i}, \lambda u(z) + au^3 \rangle
\end{pmatrix}
& \text{for } i = M + N
\end{array}
\right\} = 0,
$$

where $a := \mu_0 f'''(0)/6$. We have that the eigenfunctions are an orthonormal basis for the null space;

$$\langle \psi_{i,j}, \psi_{k,l} \rangle = \delta_{i,k} \cdot \delta_{j,l}$$

Therefore, the only nonzero third order terms in the inner product are of the form

$$\langle e^{i(kx+ly)}, e^{i(kx+ly)} e^{i(k'x+l'y)} e^{i(k'x+l'y)} \rangle$$

and counting the multiplicity of the various terms on expanding the cubic term, we get that $B_3$ is given by

(13)

$$
B_3(z) = \left\{
\begin{array}{ll}
\begin{pmatrix}
z_{1,i}\left(\lambda + 3a|z_{1,i}|^2 + 6a \sum_{(j,k)\neq(1,i)} |z_{j,k}|^2\right) \\
z_{2,i}\left(\lambda + 3a|z_{2,i}|^2 + 6a \sum_{(j,k)\neq(2,i)} |z_{j,k}|^2\right) \\
z_{3,i}\left(\lambda + 3a|z_{3,i}|^2 + 6a \sum_{(j,k)\neq(3,i)} |z_{j,k}|^2\right) \\
z_{4,i}\left(\lambda + 3a|z_{4,i}|^2 + 6a \sum_{(j,k)\neq(4,i)} |z_{j,k}|^2\right)
\end{pmatrix}
& , \; i = 1 \cdots M \\[6ex]
\begin{pmatrix}
z_{1,i}\left(\lambda + 3a|z_{1,i}|^2 + 6a \sum_{(j,k)\neq(1,i)} |z_{j,k}|^2\right) \\
z_{2,i}\left(\lambda + 3a|z_{2,i}|^2 + 6a \sum_{(j,k)\neq(2,i)} |z_{j,k}|^2\right)
\end{pmatrix}
& , \; i = M + N
\end{array}
\right\}
$$

Note that this truncated bifurcation equation has $T^{4M+2N} \times_s S_{4M+2N}$ symmetry, i.e. if $z_{ij} = r_{ij} e^{i\theta_{ij}}$, any translation of the $\theta_{ij}$s and any permutation of the $r_{ij}$s is a symmetry of the truncated problem. This is much more symmetric than the original problem. If the problem is generic, this symmetry must be broken at a higher order.

**3.1. Solution of the third order equations.** If we choose one single block $i$ in $1, \ldots, M$ and let all other $z_{j,k} = 0$, then the solutions within of this block $i =$ constant are of the following types:

$(0, 0, 0, 0)$,

this corresponds to the trivial solution branch.

$(\sqrt{-\lambda/3a}e^{i\theta}, 0, 0, 0)$,

this is a circle of steady states parametrised by $\theta$: there are four of these generated by (9) and (10), hence by $D_4$.

$(\sqrt{-\lambda/9a}e^{i\theta_1}, \sqrt{-\lambda/9a}e^{i\theta_2}, 0, 0)$,

this is a two-torus of steady states parametrised by $(\theta_1, \theta_2)$: there are six of these generated by (9), (10).

$(\sqrt{-\lambda/15a}e^{i\theta_1}, \sqrt{-\lambda/15a}e^{i\theta_2}, \sqrt{-\lambda/15a}e^{i\theta_3}, 0)$,

this is a three-torus of steady states parametrised by $(\theta_1, \theta_2, \theta_3)$: there are four of these generated by (9), (10).

$(\sqrt{-\lambda/21a}e^{i\theta_1}, \sqrt{-\lambda/21a}e^{i\theta_2}, \sqrt{-\lambda/21a}e^{i\theta_3}, \sqrt{-\lambda/21a}e^{i\theta_4})$,

this is a four-torus of steady states parametrised by $(\theta_1, \theta_2, \theta_3, \theta_4)$.

The fact that the last two solutions are not isolated even when taking into account the group orbit under (11) means that only a subset of the manifold of solutions can possibly exist in the actual equations. To find out which they are, we look at the higher order terms.

The corresponding solutions within the block $i = M + N$ are obtained from the first three types above by dropping the last two zeros. Types two and three again represent circles and two-tori of steady states parametrised by $\theta_1$ and $\theta_1, \theta_2$, respectively. Since for $\mu_0 = 2m^2$ or $\mu_0 = k^2$ the corresponding $\kappa_1, \kappa_2$ do not generate new solutions, but only a new parametrisation for the circle, we have only one solution for each type.

**3.2. A general solution.** A general solution of the system (13) is a $p$-torus with $0 \le p \le 4M + 2N$ and consists of

$$z_{i,j} = \begin{cases} 0 & for (4M+2N-p) distinct pairs of the form (i,j); \\ re^{i\theta_{i,j}} & otherwise, \end{cases}$$

and $r$ is given by

$$r^2 = -\frac{\lambda}{(6p - 3)a}.$$

The $p$ torus is parametrised by the $p$ angles $\{\theta_{i,j} \mid z_{i,j} \ne 0\}$.

## 4. An example calculation; (1,2):(2,1) to fifth order

In this section we calculate and solve the bifurcation equation for a special example

(14)
$$f(u) = \frac{u^3}{5}$$

with the case that

$$\mu_0 = 5, \ k = 1, \ l = 2.$$

Note that, $M = 1$ and $N = 0$, and so we drop the block subscript $i$ on the $z$ and $\psi$. Note that Crawford [6] has shown that the representation of $D_4 \times_s T^2$ is absolutely irreducible in the case of the bifurcating modes have parity (odd,even).

To find the 5th order terms, we have to calculate $w_3(z)$ and thus $B_5(z)$. We calculate the former using

$$w_3(z) := (\partial_u G)^{-1}[-P(\lambda u(z) + u(z)^3)]$$

which, because $u \in \text{Ker}(\partial_u G)$, is equal to

$$w_3(z) = -(\partial_u G)^{-1} P u(z)^3.$$

To invert $(\partial_u G)^{-1}$ we can use the fact that

$$(\partial_u G)e^{i(px+qy)} = (\mu_0 - p^2 - q^2)e^{i(px+qy)}.$$

We then find that the bifurcation equation to fifth order is given by

$$B_5(z) := (I - P)j_5\{\lambda u(z) + (\mu_0 + \lambda)[w_3(z)) + \frac{1}{5}(u(z) + w_3(z))^3]\} = 0,$$

which implies that

$$
\begin{aligned}
B_5(z) &= B_3(z) + (I - P)[3u^2(z)w_3(z) + \frac{\lambda}{5}u^3] \\
&= B_3(z) + \begin{pmatrix} \langle \psi_1, 3u^2(z)w_3(z) + \frac{\lambda}{5}u^3 \rangle \\ \langle \psi_2, 3u^2(z)w_3(z) + \frac{\lambda}{5}u^3 \rangle \\ \langle \psi_3, 3u^2(z)w_3(z) + \frac{\lambda}{5}u^3 \rangle \\ \langle \psi_4, 3u^2(z)w_3(z) + \frac{\lambda}{5}u^3 \rangle \end{pmatrix}.
\end{aligned}
$$

Using the symbolic algebra program MAPLE to calculate $w_3$ and $B_5$ gives the following expression for the first component of $B_5$:

$$
\begin{aligned}
B_5(z)_1 &= z_1 \left( \lambda + 3(1 + \frac{\lambda}{5})|z_1|^2 + 6(1 + \frac{\lambda}{5})(|z_2|^2 + |z_3|^2 + |z_4|^2) \right) \\
&+ z_1 \left( \frac{3}{40}|z_1|^4 + \frac{45}{32}|z_2|^4 + \frac{5}{2}|z_3|^4 + \frac{9}{10}|z_4|^4 \right) \\
&+ z_1|z_1|^2 \left( \frac{45}{16}|z_2|^2 + 5|z_3|^2 + \frac{9}{5}|z_4|^2 \right) \\
&- \frac{18}{5}\bar{z}_1\bar{z}_2^2 z_3 z_4 - \frac{9}{5}z_3^2 z_2 \bar{z}_4^2.
\end{aligned}
$$

The other components of $B_5$ can be obtained by applying the symmetries of the square (9), (10).

**4.1. Solution of the fifth order equations.** We now solve the system of equations

$$(15) \qquad\qquad B_5(z) = 0.$$

Note that $z = 0$ is always a trivial solution.

**4.1.1.** *One component of $z$ nonzero.* For example, assume that $z_2 = z_3 = z_4 = 0$. This gives that

$$z_1(\lambda + 3(1 + \frac{\lambda}{5})|z_1|^2 + \frac{3}{40}|z_1|^4) = 0$$

and so if $z_1 \neq 0$ we have that

$$\begin{aligned} |z_1|^2 &= -20 - 4\lambda + \frac{2}{3}\sqrt{900 + 330\lambda + 36\lambda^2} \\ &= -\frac{\lambda}{3} + \frac{23\lambda^2}{360} + O(\lambda^3), \; z_2 = z_3 = z_4 = 0, \end{aligned}$$

which is a one dimensional manifold of solutions. There are four of these manifolds of solutions corresponding to the group orbit under the discrete symmetries (9), (10). An application of (11) amounts to a shift of the solution along the manifold.

**4.1.2.** *Two components of $z$ nonzero.* If two of the components of $z$ are zero and two are nonzero, what happens depends on which are zero. In all cases, we can use the $T^2$ symmetry to restrict to the case of real $z$, and also we find that the only possibilities have $|z_i|^2 = |z_j|^2$ for the nonzero components.

$z_3 = z_4 = 0$:
For this case we get

$$z_1\left(\lambda + 3(1 + \frac{\lambda}{5})|z_1|^2 + 6(1 + \frac{\lambda}{5})|z_2|^2 + \frac{3}{40}|z_1|^4 + \frac{45}{32}|z_2|^4 + \frac{45}{16}|z_1|^2|z_2|^2\right) = 0$$

$$z_2\left(\lambda + 3(1 + \frac{\lambda}{5})|z_2|^2 + 6(1 + \frac{\lambda}{5})|z_1|^2 + \frac{3}{40}|z_2|^4 + \frac{45}{32}|z_1|^4 + \frac{45}{16}|z_1|^2|z_2|^2\right) = 0$$

which gives that $|z_1|^2 = |z_2|^2 = r \in R$ with

$$r\left(\lambda + 9(1 + \frac{\lambda}{5})r^2 + \frac{687}{160}r^4\right) = 0.$$

Thus, the nontrivial solutions are:

$$|z_1|^2 = |z_2|^2 = -\frac{\lambda}{9} + \frac{127}{7776}\lambda^2 + O(\lambda^3)$$

which are for fixed $\lambda$ two dimensional manifolds of solutions corresponding to translations in $T^2$.

$z_2 = z_4 = 0$:

Similarly, for this case we get $|z_1|^2 = |z_3|^2 = r \in R$ with

$$r = -\frac{\lambda}{9} + \frac{23}{1944}\lambda^2 + O(\lambda^3).$$

$z_2 = z_3 = 0$:

For this case we get that $|z_1|^2 = |z_4|^2 = r \in R$ so that

$$r = -\frac{\lambda}{9} + \frac{179}{1944}\lambda^2 + O(\lambda^3).$$

All of these two mode solutions have the same linear dependence on $\lambda$, but note that the second order terms are distinct! This indicates that the fifth order terms have broken the extra $T^4$ symmetry of the cubic terms to give the $D_4 \times_s T^2$ symmetry.

Each of the above cases gives rise to another by using the symmetry (9), (10) of the problem.

**4.2. Three components of $z$ nonzero.** This is not possible, because if e.g. $z_4 = 0$ then the fourth component of $B_5(z) = 0$ implies that

$$z_1^2 z_2 \bar{z}_3 = 0$$

implying that one of the other components of $z$ is zero. Thus it only remains to find solutions where all components of $z$ are nonzero:

**4.3. All components nonzero.** Using the $T^2$ symmetry (10), it is always possible to translate any solution of $B_5(z) = 0$ such that both

$$z_1 \in R^+ \text{ and } z_2 \in R^+.$$

If we write $z_j = r_j e^{i\theta_j}$ with $\theta_1 = \theta_2 = 0$ and look at the imaginary parts of the first two components of $B_5 = 0$, we obtain that

$$-\frac{18}{5}r_1 r_2^2 r_3 r_4 \sin(\theta_3 + \theta_4) - \frac{9}{5}r_2 r_3^2 r_4^2 \sin 2(\theta_3 - \theta_4) = 0$$

$$-\frac{18}{5}r_1^2 r_2 r_3 r_4 \sin(\theta_3 + \theta_4) - \frac{9}{5}r_1 r_3^2 r_4^2 \sin 2(\theta_4 - \theta_3) = 0$$

from which we have, as solution of this homogeneous system,

(16)                    $$\sin 2(\theta_3 - \theta_4) = \sin(\theta_3 + \theta_4) = 0.$$

This implies that, (working mod $2\pi$),

$$(2\theta_3, 2\theta_4) = \left\{ \begin{array}{l} (0,0) \\ (\pi, \pi) \end{array} \right.$$

giving that $z_3$ and $z_4$ are either both real or both pure imaginary. We treat each of these cases below, but note that already at this stage, we can see we have removed all continuous symmetries from the problem, and so our fifth order terms have broken the $T^4$ symmetry of the third order truncation to give the original symmetry group containing a maximal 2-torus.

What is more, we know that any solutions with all components nonzero must become tangent to solutions of the third order equations with all components nonzero. As the only solutions of the third order equations have $|z_i| = r$ independent of $i$, we can apply for the fifth order equations the ansatz that

$$r_i^2 = -\frac{\lambda}{21} + b_i \lambda^2 + O(\lambda^3), \; i = 1, 2, 3, 4.$$

Case of $z_3$ and $z_4$ real:

In the first case, $\theta_1 = \theta_2 = \theta_3 = \theta_4 = 0$ giving that

$$b_1 = b_2 = b_3 = b_4 = \frac{4219}{493920}$$

and the corresponding solution is thus

$$z = (1, 1, 1, 1)r$$
$$r^2 = -\frac{\lambda}{21} + \frac{4219}{493920}\lambda^2 + O(\lambda^3).$$

Secondly, we can have $\theta_1 = \theta_2 = \theta_3 = 0$, $\theta_4 = \pi$, giving that

$$(b_1, b_2, b_3, b_4) = (\frac{4603}{493920}, \frac{4603}{493920}, \frac{3259}{493920}, \frac{3259}{493920}).$$

Thirdly, with $\theta_1 = \theta_2 = 0$, $\theta_3 = \theta_4 = \pi$ we get again

$$b_1 = b_2 = b_3 = b_4 = \frac{4219}{493920}.$$

Case of $z_3$ and $z_4$ both pure imaginary:

We now assume that $\theta_1 = \theta_2 = 0$, $\theta_3 = \pi/2$, $\theta_4 = -\pi/2$ and obtain again that

$$b_1 = b_2 = b_3 = b_4 = \frac{4219}{493920}.$$

Finally we assume that $\theta_1 = \theta_2 = 0$, $\theta_3 = \pi/2$ and $\theta_4 = \pi/2$ and obtain again that

$$(b_1, b_2, b_3, b_4) = (\frac{4603}{493920}, \frac{4603}{493920}, \frac{3259}{493920}, \frac{3259}{493920}).$$

This exhausts all the possibilities of solutions $(\theta_3, \theta_4)$ to equation (16), up to the action of the group (10).

We see that the fifth order terms determine this problem, but note that the solution of the third order terms helps to organise the problem of solving the four coupled fifth order complex equations. Each isolated solution parametrised by $\lambda$ of (15) induces a solution branch of the exact bifurcation equations, and consequently a solution branch of (1) bifurcating at $(0, \mu_0) = (0, 5)$. Since the truncated bifurcation equations (15) is determined, its solutions obtained above yield the bifurcation scenario of (1) and (14) at $\mu_0 = 5$.

## References

1. E.L. Allgower, K. Böhmer and Z. Mei: Exploiting symmetry in the reduced bifurcation equations. In *Bifurcation and Symmetry*, eds. E. L. Allgower, K. Böhmer and M. Golubitsky. ISNM **104**, Birkhäuser, Basel, 1-10, 1992

2. P. Ashwin: High corank mode interactions on a rectangle. In *Bifurcation and Symmetry*, eds. E. L. Allgower, K. Böhmer and M. Golubitsky. ISNM **104**, Birkhäuser, Basel, 23-33, 1992.

3. P. Ashwin: *PhD thesis, Maths Institute, University of Warwick* 1991.

4. P. Ashwin, K. Böhmer and Z. Mei: A numerical Liapunov-Schmidt method with applications to Hopf bifurcation on a square. Preprint, 1992.

5. P.J. Aston: Analysis and computation of symmetry-breaking bifurcation and scaling laws using group theoretic methods. *SIAM J. Math. Anal.* **22**:181-212, 1991.

6. J.D. Crawford: Normal forms for driven surface waves. *Physica D*, **52**:429-457, 1991.

7. J.D. Crawford, J.P. Gollub and D. Lane: *Hidden symmetries of parametrically forced waves*. Preprint, Dept of Physics and Astronomy, University of Pittsburgh, 1992.

8. G. Dangelmayr and D. Armbruster: Steady-state mode interactions in the presence of $O(2)$ symmetry and in non-flux boundary value problems. AMS Contemp. Maths **56**, 1986.

9. M. Golubitsky and D. Schaeffer: *Singularities and groups in bifurcation theory. Volume 1*, App. Math. Sci. **51** Springer, New York, 1986.

10. M.G.M. Gomes: *PhD thesis, Mathematics Institute, University of Warwick.* 1992.

11. M.G.M. Gomes and I.N. Stewart: *Steady pdes on generalised rectangles: a change of genericity in mode interactions.* Preprint 8/1993, Mathematics Institute, University of Warwick, 1993.

12. Z. Mei: Utilization of scaling laws and symmetries in the path following of a semilinear elliptic problem. In *Bifurcation and Symmetry*, eds. E. L. Allgower, K. Böhmer and M. Golubitsky. ISNM **104**, Birkhäuser, Basel, pp. 263-273, 1992.

13. T. J. Healey and H. Kielhöfer: Symmetry and nodal properties in global bifurcation analysis of quasi-linear elliptic equations, *Arch. Rational Mech. Anal.* **113**:299-311, 1991.

E. L. Allgower

Dept. Math., Colorado State University, Fort Collins, CO 80523, USA.

P. Ashwin

Mathematics Institute, University of Warwick, Coventry CV4 7AL, UK.

K. Böhmer

FB Math., Universität Marburg, Lahnberge, D-3550 Marburg, FRG.

Z. Mei

FB Math., Universität Marburg, Lahnberge, D-3550 Marburg, FRG and Dept. Math., Xian Jiaotong University, Xian 710049, PRC.

Lectures in Applied Mathematics
Volume **29**, 1993

# Exploiting Permutation Symmetries with Fixed Points in Linear Equations

EUGENE ALLGOWER, KURT GEORG, AND RICK MIRANDA

April 22, 1993

ABSTRACT. The basic problem discussed here is the exploitation of symmetry in solving linear systems of equations which are equivariant under a permutation of the indices. It is essentially an overview of the method given in [**11**] and [**10**], augmented by a careful discussion of the case of fixed points, which was left untreated in these articles.

## 1. Introduction

Let $V$ be an $N$-dimensional vector space over $\mathbb{C}$, and let $\mathcal{L} : V \to V$ be a nonsingular linear map. Any vector $v \in V$ gives rise to a linear system

$$(1.1) \qquad \mathcal{L}u = v$$

which typically one wants to solve for the unknown vector $u$.

Suppose that $G$ is a finite group, acting linearly on $V$. We denote the action of a group element $g$ on a vector $v$ by $gv$. We say that $\mathcal{L}$ is *G-equivariant* if

$$\mathcal{L}(gv) = g\mathcal{L}(v)$$

for all $g \in G$ and all $v \in V$.

The main problem we wish to address in this article is how to use the equivariance of $\mathcal{L}$ to solve the system (1.1) more efficiently.

This problem has been addressed previously is such works as [**1**], [**2**], [**3**], [**5**], [**6**] [**7**], [**9**], [**11**], and [**10**]. Some of these, dealing with finite and boundary element methods, treat fundamental domain decomposition; others discuss the method of symmetry-adapted bases.

---

1991 *Mathematics Subject Classifications.* 65R20, 65F05, 20C15, 45L10, 31C20

*Key words and phrases.* symmetry, group, representations.

Research supported in part by the NSF under grant DMS-9104058

This paper is in final form and no version of it will be submitted elsewhere.

In this article we will concentrate on the case of permutation symmetry. This is the assumption that the vector space $V$ has a basis $\{e_i\}$ which is simply permuted by the action of $G$. That is, we have an index set $\mathcal{I}$, indexing the basis of $V$, and for every group element $g \in G$, we have a permutation (which we will also call $g$) of the index set. This induces a linear action of $G$ on $V$ by setting

$$ge_i = e_{gi}$$

and extending by linearity. This is called a *permutation representation* of the group $G$ on $V$.

The basis $\{e_i\}$ for $V$ allows us to identify $V$ with a space of column vectors of length $N$. We will denote the $i^{th}$ coordinate of a column vector $v$ by $v[i]$. Note that this implies that the permutation action of $G$ on $V$ can be expressed as

$$(gv)[m] = v[g^{-1}m].$$

We will denote the $ij^{th}$ entry of a matrix $L$ by $L[i,j]$. If the operator $\mathcal{L}$ has the matrix $L$ with respect to this basis $\{e_i\}$, then the $G$-equivariance of $\mathcal{L}$ when the representation of $G$ is a permutation representation as described above is equivalent to the formula

$$L[m,n] = L[gm, gn]$$

holding for all $g \in G$ and all indices $m,n$.

We say that a permutation representation is *without fixed points* if no index $m$ is fixed by any non-trivial group element $g$. In general, the *isotropy subgroup* of an index $m$ is the subgroup $G_m = \{g \in G \mid gm = m\}$. If $G$ acts without fixed points, then every isotropy subgroup is trivial. The *orbit* of an index $n$ is the set $Gn = \{gn \mid g \in G\}$ of indices to which $n$ is moved by the group elements in $G$.

If $G$ acts without fixed points, we see that there are exactly $\#G$ elements in each orbit. In this case any two orbits are either identical or disjoint, so that there are exactly $M = N/\#G$ orbits.

A *selection* of indices is a subset $\mathcal{J}$ of $\mathcal{I}$ which consists of one index in each orbit of $G$.

## 2. Irreducible Representations and Projectors

Let $\rho : G \to \mathrm{GL}(d)$ be a homomorphism from $G$ to the group $\mathrm{GL}(d)$ of invertible $d \times d$ matrices over $\mathbb{C}$. Such a homomorphism is called a *complex matrix representation* of $G$, of dimension $d$. It is said to be *irreducible* if the matrices of $\rho$ cannot be simultaneously put into an upper triangular block form. We will also denote the matrix $\rho(g)$ by $\mathbf{A}^\rho(g)$.

Recall that two elements $g_1$ and $g_2$ of a group $G$ are *conjugate* if $g_1 = gg_2g^{-1}$ for some $g \in G$; this is an equivalence relation on $G$, and the equivalence classes are called the *conjugacy classes* of $G$.

In the context of complex matrix representations, one usually uses the terminology that two complex matrix representations $\rho_1$ and $\rho_2$ are em equivalent if there is a fixed invertible matrix $A$ such that $A\rho_1(g)A^{-1} = \rho_2(g)$ for all $g \in G$.

The following theorem contains some basic facts about irreducible complex matrix representations of a finite group; for a reference, see [12] or [4].

THEOREM 2.1. *Let $G$ be a finite group of order $\#G$, with $R$ conjugacy classes.*

2.1.1: *There are $R$ non-equivalent irreducible complex matrix representations $\rho_1, \ldots, \rho_R$ of dimensions $d_1, \ldots, d_R$ respectively, such that any irreducible matrix representation of $G$ is equivalent to exactly one of the $\rho_i$'s.*

2.1.2: *Any complex matrix representation is equivalent to a direct sum of the $\rho_i$'s.*

2.1.3: $\sum_i d_i^2 = \#G$.

2.1.4: *The representations $\rho_i$ can be taken to be unitary.*

We call a set of representations $\{\rho_i\}$ as in the above theorem a *full set of irreducible representations* for $G$.

Given an irreducible matrix representation $\rho$ of $G$, we will write the $ij^{th}$ element of the matrix $\mathbf{A}^\rho(g)$ as $A_{ij}^\rho(g)$.

Let $V$ be any representation of $G$, and let $\rho$ be an irreducible matrix representation of $G$. Following [12, section 2.7], define maps

$$P_{ij}^\rho : V \to V$$

by the formula

$$P_{ij}^\rho(v) = \frac{\dim \rho}{\#G} \sum_{g \in G} A_{ji}^\rho(g^{-1})gv.$$

Note the reversal of indices and the use of the inverse element in the sum. Since the action operation sending $v$ to $gv$ is linear for each $g$, the map $P_{ij}^\rho$ is a linear operator from $V$ to $V$, and it is defined for each $i$ and $j$ between 1 and $\dim \rho$.

We have the following orthogonality relation (see [12, section 2.7])

PROPOSITION 2.2. *Let $\rho$ and $\tau$ be two irreducible complex matrix representations of $G$ from a full set of irreducible representations. (I.e., $\rho$ and $\tau$ are either identical or non-equivalent.) Then*

$$P_{ij}^\rho \circ P_{kl}^\tau = P_{il}^\tau \delta_{jk}\delta_{\rho\tau}.$$

Note that this implies that the "diagonal" operators $P_{ii}^\rho$ are a set of mutually orthogonal projectors, defined on any vector space $V$ on which $G$ acts, as $\rho$ ranges over a full set of irreducible representations for $G$, and as $i$ runs from 1 to $\dim \rho$. Indeed, they form a complete set of projectors (see [12, section 2.7]):

PROPOSITION 2.3. *Let $G$ act linearly on a vector space $V$. Then*

$$\sum_\rho \sum_{i=1}^{\dim \rho} P_{ii}^\rho = \mathrm{Id}_V.$$

*In the above sum, $\rho$ ranges over a full set of irreducible representations for $G$.*

We denote by $V_i^\rho$ the image of the projector $P_{ii}^\rho$. The previous Proposition implies that the subspaces $V_i^\rho$ completely decompose the original space $V$:

$$(2.4) \qquad V = \bigoplus_\rho \bigoplus_{i=1}^{\dim \rho} V_i^\rho.$$

Given a vector $v \in V$, we denote by $v_{ij}^\rho$ its image under $P_{ij}^\rho$:

$$v_{ij}^\rho = P_{ij}^\rho(v).$$

This gives the formula

$$(2.5) \qquad v = \sum_\rho \sum_{i=1}^{\dim \rho} v_{ii}^\rho$$

for any vector $v \in V$, by Proposition 2.3.

Using this notation we may re-write the definition of the $P_{ij}^\rho$'s as follows:

$$(2.6) \qquad v_{ij}^\rho[m] = \frac{\dim \rho}{\#G} \sum_{g \in G} A_{ji}^\rho(g^{-1}) v[g^{-1}m],$$

## 3. The Reduction to the Sub-problems

Let the finite group $G$ act as a group of permutations on the indices of a basis for the vector space $V$. Let $\mathcal{L}$ be a $G$-equivariant nonsingular operator on $V$, with matrix $L$ in the permuted basis. Let $\rho$ be an irreducible matrix representation of $G$. Since $\mathcal{L}$ commutes with the action of each group element $g \in G$, and the operators $P_{ij}^\rho$ are simply linear combinations of the group elements, we see that $\mathcal{L}$ commutes with each $P_{ij}^\rho$; in particular, we have

$$\mathcal{L}(P_{ii}^\rho(v)) = P_{ii}^\rho(\mathcal{L}(v))$$

for every $v \in V$. Therefore we see that $\mathcal{L}$ maps $V_i^\rho$ to itself, i.e., the subspace $V_i^\rho$ is stable under the operator $\mathcal{L}$. Let us denote by $\mathcal{L}_i^\rho$ the restriction of $\mathcal{L}$ to $V_i^\rho$:

$$\mathcal{L}_i^\rho = \mathcal{L}|_{V_i^\rho} : V_i^\rho \to V_i^\rho.$$

Note that since $\mathcal{L}$ is nonsingular, so is each $\mathcal{L}_i^\rho$.

We now have the following plan for reducing the original system (1.1) into sub-problems:

**To Solve $\mathcal{L}u = v$:**

(i) Choose a full set $\{\rho\}$ of irreducible representations for $G$.

(ii) For each $\rho$ and each $i = 1, \ldots, \dim \rho$, apply the projectors $P_{ii}^{\rho}$ to both sides of the system, obtaining the sub-system $P_{ii}^{\rho}(v) = P_{ii}^{\rho}(\mathcal{L}(u)) = \mathcal{L}(P_{ii}^{\rho}(u))$ which we write as

(3.1)
$$\mathcal{L}_i^{\rho}(u_{ii}^{\rho}) = v_{ii}^{\rho}$$

(iii) Solve these sub-problems for the unknowns $u_{ii}^{\rho}$.

(iv) Reconstruct the solution

$$u = \sum_{\rho} \sum_{i=1}^{\dim \rho} u_{ii}^{\rho}.$$

## 4. The Method Without Fixed Points

It is now time to get specific about how to write down the matrices for the sub-problems, in the case of permutation symmetry.

There are two basic formulas which govern the construction of the matrices for the sub-operators $\mathcal{L}_i^{\rho}$. The first comes from the following equation relating the group action and the operators $P_{ij}^{\rho}$ (see [12, section 2.7]):

LEMMA 4.1. *If $\rho$ is an irreducible matrix representation of $G$, then*

$$g P_{ij}^{\rho} = \sum_{k} A_{ki}^{\rho}(g) P_{kj}^{\rho}.$$

The above formula has as a special but important case the following, expressed in terms of the $v_{ij}^{\rho}$'s:

(4.2)
$$v_{kk}^{\rho}[gm] = \sum_{i=1}^{\dim \rho} A_{ik}^{\rho}(g^{-1}) v_{ik}^{\rho}[m].$$

¿From this we see the following principle: to determine all of the coordinates of the vector $v_{kk}^{\rho}$, it suffices to know the coordinates $v_{ik}^{\rho}[m]$, *for one $m$ in each $G$-orbit of indices.*

Hence we have natural "coordinates" for a vector $v$ in the space $V_k^{\rho}$ consisting of the numbers $v_{ik}^{\rho}[m]$, as $i$ goes from 1 to $\dim \rho$, and $m$ ranges over a selection of indices, i.e., one from each $G$-orbit.

If $G$ acts without fixed points, then there are $M = N/\#G$ elements in any selection; hence the number of these "coordinates" is $M \dim \rho$. This is exactly the dimension of the space $V_k^{\rho}$, and hence we have no duplication or relations among these "coordinates".

To see this, we view the introduction of the coordinates $v_{ik}^{\rho}[m]$ as showing that the dimension of $V_k^{\rho}$ is at most the number of these coordinates, which is $M \dim \rho$:

(4.3)
$$\dim(V_k^{\rho}) \leq M \dim \rho = N \dim \rho / \#G.$$

Now the total space $V$ has dimension $N$, and decomposes as the direct sum of the spaces $V_k^\rho$, by (2.4). Therefore

$$N = \dim V \le \sum_\rho \sum_{k=1}^{\dim \rho} (N \dim \rho / \#G) = \frac{N}{\#G} \sum_\rho \dim^2 \rho.$$

Now by Theorem 2.1.3, $\sum_\rho \dim^2 \rho = \#G$; hence the inequality in the previous formula must be an equality, which forces $\dim(V_k^\rho) = M \dim \rho = N \dim \rho / \#G$ for every $\rho$ and $k$, in the case where $G$ acts without fixed points.

Choose therefore once and for all a selection $\mathcal{J} \subset \mathcal{I}$ of $M$ indices. We now want to write the matrix equation for the sub-operator $\mathcal{L}_k^\rho$ using these new "coordinates". Thus we seek a matrix $L_k^\rho$, whose rows and columns are indexed by pairs $(i, m)$, ($i$ goes from 1 to $\dim \rho$, and $m$ ranges over the selection $\mathcal{J}$ of indices), which represents the matrix of the sub-operator $\mathcal{L}_k^\rho$. That is, we seek the entries $L_k^\rho[(i, m), (j, n)]$, such that

$$P_{ik}^\rho(\mathcal{L}_k^\rho(v_{kk}^\rho))[m] = \sum_{(j,n)} L_k^\rho[(i, m), (j, n)] v_{jk}^\rho[n].$$

These entries are readily worked out from the formulas for $P_{ik}^\rho$; we obtain the following.

(4.4)            $$L_k^\rho[(i, m), (j, n)] = \sum_{g \in G} L[g^{-1} m, n] A_{ji}^\rho(g^{-1})$$

Notice now that these entries are independent of the index $k$! This means that we actually have a single matrix $L^\rho = L_k^\rho$. This shows that our method is a variant of the "symmetry-adapted basis" technique described in [9].

In addition, note that to determine these matrices, we do not need to know the entire original matrix $L$; we only need to know those columns of $L$ indexed by elements in the given selection $\mathcal{J}$.

Now the specific application of the method in the case of permutation symmetry without fixed points can be described with the following algorithm.

**To Solve $\mathcal{L}u = v$:**
  (i) Choose a full set $\{\rho\}$ of irreducible representations for $G$.
  (ii) Choose a selection $\mathcal{J}$ of indices, one from each $G$-orbit.
  (iii) Determine the entries $L[m, n]$ of the matrix for the operator $\mathcal{L}$, for all indices $m \in \mathcal{I}$ and all selected indices $n \in \mathcal{J}$.
  (iv) For each irreducible matrix representation $\rho$:
      a: Form the doubly-indexed matrices $L^\rho$ using (4.4).
      b: Using (2.6), determine the coordinates $v_{ik}^\rho[m]$ for all $i$ and $k$ from 1 to $\dim \rho$ and all selected indices $m \in \mathcal{J}$.
      c: Solve the sub-problems

$$L^\rho u_{ik}^\rho = v_{ik}^\rho$$

for the unknowns $u_{ik}^\rho[m]$, again for $i, k = 1, \ldots, \dim \rho$ and all selected indices $m \in \mathcal{J}$.

d: For each $k = 1, \ldots, \dim \rho$, reconstruct every coordinate of the vector $u_{kk}^\rho$ using (4.2).

(v) Reconstruct the solution $u = \sum_\rho \sum_{k=1}^{\dim \rho} u_{kk}^\rho$.

## 5. Complexity Reduction Factors

Suppose that one is using a direct linear system solver which has complexity $O(n^3)$ for an $n \times n$ problem, as is suitable for full problems. To solve our original system of size $N$ which is considered to be full, would then require approximately $cN^3$ flops, for some constant $c$.

In using the above method, what is the reduction factor in the complexity? The work is done in the step 4(c) in the outline in the previous section, where the linear sub-problems are solved. For each irreducible matrix representation $\rho$, we must solve the system of step 4(c), which is of size $\dim \rho N / \#G$. This has computational cost approximately $c(\dim \rho N / \#G)^3$. Hence the total cost is approximately $\sum_\rho c(\dim \rho N / \#G)^3$; dividing by the cost of the original problem $cN^3$ gives a reduction factor

$$\frac{1}{(\#G)^3} \sum_\rho \dim^3 \rho.$$

Note that in the above computation we have neglected the overhead in the method, which is the work involved in constructing the matrices and right-hand sides of the sub-problems in steps 4(a) and 4(b) of the outline in the previous section. This overhead is not insignificant; we have computed it to be approximately $N^2$ flops, and the reader should consult [13] for numerical results using this method in a boundary element problem involving the unit cube.

## 6. Matrix Notation for the Projectors and Transfers

The double-indexing of the projectors $P_{jj}^\rho$ and tranfers $P_{ij}^\rho$ indicate that we may benefit from grouping these functions together and viewing the output as a matrix. In this section we would like to lay out the notation for this point of view.

Fix an irreducible representation $\rho$, and let $d = \dim \rho$ be its dimension. For any vector $v \in V$ and any index $n \in \mathcal{I}$, let $\mathbf{v}^\rho[n]$ denote the $d$-by-$d$ matrix whose $ij^{th}$ entry is $v_{ij}^\rho[n]$:

(6.1)
$$\mathbf{v}^\rho[n] = \begin{pmatrix} v_{11}^\rho[n] & \cdots & v_{1d}^\rho[n] \\ \vdots & \ddots & \vdots \\ v_{d1}^\rho[n] & \cdots & v_{dd}^\rho[n] \end{pmatrix}$$

We will also use the notation $\mathbf{P}^\rho$ to denote the matrix of operators

$$\mathbf{P}^\rho = \left( P_{ij}^\rho \right),$$

so that for a vector $v \in V$, $\mathbf{P}^\rho(v)[n] = \mathbf{v}^\rho[n]$. In particular, we think of $\mathbf{P}^\rho(v)$ as a column vector of $d \times d$ matrices, not as a $d \times d$ matrix of vectors.

Using the formula (2.6), we see that the definition of $\mathbf{v}^\rho[m]$ can be expressed as a linear combination of transposes of the matrices $\mathbf{A}^\rho$ which define the representation $\rho$:

$$(6.2) \qquad \mathbf{v}^\rho[m] = \frac{\dim \rho}{\#G} \sum_{g \in G} v[g^{-1}m] \mathbf{A}^\rho(g^{-1})^\top.$$

Proposition 2.2 and Lemma 4.1 can be expressed in the following succinct way.

PROPOSITION 6.3.

6.3.1: *Let* $g \in G$, $n \in \mathcal{I}$, *and* $v \in V$. *If* $\rho$ *is an irreducible matrix representation of* $G$, *then*

$$\mathbf{v}^\rho[gn] = \mathbf{A}^\rho(g^{-1})^\top \mathbf{v}^\rho[n].$$

6.3.2: *Let* $\rho$ *and* $\tau$ *be two irreducible complex matrix representations of* $G$ *from a full set of irreducible representations. (I.e.,* $\rho$ *and* $\tau$ *are either identical or non-equivalent.) Then*

$$\mathbf{P}^\rho \circ \mathbf{P}^\tau = \dim \tau \, \mathbf{P}^\tau \delta_{\rho\tau}.$$

The formula (6.2) expresses the matrices $\mathbf{v}^\rho[m]$ in terms of the original vector $v$. Inverting this procedure is also straightforward. Fix an index $n$. Then by (2.5), we have

$$(6.4) \qquad v[n] = \sum_\rho \operatorname{trace}(\mathbf{v}^\rho[n]).$$

In fact, the image $v^\rho$ of $v$ in the canonical space $V^\rho$ is

$$(6.5) \qquad v^\rho[n] = \operatorname{trace}(\mathbf{v}^\rho[n]).$$

The final bit of organization for these matrices $\mathbf{v}^\rho[n]$ is to make a "vector of matrices" out of them. Specifically, let $\mathbf{v}^\rho$ be defined as the column vector whose $n^{th}$ entry is the matrix $\mathbf{v}^\rho[n]$:

$$(6.6) \qquad \mathbf{v}^\rho = \begin{pmatrix} \mathbf{v}^\rho[1] \\ \mathbf{v}^\rho[2] \\ \vdots \\ \mathbf{v}^\rho[N] \end{pmatrix}.$$

Denote by $\mathbf{V}_{\mathcal{I}}^{\rho}$ the vector space of such columns of matrices. It is a complex vector space of dimension $Nd^2$, where $d = \dim \rho$.

Given a selection $\mathcal{J}$ of indices, we will denote by $\mathbf{V}_{\mathcal{J}}^{\rho}$ the vector space of columns of matrices $\mathbf{v}^{\rho}[m]$ where we only have selected coordinates $m \in \mathcal{J}$. The corresponding "vector" in $\mathbf{V}_{\mathcal{J}}^{\rho}$ will be denoted by $\mathbf{v}_{\mathcal{J}}^{\rho}$:

$$(6.7) \qquad \mathbf{v}_{\mathcal{J}}^{\rho} = \begin{pmatrix} \mathbf{v}^{\rho}[m_1] \\ \mathbf{v}^{\rho}[m_2] \\ \vdots \\ \mathbf{v}^{\rho}[m_M] \end{pmatrix}.$$

The space $\mathbf{V}_{\mathcal{J}}^{\rho}$ is a complex vector space of dimension $Md^2$, where $d = \dim \rho$ and $M$ is the number of orbits of the action of $G$ on the indices $\mathcal{I}$.

Finally fix a selection $\mathcal{J}$ and let $d = \dim \rho$. Denote by $\mathbf{U}_{\mathcal{J}}^{\rho}$ the vector space with coordinates indexed by $\mathcal{J}$ and whose entries are column vectors of length $d$. We are thinking of this space as the space of "column vectors" for elements of $\mathbf{V}_{\mathcal{J}}^{\rho}$. The space $\mathbf{U}_{\mathcal{J}}^{\rho}$ is a complex vector space of dimension $Md$.

The use of the "coordinates" $v_{ik}^{\rho}[m]$ for vectors in the subspace $V_k^{\rho}$ can be expressed in the matrix notation very nicely. View the matrix of operators $\mathbf{P}^{\rho}$ as a linear map from the space of ordinary vectors $V$ to the space of vectors of matrices $\mathbf{V}_{\mathcal{I}}^{\rho}$:

$$\mathbf{P}^{\rho} : V \to \mathbf{V}_{\mathcal{I}}^{\rho}$$

sending a vector $v \in V$ to the column vector $\mathbf{v}^{\rho}$ of matrices. Given the selection $\mathcal{J}$, we have the natural projection of $\mathbf{V}_{\mathcal{I}}^{\rho}$ onto $\mathbf{V}_{\mathcal{J}}^{\rho}$ given by keeping only those $M$ coordinates in $\mathcal{J}$; let us denote this projection by $\pi_{\mathcal{J}}^{\rho}$:

$$\pi_{\mathcal{J}}^{\rho} : \mathbf{V}_{\mathcal{I}}^{\rho} \to \mathbf{V}_{\mathcal{J}}^{\rho}.$$

Finally let $\mathbf{U}_{\mathcal{J}}^{\rho}$ be the space of "vectors of vectors" as described above. For any fixed $j = 1, \ldots, \dim \rho$, we have a natural projection $\pi_{\mathcal{J},j}^{\rho}$ of $\mathbf{V}_{\mathcal{J}}^{\rho}$ onto $\mathbf{U}_{\mathcal{J}}^{\rho}$, sending a vector of matrices $\mathbf{v}_{\mathcal{J}}^{\rho}$ to the vector of $j^{th}$ columns $\mathbf{v}_{\mathcal{J},j}^{\rho}$ (i.e., the $m^{th}$ entry of $\mathbf{v}_{\mathcal{J},j}^{\rho}$ is the $j^{th}$ column of the $m^{th}$ entry of $\mathbf{v}_{\mathcal{J}}^{\rho}$):

$$\pi_{\mathcal{J},j}^{\rho} : \mathbf{V}_{\mathcal{J}}^{\rho} \to \mathbf{U}_{\mathcal{J}}^{\rho}.$$

The composition $\pi_{\mathcal{J},j}^{\rho} \circ \pi_{\mathcal{J}}^{\rho} \circ \mathbf{P}^{\rho}$ can be restricted to the symmetry subspace $V_j^{\rho}$ to give a linear map

$$\phi_j^{\rho} = \pi_{\mathcal{J},j}^{\rho} \circ \pi_{\mathcal{J}}^{\rho} \circ \mathbf{P}^{\rho} : V_j^{\rho} \to \mathbf{U}_{\mathcal{J}}^{\rho}.$$

PROPOSITION 6.8. *With the above notation, if $G$ acts without fixed points, the linear map* $\phi_j^{\rho} : V_j^{\rho} \to \mathbf{U}_{\mathcal{M}}^{\rho}$ *is an isomorphism of complex vector spaces.*

PROOF. We can easily write down the inverse map, guided by the formula (4.2). Fix an index j, and let $(\underline{w})$ be a vector of vectors in $\mathbf{U}^\rho_{\mathcal{J}}$ (so that $\underline{w}[m]$ is a column vector of length $d = \dim \rho$, whose $i^{th}$ coordinate is $w_i[m]$, say). Define $\psi : \mathbf{U}^\rho_{\mathcal{J}} \to V^\rho_j$ by

$$\psi(\underline{w})[n] = \sum_{k=1}^{\dim \rho} A^\rho_{kj}(g^{-1})w_k[m]$$

if $n = gm$. (Under the assumption of no fixed points, there is a unique selected index $m \in \mathcal{J}$ and a unique group element $g \in G$ such that $n = gm$.) We leave it to the reader to check that $\psi$ maps $\mathbf{V}^\rho_{\mathcal{J},j}$ into $V^\rho_j$, and that it is an inverse to $\phi^\rho_j$. □

An alternate proof of the above proposition is obtained by noting that, in the case where $G$ acts without fixed points, these spaces have the same dimension. Hence it suffices to show that the linear map $\phi^\rho_j$ is 1-1, which is elementary.

As a corollary, we see that we have a similar isomorphism for the canonical subspaces. Let $\phi^\rho$ be the composition $\pi^\rho_{\mathcal{J}} \circ \mathbf{P}^\rho$ restricted to the canonical subspace $V^\rho = \bigoplus_i V^\rho_i$.

COROLLARY 6.9. *With the above notation, if $G$ acts without fixed points, the linear map $\phi^\rho : V^\rho \to \mathbf{V}^\rho_{\mathcal{J}}$ is an isomorphism of complex vector spaces.*

With this use of matrix notation, the linear system for the subproblem (3.1) can now be written as

(6.10)                           $\mathbf{L}^\rho \mathbf{u}^\rho = \mathbf{v}^\rho$

where $\mathbf{L}^\rho$ is an appropriately-indexed version of the matrix $L^\rho[(i,m),(j,n)]$ constructed in Section 4. This matrix $\mathbf{L}^\rho$ is an $Md$-by-$Md$ matrix, which because of the double-indexing, can be viewed naturally as an $M$-by-$M$ matrix whose $mn^{th}$ entry is a $d$-by-$d$ matrix. Indeed, using the definition (4.4) of $L^\rho[(i,m),(j,n)]$, we see that the $mn^{th}$ entry is the matrix $\sum_{g \in G} L[g^{-1}m,n]A^\rho(g^{-1})^\top$. Thus $\mathbf{L}^\rho$ in this form is naturally *a sum of tensor product matrices.*

Recall that the tensor product of two matrices $\mathbf{X}$ and $\mathbf{Y}$ which are $d$-by-$d$ and $e$-by-$e$ respectively is the $de$-by-$de$ matrix $\mathbf{X} \otimes \mathbf{Y}$ which breaks naturally into a $d$-by-$d$ matrix with $e$-by-$e$ matrix entries: the matrix in the $rs^{th}$ block is $X_{rs}\mathbf{Y}$.

To write our matrix $\mathbf{L}^\rho$ with this notation, one more definition is useful. Specifically, for $g \in G$ define the $M$-by-$M$ matrix $\mathbf{L}_g$ by

(6.11)                          $\mathbf{L}_g[m,n] = L[g^{-1}m,n]$.

Then

(6.12)                          $\mathbf{L}^\rho = \sum_{g \in G} \mathbf{L}_g \otimes \mathbf{A}^\rho(g^{-1})^\top$.

The overview of the method given in this article can now be written with this notation as follows:

Given: a vector $v$ of $V$.

To Solve: $\mathcal{L}(u) = v$ for the unknown vector $u$.

1: Choose a selection $\mathcal{J}$ of indices.

2: Find a full set of irreducible complex matrix representations for $G$.

3: Determine the entries $L[m, n]$ of the matrix for $\mathcal{L}$ in the selected columns, find the matrices $\mathbf{L}_g$ for each group element $g$, and form the matrix $\mathbf{L}^\rho$ for each complex matrix representation $\rho$.

4: For each $\rho$ in the full set of irreducible complex matrix representations, perform steps 4a - b:

    4a: Form the column vector (with matrix entries) $\mathbf{v}^\rho$ from the given vector $v$, using (6.2):

$$\mathbf{v}^\rho[m] = \frac{\dim \rho}{\#G} \sum_{g \in G} v[g^{-1}m]\mathbf{A}^\rho(g^{-1})^\top$$

    4b: Solve the subproblems (6.10) for the column vector (with matrix entries) $\mathbf{u}^\rho$:

$$\mathbf{L}^\rho \mathbf{u}^\rho = \left[ \sum_{g \in G} \mathbf{L}_g \otimes \mathbf{A}^\rho(g^{-1})^\top \right] \mathbf{u}^\rho = \mathbf{v}^\rho.$$

5: Reconstruct the solution vector $u$ from the $\mathbf{u}^\rho$'s using (6.5) and (2.5):

$$u[gm] = \sum_\rho \operatorname{trace}(\mathbf{A}^\rho(g^{-1})^\top \mathbf{u}^\rho[m]).$$

## 7. Variations in the Case of Fixed Points

How does the preceding discussion change if there are fixed points for the permutation action of $G$ on the indices? The alteration in the method comes from the fact that the coordinates $v_{ik}^\rho[m]$ no longer are independent coordinates for the subspaces $V_k^\rho$. One sees this by analyzing the formula (4.2) in the case where $m$ is a fixed index under a group element $g$, i.e., when $gm = m$; one obtains

$$(7.1) \qquad v_{kk}^\rho[m] = \sum_{i=1}^{\dim \rho} \mathbf{A}_{ik}^\rho(g^{-1})v_{ik}^\rho[m].$$

which gives a relation among the coordinates $v_{ik}^\rho[m]$ whenever we have a non-trivial element $g$ fixing an index $m$.

Expressed in the matrix notation, using Proposition 6.3.1 this becomes

$$(7.2) \qquad \mathbf{v}^\rho[m] = \mathbf{A}^\rho(g^{-1})^\top \mathbf{v}^\rho[m]$$

for any group element $g$ in the isotropy subgroup $G_m$.

We claim that these are the only conditions on the coordinates. To be precise, let $\widehat{\mathbf{V}^\rho_{\mathcal{J}}}$ be the subspace of $\mathbf{V}^\rho_{\mathcal{J}}$ consisting of those "vectors of $d \times d$ matrices" $(\mathbf{w}[-])$ indexed by $m \in \mathcal{J}$ satisfying (7.2), i.e.,

$$(7.3) \qquad \mathbf{w}[m] = \mathbf{A}^\rho(g^{-1})^\top \mathbf{w}[m]$$

for every $m \in \mathcal{J}$ and every $g$ in the isotropy subgroup $G_m$. Then we have immediately that the linear map $\phi^\rho$ defined above maps the canonical space $V^\rho$ into the subspace $\widehat{\mathbf{V}^\rho_{\mathcal{J}}}$.

PROPOSITION 7.4. *With the above notation, the linear map*

$$\phi^\rho : V^\rho \to \widehat{\mathbf{V}^\rho_{\mathcal{J}}}$$

*is an isomorphism of complex vector spaces.*

PROOF. Again we will proceed by constructing an inverse map $\psi$ for $\phi^\rho$. Fix an element $(\mathbf{w}[-])$ in $\widehat{\mathbf{V}^\rho_{\mathcal{J}}}$, i.e., $(\mathbf{w}[-]) \in \mathbf{V}^\rho_{\mathcal{J}}$ and satisfies (7.3) for every $m \in \mathcal{J}$ and every $g \in G_m$. Fix an index $n \in \mathcal{I}$, and write $n = gm$ for some $g \in G$ and $m \in \mathcal{J}$. Define $\psi((\mathbf{w}[-]))$ to be the vector $v$ with $n^{th}$ coordinate

$$v[n] = \text{trace}(\mathbf{A}^\rho(g^{-1})^\top \mathbf{w}[m]).$$

The reader can easily check that this assignment is well-defined, independent of the choice of the group element $g$. (In the case of non-trivial isotropy, more than one group element can send $m$ to $n$.) Moreover, one next checks that $\psi$ actually maps to the canonical subspace $V^\rho$.

Finally it is an easy matter to verify that $\psi$ is an inverse for $\phi^\rho$. $\square$

The corresponding statement for the symmetry subspaces $\mathbf{V}^\rho_j$ can be obtained as a corollary of the above, since the $j^{th}$ symmetry space $V^\rho_j$ exactly corresponds under the above isomorphism to the $j^{th}$ columns of the associated vector of matrices.

To be specific, let $\widehat{\mathbf{U}^\rho_{\mathcal{J}}}$ be the subspace of $\mathbf{U}^\rho_{\mathcal{J}}$ consisting of those "vectors of vectors" $(\mathbf{u}[-])$ indexed by $m \in \mathcal{J}$ satisfying the vector version of (7.2), i.e.,

$$(7.5) \qquad \mathbf{u}[m] = \mathbf{A}^\rho(g^{-1})^\top \mathbf{u}[m]$$

for every $m \in \mathcal{J}$ and every $g \in G_m$. Then we have immediately that the linear map $\phi^\rho_j$ defined above maps the symmetry space $V^\rho_j$ into the subspace $\widehat{\mathbf{U}^\rho_{\mathcal{J}}}$.

COROLLARY 7.6. *With the above notation, the linear map*

$$\phi^\rho_j : V^\rho_j \to \widehat{\mathbf{U}^\rho_{\mathcal{J}}}$$

*is an isomorphism of complex vector spaces.*

If indeed $G$ acts without fixed points, then $\widehat{\mathbf{V}^\rho_{\mathcal{J}}} = \mathbf{V}^\rho_{\mathcal{J}}$ and $\widehat{\mathbf{U}^\rho_{\mathcal{J}}} = \mathbf{U}^\rho_{\mathcal{J}}$, so that Proposition 6.8 and Corollary 6.9 are actually trivial consequences of Proposition 7.4 and Corollary 7.6.

Using a bit of character theory, we can develop a formula for the dimension of the vector spaces $\widehat{\mathbf{U}^\rho_{\mathcal{J}}}$ and $\widehat{\mathbf{V}^\rho_{\mathcal{J}}}$ (and therefore, by the above results, for the symmetry spaces $V^\rho_j$ and the canonical spaces $V^\rho$).

PROPOSITION 7.7. *Let the irreducible matrix representation $\rho$ have character function $\chi^\rho$. Then*

$$\dim V^\rho_j = \dim \widehat{\mathbf{U}^\rho_{\mathcal{J}}} = \sum_{m \in \mathcal{J}} \frac{1}{\#G_m} \sum_{g \in G_m} \chi^\rho(g)$$

*and*

$$\dim V^\rho = \dim \widehat{\mathbf{V}^\rho_{\mathcal{J}}} = \dim \rho \sum_{m \in \mathcal{J}} \frac{1}{\#G_m} \sum_{g \in G_m} \chi^\rho(g)$$

PROOF. By considering every component of a "vector of vectors" in $\widehat{\mathbf{U}^\rho_{\mathcal{J}}}$ separately, it is clear that

$$\dim \widehat{\mathbf{U}^\rho_{\mathcal{J}}} = \sum_{m \in \mathcal{J}} \dim\{u \in \mathbb{C}^{\dim \rho} \mid u = \mathbf{A}^\rho(g^{-1})^\top u \text{ for all } g \in G_m\}$$

and so the task is to compute the dimension of this last vector space. This space is exactly the subspace of vectors, in the space of the inverse transpose representation $\tau$ to $\rho$, where $G_m$ acts trivially. I.e., it is the "trivial part" of the restricted representation $\tau|_{G_m}$. The restricted representation $\tau|_{G_m}$ has the character function $\overline{\chi^\rho|_{G_m}}$. Moreover, the trivial part of any representation of any group $H$ has dimension $\sum_{h \in H} \chi(h)/\#H$, where $\chi$ is the character of the representation. Therefore the desired dimension in our case is the quantity $(1/\#G_m) \sum_{g \in G_m} \overline{\chi^\rho(g)}$; since this quantity must be an integer, we may dispense with the complex conjugation in the sum, proving the first part of the proposition. The second equation follows since all the spaces $V^\rho_j$ are isomorphic to each other, and so by Corollary 7.6, so are the spaces $\widehat{\mathbf{U}^\rho_{\mathcal{J}}}$; hence their direct sum $\widehat{\mathbf{V}^\rho_{\mathcal{J}}}$ has dimension equal to $\dim \rho$ times the dimension for $\widehat{\mathbf{U}^\rho_{\mathcal{J}}}$. $\square$

It is instructive to check that in the case where $G$ acts without fixed points, the above formulas give the expected answer. Specifically, in this case, for every $m \in \mathcal{J}$, $G_m = \{I\}$; since $\chi^\rho(I) = \dim \rho$, the above reduces to the value $M \dim \rho$ found in the discussion following (4.3).

Now the method proceeds in a similar manner to the fixed-point-free case. First, one chooses, for each $\rho$, a basis for the space $\widehat{\mathbf{U}^\rho_{\mathcal{J}}}$. Then one transports the linear operator $L$ from $V^\rho_j$ to $\widehat{\mathbf{U}^\rho_{\mathcal{J}}}$ via the isomorphism $\phi^\rho_j$. Actually, at this point one can transport the entire canonical space $V^\rho$ to $\widehat{\mathbf{V}^\rho_{\mathcal{J}}}$ via $\phi^\rho$; it is the same in principle.

The isomorphisms $\phi^\rho$ and $\phi^\rho_j$ are sufficiently explicit to easily transport the right-hand-side vector $v$ into the space. Thus a good choice of basis for the fixed-point space $\widehat{U^\rho_{\mathcal{J}}}$ essentially finishes the variation in the method. This we have done by hand in simple cases. The interested reader should also consult [8] for a classification and construction of the symmetry-adapted bases for all permutation representations of the point groups in dimensions 2 and 3. It is a subject of ongoing study to efficiently determine a useful basis for this space in general, and thus to give a complete algorithm in the fixed-point case. The interested reader should also consult [8] for a classification and construction of the symmetry-adapted bases for all permutation representations of the point groups in dimensions 2 and 3.

## References

1. E. L. Allgower, K. Böhmer, K. Georg, and R. Miranda, *Exploiting symmetry in boundary element methods*, SIAM J. Numer. Anal. **29** (1992), 534–552.

2. E. L. Allgower, K. Böhmer, and Z. Mei, *On a problem decomposition for semi-linear nearly symmetric elliptic problems*, Parallel Algorithms for PDE's (Braunschweig, Fed. Rep. Germany) (W. Hackbusch, ed.), Notes on Numerical Fluid Mechanics, vol. 31, Vieweg Verlag, 1991, pp. 1–17.

3. A. Bossavit, *Symmetry, groups, and boundary value problems. A progressive introduction to noncommutative harmonic analysis of partial differential equations in domains with geometric symmetry*, Computer Methods in Applied Mechanics and Engineering **56** (1986), 165–215.

4. C. W. Curtis and I. Reiner, *Representation theory of finite groups and associative algebras*, John Wiley and Sons, 1962.

5. M. Dellnitz and B. Werner, *Computational methods for bifurcation problems with symmetries — with special attention to steady state and Hopf bifurcation points*, J. Comput. Appl. Math. **26** (1989), 97–123.

6. C. C. Douglas and J. Mandel, *The domain reduction method: High way reduction in three dimensions and convergence with inexact solvers*, (Philadelphia), Fourth Copper Mountain Conference on Multigrid Methods, SIAM, 1989, pp. 149–160.

7. _____, *A group theoretic approach to the domain reduction method: The commutative case*, Computing **47** (1990).

8. A. Fässler and R. Mäder, *Symmetriegerechte Basisvektoren für Permutationsdarstellungen aller endlichen Punktgruppen der Dimensionen 3 und 2*, ZAMP **31** (1980), 277–292.

9. A. Fässler and E. Stiefel, *Group theoretical methods and their applications*, Birkhäuser, Boston, 1992.

10. K. Georg and R. Miranda, *Exploiting symmetry in solving linear equations*, Bifurcation and Symmetry (Basel, Switzerland) (E. L. Allgower, K. Böhmer, and M. Golubitsky, eds.), ISNM, vol. 104, Birkhäuser Verlag, 1992, pp. 157–168.

11. _____, *Symmetry aspects in numerical linear algebra with applications to boundary element methods*, 1993.

12. J.-P. Serre, *Linear representations of finite groups*, Graduate Texts in Mathematics, vol. 42, Springer Verlag, Berlin, Heidelberg, New York, 1977.

13. J. Walker, *Numerical experience with exploiting symmetry groups for bem*, Exploiting Symmetry in Applied and Numerical Analysis (Providence, RI) (E. L. Allgower, K. Georg, and R. Miranda, eds.), Lectures in Applied Mathematics, vol. 28, American Mathematical Society, 1993.

Dept. of Mathematics, Colorado State University, Ft. Collins, CO 80523
*E-mail address*: allgower(georg,miranda)@math.colostate.edu

Lectures in Applied Mathematics
Volume **29**, 1993

# Topological Constraints for Explicit Symmetry Breaking

D. ARMBRUSTER AND E. IHRIG

February 8, 1993

ABSTRACT. Bifurcation problems with O(2) and O(3) symmetry which contain small non-symmetric terms in the bifurcation equations are considered. The topological properties of all orbits of fixed points for those two symmetry groups are determined. It is shown how their dimension, orientability and Euler characteristic restricts the possible vector fields that result from the small symmetry breaking terms. The occurence of periodic orbits is predicted and the possibility of chaotic behavior is restricted to bifurcations of non-axisymmetric solutions in O(3)-symmetry breaking.

## 1. Introduction

A typical model for a physical experiment identifies the physical laws and geometrical constraints which are relevant for the description of the experiment. Typically such a model neglects forces that are considered small and the geometrical peculiarities of a specific experiment. Therefore a mathematical formulation of a physical experiment often will show some degree of symmetry. A typical example is the description known as Euler buckling for the stationary behaviour of a beam under axial load. After reducing the problem to a single degree of freedom, i.e. to a configuration in a plane, a reflection symmetry is introduced by claiming that the beam has equal probability to buckle right or left. In such a way bifurcation problems are generated that are invariant under a symmetry group. Following [5] these problems have been thoroughly studied in recent years

1991 *Mathematics Subject Classification.* 34C35, 58F14.

*Key words and phrases.* dynamical system, symmetry breaking, topology.

D.A. would like to thank Ian Melbourne for bringing the reference [9] to his attention. Subsequent stimulating discussions with Reiner Lauterbach are also greatfully acknowledged. This work was partially supported through the Airforce Office of Scientific Research and NSF grants DMS 9017174 and DMS 9101964

This paper is in final form and no version of it will be submitted for publication elsewhere.

and are very well understood. Starting with a trivial solution that has the full symmetry of the problem (e.g. the straight beam) the bifurcated solutions break that symmetry and are invariant only under a subgroup of the original group of symmetries. This breaking of the symmetry due to a change in a bifurcation parameter is called *spontaneous symmetry breaking*.

Comparing the predictions of models which exhibit spontaneous symmetry breaking with the original physical experiment it turns out that those experiments often do not agree with the predictions. This was pointed out in [5] for the Euler buckling which theoretically should lead to an equal probability for right and left buckling. Practically however, due to imperfections in a given beam it will buckle with high probability to a prefered state. An incorporation of such imperfections leads to a different kind of symmetry breaking which is characterized by small terms in the governing equations that are not invariant under the original symmetry group. This is called *explicit symmetry breaking*, *system symmetry breaking* or *forced symmetry breaking*.

This paper is concerned with the following question: Let us consider a bifurcation problem for a vector field which is equivariant under a continuous group $\Gamma$. We assume a spontaneous symmetry breaking of the trivial solution at a certain value of a bifurcation parameter $\lambda$, say at $\lambda_c = 0$. What are the structurally stable phase portraits that are consistent with the topological constraints coming from a complete explicit symmetry breaking, for $\lambda$ bounded away from zero. The symmetry groups that we will analyze are $O(2)$ and $O(3)$. While there is no new information coming from $O(2)$ symmetry breaking it is still instructive to look at known results form this topological angle. Explicit symmetry breaking for $O(3)$ is much more complicated and leads to very interesting constraints for the possible vector fields. In particular we can predict for which kind of representation of $O(3)$ we can expect periodic, quasiperiodic or chaotic behaviour.

Our approach complements the one taken by Lauterbach and Roberts [9]. They are concerned with explicit symmetry breaking by perturbations of lower symmetry. Then, using group theoretic methods they discuss possible phase portraits that have residual symmetry. We assume that the physical problem has no symmetry left and therefore our methods do not rely on group theory but on differential geometry. In fact, as we will see in Section 4 we can recover some of results of [9] by appropriately adding symmetry to the asymmetric phase portraits that we predict on topological grounds.

Specifically we consider a spontaneous symmetry breaking of an $O(n)$ symmetric vector field F. As a result we will get bifurcation branches which describe solutions x that are invariant under a subgroup of $O(n)$ say $\Sigma_x$. Hence $\Sigma_x$ is the isotropy subgroup of the nontrivial solution x. Acting with the full group $O(n)$ on x transforms x into equivalent solutions, i.e. there is an orbit X associated with x. This orbit is the quotient space $\Gamma/\Sigma_x$ which in turn is isomorphic to a compact manifold M. If x is hyperbolic then, in terms of a phase space description, M is a normally hyperbolic invariant manifold of solutions with a trivial dynamics. E.g. if x is a fixed point then M is an invariant manifold of fixed points, if x is a limit cycle then M is a manifold with a trivial dynamics in all but one dimension. After perturbing the vector field F to $F_\epsilon$ which has no symmetries, we may use the following Theorem, [9].

THEOREM 1.1. *a) There exists a normally hyperbolic invariant manifold $M_\epsilon$ of solutions to the perturbed problem for $\epsilon$ small enough.*
*b) $M_\epsilon$ is diffeomorphic to M, in particular its topological properties are the same.*

The problem that we want to analyze in the following sections is: What are the topological constraints on the vector fields for all possible manifolds M of $\Gamma = O(2)$ and $\Gamma = O(3)$? What are the simplest structurally stable vector fields satisfying these constraints and what are the typical codimension 1 bifurcations between them?

The rest of the paper is organized as follows: Section 2 discusses O(2) symmetry breaking, Section 3.1 recalls all possible isotropy subgroups for O(3)-equivariant bifurcations and determines the Euler characteristic for all orbits M $= O(3)/\Sigma_x$, Section 3.2 discusses the Poincaré-Hopf theorem and the resulting structurally stable vector fields. Section 4 adds residual symmetry to some of the vector fields of 3.2.

## 2. Explicit symmetry breaking from O(2) symmetry

Consider a steady state bifurcation in an O(2) invariant vector field. The normal form for that bifurcation can be written as

$$(2.1) \qquad \frac{dA}{dt} = \mu A + cA|A|^2.$$

where $\mu, c$ are real coefficients and A is a complex amplitude for an unstable mode. Reduction to polar coordinates

$$(2.2) \qquad \frac{dr}{dt} = \mu r + cr^3$$

$$(2.3) \qquad \frac{d\phi}{dt} = 0$$

reveals the manifold M of equivalent steady states: $r_s = \sqrt{-\frac{\mu}{c}}, \phi = \phi_0$ with $\phi_0 \in [0, 2\pi]$. Assuming the symmetry broken steady state bifurcation is described by

$$(2.4) \qquad \frac{dA}{dt} = \mu A + cA|A|^2 + \epsilon f(A).$$

with $\epsilon$ measuring the symmetry breaking and $\mu >> \epsilon$ we can reduce the perturbed dynamics to a dynamics on the perturbed manifold $M_\epsilon$ which is a closed curve parametrized by $\phi$:

(2.5) $$\frac{d\phi}{dt} = \epsilon \tilde{f}(\phi)$$

Depending on the zeros of $\tilde{f}$ there are two different types of flows on the circle: $\tilde{f} \neq 0$ introduces a drift along the circle, with $\tilde{f} = const$ being the special case of a symmetry breaking from $O(2)$ to $SO(2)$ leading to a constant speed along the circle. $\tilde{f}(\phi_n) = 0$ defines fixed points $\phi_n$ which are created in pairs in a saddle node bifurcation. Note that the simplest vector field that shows these two phase portraits is given by

(2.6) $$\frac{dA}{dt} = \mu A + cA|A|^2 + \epsilon_1 + \epsilon_2 \bar{A}$$

which has been suggested by [4]. It is also now obvious why the $O(2)$ symmetric bifurcation has codimension infinity relative to all possible symmetry breaking perturbations: There are infinitely many perturbations with different symmetries (corresponding to the Fourier modes along a circle) that break the rotation symmetry.

Notice also that the standard result for forced vibrations fits into this scheme: The averaged vector field for the forced van der Pol oscillator is given as

(2.7) $$\dot{u} = u - \sigma v - u(u^2 + v^2)$$
(2.8) $$\dot{v} = v + \sigma u - v(u^2 + v^2) - \gamma$$

with $\gamma$ proportional to the forcing amplitude and $\sigma$ describing the detuning [6]. It is well known that for $(\gamma, \sigma)$ going to zero there exist two regimes for the $\omega$ limit sets of these equations. One regime has a source surrounded by a stable limit cycle, called the drifting solution. This limit cycle is destroyed through a saddle node bifurcation which leads to a regime of three fixed points, the source at the origin and a saddle and a sink close to the former limit cycle. The latter fixed points are called phase locked. The symmetry that is broken here by the explicit time dependency of the forcing is the phase invariance of the periodic solution of the autonomous differential equation.

For an application of that approach to the Taylor Couette problem see [1] and [11].

## 3. Explicit symmetry breaking from O(3) symmetry

**3.1. Spontaneous symmetry breaking.** We review here the work by Ihrig and Golubitsky [8] and Chossat et al [2]. Let us consider $O(3)$ as the direct sum

(3.1) $$O(3) = SO(3) \oplus Z_2^c$$

where $Z_2^c = \{Id, -Id\}$. All subgroups of $O(3)$ fall into three classes: Class I contains those subgroups that are subgroups of $SO(3)$, class II constitutes subgroups that contain -Id, and class III are all those subgroups that are neither in class I nor in class II, i.e. they are not subgroups of SO(3) and they do not contain -Id. In the terminology of [5] they are $O(2)^-, O^-, D_m^z, D_{2m}^d, Z_{2m}^-$ which are subgroups of O(3) isomorphic to the subgroups with the same name without the superscripts but are not conjugate to them.

In order to analyze the orbit structure of fixed points that are invariant under some subgroup of O(3) we need a few facts:

LEMMA 3.1. *Let X denote an orbit of O(3). Then a) X is either connected or has two connected components. Each connected component of X is diffeomorphic to $SO(3)/(SO(3) \cap \Sigma_x)$ where $\Sigma_x$ is the isotropy subgroup of a point x in X; b) X has two components iff $\Sigma_x \subset SO(3)$.*

PROOF. a) Let $x \in X$ be a point on the orbit and $g \in O(3) \sim SO(3)$. Then by the coset decomposition of O(3) the orbit X is generated as

$$X = \quad O(3)x \quad = (SO(3) \cup SO(3)g)x$$
$$= SO(3)x \cup SO(3)(g(x))$$

We can have two cases:

Case 1: X is connected. Then $SO(3)x \cap SO(3)g(x) \neq \emptyset$. Hence they are identical since they are both SO(3)-orbits. Therefore X itself is an SO(3)-orbit.

Case 2: X is not connected. Since the SO(3) orbits are both connected X is the union of two connected sets each of which is a diffeomorphic SO(3) orbit.

b) If X has two components then $SO(3)x \neq SO(3)g(x)$ i.e. $g \in O(3) \sim SO(3)$ cannot be in $\Sigma_x$, the isotropy subgroup of x. Therefore the isotropy subgroup must be contained in SO(3). Conversely, if $\Sigma_x \subset SO(3)$ then the two cosets SO(3) and SO(3)g act differently on x, and therefore $SO(3)x \cap SO(3)g(x) = \emptyset$ i.e. X is not connected. $\square$

The next Lemma discusses the orientability of the orbits:

LEMMA 3.2. *a) If X is an orbit of SO(3) then X is orientable iff $\Sigma_x \neq O(2)$. b) If X is an orbit of O(3) then X is orientable iff $\Sigma_x \not\supset O(2)$ or $\Sigma_x$ is either SO(3) or O(3).*

PROOF. First we consider the case in which $\Sigma_x$ is finite. If $\Gamma$ is any Lie group, and $\Delta$ is any finite subgroup of $\Gamma$, then $\Gamma/\Delta$ is orientable. This is true because $\Gamma$ has a non-zero n-form (where n = dim($\Gamma$)) which is invariant under right translation by all the elements of $\Gamma$. Because it is invariant under right translation it is never 0. Also, it is invariant under right translation by the elements of $\Delta$, so that it will induce an n-form on the space of left cosets, $\Gamma/\Delta$. This n-form is never 0, and hence is an orientation of $\Gamma/\Delta$.

To finish we consider the case in which $\Sigma_x$ is not finite. In this case the dimension of $\Gamma/\Sigma_x$ is not the same as the dimension of $\Gamma$ and the induced n-form on $\Gamma/\Sigma_x$

will become zero and does not define an orientation. Thus, when $\Gamma = SO(3)$, then $\Sigma_x$ must be either SO(2), O(2) or SO(3). $\Gamma/\Sigma_x$ will be $S^2$ ( the two dimensional sphere), RP(2) (two dimensional real projective space) or 0. RP(2) is the only one of these manifolds which is not orientable. This completes a). Using Lemma 1, and part a), we see that an O(3)-orbit X is not orientable if and only if $\Sigma_x \cap SO(3) = O(2)$. This holds if and only if $\Sigma_x$ contains O(2) and is not three dimensional. $\square$

Intuitively Lemma 2 reflects the fact that every fixed point x that has isotropy subgroup SO(2) or O(2) represents an axisymmetric solution. Given a full SO(3)-symmetric physical problem the orientation of that axis is arbitrary. In case of the isotropy subgroup SO(2) the rotation axis has a polar orientation and the action of SO(3) on this axis sweeps out a whole sphere, so $SO(3)\backslash SO(2) \simeq S^2$. In the case of an isotropy subgroup O(2) north and south pole of the rotation axis are equivalent. Hence when sweeping out $S^2$ under the action of SO(3) on such a configuration antipodal points on the sphere $S^2$ have to be identified. This is the standard visualization of the real projective plane.

LEMMA 3.3. *The dimension of the orbit X is 3 for the finite isotropy subgroups $\Sigma_x$ and two for $\Sigma_x$ isomorphic to SO(2) or O(2).*

PROOF. A trivial application of Proposition 1.2 in [5] , Chapter XIII. $\square$

The following theorem ties all information together:

THEOREM 3.1. *Let X be an O(3) orbit, i.e.* $X \simeq O(3)/G$ *where* $G = K$, $G = K \oplus Z_2^c$ *or* $G = O(2)^-, O^-, D_n^z, D_{2n}^d$ *or* $Z_{2n}^-$ *where* $K = SO(3), O(2), SO(2)$, $T, O, I, D_n, Z_n$. *Then*
*Case 1: If G is finite then dim X =3, X is orientable and the Euler characteristic* $\chi(X) = 0$. *If* $G \subset SO(3)$ *then X has two components, otherwise it is connected.*
*Case 2: If* $G = O(2)^-, SO(2) \oplus Z_2^c$ *then* $X \simeq S^2$, *so dim X = 2, X is orientable and* $\chi(X) = 2$.
*Case 3: If* $G = SO(2)$ *then* $X \simeq S^2 \cup S^2$, *so dim X = 2, X is orientable and* $\chi(X) = 2 + 2$.
*Case 4:* $G = O(2) \oplus Z_2^c$ *then* $X \simeq RP(2)$, *so dim X = 2, X is nonorientable and* $\chi(X) = 1$.
*Case 5:* $G = O(2)$ *then* $X \simeq RP(2) \cup RP(2)$, *so dim X = 2, X is nonorientable and* $\chi(X) = 1 + 1$.

Note: The Euler characteristic of the form n+n denotes that each component of the orbit X has the characteristic n.

PROOF. Dimension and orientablity are direct consequences of the previous Lemmas. The Euler characteristic of odd dimensional orientable compact manifolds is zero, the Euler characterisic of $S^2 = 2$ and that of $\mathrm{RP}(2) = 1$ [12]. $\square$

**3.2. Consequences for vector fields.** The topological properties discussed in the previous section generate constraints for structurally stable vector fields on the manifolds of fixed points. These constraints follow from a direct application of the Poincaré-Hopf Theorem:

THEOREM 3.2. *Consider a compact manifold M and a smooth vector field w on M with isolated zeroes. Then the sum $\Sigma i$ of the indices of all zeroes of such a vector field is equal to the Euler characteristic of the manifold* [10],

$$\Sigma i = \chi(M)$$

Note that in the following we are mainly interested in structurally stable vector fields. In that case the index of a vector field $w$ at a zero $x_0$ is simply given by the sign $\det Dw(x_0)$. Let us inspect some of the consequences of this theorem:

Consider O(2)-symmetry breaking. We can recover our intuitive classification of structurally stable vector fields by noting that $\chi(S^1) = 0$. The index of a source on $S^1$ is one, the index of a sink is -1. Therefore the structurally stable vector fields on $S^1$ are a) the parallel flow along the circle, i.e. a periodic solution, b) an even number of fixed points.

Consider the consequences of the Theorem for flows that arise from O(3)-symmetry breaking:

    i) Explicit symmetry breaking of steady state solutions with $\Sigma_x \subset SO(3)$ leads to vector fields on two disconnected manifolds $X_i$ i = 1,2. Assuming a complete symmetry breaking there is no reason why the symmetry breaking perturbations should act in the same way on both manifolds $X_i$. Hence such flows would exhibit multistability.

    ii) Explicit symmetry breaking of steady state solutions with finite $\Sigma_x$ leads to 3-d flows whose Euler characteristic is zero. Hence, according to the Poincaré-Hopf Theorem we can have flows on X without singularities. It is unclear what the generic flows in such a situation are but the possibility of chaotic flows as well as of 3-tori exists.

    iii) Explicit symmetry breaking with a continuous isotropy subgroup leads to 2-d flows. For them the Poincaré-Bendixson Theorem applies which leads to further restrictions. Consider e.g. $X \simeq S^2$, then for the structurally stable vector fields the index of a sink or source is +1, the index for a saddle is -1. Then the number of fixed points of source, sink and saddle type is related according to Theorem 3.2 by

$$n_{sink} + n_{source} - n_{saddle} = 2$$

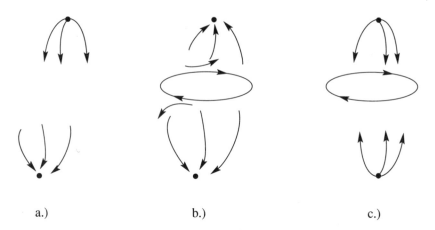

a.)                        b.)                        c.)

FIGURE 1. Schematic vector fields on the sphere with a) one
sink and one source, b) two sinks and c) two sources. The limit
cycles in b) is unstable the one in c) is stable

Under the assumption that perturbations have large wavelengths and
therefore do not create many fixed points it makes sense to examine a
few typical phase portraits for small integer numbers of fixed points.
Taking $(n_{sink}, n_{source}, n_{saddle}) = (1, 1, 0)$ we find a vector field with two
fixed points on the sphere, one stable the other one unstable (Figure
1a). $(n_{sink}, n_{source}, n_{saddle}) = (2, 0, 0)$ does not lead to a phase portrait
consistent with the Poincaré-Bendixson Theorem. Therefore we require
the existence of an unstable periodic orbit as shown in Figure 1b. Similarly $(n_{sink}, n_{source}, n_{saddle}) = (0, 2, 0)$ leads to the existence of a stable
limit cycle. Note that the generic bifurcations from the vector field in
Figure 1a to Figure 1b or Figure 1c are Hopf bifurcations. Additional
bifurcations might destroy the limit cycle via a saddle node bifurcation
similarly to the generic bifurcations in the O(2) case. The corresponding vector fields would have indices $(n_{sink}, n_{source}, n_{saddle}) = (1, 2, 1)$ or
$(2, 1, 1)$.

Consider a vector field on $X \simeq RP(2)$ which we visualize as a flow on
a sphere with antipodal points identified. With the same argument as
above we have the relation between the number of fixed points:

$$n_{sink} + n_{source} - n_{saddle} = 1$$

Therefore even the simplest vector field with $(n_{sink}, n_{source}, n_{saddle}) =
(1, 0, 0)$ automatically has to generate an unstable limit cycle which is

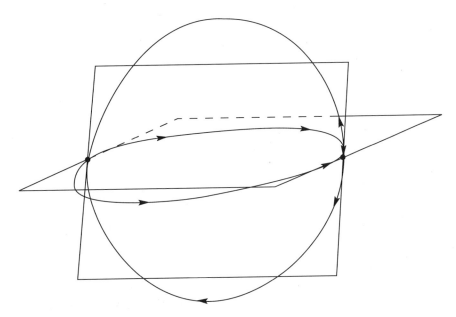

FIGURE 2. Structurally stable heteroclinic cycle on a sphere

symmetric with respect to inversion. Note that, as a perturbation moves
the unstable limit cycle closer to the fixed point, the cycle splits up in a
saddle source bifurcation and re-forms as a stable limit cycle in a saddle
sink bifurcation.

## 4. Residual symmetry

So far we have assumed complete symmetry breaking. Assuming that our
perturbations have some residual symmetry of a subgroup of O(3) we can recover some of the results of [9] in a very intuitive way. This is true for the 2-d
flows on $S^2$ or $RP(2)$ which are easy to visualize. Consider the orbit $X \simeq S^2$. If
we require that the perturbations that break the O(2) symmetry preserve some
residual subgroup $\Delta$ then the topologically possible vector fields are additionally
constrained to be invariant under H. In particular it is easy to construct heteroclinic orbits in some of these cases. Consider e.g. $\Delta = Z(2) \times Z(2)(\simeq D_2)$
and an action on $S^2$ that creates two orthogonal invariant planes (cf. Figure 2).
This forces two fixed points at the intersection of the planes with $S^2$ ("North"
and "South" pole). Since each of the reflections in $Z(2) \times Z(2)$ are independent
of each other we can arrange the vector field such that the North pole is stable
in one subspace and unstable in the other and vice versa for the South pole.
Hence we have created structurally stable heteroclinic cycles connecting these
two fixed points. Clearly for topological reasons we have four fixed points in
the four sections of the sphere between the heteroclinic connections. If those
are asymptotically stable (unstable) the heteroclinic cycle is asymptotically un-

stable (stable). Obviously more complicated heteroclinic arrangements can be constructed with more complicated residual symmetries $\Delta$.


## 5. Conclusion

We have shown that the topological properties of the orbits of fixed points that occur in spontaneous symmetry breaking bifurcations with $O(2)$ and $O(3)$ symmetry restrict the possible flows along those orbits once the symmetry of a given system is broken by small perturbations in the bifurcation equations. These properties allow us to make certain predictions for the behavior of bifurcations in $O(3)$-symmetric models for not quite perfectly symmetric physical problems like e.g. spherical convection models for planets or stars. Specifically we expect axisymmetric solutions to behave rather stably: The worst that can happen to them is that they aquire a slow rotation of their axis. On the other hand truly three-dimensional stationary pattern (characterized by a discrete isotropy sub-group) are very likely to exhibit complicated time-dependent behavior.

Since our approach is topological and perturbative the results are global in the sense that they apply for all fixed points of a problem whose mathematical model leads to a dynamical system that is equivariant under the symmetry groups $O(2)$ or $O(3)$. However the results are local in the size of the explicit symmetry breaking governed by the parameter $\epsilon$. Increasing $\epsilon$ will eventually lead to the destruction of the hyperbolicity transverse to the group orbits. The same problem occurs as we reduce our bifurcation parameter $\lambda$ which governs the spontaneous symmetry breaking. As soon as $\lambda$ and $\epsilon$ become comparable the orbit of fixed point is no longer hyperbolic.

Our topological approach seems to have many more applications. For instance Dangelmayr and Wegelin [3] discuss spontaneous symmetry breaking through Hopf bifurcations in interacting clusters of oscillators. It turns out that the relevant Birkhoff normal form has the structure of an $O(2) \times O(2)$ symmetric problem up to order 3 while higher order terms (fourth and fifth order) are determined by the specific discrete symmetry of the problem considered. A classification of the solution types of these bifurcation equations could start with the $O(2) \times O(2)$-solution orbits and then determine the possible flows on those orbits due to explicit symmetry breaking from the higher order terms.

Another interesting question that we are pursuing is the applicability of this approach to other limit sets than just fixed points. In particular nothing in this paper really requires these invariant orbits to be orbits of fixed points. They could equally well be group orbits of periodic solutions, homoclinic cycles or more complicated objects. However here the possibility of resonances arises which needs further study.

## References

1. D.Armbruster, A.Mahalov, On the explicit symmetry breaking in the Taylor-Couette problem, Physics Lett. A 167, (1992), p.251.
2. P.Chossat, R.Lauterbach, I.Melbourne, Steady-state bifurcation theory with O(3)-symmetry, Arch. Rat. Mech. Anal. 113, (1990), p 313-376.
3. G.Dangelmayr, M.Wegelin, Hopf bifurcation with $D_3 \times D_3$-symmetry, to appear in ZAMP. (1992)
4. M.Golubitsky, D.Schaeffer, A Discussion of Symmetry and Symmetry Breaking, Proc. of Symposia in Pure Mathematics Vol. 40 Part 1, (1983), p.499.
5. M.Golubitsky, I.Stewart, D.Schaeffer, *Singularities and Groups in Bifurcation Theory Vol. II*, Springer Verlag, 1988
6. J.Guckenheimer, Ph.Holmes, *Nonlinear oscillations, Dynamical Systems and Bifurcations of Vector Fields*, Springer Verlag 1983
7. M.W.Hirsch, C.C.Pugh, M.Shub, *Invariant Manifolds, Lecture Notes in Mathematics 583*, Springer Verlag 1977
8. E.Ihrig, M.Golubitsky, Pattern selection with O(3) symmetry, Physica 12D, (1984), p. 1
9. R.Lauterbach, M.Roberts, Heteroclinic Cycles in Dynamical Systems with Broken Spherical Symmetry, preprint (1990)
10. J.Milnor, *Topology from the differentiable viewpoint*, University Press of Virginia 1965
11. R.Raffai, P.Laure, Influence of an axial mean flow on the Couette-Taylor problem, to appear in Europhysics Letters (1992)
12. M.Spivak, *Differential Geometry*, Publish or Perish Inc. 1979

DEPARTMENT OF MATHEMATICS, ARIZONA STATE UNIVERSITY, TEMPE, AZ 85287

DEPARTMENT OF MATHEMATICS, ARIZONA STATE UNIVERSITY, TEMPE, AZ 85287

Lectures in Applied Mathematics
Volume **29**, 1993

# A Numerical Liapunov-Schmidt Method for Finitely Determined Problems

P. Ashwin,* K. Böhmer, Z. Mei[†]

## March 28, 1993

For bifurcation problems with symmetry, we present a numerical method to approximate the Liapunov-Schmidt reduced bifurcation equation. Using computer algebra and symmetry respecting discretisation, we obtain the correct branching behaviour for finitely determined generic problems.

## 1 Introduction

Our goal is to obtain branching information at steady state and Hopf bifurcations of nonlinear problems with symmetry, whose bifurcation equation is finitely determined. Let $E \hookrightarrow \hat{E}$ be Banach spaces embedded into a Hilbert space with scalar product $\langle \cdot, \cdot \rangle$ and $G : E \times \mathbf{R} \to \hat{E}$ be smooth and equivariant under some action of a compact symmetry group $\Gamma$. We consider the bifurcation scenario of the problem

$$\frac{\partial u}{\partial t} = G(u, \lambda) \quad \text{with } (u, \lambda) \in E \times \mathbf{R} \tag{1}$$

at a given bifurcation point $(u_0, \lambda_0)$. In particular, we are interested in partial differential operators. The Liapunov-Schmidt method is a well-known tool for steady state bifurcations with finite dimensional $\mathrm{Ker}(DG(u_0, \lambda_0))$. Under appropriate assumptions on $DG(u_0, \lambda_0)$ and its kernel, the method gives a reduction of the bifurcation phenomena of (1) to a low dimensional bifurcation equation

$$B(\alpha, \lambda) = 0.$$

Here $(\alpha, \lambda)$ is an element in $\mathrm{Ker}(DG(u_0, \lambda_0))$ with respect to a given basis. Usually $B(\alpha, \lambda)$ is highly nonlinear and moreover is not obtainable by analytical methods. We study the local behaviour of the nonlinear problem near the bifurcation point $(u_0, \lambda_0)$, so the truncated Taylor expansions of $B(\alpha, \lambda)$ and the formalism of germs will be appropriate tools. We will combine numerical and computer algebraic methods to perform the Liapunov-Schmidt reduction. Instead of the original bifurcation equation $B(\alpha, \lambda) = 0$ we introduce its polynomial truncations $B_k(\alpha, \lambda) = 0$. For most differential equations, neither $B$ nor $B_k$ are

1991 Mathematics Subject Classification. Primary 58G28, 35B32, 65M12, 65M99
*Supported by the Royal Society European Exchange Program, UK.
†Supported by Deutsche Forschungsgemeinschaft, FRG.

directly available and thus have to be approximated by discretisation methods. In fact, only $B_k^h$, the discretisation of $B_k$ with some non-zero discretisation parameter $h$, is really computable. It is related to $B$ via the following approximately commutative diagram:

$$
\begin{array}{ccc}
B(\alpha,\lambda)=0 & \xrightarrow[\;truncation\;]{\;polynomial\;} & B_k(\alpha,\lambda)=0 \\[2pt]
\Big\downarrow\; discretisation & & discretisation\;\Big\downarrow \\[2pt]
B^h(\alpha,\lambda)=0 & \xrightarrow[\;truncation\;]{\;polynomial\;} & B_k^h(\alpha,\lambda)=0
\end{array}
$$

In this paper we consider symmetry respecting calculations of $B_k^h(\alpha,\lambda)$ and generic conditions implying that the bifurcation equations (1) and $B_k^h(\alpha,\lambda)=0$ describe the same scenario. To achieve this goal we introduce some concepts from the theory of germs of equivariant bifurcation problems. By assuming the bifurcation scenario of $B=0$ is determined by its Taylor expansion truncated to high enough order (finite determinacy), and assuming it is stable with respect to small equivariant perturbations (structural stability) we can get the desired equivalence. We apply a truncated Liapunov-Schmidt method to obtain the reduced bifurcation equations, see Ashwin [6, 7].

Regarding the $\Gamma$–equivariance of our problem, we split the group $\Gamma$ into a semi–product of continuous and finite groups; $\Gamma = \Gamma_c \times_s \Gamma_d$. These two components have to be treated differently due to the very nature of discretisations. As demonstrated by Golubitsky-Schaeffer [19], Golubitsky-Stewart-Schaeffer [21], we need to incorporate the symmetry at every stage in order to get any sort of stable behaviour. For finite groups, a symmetry respecting discretisation can be defined. For continuous groups, this is not possible except in special cases. In general we must choose a basis for the discretisation that contains its own group orbit, for example, Fourier series for periodic problems. If we miss some symmetry in the discretisation of a symmetric problem then generally we cannot expect that the actual bifurcation scenario is reflected in the discretised problem. Instead we expect a one parameter deformation of the problem.

A numerical realisation relating $B$ and $B_k^h$ in the desired way is described in Section 3. We discretise the original problem with symmetry respecting methods to maintain $\Gamma_d$–equivariance. Since certain higher derivatives of the problem are important for determining the bifurcation behaviour, we have to ensure that the derivatives of the discretised operator sufficiently resemble those of the original operator. We therefore need the concept of *consistent differentiability* introduced in Böhmer [10, 11] and Böhmer-Mei [12]. This is related to ideas of Brezzi-Rappaz-Raviart [13] and Crouzeix-Rappaz [14]. Classical stability arguments for stability of discretisations need modification as they require that $DG$ is invertible; precisely this breaks down at bifurcation points.

We discuss an application in Section 4 and more fully in a forthcoming paper [8]: we use a standard technique to apply the Liapunov-Schmidt method to Hopf bifurcations. We handle $\Gamma_c$ in an analytical way whilst $\Gamma_d$ is treated numerically. These two approaches are unified by using computer algebraic tools. This allows the discussion of bifurcations, that cannot be solved directly, using a hybrid numerical/computer algebraic approach.

# 2   Bifurcation, singularities and truncated Liapunov-Schmidt reduction

In order to identify the bifurcation scenario for the true bifurcation equation $B(\alpha, \lambda)$ from its truncated discretisations $B_k^h(\alpha, \lambda)$, we have to make assumptions about $B(\alpha, \lambda)$. After detailing these, we briefly review an iterative Liapunov-Schmidt method that generates the jets (truncated Taylor series) of the bifurcation equation to desired order, by successively solving linear problems. Numerical approximation of this process by symmetry respecting discretisation is discussed in Section 3.

## 2.1   Singularities and bifurcation equations

The natural language for local steady state bifurcation problems is that of germs of singularities [24, 19, 21]. Let $E_{m,1}^\nu$ denote the module of germs at $(0,0)$ of $C^\infty$ vector– or matrix–valued functions $\mathbf{R}^m \times \mathbf{R} \to \mathbf{R}^\nu$, ($\nu = m$ or $m \times m$ respectively) over the ring $E_{m,1}^1$ of germs of scalar–valued functions. The set of $\Gamma$–equivariant bifurcation problems is defined by

$$\mathcal{F}_\Gamma := \{f \in E_{m,1}^m \ : \ f(0,0) = 0, \ Df(0,0) = 0, \ \gamma f(u, \lambda) = f(\gamma u, \lambda) \ \forall \ \gamma \in \Gamma\}. \quad (2)$$

To identify elements in $\mathcal{F}_\Gamma$ that are merely deformations of each other, we define an equivalence of germs in $\mathcal{F}_\Gamma$.

**Definition 1** *([21]) Consider germs $f, g$ in $\mathcal{F}_\Gamma$. If there exist germs $X \in E_{m,1}^m$, and $S \in E_{m,1}^{m \times m}$ with $\partial_u X(0,0)$, $S(0,0)$ regular, and $\Lambda \in E_1^1$ with $\Lambda : \mathbf{R} \to \mathbf{R}$, $\Lambda'(0) > 0$ such that*

$$f(u, \lambda) = S(u, \lambda)g(X(u, \lambda), \Lambda(\lambda)) \quad and$$
$$X(\gamma u, \lambda) = \gamma X(u, \lambda), \quad S(\gamma u, \lambda)\gamma = \gamma S(u, \lambda) \ \forall \ \gamma \in \Gamma,$$

*then $f$ and $g$ are said to be equivalent (we write $f \sim g$).*

An important consequence of this definition is that if $f \sim g$ then for fixed $\lambda$

$$f(u, \lambda) \quad \text{and} \quad g(X(u, \lambda), \Lambda(\lambda)) \text{ have the same number of zeros.}$$

We write $j_k g$ for the truncated Taylor expansion of $g$ with respect to all its arguments up to and including the homogeneous terms of order $k$ (the so called $k$-jet).

**Definition 2** *For $g \in \mathcal{F}_\Gamma$ and $l \in \mathbf{N}$ suppose $j_l g \sim g$, and let $k \in \mathbf{N}$ be the minimum such that $j_k g \sim g$. Then $g$ is said to be $k$-determined. If $g$ is $k$-determined for some $k$, $g$ is said to be finitely determined.*

**Definition 3** *For any* $g \in \mathcal{F}_\Gamma$ *define the pseudo-norm* $\| \cdot \|_k^0$ *(a norm on the subspace of* $k-$*jets) by*

$$\|g\|_k^0 = \sum_{|i|+j \leq k} \left| \frac{\partial^{|i|+j} g}{\partial u^i \partial \lambda^j}(0,0) \right|, \qquad (3)$$

*using a multi-index* $i$. *If there exists* $\varepsilon > 0$ *such that all perturbations* $f \in \mathcal{F}_\Gamma$ *of* $g$ *with*

$$\|f - g\|_k^0 < \varepsilon$$

*satisfy* $f \sim g$, *then we say* $g$ *is stable with respect to* $\| \cdot \|_k^0$.

A consequence of this is that $\Gamma$ must be absolutely irreducibly represented on $\mathbf{R}^m$ (see [21]). We shall work with the set of germs of bifurcation problems that are $k-$determined and stable with respect to $\| \cdot \|_k^0$ (or just simply *stable*).

To justify this restriction, firstly note that the set of germs of bifurcations which are *not* $k-$determined is of codimension infinity in $\mathcal{F}_\Gamma$. Secondly, we examine only generic structurally stable bifurcations that persist under sufficiently small $\Gamma$−equivariant perturbations of $g(u, \lambda)$, e.g. pitchfork bifurcations under $\mathbf{Z}_2$-symmetry or corank-1 or 2 bifurcations of nonlinear elliptic problems under $\mathbf{D}_4$-equivariance. Note that for both these examples, singularity theory can be used to show that any stable problems must be $3-$determined. The finite determinacy and stability assumptions together imply that a bifurcation problem with germ $g$ is in fact stable with respect to the smooth norm in the following sense: let $\bar{f}$, $\bar{g}$ be two functions whose corresponding germs $f, g$ are in $\mathcal{F}_\Gamma$ with $\|f - g\|_k^0 < \varepsilon$. Then there is a neighbourhood $U$ of the origin such that in the Sobolev-norm

$$\|\bar{g}\|_k := \sum_{|i|+j \leq k} \max_{(u,\lambda) \in U} \left| \frac{\partial^{|i|+j} \bar{g}}{\partial u^i \partial \lambda^j}(u, \lambda) \right| \qquad (4)$$

we have $\|\bar{f} - \bar{g}\|_k < 2\varepsilon$. Conversely, we always have $\|\bar{g}\|_k^0 \leq \|\bar{g}\|_k$.

## 2.2   The truncated Liapunov-Schmidt method

In this section we reduce the operator equation (1) to the truncated bifurcation equation (2), see [6, 7]. To this end, we consider the $\Gamma$-equivariant operator equation

$$\Phi : E \times \mathbf{R} \to \hat{E}, \quad \Phi(\gamma u, \lambda) = \gamma \Phi(u, \lambda) \ \forall \ \gamma \in \Gamma,$$
$$0 = \Phi(u, \lambda) =: D\Phi(0,0)(u, \lambda) + R(u, \lambda), \qquad (5)$$

where $\Phi$ is a smooth 1-parameter mapping between the Banach spaces $E \hookrightarrow \hat{E}$ with $\Phi(0,0) = 0$. Note that by definition, $R(z) = O(\|z\|^2)$, where we write $z = (u, \lambda)$. We define

$$\Phi_0' := D\Phi(0,0) = (\partial_u \Phi(0,0), \partial_\lambda \Phi(0,0))$$

and assume that $\Phi_0'$ is a Fredholm operator of index 1 with a kernel of dimension $m + 1$, $m \geq 1$. Furthermore, our definition of $\mathcal{F}_\Gamma$ excludes turning point bifurcations and so we assume, reparameterising if necessary, that

$$\partial_\lambda \Phi(0,0) = 0.$$

This corresponds to assuming existence of a trivial solution from which the branches bifurcate. Using the Fredholm condition we split the spaces in the standard way:

$$E \times \mathbf{R} = \mathrm{Ker}(\Phi_0') \oplus \mathrm{Im}(\Phi_0'^*) \quad \text{and} \quad \hat{E} = \mathrm{Ker}(\Phi_0'^*) \oplus \mathrm{Im}(\Phi_0') \qquad (6)$$

with $\Phi_0'^*$ the adjoint operator and with $\oplus$ indicating orthogonality w.r.t. $\langle \cdot, \cdot \rangle$. We define the projections

$$\begin{aligned} Q : E \times \mathbf{R} \to \mathrm{Im}(\Phi_0'^*), \quad & \mathrm{Ker}(Q) = \mathrm{Ker}(\Phi_0') = \mathrm{Ker}(\partial_u \Phi_0) \times \mathbf{R}; \\ \hat{Q} : \hat{E} \to \mathrm{Im}(\Phi_0'), \quad & \mathrm{Ker}(\hat{Q}) = \mathrm{Ker}(\Phi_0'^*). \end{aligned} \qquad (7)$$

Using these projectors we write $z = \eta + w$ with $\eta \in \mathrm{Ker}(\Phi_0')$ and $w \in \mathrm{Im}(\Phi_0'^*)$ and note that (5) holds if and only if

$$\hat{Q}\Phi(\eta + w) = 0 \quad \text{and} \qquad (8)$$
$$(I - \hat{Q})\Phi(\eta + w) = 0. \qquad (9)$$

Equation (8) is solvable in terms of $\eta$ (for small $\eta$) and gives $w(\eta) \in \mathrm{Im}(\Phi_0'^*)$ uniquely; this is substituted into Equation (9) to give the *bifurcation equation*:

$$B(\eta) := (I - \hat{Q})\Phi(\eta + w(\eta)) = (I - \hat{Q})R(\eta + w(\eta)) = 0. \qquad (10)$$

If symmetries of $\Phi$ are to be inherited by $B$, we must (and for finite symmetry groups are able to) choose projectors $Q$, $\hat{Q}$ commuting with the symmetries, see e.g. [21, 36]. We now give an iterative method to solve Equation (8) that develops the Taylor expansion of $B$. The method is presented in Ashwin [6, 7] and a variation of this is applied in [1]. The functions $w(\eta)$ and $B(\eta)$ above are related to $w_k$ and $B_k$ by Theorem 5.

**Algorithm 4 :** *Truncated Liapunov-Schmidt Method:*
   **Initialisation:** *Pick an initial function*

$$w_1(\eta) = 0, \qquad (11)$$

*a mapping from* $\mathrm{Ker}(\Phi_0')$ *to* $\mathrm{Im}(\Phi_0'^*)$.
   **Iteration:** *For* $k = 2, 3, \ldots$ *until determinacy do:*
   *At each stage define the truncated bifurcation equation*

$$B_k(\eta) := (I - \hat{Q})j_k R(\eta + w_{k-1}(\eta)) \qquad (12)$$

*where* $B_k$ *is a polynomial map of $k$-th order from* $\mathrm{Ker}(\Phi_0')$ *to* $\mathrm{Ker}(\Phi_0'^*)$. *We usually identify* $\eta \in \mathrm{Ker}(\Phi_0')$ *with* $(\alpha, \lambda) \in \mathbf{R}^{m+1}$ *and* $\mathrm{Ker}(\Phi_0'^*)$ *with* $\mathbf{R}^m$. *(By splitting $\eta$ into $(\alpha, \lambda)$ with fixed $\lambda \in \mathbf{R}$ the equivariance is maintained.) Generate the next $w_k$ by*

$$w_k(\eta) := Q(\Phi_0')^{-1}\hat{Q}j_k R(\eta + w_{k-1}(\eta)). \qquad (13)$$

*The iteration creates a sequence of polynomial functions*

$$w_k : \mathrm{Ker}(\Phi_0') \to \mathrm{Im}(\Phi_0'^*). \qquad (14)$$

**Theorem 5 :** *([8]) For the iteration defined in (13), we have*

$$\begin{aligned} w_k(\eta) &= j_k w(\eta), \\ B_k(\eta) &= j_k B(\eta). \end{aligned}$$

## 2.3  Hopf bifurcation

Algorithm 4 can be applied to find branches of periodic solution in the neighbourhood of a Hopf bifurcation. Assume that the linearisation $L = \partial_u G(0,0)$ of Equation (1) *has $2l$ imaginary pairs of eigenvalues $\pm i\omega_0$ (that are semi-simple) and no other eigenvalues at $mi\omega_0$ for all $m \in \mathbf{Z}$, $m \neq \pm 1$.*

Using a standard trick (see Vanderbauwhede [36]) we define a steady-state problem from the original 1-parameter Hopf bifurcation problem

$$\Phi(u,\tau,\lambda) := G(u,\lambda) - \omega_0(1+\tau)\partial_s u = 0 \qquad (15)$$

that is a mapping from $C^1_{2\pi} \times \mathbf{R} \times \mathbf{R}$ to $C^0_{2\pi}$ where

$$
\begin{aligned}
C^0_{2\pi} &:= \{u(s) : u(s+2\pi) = u(s), u \in L^2(\mathbf{R}, \hat{E})\}, \\
C^1_{2\pi} &:= \{u(s) \in C^0_{2\pi} : \frac{du}{ds} \in L^2(\mathbf{R}, E)\}.
\end{aligned}
$$

We treat the parameter $\tau$ as a state variable (i.e. we allow changes of coordinates mixing $\tau$ and $u(t)$), but we keep $\lambda$ as the parameter. We introduce the inner product

$$\langle u, v \rangle_{C^i_{2\pi}} := \frac{1}{2\pi} \int_{s=0}^{2\pi} \sum_{j=0}^{i} \left\langle \frac{\partial^j u}{\partial s^j}, \frac{\partial^j v}{\partial s^j} \right\rangle \, ds.$$

The hypotheses for Hopf bifurcation of $G$ imply that $\Phi'_0$ satisfies the conditions for Liapunov-Schmidt reduction: fixed points in $C^1_{2\pi}$ correspond to periodic solutions of Equation (1).

By using a Fourier basis for the periodic function space, we can re-write the problem so that it has just a discrete symmetry group $\Gamma$. In fact, as discussed in [8] we need only consider finite Fourier sums to get the bifurcation equation to any finite order.

# 3  Discretising the truncated Liapunov-Schmidt method

If we want to reproduce the bifurcation scenario for $B_k(\alpha, \lambda) = 0$ numerically, the discretisation must be $\Gamma$-equivariant and the higher derivatives have to be well enough approximated in a sense we shall define. We use techniques developed in [10, 11, 12].

## 3.1  Consistently differentiable discretisation methods

To define a discretisation of (1) or (5), see e.g. Stetter [31], Stummel [33], we consider sequences of problems indexed by $h \in H$. We need (linear) bounded restriction operators $\pi^h$ and $\hat{\pi}^h$ and finite dimensional discrete spaces $E^h \times \mathbf{R}$ and $\hat{E}^h$. Usually $\pi^h$ acts on $z = (u, \lambda) \in E \times \mathbf{R}$ by transforming $u \in E$ into a discrete $u^h = \pi^h u \in E^h$, leaving $\lambda$ unchanged. For simplicity, we restrict our discussion to this situation and assume the following approximation properties of $\pi^h$, $\hat{\pi}^h$:

$$\pi^h : E \times \mathbf{R} \to E^h \times \mathbf{R}, \quad \hat{\pi}^h : \hat{E} \to \hat{E}^h$$
$$\pi^h z = \pi^h(u, \lambda) = (\pi^h u, \lambda) \text{ and } \|\pi^h z\| = \|z\| + O(h^p \|z\|_{E_s}), \qquad (16)$$
$$\|\hat{\pi}^h \hat{y}\| = \|\hat{y}\| + O(h^p \|\hat{y}\|_{\hat{E}_s})$$

for fixed $z, \hat{y}$ in appropriate (smooth) subspaces $E_s (\subset E)$, $\hat{E}_s (\subset \hat{E})$. The $O(h^p)$ terms in $\pi^h z$ are due to $\pi^h u$. Usually, if $E$ is defined for a Sobolev-norm (4) with derivatives of order $|i| + j \le n$, then $E_s$ requires a norm with derivatives of orders $|i| + j \le n + p$. In such a case we choose $n$ such that $\Phi$ is continuous on $E$. For simplicity, we do not distinguish between $E, \hat{E}$ and their smooth subspaces $E_s$, $\hat{E}_s$. Strictly speaking, the latter are needed for consistency and convergence in (16) and (17). These conditions may be relaxed by using weak discrete Sobolev-norms as indicated in Hackbusch [23]. The operator $\Phi$ in (5) is transformed into a discrete operator $\Phi^h : D(\Phi^h) \subset E^h \to \hat{E}^h$.

For many discretisation methods we find $\Phi^h$ is $r$ *times consistently differentiable* with $\Phi$ in the sense:

$$\Phi^h(\pi^h z) = \hat{\pi}^h \Phi z + O(h^p \|z\|_{E_s}) \quad \text{for } z \in E_s \quad \text{and for } j = 1, \cdots, r,$$
$$\Phi^{h(j)}(\pi^h z)\pi^h z_1 \cdots \pi^h z_j = \hat{\pi}^h (\Phi^j(z) + \varepsilon(h^p, z)) z_1 \cdots z_j \text{ for } z, z_1, \cdots, z_j \in E_s,$$
$$(17)$$

where $\varepsilon(h^p, z)$ represents a $j$-linear operator with $\|\varepsilon(h^p, z)\| \le O(h^p \|z\|_{E_s})$.

We give some examples for difference methods; similar considerations are possible for finite element methods, see e.g. Brezzi-Raviart-Rappaz [13] and Crouzeix-Rappaz [14]. We use the Taylor formula to yield remainders for divided differences. For $f \in C^{n+1}(\mathbf{R})$ we have, defining the shift operator $(s_\alpha u)(x) := u(x + \alpha)$,

$$f(x + t) - f(x) - \cdots - \tfrac{1}{n!} f^{(n)}(x) t^n$$

$$= \int_x^{x+t} \tfrac{1}{n!}(x + t - \tau)^n f^{(n+1)}(\tau)\, d\tau$$

$$= \tfrac{1}{n!} t^{n+1} \int_0^1 (1 - \sigma)^n s_{t\sigma}(f^{(n+1)}(x))\, d\sigma \qquad (18)$$

$$=: \tfrac{1}{n!} t^{n+1} Q_n(t, f^{(n+1)})(x).$$

For example, we obtain for appropriately smooth $u$

(i)    $\tfrac{1}{2}(u(x + h) + u(x - h)) = u(x) + \tfrac{1}{2} h^2 P_1(h, u'')(x),$

(ii)    $\tfrac{1}{2h}(u(x + h) - u(x - h)) = u'(x) + \tfrac{1}{4} h^2 P_2(h, u''')(x), \qquad (19)$

(iii)    $\tfrac{1}{h^2}(u(x + h) - 2u(x) + u(x - h)) = u''(x) + \tfrac{1}{6} h^2 P_3(h, u^{(4)})(x)$

with $P_n(h, u)(x) := \int_0^1 (1 - \sigma)^n (s_{h\sigma} + s_{-h\sigma}) u(x)\, d\sigma$. The $P_n$ and $Q_n$ are linear operators in $u$. For functions $u$ defined in $\mathbf{R}^m$ similar results are available. As an example we use the nonlinear Laplacian

$$\Phi(u, \lambda) = \Delta u + \lambda f(u, \nabla u) = 0 \quad \text{in} \quad \Omega = [0, 1]^2, \qquad (20)$$

with $u|_{\partial\Omega} = 0$ or other boundary conditions. Introducing an equidistant grid in both $x$ and $y$ directions

$$\mathbf{G}_h := \{(x_a, y_b) \in \Omega \mid x_a := ah, \ y_b := bh, \ 0 \le a, b \le 1/h = N, \ a, \ b \in \mathbf{N}\},$$

$$\mathbf{G}_h^i := \{(x_a, y_b) \mid 1 \le a, b \le N - 1\}$$

we choose

$$E_s = H_0^4(\Omega) \subset E = H_0^2(\Omega), \quad \hat{E}_s = H^2(\Omega) \subset \hat{E} = L^2(\Omega);$$
$$E^h := \{u : \mathbf{G}_h \to \mathbf{R}, \ u = 0 \text{ on } \mathbf{G}_h \backslash \mathbf{G}_h^i\}, \quad \hat{E}^h := \{u : \mathbf{G}_h^i \to \mathbf{R}\};$$
$$\pi^h u := u|_{G_h}, \quad \hat{\pi}^h v := v|_{G_h^i}.$$

For a point $(x_a, y_b) \in \mathbf{G}_h^i$ we denote the values of $u$ in the five-point star neighbours as follows:

This yields

$$\Delta^h u := \tfrac{1}{h^2}(u_{a-1} + u_{a+1} + u_{b-1} + u_{b+1} - 4u_{ab}),$$
$$\text{and}$$
$$\nabla^h u := \tfrac{1}{2h}(u_{a+1} - u_{a-1}, u_{b+1} - u_{b-1}) =: (\partial_x^h u, \partial_y^h u)_{ab}.$$

We define, for the Dirichlet boundary conditions,

$$\begin{aligned}\Phi^h(u^h, \lambda) := \Delta^h u^h + \lambda f(u, \nabla^h u) &= 0 \quad \text{in } \mathbf{G}_h^i, \\ u^h &= 0 \quad \text{on } \mathbf{G}_h \backslash \mathbf{G}_h^i.\end{aligned}$$

To verify (17) we compare

$$\Phi^{(j)}(u, \lambda)(v, \mu)^j = \Delta(\delta_{j,0}u + \delta_{j,1}v) + \sum_{|\alpha|=j} D^\alpha\big(f(u, \nabla u)\lambda\big)(v, \nabla v, \mu)^\alpha \quad \text{in } \mathbf{G}_h^i$$

with

$$\begin{aligned}(\Phi^h)^{(j)}(\pi^h u, \lambda)(\pi^h v, \mu)^j &= \Delta^h(\delta_{j,0}\pi^h u + \delta_{j,1}\pi^h v) \hspace{2cm} (21)\\ &\quad + \sum_{|\alpha|=j} D^\alpha\big(f(\pi^h u, \nabla^h \pi^h u)\lambda\big)(\pi^h v, \nabla^h \pi^h v, \mu)^\alpha,\end{aligned}$$

where we have used the Kronecker $\delta_{i,j}$ and multi-index notation. Now we have to estimate the errors for the different terms in Equation (21), see (19), by considering

$$(\Delta u - \Delta^h u)_{ab} = \frac{h^2}{6} P_3(h, \Delta^4 u)(ab) \quad \text{for } u \in C^4(\Omega),$$

using (18), (19) and $(\alpha_1, \alpha_2) = (i, j, k)$ for $\alpha_3 = 0$. Thus

$$f^{(i,j,k)}(u, \nabla^h u) v^i (\partial_x^h v)^j (\partial_y^h v)^k - f^{(i,j,k)}(u, \nabla u) v^i (\partial_x v)^j (\partial_y v)^k$$

$$= \left[ f^{(i,j,k)}(u, \nabla^h u) - f^{(i,j,k)}(u, \nabla u) \right] v^i (\partial_x^h v)^j (\partial_y^h v)^k$$

$$+ f^{(i,j,k)}(u, \nabla u) \left[ v^i (\partial_x^h v)^j (\partial_y^h v)^k \right.$$

$$\left. - v^i (\partial_x^h v)^j (\partial_y v)^k + v^i (\partial_x^h v)^j (\partial_y v)^k - v^i (\partial_x v)^j, (\partial_y v)^k \right]$$

$$= \left\{ Q_0 \left( \frac{h^2}{4} P_2(h, \partial_x^3 u), f^{(i,j+1,k)}(u, \nabla u) \frac{h^2}{4} P_2(h, \partial_x^3 u) \right) \right.$$

$$\left. + Q_0 \left( \frac{h^2}{4} P_2(h, \partial_y^3 u), f^{(i,j,k+1)}(u, \nabla u) \frac{h^2}{4} P_2(h, \partial_y^3 u) \right) \right\} v^i (\partial_x^h v)^j (\partial_y^h v)^k$$

$$+ f^{(i,j,k)}(u, \nabla u) \left\{ v^i (\partial_x^h v)^j (\frac{h^2}{4} P_2(h, \partial_y^3 v))^k + v^i (\frac{h^2}{4} P_2(h, \partial_x^3 v))^j (\partial_y^h v)^k \right\}$$

$$=: h^2 \varepsilon^{(i,j,k)}(h^2, u) v^{(i,j,k)},$$

where $\varepsilon^{(i,j,k)}$ is a $(i, j, k)$-linear operator acting on $v$. Analogous results hold for derivatives with respect to $\lambda$. By presenting this derivation of $\varepsilon^{(i,j,k)}(h^2, z)$ we have gained some insights into the structure of the operator $\varepsilon(h^p, z)$ in (17), in particular the $h^p = h^2$ factor is verified. If $u \in C^4(\Omega)$ is replaced by a weaker condition, then $h^2$ has to be replaced with $h^p$ for some $0 < p < 2$. A combination of (17) with the usual stability

$$\| z^h - \bar{z}^h \| \le S \| \Phi^h(z^h) - \Phi^h(\bar{z}^h) \|$$

for $z^h$, $\bar{z}^h$ near the solution $z_0^h$ for $\Phi^h(z_0^h) = 0$, yields convergence of $z_0^h$ to $z_0$ with $\Phi(z_0) = 0$. Since the operators $\Phi^{h(j)}(\pi^h z) \pi^h \cdots \pi^h$ and $\hat{\pi}^h \Phi^{(j)}(z)$ in (17) act multilinearly on $z_1, \ldots, z_j$ the same is true for $\varepsilon(h^p, z)$, working from its definition.

## 3.2   Equivariant discretisation methods

In order that the discretisation of (1) inherits $\Gamma$–equivariance, we need reflect this in the discretisation method. In particular, $E^h$ and $\hat{E}^h$ need to be closed under the action of $\Gamma$. Moreover, $\Phi^h, \pi^h$ and $\hat{\pi}^h$ have to commute with the action of $\Gamma$, i.e.

$$\begin{aligned} &\Phi^h(\gamma u^h, \lambda) = \gamma \Phi^h(u^h, \lambda), \\ &\pi^h \gamma u^h = \gamma \pi^h u^h, \quad \hat{\pi}^h \gamma \hat{u}^h = \gamma \hat{\pi}^h \hat{u}^h \\ &\text{for all } \gamma \in \Gamma \text{ and } u^h \in E^h, \hat{u}^h \in \hat{E}^h. \end{aligned} \tag{22}$$

For difference methods, e.g. for the problem (20), a $\mathbf{D}_4$ invariant grid in $[0, 1]^2$ and the $\Delta^h$, $\nabla^h u^h$ in Section 3.1 will fulfill these criteria. For finite element methods, $\Gamma$–invariant approximating subspaces have to be chosen such that (22) is satisfied. This nontrivial task can be performed by using group representation theory and a basis that does not destroy the sparseness structure of the original problem. Suitable bases, e.g. the a so-called *symmetry respecting bases*, are studied in Stiefel-Fässler [32], Allgower-Böhmer-Mei [4, 5], Allgower-Böhmer-Georg-Miranda [3], Georg-Miranda [18] and Douglas-Mandel [16]. For a given problem, $\Phi^h$ has to be formulated w.r.t. a symmetry respecting basis satisfying (22). Then fixed point spaces within $E^h$ and $\hat{E}^h$ can be defined for subgroups $\Sigma \preceq \Gamma$ in the following way

$$E^{h,\Sigma} := \mathrm{Fix}^\Sigma(E^h) := \{z^h \in E^h \ : \ z^h = \sigma z^h \ \forall \ \sigma \in \Sigma\}. \tag{23}$$

As a consequence of (22) and (23) we are able to consider the discrete problem $\Phi^h(z^h)$ on the fixed point space (23) for subgroups $\Sigma \preceq \Gamma$. The following proposition formalises this result.

**Proposition 6** *Let (22) be satisfied. Then for any subgroup $\Sigma \preceq \Gamma$ we have*

$$\Phi^{h,\Sigma} := \Phi^h|_{E^{h,\Sigma}} : E^{h,\Sigma} \to \hat{E}^{h,\Sigma},$$

*and $\Phi^h$ and all its derivatives (and remainder terms) evaluated at $z^h$ are equivariant with respect to the isotropy subgroup $\Sigma_{z^h}$ of the point $z^h$. If $\Phi^h$ is stable or $r$-times consistently differentiable, the same is true for $\Phi^{h,\Sigma}$.*

Throughout we have assumed that we study the singularity of the original problem at $(0,0) = z_0$, i.e. $\Phi(0,0) = \Phi(z_0) = 0$. We shall restrict to discussing bifurcation problems at $(0,0)$ with

$$\begin{aligned}
\partial_\lambda \Phi(0,0) &= 0, \\
\mathrm{Ker}(\Phi_0') &= \mathrm{Ker}(\partial_u \Phi(0,0)) \times \mathbf{R} = [\psi_1, \ldots, \psi_m] \times \mathbf{R} \quad \text{and} \\
\mathrm{Ker}(\Phi_0'^*) &= \left[\hat{\psi}_1, \ldots, \hat{\psi}_m\right].
\end{aligned} \tag{24}$$

Let

$$\Phi^h(z_0^h) = 0, \ z_0^h \approx (0,0) \ \text{and} \ z_0^h \in \mathrm{Fix}^\Gamma(E^h) = E^{h,\Gamma}. \tag{25}$$

For a simple bifurcation point of $\Phi$ ($m = 1$ in (24)) Brezzi-Rappaz-Raviart [13] have shown that $\Phi^h$ has a simple bifurcation point in $z_0^h = (0,0) + O(h^p)$. We introduce a translated operator

$$\hat{\Phi}^h(z^h) := \Phi^h(z_0^h + z^h) - \Phi^h(z_0^h), \tag{26}$$

and drop the $\hat{\ }$ in the following. Thus, without loss of generality, $(0,0)$ is also a singular point of $\Phi^h$. For higher singularities of $\Phi$, Allgower-Böhmer [2] and Böhmer [11] have shown that under the additional hypothesis

$$\partial_\lambda \Phi^h(0,0) = 0$$

by discretising extended systems for $\Phi$ there exist $\psi_i^h$ and $\hat{\psi}_i^h$ such that, for $i = 1, \ldots, m$,

$$\begin{aligned}
\partial_{u^h} \Phi^h(0,0)\psi_i^h &= O(h^p), & \psi_i^h &= \pi^h \psi_i + O(h^p), \\
\left(\partial_{u^h} \Phi^h(0,0)\right)^* \hat{\psi}_i^h &= O(h^p), & \hat{\psi}_i^h &= \pi^h \hat{\psi}_i + O(h^p).
\end{aligned} \tag{27}$$

So generally the kernels of the discrete operator $\Phi^{h'}$ and $\Phi^{h'*}$ have smaller dimensions than those of the operators $\Phi'$ and $\Phi'^*$. In this paper we assume, even more stringently than (27), that

$$
\begin{aligned}
\mathrm{Ker}(\Phi_0^{h\prime}) &= \mathrm{Ker}(\partial_{u^h}\Phi^h(0,0)) \times \mathbf{R} = \left[\psi_1^h, \dots, \psi_m^h\right] \times \mathbf{R}, \\
\mathrm{Ker}(\Phi_0^{h\prime *}) &= \left[\hat{\psi}_1^h, \dots, \hat{\psi}_m^h\right] \quad \text{with} \\
\psi_i^h &= \pi^h \psi_i + O(h^p), \quad \hat{\psi}_i^h = \hat{\pi}^h \hat{\psi}_i + O(h^p), \ i = 1, \dots, m.
\end{aligned}
\tag{28}
$$

Since we work with an equivariant discretisation method, we may assume that $\psi_i^h$ and $\hat{\psi}_i^h$ have been determined in the same fixed point spaces as the $\psi_i$ and $\hat{\psi}_i$ respectively. The strong condition (28) is a generic situation. Our assumption that the problem is a generic $\Gamma$–equivariant problem means that in particular the kernel of the linearised operator is of generic dimension, that is, the dimension of an absolutely irreducible representation of $\Gamma$ for one parameter. Here we only give hints of how to verify (28). For many interesting problems, in particular, problems with a trivial solution branch, e.g. the Brusselator equations in Section 4 or the nonlinear Laplacian on the square, one can immediately show that (28) is satisfied at the bifurcation points where the singularity is resulted merely by symmetries of the domain (cf. [25]). In these cases, it is often possible that the symmetries of $\Phi$ and the domain $\Omega$ may be used to generate $\psi_2, \dots, \psi_m$ from $\psi_1$. For equivariant discretisations, the same strategy may be applied to generate the $\psi_i^h$. Alternatively, one can sometimes restrict to one dimensional fixed point subspaces and apply the Equivariant Branching Lemma [19] in the following way: choose a basis such that

$$
[\psi_1, \dots, \psi_m] = \mathrm{Ker}(\partial_u \Phi(0,0))
$$

and for $i \neq j$ there exist bifurcation subgroups $\Sigma_i \neq \Sigma_j$ with $\psi_i \in E^{\Sigma_i}$, $i = 1, \dots, m$. In this case, Proposition 6 implies that

$$
\mathrm{Ker}(\partial_{u^h}\Phi^h(0,0)) = [\psi_1^h, \dots, \psi_m^h] \quad \text{with } \psi_i^h \in E^{h,\Sigma_i}, \ \Sigma_i \neq \Sigma_j \text{ for } i \neq j,
$$

hence, $\mathrm{Ker}(\partial_{u^h}\Phi^h(0,0)) \cap E^{h,\Sigma_i} = [\psi_i^h]$. If (24) is satisfied we have

$$
(\Phi'(0,0))^* = (\partial_u\Phi(0,0), 0)^* = ((\partial_u\Phi)^*(0,0), 0)^T
$$

and hence the decomposition of $\mathrm{Ker}(\partial_u\Phi(0,0))$ into basis vectors with different symmetries will apply to $(\Phi_0')^*$ as well, especially in the cases where the corresponding bifurcation subgroup or the intrinsic isotropy subgroup has nontrivial conjugatcy classes, see e.g. Dellnitz-Werner [15] and Rabier [28].

Finally we need some consequences of Allgower-Böhmer [2] and Böhmer [11]. Let $\phi$ and $\hat{\psi}$ be solutions of

$$
\Phi_0'\phi = g \quad \text{and} \quad \Phi_0'^*\hat{\psi} = \hat{g},
$$

then there exist discrete solutions $\phi^h, \hat{\psi}^h$ of

$$
\begin{aligned}
\Phi_0^{h\prime}\phi^h &= \pi^h g \quad \text{and} \quad \Phi_0'^*\hat{\psi}^h = \hat{\pi}^h \hat{g} \quad \text{such that} \\
\phi^h &= \pi^h \phi + O(h^p), \quad \text{and} \quad \hat{\psi}^h = \hat{\pi}^h \hat{\psi} + O(h^p).
\end{aligned}
$$

If $g$ and hence $\phi$ is in $E^\Sigma$, then the same is true for the $\phi^h$ and the $O(h^p)$ error terms in $E^{h,\Sigma}$. An analogous result holds for $\hat{g}$.

## 3.3   Equivariant discrete projections

We need the results in 3.1 and 3.2 to be able to formulate an appropriate discretisation reflecting the bifurcation scenario. The singularity of (1) may be either analytically or numerically determined in the sense that we either know the elements in the kernels of $\Phi_0'$, $\Phi_0'^*$ or their approximations (28). To re-write Algorithm 4 in a discrete setting we need discrete projectors $Q^h$ and $\hat{Q}^h$. The severe stability problems discussed in [10, 11, 12] do not occur here due to the continuous and discrete kernels having the same dimensions. We use the discrete pairing or inner product $\langle .,.\rangle^h$ related to $\langle .,.\rangle$ by

$$\langle \pi^h u, \pi^h v\rangle^h = \langle u, v\rangle + O(h^p \|u\|_s \|v\|_s), \tag{29}$$

where $\|.\|_s$ indicates the smooth norm. Corresponding results hold in $\hat{E}$ and $\hat{E}^h$ as well. We can assume that the original and discrete pairings $\langle \cdot, \cdot\rangle$ and $\langle \cdot, \cdot\rangle^h$ are $\Gamma$–invariant

$$\langle u, v\rangle = \langle \gamma u, \gamma v\rangle, \quad \langle u^h, v^h\rangle = \langle \gamma u^h, \gamma v^h\rangle \quad \forall \, \gamma \in \Gamma \tag{30}$$

by defining, if necessary, a symmetrised pairing

$$[u, v] := \frac{1}{|\Gamma|} \sum_{\gamma \in \Gamma} \langle \gamma u, \gamma v\rangle$$

for a finite group $\Gamma$. Now we may split, as in Section 2,

$$\begin{aligned} E^h &= \mathrm{Ker}(\Phi_0^{h\prime}) \oplus \mathrm{Im}(\Phi_0^{h\prime*}) =: N^h \oplus M^h \\ \hat{E}^h &= \mathrm{Ker}(\Phi_0^{h\prime*}) \oplus \mathrm{Im}(\Phi_0^{h\prime}) =: \hat{N}^h \oplus \hat{M}^h. \end{aligned} \tag{31}$$

Since kernels and images of $\Gamma$–equivariant linear operators in (31) and in Section 2 are $\Gamma$–invariant subspaces, the orthogonality indicated by $\oplus$ is valid for the $\Gamma$–invariant pairing $\langle \cdot, \cdot\rangle$. The following Proposition relates the continuous projectors

$$\begin{aligned} Qu &:= u - \sum_{i=1}^m \langle \psi_i, u\rangle \psi_i, \quad Q(u, \lambda) := Qu \quad \text{for } (u, \lambda) \in E \times \mathbf{R}, \\ \hat{Q}\hat{u} &:= \hat{u} - \sum_{i=1}^m \langle \hat{\psi}_i, \hat{u}\rangle \hat{\psi}_i \quad \text{for } \hat{u} \in \hat{E}, \end{aligned} \tag{32}$$

with their discrete counterparts

$$\begin{aligned} Q^h u^h &:= u^h - \sum_{i=1}^m \langle \psi_i^h, u^h\rangle^h \psi_i^h, \quad Q^h(u^h, \lambda) := Q^h u^h \quad \text{for } (u^h, \lambda) \in E^h \times \mathbf{R}, \\ \hat{Q}^h \hat{u}^h &:= \hat{u}^h - \sum_{i=1}^m \langle \hat{\psi}_i^h, \hat{u}^h\rangle^h \hat{\psi}_i^h \quad \text{for all } \hat{u}^h \in \hat{E}^h. \end{aligned}$$

**Proposition 7** *If (28)-(30) are satisfied, then the projectors $Q$, $\hat{Q}$, $Q^h$, $\hat{Q}^h$ are $\Gamma$–equivariant projections satisfying*

$$\begin{aligned} \pi^h Q z - Q^h \pi^h z &= O(h^p \|z\|) \quad \text{and} \\ \hat{\pi}^h \hat{Q}\hat{z} - \hat{Q}^h \hat{\pi}^h \hat{z} &= O(h^p \|\hat{z}\|) \end{aligned}$$

*with error terms also equivariant under $\Gamma$.*

Summarising, we are able to reproduce the operators $\Phi$, $Q$, $\hat{Q}$ in (32), (8), (9) by $\Phi^h$, $Q^h$, $\hat{Q}^h$ up to $\Gamma$–equivariant perturbations of (relative) uniformly bounded size $O(h^p)$, see (18), (19) and (26).

## 3.4   Discrete finite determinacy

We want to ensure that the $\Gamma$–equivariant discretisation merely causes small perturbations of the reduced bifurcation equation, using the pseudo-norm in Definition 3. Let the original problem be $k$-determined, hence $B_k$ reflects the bifurcation scenario. The assumption of stability then gives that $B_k^h$ contains the whole bifurcation scenario of the original problem. Because of this, we need only show that if the $\Gamma$-equivariant problem $\Phi$ and discretisation $\Phi^h$ satisfy certain conditions, then $B_k^h$ coincides with $B_k$ up to small $\Gamma$-equivariant perturbations of $k$-th order.

To solve (12), (13) with $w_k \in \text{Im}(\Phi_0'^*)$ numerically, we need a *modified stability* of the form, see (31),

$$\Phi_0^{h'}|_{M^h} : M^h \to \hat{M}^h := \Phi_0^{h'}(M^h) \tag{33}$$
$$\text{is boundedly invertible, uniformly for small } h,$$

with appropriate norms chosen in $E^h$ and $\hat{E}^h$. Stability results of this type follow for many discretisation methods, see Brezzi-Rappaz-Raviart [13], Esser [17], Grigorieff [22], Reinhard [29], Stummel [33] and Vainikko [35].

Before stating the main results on the difference of $B_k^h$ and $B_k$ in Theorem 9, we formulate the corresponding numerical algorithm for the truncated Liapunov-Schmidt reduction:

**Algorithm 8** *Discrete Truncated Liapunov-Schmidt Method.*
   **Initialisation:** *Define, for $\eta^h \in \text{Ker}(\Phi_0^{h'})$, the function $w_1^h(\eta^h)$ by*

$$w_1(\eta^h) = 0.$$

**Iteration:** *For $k = 2, 3, \ldots$, until determinacy do:*

$$B_k^h(\eta^h) := (I - \hat{Q}^h)j_k R^h(\eta^h + w_{k-1}^h(\eta^h)) \tag{34}$$

*where $B_k^h$ is a polynomial of $k$-th order in $\eta^h$ from $\text{Ker}(\Phi_0^{h'})$ to $\text{Ker}(\Phi_0^{h'*})$. We generate the next function $w_k^h$ with the formula:*

$$w_k^h(\eta^h) := Q^h(\Phi_0^{h'})^{-1}\hat{Q}^h j_k R^h(\eta^h + w_{k-1}^h(\eta^h)) \tag{35}$$

*where $w_k^h$ is a $k$-th order polynomial from $\text{Ker}(\Phi_0^{h'})$ to $\text{Im}(\Phi_0^{h'*})$.*

The actual computation for the proof of the next theorem uses the modified stability in (33) and is discussed in detail in [10, 11].

**Theorem 9** *Let $\Phi$ and its discretisation $\Phi^h$ satisfy the following conditions:*

   (i)   *$r$-times consistent differentiability, see (17),*
   (ii)   *$\Phi^h$ is a $\Gamma$-equivariant discretisation for $\Phi$, see (22),*
   (iii)   *the singularity satisfies (24), (28)*
   (iv)   *the pairings are chosen to satisfy (29), (30),*
   (v)   *$\Phi_0^{h'}$ satisfies the modified stability, see (33).*

*Then $w_k$, $B_k$ from Algorithm 4 and their discrete approximations from Algorithm 8 satisfy*

$$\|w_k^h(\pi^h\eta) - \pi^h w_k(\eta)\| = O(h^p\|\eta\|), \quad \|B_k^h(\pi^h\eta) - \pi^h B_k(\eta)\| = O(h^p\|\eta\|) \quad (36)$$

*with the differences being $\Gamma$-equivariant. The norm for the $w_k^h$ approximation is given by (4), for $B_k^h$ we use (3). Furthermore, $B_k$ and $B_k^h$ are elements in $\mathcal{F}_\Gamma$.*

As a consequence of Theorem 9 and the genericity assumption in Section 1 we know that $B_{k+1}$ and $B_{k+1}^h$ represent the same bifurcation scenario for small enough $h$ and large enough $k$. The validity of the stability assumption can be monitored by comparing results for different values of $h$ and seeing that the coefficients of $B_k^h$ do not converge to values giving an unstable bifurcation problem.

# 4    Hopf Bifurcation for Brusselator Equations

In this section we consider an example from the dynamics of chemical systems as an application of our numerical Liapunov-Schmidt method.

## 4.1    The Brusselator equations with Neumann boundary conditions

Let us consider a simplified model for the Belusov-Zhabotinskii reaction:

$$\begin{aligned}
\partial_t u_1 &= \Delta u_1 + A - (B+1)u_1 + u_1^2 u_2, \quad \text{in } \Omega := [0,1] \times [0,1], \\
\partial_t u_2 &= d\Delta u_2 + Bu_1 - u_1^2 u_2,
\end{aligned} \quad (37)$$

where $A, d \in (0, +\infty)$, $B \in [0, \infty)$ are parameters, see Prigogine-Glansdorff [27]. We choose Neumann boundary conditions

$$\partial_n u_1 = \partial_n u_2 = 0, \quad \text{on } \partial\Omega,$$

($\partial_n$ represents the normal derivatives along the boundary of $\Omega$), see also Rothe [30]. This choice allows a fully analytical approach and thus a direct comparison between analytical and numerical methods. This problem has a (constant) trivial solution at $(u_1, u_2) = (A, B/A)$. By fixing the parameters $A$ and $d$ we consider the bifurcation of (37) from the trivial solution curve by varying $B$. To this end, we carry out the substitutions $(u_1, u_2) \longleftrightarrow (A, B/A) + (u_1, u_2)$, $t \longleftrightarrow \omega_0(1+\tau)t$, with the frequency $\omega_0$ to be specified later, and derive from (37) the equivalent form:

$$\Phi(u, B, \tau) := L(B)u - \omega_0\partial_t u + R(u, B) = 0, \quad (38)$$

where $u := (u_1, u_2)^T$,

$$L(B) := \begin{pmatrix} \Delta + B - 1 & A^2 \\ -B & d\Delta - A^2 \end{pmatrix}; \quad E := (C_0^{2,s}(\Omega))^2 \to \hat{E} := (C(\Omega))^2 \quad (39)$$

(with $C_0^{2,s}(\Omega)$ indicating homogeneous Neumann boundary conditions) and

$$R(u, B, \tau) := \begin{pmatrix} 1 \\ -1 \end{pmatrix}(\frac{B}{A}u_1^2 + 2Au_1u_2 + u_1^2u_2) - \omega_0\tau\partial_t u. \quad (40)$$

Let us consider the spaces $C_{2\pi}^0$ and $C_{2\pi}^1$ as in Section 2.3 with respect to the following basis with coefficients in $E, \hat{E}$ respectively,

$$\{1, \cos lt, \sin lt, \quad l = 1, 2, \ldots\}.$$

The $\mathbf{D_4} \times \mathbf{S}^1$-equivariant operator $\Phi$ maps $C_{2\pi}^1$ into $C_{2\pi}^0$ and

$$\Phi(0, B, \tau) = 0 \quad \text{for all} \quad B, \ \tau \in [0, \infty).$$

If, for a given constant $d_0 > 0$, the system

$$B - 1 - A^2 - (1 + d_0)c = 0, \tag{41}$$
$$(1 + c)(A^2 + d_0 c) - d_0 B c > 0 \tag{42}$$

has solutions $(A_0, B_0) \in \mathbf{R}_+^2$ for exactly one eigenvalue $c := (k^2 + l^2)\pi^2, k, l \in \mathbf{N}$ of $-\Delta$ with multiplicity 2, then the operator

$$\partial_u \Phi_0 := \partial_u \Phi(0, B_0, 0) = -\omega_0 \partial_t + L(B_0) \tag{43}$$

with $\omega_0 = \sqrt{(1 + c)(A_0^2 + d_0 c) - d_0 B_0 c}$ has a four dimensional null space:

$$\text{Ker}(\partial_u \Phi_0) = [\psi_1, \psi_2, \psi_3, \psi_4], \tag{44}$$

where with $e_1 := (1, 0)^T$, $e_2 := (0, 1)^T$,

$$P = (P_{ij})_{i,j=1}^2 := \frac{1}{\omega_0} \begin{pmatrix} B_0 - 1 - c & A_0^2 \\ -B_0 & -d_0 c - A_0^2 \end{pmatrix}, \quad \text{and}$$
$$\psi_1 := (e_1 \cos t + P e_1 \sin t)\phi_{kl}, \quad \psi_2 := (-P e_1 \cos t + e_1 \sin t)\phi_{kl},$$
$$\psi_3 := (e_1 \cos t + P e_1 \sin t)\phi_{lk}, \quad \psi_4 := (-P e_1 \cos t + e_1 \sin t)\phi_{lk}$$

with $\phi_{kl}(x, y) := 2\cos k\pi x \cos l\pi y$, $\phi_{lk}(x, y) := 2\cos l\pi x \cos k\pi y$, see Ashwin-Mei [9]. Similarly, the null space of the adjoint operator $\partial_u \Phi^*(0, B_0, 0)$ is

$$\text{Ker}(\partial_u \Phi_0^*) = [\psi_1^*, \ \psi_2^*, \ \psi_3^*, \ \psi_4^*] \tag{45}$$

with $P_{21} = -B/\omega_0$ and

$$\psi_1^* := (P^T e_2 \cos t + e_2 \sin t)\phi_{kl}/P_{21}, \quad \psi_2^* := (-e_2 \cos t + P^T e_2 \sin t)\phi_{kl}/P_{21},$$
$$\psi_3^* := (P^T e_2 \cos t + e_2 \sin t)\phi_{lk}/P_{21}, \quad \psi_4^* := (-e_2 \cos t + P^T e_2 \sin t)\phi_{lk}/P_{21}.$$

We have

$$\langle \psi_i^*, \ \psi_j \rangle = \delta_{ij}, \quad i, j = 1, \ldots, 4 \tag{46}$$

using the $\mathbf{D_4} \times \mathbf{S}^1$-invariant $L^2$-product

$$\langle u, \ v \rangle = \frac{1}{2\pi} \int_0^{2\pi} \int_\Omega (u, v) \, dx \, dy \, dt \tag{47}$$

in $C_{2\pi}^0$, where $(\cdot, \cdot)$ represents the Euclidean product in $\mathbf{R}^2$. Because $\partial_u \Phi(0, B_0, 0)$ is a Fredholm operator with index 0, and the zero eigenvalues are of semi-simple type, we have the direct, but not orthogonal decompositions in invariant subspaces:

$$C^0_{2\pi} = \mathrm{Ker}(\partial_u\Phi_0) \oplus \mathrm{Im}(\partial_u\Phi_0), \quad C^1_{2\pi} = \mathrm{Ker}(\partial_u\Phi_0) \oplus (\mathrm{Im}(\partial_u\Phi_0) \cap C^1_{2\pi}). \quad (48)$$

Define the $\mathbf{D}_4 \times \mathbf{S}^1$-equivariant projection $\hat{Q} : C^0_{2\pi} \to \mathrm{Im}(\partial_u\Phi_0)$ with

$$\hat{Q}w := w - \sum_{i=1}^{4} \langle \psi_i^*, w \rangle \psi_i, \quad w \in C^0_{2\pi}. \quad (49)$$

Equation (48) shows that $(I - \hat{Q})$ is a projection from $C^0_{2\pi}$ into $\mathrm{Ker}(\partial_u\Phi_0)$. The advantage of (48) as compared with (6) is the fact that $Q = \hat{Q}|C^1_{2\pi}$.

## 4.2   A second order truncated Liapunov-Schmidt method

Having set up the splitting and the null space projections, we now progress with the iterative Liapunov-Schmidt reduction of Section 2. For $B = B_0 + \lambda$ and $\tau$ in (38), and for any

$$\eta := (z, \lambda, \tau) = (\sum_{i=1}^{4} \alpha_i \psi_i, \lambda, \tau) \in \mathrm{Ker}(\partial_u\Phi_0) \times \mathbf{R}^2 = \mathrm{Ker}(\Phi_0'), \quad (50)$$

we find that the truncated bifurcation equation to second order is given by:

$$
\begin{aligned}
B_2(z, \lambda, \tau) &= (I - \hat{Q})\left[ \begin{pmatrix} 1 & 0 \\ -1 & 0 \end{pmatrix} \lambda \sum_{i=1}^{4} \alpha_i\psi_i + j_2 R(\sum_{i=1}^{4} \alpha_i\psi_i, B_0 + \lambda, \tau) \right] \\
&= (I - \hat{Q})\left\{ \begin{pmatrix} 1 & 0 \\ -1 & 0 \end{pmatrix} \lambda \sum_{i=1}^{4} \alpha_i\psi_i + \begin{pmatrix} 1 \\ -1 \end{pmatrix} \left[ \frac{B_0}{A_0}(\sum_{i=1}^{4} \alpha_i\psi_i^1)^2 \right. \right. \\
&\quad \left. \left. + 2A_0(\sum_{i=1}^{4} \alpha_i\psi_i^1)(\sum_{i=1}^{4} \alpha_i\psi_i^2) \right] - \omega_0\tau\partial_t \sum_{i=1}^{4} \alpha_i\psi_i \right\} = 0.
\end{aligned}
$$

We have used the notation $\psi_i = (\psi_i^1, \psi_i^2)^T$ and $j_2$ truncates to the polynomial of degree 2 in the Taylor expansion of $R$ at $(0, B_0, 0)$ with respect to $\alpha_i, \lambda, \tau$. Due to the orthogonality properties of $\{\phi_{k,l}\}$, the equation $B_2(z, \lambda, \tau) = 0$ becomes

$$
\begin{pmatrix} \lambda & -(P_{11} - P_{12})\lambda - 2\tau\omega_0 \\ (P_{11} - P_{12})\lambda + 2\tau\omega_0 & \lambda \end{pmatrix} \begin{pmatrix} \alpha_{i+1} \\ \alpha_{i+2} \end{pmatrix} = 0, \ i = 0, 2.
$$

Evidently, $B_2(z, \lambda, \tau)$ is not stable as it has vertically branching bifurcating solutions: we have a linear equation for $(\alpha_i)$ that has only zero as a solution unless both $\lambda$ and $\tau$ are zero. To investigate the structure of solution branches of (38) at $(0, B_0, 0)$, the next step of the iterative Liapunov-Schmidt method is needed.

## 4.3  A third order truncated Liapunov-Schmidt method

To find the third order truncation, we first need to compute $w_2$ from the equation

$$\partial_u \Phi(0, B_0, 0) w_2 = \hat{Q}\left[ \begin{pmatrix} 1 & 0 \\ -1 & 0 \end{pmatrix} \lambda \sum_{i=1}^{4} \alpha_i \psi_i + j_2 R(\sum_{i=1}^{4} \alpha_i \psi_i, B_0 + \lambda, \tau) \right]$$

$$=: f(\alpha, \lambda, \tau).$$

With Neumann boundary conditions we can do this combining computer analysis with either a numerical approximation, as discussed in Section 3, or a direct analytical approach as in [9]. In particular, the analytical results obtained by computer algebra yield a verification of the numerical Liapunov-Schmidt reduction, the latter being faster than analytical computations. Since

$$f(\alpha, \lambda, \tau) = f_0 + f_1 \cos t + f_2 \sin t + f_3 \cos 2t + f_4 \sin 2t, \tag{51}$$

where $f_i, i = 0, 1, \ldots, 4$ are functions of $\alpha, \lambda, \tau$ and $x, y$, we decompose the expected solution into the form:

$$w_2 = w_2^0 + w_2^1 \cos t + w_2^2 \sin t + w_2^3 \cos 2t + w_2^4 \sin 2t. \tag{52}$$

These $w_2^i$ are independent of $t$ and solutions of the corresponding $\mathbf{D}_4$-equivariant equations, for example

$$\begin{pmatrix} L(B_0) & -\omega_0 I \\ \omega_0 I & L(B_0) \end{pmatrix} \begin{pmatrix} w_2^1 \\ w_2^2 \end{pmatrix} = \begin{pmatrix} f_1(\alpha, \lambda, \tau) \\ f_2(\alpha, \lambda, \tau) \end{pmatrix}. \tag{53}$$

In these equations $\mathbf{S}^1$ is eliminated and hence the conditions in Section 3 are satisfied. The expressions $f_0, \ldots f_4$ are quadratic polynomials of $\alpha_i, i = 1, \ldots, 4$. Thus picking the coefficients of the various monomials in the $\alpha_i$ gives us a set of problems on the square. We find the solutions $w_2^k$ of (53) in the form:

$$w_2^k = \sum_{i,j=1}^{4} \alpha_i \alpha_j w_2^{kij}. \tag{54}$$

Due to the $\mathbf{D}_4 \times \mathbf{S}^1$–equivariance of (38), the terms $w_2^{kij}$ in the solution of (54) are related by group actions. This fact will be utilised to reduce the computational work. Furthermore, depending on the symmetries of the right hand side of (53), the problem (53) need only to be solved in the appropriate subdomain of $\Omega$, and thus the computational effort is once more reduced, see e.g. Mei [26]. The action of $\mathbf{D}_4$ on the domain $\Omega = [0, 1] \times [0, 1]$ is generated by action of the reflections $R$ and $S$,

$$R(x, y) = (y, x), \quad S(x, y) = (1 - x, y) \quad \forall \ (x, y) \in \Omega. \tag{55}$$

In the function space $C_{2\pi}^0$, the action of $\mathbf{D}_4 \times \mathbf{S}^1$ is defined to be diagonal, i.e.,

$$(\delta, \theta) \begin{pmatrix} u \\ v \end{pmatrix} = \begin{pmatrix} u(\delta(x, y), t - \theta) \\ v(\delta(x, y), t - \theta) \end{pmatrix} \quad \forall (\delta, \theta) \in \mathbf{D}_4 \times \mathbf{S}^1.$$

Let $(\alpha_1, \ldots, \alpha_4)$ denote the coordinates of an arbitrary element $\sum \alpha_i \psi_i$ in the null space $\text{Ker}(\partial_u \Phi_0)$, since

$$S\phi_{kl} = (-1)^k \phi_{kl}, \quad R\phi_{kl} = \phi_{lk},$$
$$S\phi_{lk} = (-1)^l \phi_{lk}, \quad R\phi_{lk} = \phi_{kl}$$

hold for Neumann boundary conditions, the induced action of $\mathbf{D}_4$ on $(\alpha_1, \ldots, \alpha_4)$ is generated by

$$S(\alpha_1, \alpha_2, \alpha_3, \alpha_4)^T = ((-1)^k \alpha_1, (-1)^k \alpha_2, (-1)^l \alpha_3, (-1)^l \alpha_4)$$

and

$$R(\alpha_1, \alpha_2, \alpha_3, \alpha_4)^T = (\alpha_3, \alpha_4, \alpha_1, \alpha_2).$$

The induced action of $\mathbf{S}^1$ on $\alpha$ is as follows:

$$T_\theta \begin{pmatrix} \alpha_1 \\ \alpha_2 \\ \alpha_3 \\ \alpha_4 \end{pmatrix} := \begin{pmatrix} \cos\theta & \sin\theta & 0 & 0 \\ -\sin\theta & \cos\theta & 0 & 0 \\ 0 & 0 & \cos\theta & \sin\theta \\ 0 & 0 & -\sin\theta & \cos\theta \end{pmatrix} \begin{pmatrix} \alpha_1 \\ \alpha_2 \\ \alpha_3 \\ \alpha_4 \end{pmatrix} =: \begin{Bmatrix} t_\theta \begin{pmatrix} \alpha_1 \\ \alpha_2 \end{pmatrix} \\ t_\theta \begin{pmatrix} \alpha_3 \\ \alpha_4 \end{pmatrix} \end{Bmatrix}.$$

In particular,

$$T_\pi \alpha = -\alpha, \quad T_{\pi/2} \alpha = (\alpha_2, -\alpha_1, \alpha_4, -\alpha_3)^T. \tag{56}$$

Hence, the symmetry group of the (reduced) bifurcation equation depends on the parity of $k, l$. The symmetries and the equivariance of the equations for $w_2^i$ lead via (38), (51)-(56) to the relations

$$\gamma w_2^i(\alpha, \lambda, \tau) = w_2^i(\gamma \alpha, \lambda, \tau) \quad \forall \gamma \in \mathbf{Z}_2 \times \mathbf{Z}_2 \times \mathbf{S}^1 \text{ or } \mathbf{D}_4 \times \mathbf{S}^1, \ i = 0, 1, \ldots, 4.$$

The different cases $\mathbf{Z}_2 \times \mathbf{Z}_2$ or $\mathbf{D}_4$ depend on the chosen values of $(k, l)$. Once the coefficients $w_2^{i11}, w_2^{i12}, w_2^{i13}, w_2^{i14}$ of the terms $\alpha_1^2, \alpha_1\alpha_2, \alpha_1\alpha_3, \alpha_1\alpha_4$ in the expansion of $w_2^i, i = 0, \ldots, 4$ are known, all the other terms can be derived via group actions. For example, to solve the equations (53), we solve four equations corresponding to $(i, j) = (1, 1), (1, 2), (1, 3), (1, 4)$ in the expansion of the right hand side of (53) w.r.t. $\alpha_1, \ldots, \alpha_4$. Even in the computation of these coefficients, the special structure of the systems allows further decomposition of the solutions and reduction of computational effort.

After these preparations we are able to compute $B_3$. For $u = (u_1, u_2)^T$, $v = (v_1, v_2)^T \in C_{2\pi}^1$, we define

$$R_1(u, v) := \begin{pmatrix} 1 \\ -1 \end{pmatrix} \left[ \frac{B_0}{A_0} u_1 v_1 + A_0(u_1 v_2 + u_2 v_1) \right], \quad R_2(u) := \begin{pmatrix} 1 \\ -1 \end{pmatrix} u_1,$$

$$R_3(u, u, u) := \begin{pmatrix} 1 \\ -1 \end{pmatrix} u_1^2 u_2, \quad R_4(u, v) := \begin{pmatrix} 1 \\ -1 \end{pmatrix} u_1 v_1 / A_0.$$

Hence,

$$B_3(z, \lambda, \tau)$$

$$
\begin{aligned}
= \ & (I - \hat{Q})\left[\begin{pmatrix} 1 & 0 \\ -1 & 0 \end{pmatrix}\lambda\Big(\sum_{i=1}^{4}\alpha_i\psi_i + w_2\Big) + j_3 R\Big(\sum_{i=1}^{4}\alpha_i\psi_i + w_2, B_0 + \lambda, \tau\Big)\right] \\
= \ & B_2(z, \lambda, \tau) + (I - Q)(2R_1(z, w_2(\eta)) + \lambda R_2(w_2(\eta)) \qquad\qquad (57) \\
& + R_3(z, z, z) + \lambda R_4(z, z) - \omega_0\tau\partial_t w_2(\eta)).
\end{aligned}
$$

Since all the second order terms are already known from the calculation of $B_2$, we only need to consider the third order terms. The analytical computations were performed by a Maple program giving an output in Fortran code, while numerical computation was done on the corresponding fundamental domain with a finite difference method. Due to the $\mathbf{D}_4 \times \mathbf{S}^1$-equivariance of $B_3$, we need only calculate a few cubic terms to obtain the branching behaviour at bifurcation.

## 4.4 Results

As an example, we consider in [8] truncated Liapunov-Schmidt reduction at the Hopf bifurcation of the $(1, 2) : (2, 1)$ spatial mode with $A_0 = 1, d_0 = 1/5$. We have observed first order convergence in $h$ of the numerical predictions to the exact results. There are vertex oscillations, edge oscillations and rotating wave solution branches bifurcating from this Hopf bifurcation point, using the classification of Swift [34]. Moreover, we see reduced equations for which Swift predicts branching of periodic solutions with submaximal symmetry, i.e. branches not predicted by the equivariant Hopf theorem [20].

# 5 Acknowledgment

We thank Jacques Furter for helpful comments on an earlier version of the manuscript.

# References

[1] E. L. Allgower, P. Ashwin, K. Böhmer and Z. Mei: *Liapunov-Schmidt reduction for a bifurcation problem with periodic boundary conditions on a square domain.* Preprint, 1992.

[2] E. L. Allgower and K. Böhmer: Resolving singular nonlinear equations. *Rocky-Mountain J. Math.*, **18**:225-268, 1988.

[3] E. L. Allgower, K. Böhmer, K. Georg and R. Miranda: Exploiting symmetry in boundary element methods. *SIAM J. Numer. Anal.* **29**:534-552, 1992.

[4] E. L. Allgower, K. Böhmer and Z. Mei: On a problem decomposition for semi-linear nearly symmetric elliptic problems. In: *Parallel Algorithms for Partial Differential Equations*, W. Hackbusch (Ed.), pp. 1-17, Vieweg Verlag 1991.

[5] E. L. Allgower, K. Böhmer and Z. Mei: A complete bifurcation scenario for the 2d-nonlinear Laplacian with Neumann boundary conditions on the unit square. In: *Bifurcations and Chaos: Analysis, Algorithms, Applications*, R. Seydel, F. W. Schneider, T. Küpper, H. Troger (Eds.), pp. 1-18, Birkhäuser Verlag, Basel 1991.

[6] P. Ashwin: High corank mode interactions on a rectangle. In: *Bifurcation and Symmetry*, ed. E. L. Allgower, K. Böhmer and M. Golubitsky. ISNM **104**, Birkhäuser, Basel, pp. 23-33, 1992.

[7]  P. Ashwin: *PhD thesis*. Mathematics Institute, University of Warwick, 1991.

[8]  P. Ashwin, K. Böhmer and Z. Mei: *A numerical Liapunov-Schmidt method with applications to Hopf bifurcation on a square*. Preprint, 1992.

[9]  P. Ashwin and Z. Mei: *Liapunov-Schmidt reduction at Hopf bifurcation of the Brusselator equations on a square*. Preprint, 1992.

[10] K. Böhmer: *Developing a numerical Lyapunov-Schmidt method*. Bericht zur Fachbereich Mathematik der Philipps-Universität Marburg 1990.

[11] K. Böhmer: On a numerical Lyapunov-Schmidt method for operator equations. Submitted to *Computing* 1992.

[12] K. Böhmer and Z. Mei: On a numerical Lyapunov-Schmidt method. In: *Computational Solutions of Nonlinear Systems of Equations*, E. L. Allgower, K. Georg (Eds.): pp. 79-98, Lectures in Applied Mathematics **26**, AMS, Providence, 1990.

[13] F. Brezzi, I. Rappaz and P. A. Raviart: Finite Dimensional Approximation of Nonlinear problems, Part I: Branches of Nonsingular solutions. *Numer. Math.* **36**:1-25, 1980. Part II: Limit points, *Numer. Math.* **37**:1-28, 1981. Part III: Simple bifurcation points, *Numer. Math.* **38**:1-30, 1981.

[14] M. Crouzeix and J. Rappaz: *On Numerical Approximation in Bifurcation Theory*. Masson, Paris, Springer-Verlag, Berlin-New York, 1988.

[15] M. Dellnitz and B. Werner: Computational methods for bifurcation problems with symmetries-with special attention to steady state and Hopf bifurcation points, *J. Comp. Appl. Math.* **26**:97-123, 1989.

[16] C. C. Douglas and J. Mandel: An abstract theory for the domain reduction method. *Computing* **43**:75-96, 1992.

[17] H. Esser: Stabilitätsungleichungen für Diskretisierungen von Rand- wertaufgaben gewöhnlicher Differentialgleichungen. *Numer. Math.* **28**:69-100, 1977.

[18] K. Georg and R. Miranda: Exploiting symmetry in solving linear equations. In: *Bifurcation and Symmetry*, ed. E. L. Allgower, K. Böhmer and M. Golubitsky. ISNM **104**, Birkhäuser, Basel, pp. 157-168, 1992.

[19] M. Golubitsky and D. Schaeffer: *Singularities and Groups in Bifurcation Theory*. *Volume 1*, App. Math. Sci. **51** Springer, New York, 1986.

[20] M. Golubitsky and I. Stewart: Hopf bifurcation in the presence of symmetry, *Arch. Rat. Mech. Anal.* **87**:107-165, 1985.

[21] M. Golubitsky, I.N Stewart, and D. Schaeffer: *Singularities and Groups in Bifurcation Theory. Volume 2*. App. Math. Sci. **69** Springer, New York, 1988.

[22] R.D. Grigorieff: Zur Theorie linearer approximationsregulärer Operatoren, I und II. *Math. Nachr.*, **55**:233-249 and 251-263, 1973.

[23] W. Hackbusch: *Theorie und Numerik elliptischer Differentialgleichungen*. Teubner Verlag, Stuttgart 1986.

[24] J. Martinet: *Singularities of Smooth functions and Maps*. LMS lecture notes **58**, CUP, 1982.

[25] Mei, Z.: Bifurcations of a simplified buckling problem and the effect of discretizations. *Manuscripta Mathematica* **71**:225-252, 1990.

[26] Z. Mei: Path following around Corank-2 bifurcation points of a semi-linear elliptic problem with symmetry. *Computing* **47**:69-85, 1991.

[27] I. Prigogine and P. Glansdorff: *Structure, Stabilité et Fluctuations*. Masson, Paris 1971.

[28] P. J. Rabier: New aspects in bifurcation with symmetry, *Differential and Integral Equations*, **6**:1015-1034, 1990.

[29] H. J. Reinhard: *Analysis of Approximation Methods for Differential Integral Equations*. Springer-Verlag, Berlin, Heidelberg, Tokyo, New York, 1985.

[30] F. Rothe: *Global Solutions for Reaction-Diffusion Systems*. Lecture Notes in Math. **1072** Springer-Verlag, Berlin, Heidelberg, Tokyo, New York, 1984.

[31] H. J. Stetter: *Analysis of Discretization Methods for Ordinary Differential Equations*. Springer-Verlag Berlin, Heidelberg, 1973.

[32] E. Stiefel and A. Fässler: *Gruppetheoretische Methoden und ihre Anwendung*. B. G. Teubner Verlag, Stuttgart 1979.

[33] F. Stummel: Diskrete Konvergenz linearer Operatoren, I. *Math. Ann.*, **190**:45-92, 1970. II. *Math. Z.* **120**:231-264, 1971. III. *Proc. Oberwolfach* 1971, **ISNM** 20, 196-216 (1972).

[34] J.W. Swift: Hopf bifurcation with the symmetry of the square. *Nonlinearity* **1**:333-377, 1988.

[35] G. Vainikko: *Funktionsanalysis der Diskretisierungsmethoden*. Teubner Texte zur Mathematik, Teubner, Leipzig, 1976.

[36] A. Vanderbauwhede. *Local bifurcation and symmetry*. Research Notes in Math. **75**, Pitman, Boston, 1982.

P. Ashwin
Mathematics Institute, University of Warwick, Coventry CV4 7AL, UK.

K. Böhmer
FB Math., Universität Marburg, Lahnberge, D-3550 Marburg, FRG.

Z. Mei
FB Math., Universität Marburg, Lahnberge, D-3550 Marburg, FRG and
Dept. Math., Xian Jiaotong University, Xi'an 710049, PRC.

Lectures in Applied Mathematics
Volume **29**, 1993

# Exploiting and Detecting Space-Time Symmetries

NADINE AUBRY and WENYU LIAN

ABSTRACT   We use biorthogonal decompositions to investigate spatio-temporal behavior, particularly in terms of space-time symmetries. Biorthogonal decompositions consist in decomposing a spatially and temporally evolving function into orthonormal temporal modes in a Hilbert space H(T) and orthonormal spatial modes in a Hilbert space H(X) which are related by an isomorphism.   In this context, a space-time symmetry can be introduced as a pair of operators, one acting on H(X), the other one on H(T).   We show how such tools can be used to analyze symmetry related space-time bifurcations by analyzing computed solutions of the Kuramoto-Sivashinsky equation.   Another space-time symmetry exploitation is the derivation of spectral laws and self-similar modal relations for fully developed turbulence.

## 1. Introduction

While dynamical systems theory is a powerful tool for temporal systems, its usefulness for systems involving both spatial and temporal complexity, such as turbulence in open flows, is questioned (see e.g. [1], [2]).   Recently, biorthogonal decompositions have been proposed to study space-time dynamics [3], providing orthogonal spatial modes in a Hilbert space H(X) of spatial functions and orthogonal temporal modes in a Hilbert space H(T) of temporal functions for which there is a one-to-one correspondence.   The spatio-temporal dynamics can then be studied in a temporal configuration space $\chi$(T) and a spatial one $\chi$(X), these two spaces being related to each other via an isomorphism.

1992 Mathematics Subject Classification.   Primary 58F14, 20C30, 20C32.
Final version
Supported by NSF/PYI award MSS89-57462, ONR grant N00014-90-J-1554 (under the Fluid Dynamics Program (Code 1132F)) and NATO grant No. 900265.

This provides a framework to study and detect the presence of spatio-temporal symmetries defined as pairs of operators, coupled by the previous isomorphism, whose spatial (resp. temporal) component acts on $\chi(X)$ (resp. $\chi(T)$).

## 2. Biorthogonal decompositions and space-time symmetries

We consider a spatio-temporal complex valued function $u(x,t)$ where $x \in X, t \in T$ (X and T being the spatial and temporal domains) and introduce appropriate Hilbert spaces $H(X)$ and $H(T)$ of spatial and temporal functions defined on X and T, respectively. Then, each spatio-temporal function defines a linear operator $U: H(X) \to H(T)$ such that

$$(1) \qquad \forall \varphi \in H(X), \quad (U\varphi)(t) = \int_X u(x,t)\varphi(x)dm(x),$$

and its adjoint $U^*: H(T) \to H(X)$ such that

$$(2) \qquad \forall \psi \in H(T), \quad (U^*\psi)(x) = \int_T \overline{u(x,t)}\psi(t)dm(t)$$

where the bar denotes the complex conjugate and $dm(x)$ (resp. $dm(t)$) the measure defining the scalar product in $H(X)$ (resp. $H(T)$). When the operator U is compact, the spectral decomposition of U can be written as

$$(3) \qquad u(x,t) = \sum_{k=1} A_k \overline{\varphi_k(x)}\psi_k(t)$$

$$\text{with} \quad A_1 \geq A_2 \geq ... > 0,$$

$$\lim_{N \to \infty} A_N = 0$$

$$\text{and} \quad (\varphi_k, \varphi_1) = (\psi_k, \psi_1) = \delta_{k,1}.$$

The isomorphism coupling the modes $\psi_k$ and $\varphi_k$, hereafter called chronos and topos, is given by the operator U itself such that

$$(4) \qquad U\varphi_k = A_k \psi_k.$$

The subspace $\chi(X) = \text{Ker}(U)^\perp$ (resp. $\chi(T) = \text{Ker}(U^*)^\perp$) is the smallest euclidean space containing the dynamics, namely all the vectors $\xi_t \in H(X)$ (resp. $\eta_x \in H(T)$) such that

$$(5) \qquad \forall x \in X, \xi_t(x) = u(x,t)$$

(6)                  (resp. $\forall t \in T, \eta_x(t) = u(x,t)$).

The above procedure is equivalent to the spectral decomposition of the self-adjoint operator

$$V = \begin{pmatrix} 0 & U^* \\ U & 0 \end{pmatrix}$$

defined on $H(X) \oplus H(T)$. Biorthogonal decompositions have been generalized in [4] to the case of non-compact operators (eventually with a singular kernel $u(x, t)$) by applying the theory of eigenfunction expansion of self-adjoint operators and introducing generalized chronos and topos defined in Hilbert spaces different (slightly 'larger') than those on which the operator $U$ and $U^*$ act. This generalization may not be necessary for pure data analysis purposes which, in practice, always correspond to finite matrices but is a necessity for theoretical deductions and model derivations of physical situations where the spectrum has a continuous component, the domain is infinite and/or singularities occur. A well-known example of this situation is turbulence in fluids (see Section 3.3).

The introduction of space-time symmetries is similar whether the operators $U$ and $U^*$ are compact or not, except that in the latter case, the symmetry operators $S$ and $\tilde{S}$ (see the definition below) need to be defined in the 'larger' spaces. For the sake of simplicity, we limit the following discussion to compact operators $U$ and $U^*$ (otherwise, see [3]). Applications of biorthogonal decompositions to study spatio-temporal dynamics can be found in [3, 4, 5, 6, 7]. Such a decomposition is obviously not new, it has its roots in the spectral decomposition of operators with symmetric kernels which can be found early in text books (e.g. [8]) for compact operators. The generalization to operators with Carleman kernels can be found in von Neumann [9].

When $u$ is real, two operators with symmetric kernels can be considered, one is $V$ and the other one is $U^*U$: $H(X) \to H(X)$ (and similarly $UU^*$: $H(T) \to H(T)$). The decomposition of the latter can be connected with the statistical tool called Karhunen-Loève (KL) expansion [10] introduced in turbulence in [11] when the operator $U$ is compact, a space-time domain is considered (instead of a statistical ensemble) and $L^2$-norms are used in the definitions (1) and (2). Then the KL expansion corresponds to the spectral decomposition of $U^*U$ in one direction (e.g.

physical space for $U^*U$) (see [12, 4] for this connection and its consequences). Even if the time information of $u(x, t)$ can be recovered from the Fourier coefficients of the expansion, considering the decomposition of the two-point statistics (the kernel of $U^*U$) does not permit the introduction of a space-time symmetry (see below), but only that of a (statistical) spatial symmetry. The question of preserving spatial symmetries in dynamical systems based on KL modes has been addressed in [11].

As a natural generalization of the notion of symmetries in temporal (or spatial) systems in classical group theory (see e.g. [14]), space-time symmetries, introduced and defined in [5] as actions of *two* operators, S acting on $H(X)$ and $\tilde{S}$ acting on $H(T)$ which intertwine the operator U, can be written as

$$(7) \qquad\qquad US = \tilde{S}U \text{ and } US^* = \tilde{S}^*U$$

where the star denotes the adjoint operator. Equivalently, we can write VT=TV where

$$T = \begin{pmatrix} S & 0 \\ 0 & \tilde{S} \end{pmatrix}.$$

The classical notion of a symmetry acting on a single Hilbert space can be recovered by setting S (or equivalently $\tilde{S}$) to the identity operator, namely US = U (or $\tilde{S}U$ = U). The presence of a space-time symmetry (7) is equivalent to the degeneracy of the spatial and temporal eigenspaces in the following sense:

$$(8) \qquad\qquad \text{if } U\varphi_k = A_k \psi_k, \text{ then } US\varphi_k = A_k\tilde{S}\psi_k.$$

This implies that eigenspaces of topos are invariant under the action of S and eigenspaces of chronos are invariant under the action of $\tilde{S}$.


Similarly, if G is a locally compact group and $S_g$ (resp. $\tilde{S}_g$) a unitary representation of G in $H(X)$ (resp. $H(T)$), the spatio-temporal action of G can be written as:

$$(9) \qquad\qquad US_g = \tilde{S}_gU, \quad \forall g \in G$$

so that the eigenspaces of U are degenerate under the action of G in the sense of (8) and the restriction of S (resp. $\tilde{S}_g$) to each spatial (resp. temporal) eigenspace defines a unitary representation of G. These two representations, which are equivalent (due to the existence of the isomorphism U between the two eigenspaces), have the same decomposition in irreducible representations, the latter determining the (common) dimension of the corresponding eigenspaces.

A more general definition of a space-time symmetry is a pair of operators $(S,\tilde{S})$, S (resp. $\tilde{S}$) being a representation of the semi-group N on H(X) (resp. H(T)) such that

(10)                                      $US = \gamma\,\tilde{S}U$

(11)                                      $US^* = \gamma^{-1}\tilde{S}^*U$

where $\gamma$ is a real $0 < \gamma < 1$. The existence of a pair $(S,\tilde{S})$ satisfying (10) and (11) is equivalent to the following proposition:

(i) there exists a subsequence of terms in the decomposition (3) such that the spectrum satisfies the relation:

(12)                                      $A_{k+1} = \gamma\,A_k.$

(This remains valid if $\gamma$ equals one, in which case, the spectrum $A_k$ is degenerate, see above.) The generalization to representations of the group Z, useful for instance in Section 3.3, can be made with non-compact operators.  If $\gamma < 1$, the spectrum $A_k$ of (12) decays exponentially fast and all topos and chronos can be deduced from the first one in the sequence by the relation

(13)                          $\varphi_{k+1} = S\varphi_k$ and $\psi_{k+1} = \tilde{S}\psi_k$

where $\varphi_{k+1}$, $\psi_{k+1}$ and $\varphi_k$, $\psi_k$ correspond to different eigenvalues $A_{k+1} = \gamma A_k.$

Perhaps, it is worth mentioning that the commutation of U (7) with a spatio-temporal symmetry $(S, \tilde{S})$ (or equivalently the commutation of V with T) implies that $U^*U$ commutes with S and $UU^*$ with $\tilde{S}$:

$$U^*US = SU^*U$$

(14)

$$UU^*\tilde{S} = \tilde{S}UU^*$$

(and therefore the topos are invariant under the action of S and the chronos under the action of $\tilde{S}$) but the converse is not true since S and $\tilde{S}$ are (selectively) coupled through the isomorphism U in (7) but not in (14).  This emphasizes the difference between statistical space-time symmetries (14) and a deterministic space-time symmetry (7) (see also the criticism of [15] on statistical symmetries).

A simple example where (7) is satisfied is furnished by traveling waves defined as

(15)                          $u(x - x_0, t) = u(x, t + t_0)$

defined as

(15)                         $u(x - x_0, t) = u(x, t + t_0)$

for all $x, x_0 \in X$, $t, t_0 \in T$ such that $x_0 + ct_0 = 0$. In this case, defining

(16)                         $(\tilde{S}_{t_0} \psi)(t) = \psi(t - t_0)$

(17)                         $(S_{x_0} \varphi)(x) = \varphi(x - x_0)$,

we have the commutation relation

$US_{x_\bullet} = \tilde{S}_{t_\bullet} U$ for all $x_0, t_0$ such that $x_0 + ct_0 = 0$.

which is equivalent to the degeneracy of topos and chronos with respect to the group

of spatio-temporal translations of speed c: if $U\varphi = A\psi$, then $US_{x_0}\varphi = A\tilde{S}_{t_0}\psi$.

## 3. Analysis of spatio-temporal bifurcations

The two first examples in this section are solutions of the Kuramoto-Sivashinsky equation (KSE):

(18)         $$\frac{\partial u}{\partial t} + 4\frac{\partial^4 u}{\partial x^4} + \alpha\left(\frac{\partial^2 u}{\partial x^2} + \frac{1}{2}\left(\frac{\partial u}{\partial x}\right)^2\right) = 0 \quad 0 \leq x \leq 2\pi$$

which has been derived as an amplitude model equation for a variety of physical problems. It has been extensively studied numerically and theoretically for its large variety of bifurcations and the complexity of its solutions (see [16], [17], [18] and references therein). Here, the KSE, subject to periodic boundary conditions u(x, t) = u(x+2π, t), is integrated for various parameter values $\alpha$, by a spectral method using forty real Fourier modes, after checking that a higher number of Fourier modes does not alter the solutions for the ranges of parameter values considered in this paper. In all examples, we use the decomposition of the operator U defined in (1) with $L^2$-norms.

3.1 A space-time Hopf bifurcation. This example partially appeared in the review article [12]. It occurs at $\alpha = 83.75$ at which a stable fixed point whose spatial structure consists of a small hump and a large one starts oscillating mostly on the higher hump [16]. The decomposition of the fixed point consists of one term only which can be written as u(x) = $A_0 \, \varphi_0(x)$. Indeed, at $\alpha = 82$, the second eigenvalue numerically found is $10^{-10}$, which we estimate below the order of magnitude of our computational errors. After the oscillations start, the solution evolves in space and time (Figure 1); the configuration spaces $\chi(X)$ and $\chi(T)$ become of higher dimension (evaluated by eigenvalues above computational errors) and we now concentrate on the first three terms in the decomposition (3). The temporal dynamics in $(\varphi_1, \varphi_2, \varphi_3)$ is a circle showing that indeed a Hopf bifurcation took place (this point was questioned in [16]). Figures 2 and 3 show the first three temporal and spatial eigenmodes. The second and third chronos are sinusoidal, translated with respect to each other, so that we have $\psi_2 = \tilde{S}_{t_0} \psi_3$ where $\tilde{S}_{t_0}$ is the translation operator (16). Although the relation between the topos $\varphi_2$ and $\varphi_3$ is more complex, we observe the following features, which leads to an understanding of the spatio-temporal structure of the signal. The first topo is the temporal average of the function and coincides with the renormalized fixed point before the bifurcation. The second and third topos $\varphi_2$ and $\varphi_3$ intersect at the points $x_{LH}$ and $x_{SH}$ corresponding to the large and small humps of the first topo and are quasi-shifted with respect of each other, the corresponding shifts being of opposite signs on both sides of $x_{LH}$ (and $x_{SH}$). The translation coefficient varies with the spatial location, first increases with the distance from $x_{LH}$ and then decreases to reach zero at $x_{SH}$. These (irregular) shifts between topos together with the regular shift between chronos indicate the presence of left and right 'traveling' waves starting at $x_{SH}$ and reaching $x_{LH}$, as we show by summing $A_2\psi_2(t)\varphi_2(x) + A_3\psi_3(t)\varphi_3(x)$ (Figure 4). It is interesting to note that $A_2 \neq A_3$ since this is not the case of a pure traveling wave (15) corresponding to an exact translation symmetry between topos. Spatial modes for this solution have also been computed in [18] from the decomposition of the spatial two-point correlation.

Figure 1. Three-dimensional representation of a typical spatio-temporal solution at $\alpha = 84$: time increases from bottom to top, x increases from left to right;

Reprinted from *On the hidden beauty of the proper orthogonal decomposition* by Nadine Aubrey, Theoret. and Comput. Fluid Dynamics, volume 2, 1991, pp. 339–352, Springer-Verlag.

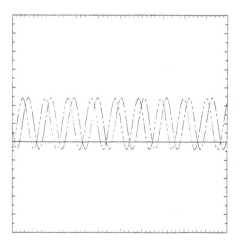

Figure 2. Chronos $\psi_1(t)$——, $\psi_2(t)$—·—, $\psi_3(t)$-···—corresponding to the solution displayed in Figure 1.

Reprinted from *On the hidden beauty of the proper orthogonal decomposition* by Nadine Aubrey, Theoret. and Comput. Fluid Dynamics, volume 2, 1991, pp. 339–352, Springer-Verlag.

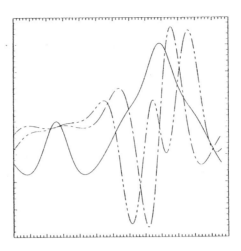

Figure 3. Topos $\varphi_1(x)$———, $\varphi_2(x)$—·—, $\varphi_3(x)$—··—corresponding to the solution displayed in Figure 1.

Reprinted from *On the hidden beauty of the proper orthogonal decomposition* by Nadine Aubrey, Theoret. and Comput. Fluid Dynamics, volume 2, 1991, pp. 339–352, Springer-Verlag.

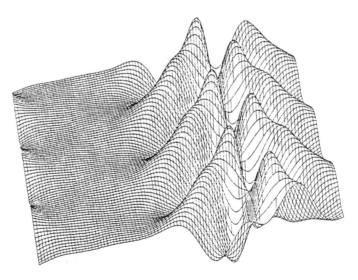

Figure 4. Three-dimensional representation of the spatio-temporal part of the solution $A_2 \, \varphi_2(x) \, \psi_2(t) + A_3 \, \varphi_3(x) \, \psi_3(t)$ corresponding to the solution shown in Figure 1 (representation conventions are the same as in Figure 1).

Reprinted from *On the hidden beauty of the proper orthogonal decomposition* by Nadine Aubrey, Theoret. and Comput. Fluid Dynamics, volume 2, 1991, pp. 339–352, Springer-Verlag.

**3.2. Spatio-temporal chaos.** As $\alpha$ is increased, a traveling wave bifurcation occurs followed by a spatio-temporal complex regime. We then analyze a typical solution obtained at the bifurcation parameter value $\alpha = 91$, for a time window [2.2, 2.4] after eliminating the transient. A certain two-by-two coupling is noticed in the spectrum, perhaps indicating the breaking of a symmetry (7). In order to understand this feature, we display topos and chronos: Topos are nearly sinusoidal, coupled and such that the topos of a same pair are shifted with respect to each other, a characteristic pattern of traveling waves. Although chronos show more irregularity and complexity, we could identify time intervals during which two consecutive chronos, with two rather well-defined frequencies, are also approximately shifted. The intermittent, chaotic regime is then divided into laminar, traveling periods alternating with turbulent or complex phases. The spatio-temporal chaos then occurs through a breaking of the temporal translation symmetry (valid before the bifurcation to 'chaos') in an intermittent fashion. Note that although topos are nearly Fourier modes, our analysis is different from a Fourier decomposition: both a spatial Fourier decomposition and a spatial and temporal Fourier decomposition would be useless from the point of view of space-time symmetries, because we would then lose the isomorphism mentioned in Section 2. Intermittency consisting of chaotic states connected by traveling waves has been observed in wall turbulence models including nonzero streamwise modes [19]. Heteroclinic orbits connecting traveling and standing waves have been identified in a rotating layer instability study [20].

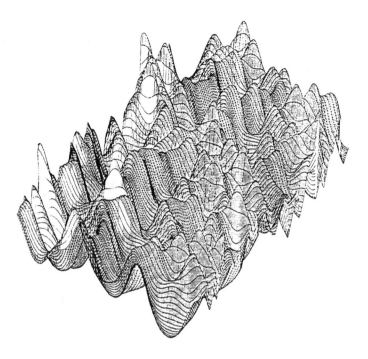

Figure 5. Three-dimensional representation of a typical chaotic spatio-temporal solution at $\alpha = 91$ (representation conventions are the same as in Figure 1).

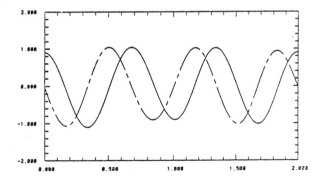

Figure 6. Topos $\varphi_3(x)$——, $\varphi_4(x)$—·— corresponding to the solution displayed in figure 5.

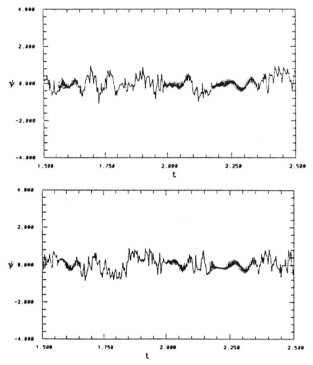

Figure 7.  Chronos $\psi_3$(t) (top), $\psi_4$(t) (bottom) corresponding to the solution
displayed in Figure 5.

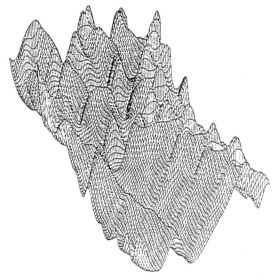

Figure 8.  Three-dimensional representation of the solution displayed in Figure 5,
enlightening the intermittent traveling of the solution (representation conventions are
the same as in Figure 1).

### 3.3. Space-time symmetries in fully developed turbulence. The compressible Navier-Stokes equations (NSE) (the incompressible equations being only a particular case) are invariant under dilation space-time symmetries. This example was treated in [4] for the incompressible case and in [21] for the compressible case. To illustrate this point, we consider the incompressible NSE:

$$\frac{\partial u}{\partial t} + (u.\nabla)u = -\nabla p + \frac{1}{Re}\nabla^2 u$$

$$\nabla.u = 0,$$

which are invariant under the transformation:

$$x' = \lambda x, t' = \lambda^{1-h}t, u' = \lambda^h u.$$

We then assume that the flow itself, in its fully developed state is invariant under this transformation, namely the flow is self-similar. This assumption is common in turbulence, at least in a 'statistical' sense, and is at the origin of the celebrated power law Kolmogorov Fourier spectrum for homogeneous incompressible turbulence. This property, which can be written in the operator form as

(19)
$$\tilde{S}_\lambda U = \lambda^{-h}US_\lambda,$$

where

(20)
$$(S_\lambda \varphi)(x) = \varphi(\lambda^{-1}x)$$

(21)
$$\text{and } (\tilde{S}_\lambda \psi)(t) = \psi(\lambda^{-(1-h)}t),$$

has many consequences, such as the spectral law of the type of (12) and the hierarchy relation between spatial and temporal eigenmodes of the type of (13) with the appropriate symmetries (20) and (21). Note that the operator U does not need to be compact, as it is required in any theory of fully developed turbulence.

## 4. Concluding remarks

We believe that space-time symmetries play an important role in the route to 'spatio-temporal' chaos (or complexity) of spatially extended systems. Traveling waves constitute the simplest example of a spatio-temporal behavior in presence of a space-time symmetry. The two first examples treated in Section 3 are bifurcations from pure traveling waves which originate in the breaking of one (either spatial or temporal) component of the symmetry. The third example shows that as complexity develops, space-time symmetries may appear (rather than break) and carry an important part of the information regarding the spatial and temporal structure of the attractor. The tools used in this paper are useful for the study of spatio-temporal symmetries, particularly in the context of non-compact operators.

## References

1. M. V. Morkovin, Recent insights into instability and transition to turbulence in open-flow system. Report *AIAA*-88-3675.

2. P. Bergé, Le chaos, théorie et experiences, Séries Synthèses, Eyrolles.

3. N. Aubry, R. Guyonnet and R. Lima, Spatio-temporal analysis of complex signals: theory and applications. *J. Stat. Phys.* 64 (1991), 683-738.

4. N. Aubry, R. Guyonnet and R. Lima. Turbulence spectra. *J. Stat. Phys.* 67 (1992), 203-228.

5. N. Aubry, R. Guyonnet and R. Lima, Spatio-temporal symmetries and bifurcations via biorthogonal decompositions. *J. Nonlinear Sci.* 2 (1992) 183-215.

6. S. Slimani, N. Aubry, P. Kolodner and R. Lima, Biorthogonal decomposition analysis of binary fluid convection. In: Bifurcation Phenomena and Chaos in Thermal Convection, H. H. Bau, L. Bertram and S. A. Korpela (ed.) New York: ASME, 1992, pp. 39-46.

7. N. Aubry, M. Chauve and R. Guyonnet, Transition to turbulence on a rotating flat disk, *Phys. Fluids A* (submitted).

8. R. Courant and D. Hilbert, Methoden der Mathematischen Physik, Interscience Publishers, Inc. New York, Verlag von Julius Springer (1924).

9. J. von Neuman, *Actualités Scientifiques et Industrielles*, Série n° 229, Paris (1935).

10. M. Loève, Probability Theory. Van Nostrand, New York (1955).

11. J. L. Lumley, The structure of inhomogeneous turbulent flows. In Atmospheric turbulence and radio wave propagation, A. M. Yaglom and V. I. Tatarski (ed.) Moscow: Nauka (1967).

12. N. Aubry, On the hidden beauty of the proper orthogonal decomposition. *Theor. and Comput. Fluid Dyn.* (1991) 339-352.

13. N. Aubry, W. Y. Lian and E. S. Titi. Preserving symmetries in the proper orthogonal decomposition, *SIAM Journal on Statistical and Scientific Computing* **14**, No. 2 (1993) 483-505.

14. M. Golubitsky and D. G. Schaeffer, Singularities and groups in bifurcation theory, vol. 1. Applied Mathematical Sciences **69**, Springer-Verlag (1985).

15. E. Barany, M. Dellnitz and M. Golubitsky, Detecting the symmetries of attractors, Research Report UH/MD-143, University of Houston, Department of Mathematics (1992).

16. J. M. Hyman, B. Nicolaenko, S. Zaleski, Order and complexity in the Kuramoto-Sivashinsky model of weakly turbulent interfaces, *Physica*18D (1985) 265-292.

17. M. Jolly, I. G. Kevrekidis, E. S. Titi, Approximate inertial manifolds for the Kuramoto-Sivashinsky equation: Analysis and computations. *PhysicaD* **44** (1990) 38-60.

18. M. Kirby, D. Armbruster and W. Guttinger, An approach for the analysis of spatially localized oscillations, *International Series of Numerical Mathematics* **97**, Birkhauser Verlag Basel (1991) 183-187.

19. S. Sanghi and N. Aubry, Mode interaction models for near wall turbulence, *J. Fluid Mech.* **247** (1993) 455-488.

20. E. Knobloch and M. Silber, Oscillatory convection in a rotating layer, preprint, 1992.

21. N. Aubry and W. Lian, Spatio-temporal structure of compressible turbulence, in Transitional and turbulent Compressible Flows, L. Kral and T. Zang (ed.), NEW-YORK: ASME, 1993.

BENJAMIN LEVICH INSTITUTE AND DEPARTMENT OF MECHANICAL ENGINEERING, THE CITY COLLEGE OF THE CITY UNIVERSITY OF NEW YORK, NEW YORK, NY 10031. *E-mail address*: menna@ccnyvme.bitnet.

Lectures in Applied Mathematics
Volume **29**, 1993

# LATTICE PERIODIC SOLUTIONS WITH LOCAL GAUGE SYMMETRY

ERNEST BARANY

February 11, 1993

ABSTRACT. Solutions that exhibit the symmetry of spatial lattices are familiar in problems with Euclidean equivariance. Such solutions show up in finite dimensional local bifurcation problems by virtue of Lyapunov-Schmidt reduction of the governing partial differential equations and a "brute force" compactification of the translations in the Euclidean group. Similar solutions are known to occur in superconductors where the situation is complicated by the existence of local gauge symmetries. We show that the standard bifurcation theory methods can be extended to include systems with local gauge symmetries, but that the physically observed lattice periodic states in superconductors arise from a different mechanism than in the conventional case.

## 1. INTRODUCTION

The work described in this article is motivated by experiment. It has long been known that when a thin layer of fluid is heated from below (Rayleigh-Benard convection) a flow pattern can emerge that has the symmetry of a planar hexagonal lattice, as well as other possibilities. More specifically, what is observed is that if the temperature difference between the top and bottom of the fluid layer is sufficiently small, a static, featureless 'conduction' state is seen that possesses the full Euclidean symmetry. (We idealize to the case of a layer of infinite horizontal extent.) If the temperature difference is increased beyond a critical value, the conduction state loses stability to a time independent convection flow pattern that breaks the Euclidean symmetry. This transition can be understood in the context of steady state bifurcation theory (see [4], Case Study 4).

Another experiment with similar results from the point of view of symmetry involves so-called 'Type II' superconductors [8] in the presence of an externally applied, spatially constant, magnetic field. When a slab of this type of material is placed in a sufficiently strong magnetic field oriented perpendicular to the slab it is observed that the superconductivity is 'quenched' and the resulting normal metal state exhibits the symmetry of the Euclidean plane. When the strength of the field is decreased below a critical value this state loses stability to a superconducting state wherein the magnetic field is confined to localized vortex tubes, and these

1991 *Mathematics Subject Classification.* Primary 81R40, 58E09; Secondary 82D55, 81T13.
This research was supported in part by NSF grant DMS-8905456

vortices form an array with the symmetry of an hexagonal lattice. Since both systems are Euclidean equivariant, it is reasonable to conjecture that the mechanisms
underlying the formation of these patterns is similar. The present work describes
why this conjecture is false. The fundamental tool is 'gauge-fixing', wherein the
gauge symmetry of the problem is exploited to simplify the form of the fields, in
addition the procedure is seen to simplify the bifurcation analysis.

This paper is expository; most of the results presented can be found in reference [1]. The presentation here is markedly different, however, and stresses the way
in which the exploitation of the symmetries of the problem simplifies the analysis.
The outline of the rest of the paper is as follows. In section 2 we present a very brief
review of the analysis for problems without gauge symmetry. The principal results
are the equivariant version of Lyapunov-Schmidt reduction and the compactification procedure that ensures the kernel is finite dimensional. In section 3 we define
what is meant by local gauge symmetry and introduce the notion of gauge fixing. In
section 4 we combine the ideas of the previous two sections and arrive at the main
results of reference [1]. Finally, in section 5 we show how this scenario is realized
in the specific example of the Ginzburg-Landau equations of superconductivity.

## 2. STANDARD BIFURCATION THEORY

The material in this section is standard, a complete discussion can be found in
[3]. We assume that we are looking at some dynamical variable $u(x)$ ($x \in \mathbf{R}^n$, $n = 2$
for planar lattice solutions) and a control parameter $\lambda$. For our particular examples,
$u(x)$ and $\lambda$ are, respectively, (a component of) the velocity and the temperature
difference for convection and the order parameter and external field strength for
superconductivity. Since we are looking for steady states, the equations are of
the form $\mathcal{G}(u(x), \lambda) = 0$, where $\mathcal{G}$ represents some nonlinear partial differential
operator that can be thought of as a Banach space map $\mathcal{G} : \mathcal{B} \times \mathbf{R} \to \mathcal{E}$ with
$\mathcal{E} \subset \mathcal{B}$ appropriate Banach spaces. We also assume we have a symmetry group $\Gamma$
(that contains the $n$-dimensional Euclidean group $\mathbf{O}(n)\dot{\times}\mathbf{R}^n$) that acts on $u(x)$.
The manifestation of symmetry in the dynamical equations is the equivariance
condition $\mathcal{G}(\gamma \cdot u, \lambda) = \gamma \cdot \mathcal{G}(u, \lambda)$ where $\gamma \in \Gamma$. We also assume that there is a fully
symmetric solution $u(x) = u_0 = 0$ so that $\mathcal{G}(u_0, \lambda) = 0$.

A *local bifurcation* occurs when the trivial solution $u_0$ loses stability to some
other solution as the control parameter $\lambda$ crosses some critical value $\lambda_0$ in such a
way that the amplitude of the new solution goes continuously to zero as $\lambda \to \lambda_0$.
By the implicit function theorem, this can happen only if $\mathcal{G}_u(u_0, \lambda_0) = 0$, where
$\mathcal{G}_u$ is the Frechet derivative of $\mathcal{G}$. We refer to such a point $(u_0, \lambda_0)$ as a *bifurcation
point* for $\mathcal{G}$.

The Lyapunov-Schmidt reduction is a procedure by which it can be shown that
only the subspace $ker(\mathcal{G}_u) \subset \mathcal{B}$ is relevant to the bifurcation problem in the sense
that the solutions to $\mathcal{G}(u, \lambda) = 0$ near a bifurcation point are in one-to-one correspondence with the solutions to a reduced set of equations

(2.1) $$ g : ker(\mathcal{G}_u(u_0, \lambda_0)) \times \mathbf{R} \to [range(\mathcal{G}_u(u_0, \lambda_0))]^\perp $$

where $[range(\mathcal{G}_u(u_0, \lambda_0))]^\perp$ is the complement of the range of $\mathcal{G}_u(u_0, \lambda_0)$ in $\mathcal{E}$. We
assume $\mathcal{G}_u(u_0, \lambda_0)$ is a Fredholm operator so the kernel and the complement of the
range in equation (2.1) have the same dimension.

The consequence of equivariance in this procedure is that (generically) the kernel and complement to the range of $\mathcal{G}$ will carry an irreducible representation of $\Gamma$, so solutions to $\mathcal{G}(u, \lambda) = 0$ near a bifurcation point are determined as singular points of irreducible vector fields. This is a very useful observation because there are results based on the geometry of the group action that allow one to conclude that solutions of certain isotropy types must exist for any vector field, independent of any specific set equations. (see e.g. chapter XIII of [4]) Such results are particularly important in this context because the Lyapunov-Schmidt reduction, being based on the implicit function theorem, does not, in general, provide a method for the explicit construction of the reduced equations (2.1).

The real utility of these techniques is in the case that $ker(\mathcal{G})$ is finite dimensional, then a choice of coordinates is always possible such that the reduced equations are polynomials. Unfortunately, if $\Gamma$ is noncompact, as it is for our case, the irreducible representations are infinite dimensional, so this whole procedure is of little value. The conventional solution, which is not particularly satisfactory, is to restrict the symmetry group to a compact subgroup and deal with the finite dimensional representations that result. For the lattice-like solutions we seek, this amounts to restricting the Euclidean group $E^2 = \mathbf{O(2)} \dot{\times} \mathbf{R}^2$ to the compact group $H_{\mathcal{L}} \dot{\times} T^2$, where $H_{\mathcal{L}} \subset \mathbf{O(2)}$ is the holohedry of the lattice $\mathcal{L}$, and $T^2$ is the two-torus of lattice translations. The kernel will then consist of a set of lattice periodic functions that are invariant under the torus of translations and irreducible under the rotations in the holohedry. When this procedure is applied to the case of hexagonal symmetry, the methods reveal the presence of solutions with the symmetry of hexagons as well as solutions with the symmetry of convection rolls, and possibly others depending on the particular irreducible representation chosen.

## 3. Gauge symmetry

Gauge symmetry is easier to introduce as an invariance than as an equivariance, and since the particular case of interest turns out to be variational we will restrict to this situation for the remainder of the paper. A typical "energy" functional for a variational problem in physics can be written

$$(3.1) \qquad \int_{\Omega} \nabla u \cdot \nabla u + \Phi(u) \qquad \Omega \subset \mathbf{R}^3$$

where $\Omega$ is the spatial domain of the problem and $\Phi(u)$ is a potential function of $u(x)$. We start out in three dimensions, and later restrict to the relevant case of two dimensions by assuming trivial dependence in the third dimension.

The group of spatial symmetries $\Gamma_s$ is the symmetry group of $\Omega$, and we assume that $u(x)$ is a spatial scalar so that the action of $\Gamma_s$ on $u(x)$ is

$$(3.2) \qquad \gamma_s \cdot u(x) = u(M_{\gamma_s^{-1}} x)$$

where $\gamma_s \in \Gamma_s$ and $M_\gamma$ is a matrix representing $\gamma$ in the defining representation of $\mathbf{O(3)}$.

An additional group of invariances, called *gauge* symmetries, can occur if $u \in V$, where V is some vector space. The gauge group $\Gamma_g$ is then a subgroup of the group of isometries of V, usually $V = \mathbf{C}^m$ so $\Gamma_g \subset \mathrm{SU}(m)$. For our model $m = 1$, so for simplicity we restrict to that case and the action of $\Gamma_g$ on $u(x)$ is then just multiplication by a complex phase,

$$(3.3) \qquad\qquad\qquad \gamma_g \cdot u = e^{i\theta} u$$

where $\gamma_g \in \Gamma_g$. With all this, the variational problem

$$(3.4) \qquad\qquad\qquad \int_\Omega |\nabla u|^2 + \Phi(|u|)$$

is gauge invariant, and the associated Euler-Lagrange equations are equivariant with respect to the actions (3.2) and (3.3).

The situation becomes more interesting if we allow *local* gauge transformations which amounts to letting $\theta$ be a function of $x$ in (3.3). In these circumstances, the $\Phi(|u|)$ term in (3.4) is still gauge invariant, but the gradient term is not, since there will be extra terms showing up that involve $\nabla\theta(x)$. To obtain a system that is invariant under local gauge symmetries, it is necessary to introduce a new dynamical field $A(x)$ that is a spatial vector $A : \mathbf{R}^3 \to \mathbf{R}^3$ with the transformation properties:

$$(3.5) \qquad\qquad \gamma_s \cdot A(x) \;=\; M_{\gamma_s} A(M_{\gamma_s^{-1}} x) \quad \gamma_s \in \Gamma_s$$
$$(3.6) \qquad\qquad \gamma_g \cdot A(x) \;=\; A(x) - \nabla\theta(x) \quad \gamma_g \in \Gamma_g$$

With these definitions, the construction $(\nabla + iA(x))u(x)$ is equivariant with respect to gauge transformations, so the new variational problem

$$\int_\Omega |(\nabla + iA)u|^2 + \Phi(|u|)$$

is locally gauge invariant and the resulting Euler-Lagrange equations obtained by varying with respect to $u(x)$ and $A(x)$ are equivariant with respect to the actions (3.2), (3.3), (3.5) and (3.6).

Before discussing the effect of the presence of such symmetries on bifurcation problems, we discuss a few relevant aspects of the lore of gauge invariant systems. The conventional doctrine holds gauge transformations to be *unobservable*, so that only gauge invariant quantities such as $|u|^2$ and $\nabla \times A$ are physical. This allows a procedure called "gauge fixing" wherein preliminary gauge tranformations are performed on the fields to select a convenient representative from each gauge orbit. A conventional choice is called Coloumb gauge, where it is assumed that the field $A(x)$ satisfies $\nabla \cdot A(x) = 0$. For simple situations, such as a problem with Dirichlet or Neumann boundary conditions on a finite region in $\mathbf{R}^3$, this condition determines $A(x)$ *uniquely* and so determines $u(x)$ up to constant phase transformations. This is very useful because the group of gauge transformations is quite large and not easy to deal with, so by selecting a representative in this way we end up with a compact group of residual gauge symmetries.

## 4. BIFURCATION WITH GAUGE SYMMETRY

At first glance, the presence of gauge symmetry would seem to greatly complicate the bifurcation theory analysis. As indicated in the previous section, the local gauge group is a large, indeed infinite dimensional, Lie group and so its irreducible representations are infinite dimensional and the Lyapunov-Schmidt reduction is again of questionable value. Fortunately, it turns out that the two procedures already presented, lattice compactification and gauge fixing, combine nicely to solve the infinite dimensionality problem, and the resulting bifurcation analysis can proceed in the usual way.

In the case with no gauge symmetry, the compactification process amounts to restricting to fields $u(x)$ that are invariant under translations in the lattice. If $\mathcal{L}$ denotes the lattice, and $t \in \mathcal{L}$ is a vector corresponding to a translation in $\mathcal{L}$, then lattice periodicity means $u(x + t) = u(x)$ for all $t \in \mathcal{L}$. For the case of gauge symmetry, the assumption that physical quantities exhibit lattice periodicity means that the fields should be required to be not strictly lattice periodic, but only lattice periodic *modulo gauge transformations*. That is, solutions to a gauge invariant problem, $(u(x), A(x))$, will have the physical symmetry of a lattice $\mathcal{L}$ if there exist functions $g_t(x)$ such that

$$(4.1) \qquad\qquad u(x + t) = e^{i g_t(x)} u(x)$$

$$(4.2) \qquad\qquad A(x + t) = A(x) - \nabla g_t(x)$$

We refer to fields that obey (4.1) and (4.2) as gauge $\mathcal{L}$-periodic.

With this definition we are in a position to state the main theorem of [1] which was proved with the help of R. Glowinski and H. Lopes (see [1] and its appendix for the proof).

**Theorem 4.1.** *Suppose $(u, A)$ is gauge $\mathcal{L}$-periodic. Then $(u, A)$ is gauge equivalent to $(v, \mathcal{P} + C)$ where:*

  **a:** *$\mathcal{P}$ is $\mathcal{L}$-periodic with mean zero.*
  **b:** *$\nabla \cdot \mathcal{P} = 0$.*
  **c:** *$C(x) = k x^{\perp}$ for some real number $k$.*
  **d:** *$v(x)$ is gauge $\mathcal{L}$-periodic with $g_t(x) \propto k \times$ (area of parallelogram spanned by $x$ and $t$).*

A few comments are in order.

- Since $\mathcal{P}$ is $\mathcal{L}$-periodic we may average $\mathcal{P}$ over any fundamental cell of the lattice and get the same answer. This number is the *mean* of $\mathcal{P}$. We denote the space of $\mathcal{L}$-periodic, divergence free, mean zero $\mathcal{P}$ by $\mathcal{X}_{\mathcal{L}}$.
- The notation $x^{\perp}$ refers to a vector in the plane of the lattice that is perpendicular to $x$ and of the same magnitude. If the Cartesian coordinates of $x$ are $(x_1, x_2)$, then $x^{\perp} = (-x_2, x_1)$.
- It can be shown that $C$ corresponds to the externally applied (spatially constant) part of the magnetic field, and the constant $k$ is determined by the strength of that field. Thus, $k$ is the control parameter of the bifurcation.
- We refer to the fields $v(x)$ of Theorem 4.1(d) as $\mathcal{L}$-theta functions, and denote this by $v \in \mathcal{T}_{\mathcal{L}}$. In light of the above comment, the phase transformation suffered by a $\mathcal{L}$-theta function under a translation $t \in \mathcal{L}$ is proportional to the magnetic flux through the parallelogram formed by $x$ and $t$.

Finally, we mention that similar gauge conditions were introduced in an *ad hoc* way on physical grounds by earlier authors (see [5] and [6]).

After implementation of Theorem 4.1 a gauge invariant functional can be thought of as

$$\mathcal{V} : \mathcal{T}_\mathcal{L} \times \mathcal{X}_\mathcal{L} \times \mathbf{R} \to \mathbf{R}$$

where the $\mathbf{R}$ on the left side of the arrow is the space of the parameters $k$.

In this context, the Lyapunov-Schmidt reduction of the Euler-Lagrange equations will produce a finite dimensional system. To see the structure of the reduced system we consider the action of some of the residual symmetries. First note that the global gauge transformations (constant phase transformations) leave the spaces $\mathcal{T}_\mathcal{L} \times \mathcal{X}_\mathcal{L} \times \mathbf{R}$ invariant. Such transformations act by

$$\theta \cdot (v, \mathcal{P}, k) = (e^{i\theta} v, \mathcal{P}, k)$$

and we denote this circle group of symmetries by $S^1$. The fact that this group acts trivially on $\mathcal{X}_\mathcal{L}$ and $k$ leads to a fundamental dichotomy in the expected bifurcation behavior. The gauge equivariant equations $\mathcal{G}(v, \mathcal{P}, k) = 0$, where $\mathcal{G} = \nabla \mathcal{V}$, must commute with the action of $S^1$, so the kernel $K$ of the linearization must be $S^1$-invariant. This requires that

$$K = K_v \oplus K_\mathcal{P}$$

where $K_v$ consists only of kernel vectors of the form $(v, 0)$ in $\mathcal{T}_\mathcal{L} \times \mathcal{X}_\mathcal{L}$, and $K_\mathcal{P}$ of kernel vectors of the form $(0, \mathcal{P})$. This decomposition follows since $S^1$ acts trivially on $K_\mathcal{P}$ and nontrivially on each nonzero vector in $K_v$. Now since all of the other residual symmetries of the gauge-fixed problem leave both spaces $K_v$ and $K_\mathcal{P}$ separately invariant, it follows that for codimension one bifurcations we expect that (generically) the kernel $K$ will equal either $K_\mathcal{P}$ or $K_v$. This is the key observation: in the former case the structure of the kernel is exactly the same as occurs in the standard case of lattice periodicity (Rayleigh-Benard convection), but the latter possibility is new and describes the type of lattice-like solution seen in superconductors. Further, when $K = K_v$, the nontrivial $S^1$ action and the expected irreducibility of the kernel at the bifurcation point require that $\dim K = 2$. Finally, it can be shown that the expected codimension one bifurcation is just a pitchfork of revolution which produces one nontrivial branch of group orbits of solutions. As such, the dynamics of this bifurcation is much simpler than the dynamics of the standard case of Rayleigh-Benard convection. In the foregoing context of Lyapunov-Schmidt reduction, this conclusion requires the observation that an additional reflection symmetry is present that does not commute with the global gauge symmetry so that the residual symmetry group contains $\mathbf{O(2)}$. Alternately, a splitting lemma reduction can be applied to the potential formalism, and the $S^1$ symmetry is sufficient to draw the same conclusion. The explicit verification of this for the specific case of the Ginzburg-Landau equations is slightly easier using the Lyapunov-Schmidt scenario.

## 5. GINZBURG-LANDAU EQUATIONS

In this final section we show how this analysis occurs in practice in the Ginzburg-Landau equations of superconductivity [2]. These equations are derived (formally) from the functional

$$(5.1) \qquad \mathcal{V}(u, A) = \int |(i\nabla + A)u|^2 - |u|^2 + \frac{1}{2}|u|^4 + |\nabla \times (A - A_0)|^2$$

where $A_0$ is the vector potential corresponding to the externally applied part of the magnetic field (see [1] and [8] for more details). In an unbounded domain the functional (5.1) is not well defined so the usual derivation is purely formal. A bonus of our gauge fixing procedure will be to rigorously define the variational problem.

To see the lattice-like solutions, first we assume a "quasi-two dimensional" situation wherein we assume that the observables are constant in the third dimension and also that the third component of the vector field $A$ vanishes. We also introduce complex coordinates in the remaining two dimensional space, and write $z \in \mathbf{C} \cong \mathbf{R}^2$ as the coordinate and $\mathcal{A} : \mathbf{C} \to \mathbf{C}$ as the complex vector field. With this notation the variational functional is

$$\mathcal{V}(u, A) = \int \quad 2|(i\partial_z + \mathcal{A})u|^2 + 2|(i\partial_{\bar{z}} + \overline{\mathcal{A}})u|^2 - |u|^2 + \frac{1}{2}|u|^4$$
$$+4|\partial_{\bar{z}}(\mathcal{A} - \mathcal{A}_0) - \partial_z(\overline{\mathcal{A}} - \overline{\mathcal{A}}_0)|^2$$

(5.2)

We now implement the gauge fixing procedure embodied in Theorem 4.1 and the energy (5.2) becomes

$$\mathcal{V}(v, \mathcal{P}, k) = \int_{\mathcal{C}} \quad 2|(i\partial_z + \mathcal{P} - ik\bar{z})v|^2 + 2|(i\partial_{\bar{z}} + \overline{\mathcal{P}} + ikz)v|^2 - |v|^2$$
$$+\frac{1}{2}|v|^4 + 4|\partial_{\bar{z}}\mathcal{P} - \partial_z\overline{\mathcal{P}} - 2i(k - k_0)|^2$$

where we may now take the domain of integration to be a fundamental cell $\mathcal{C}$ of the lattice $\mathcal{L}$, and the variational problem is rigorously defined.

For convenience we define $w = \sqrt{k}\,z$, $\psi = \sqrt{\frac{1}{k}}\,v$, $\mathcal{Q} = \sqrt{\frac{1}{k}}\mathcal{P}$, $\lambda = 1 - \frac{1}{4k}$, $L_+ = \partial_{\bar{w}} + w$ and $L_- = \partial_w - \bar{w}$, and obtain the simple form

$$\mathcal{V}(\psi, \mathcal{Q}) = \int_{\mathcal{C}} \quad |(L_- - i\mathcal{Q})\psi|^2 + |(L_+ - i\overline{\mathcal{Q}})\psi|^2 - 2(1 - \lambda)|\psi|^2$$
$$+\frac{1}{4}|\psi|^4 + 2|\partial_{\bar{w}}\mathcal{Q} - \partial_w\overline{\mathcal{Q}}|^2$$

(5.3)

The Ginzburg-Landau equations are the Euler-Lagrange equations of (5.3) which are

(5.4) $\quad (L_-L_+ - \lambda)\psi = [\frac{1}{4}|\psi|^2 + |\mathcal{Q}|^2 + i(\mathcal{Q}L_+ + \overline{\mathcal{Q}}L_-)]\psi$

(5.5) $\quad \partial^2_{w\bar{w}}\mathcal{Q} = \frac{1}{8}[\overline{\psi}(i\partial_w - i\bar{w} + \mathcal{Q})\psi + \psi(-i\partial_w - i\bar{w} + \mathcal{Q})\overline{\psi}]$

It is easy to see that the trivial state $(\psi, \mathcal{Q}) = (0, 0)$ is always a solution, and also that $\lambda = 0$ is a bifurcation point. The Frechet derivative of (5.4) and (5.5) at the trivial solution is just the linearization, obtained by setting the right hand side to zero. For $\lambda < 0$ the only solution to the linearization is the trivial one, but when $\lambda = 0$ there is a two dimensional space of solutions in the form of $(\psi, \mathcal{Q}) = (cv_0(w), 0)$, where $c$ is a complex constant and

$$v_0(w) = e^{w(w - \bar{w})}\theta_3(\frac{w}{r}; e^{i\theta})$$

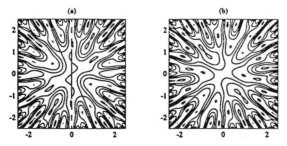

FIGURE 1. Hexagonal case: (a)real part, (b)imaginary part

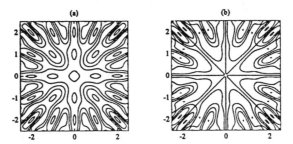

FIGURE 2. Square case: (a)real part, (b)imaginary part

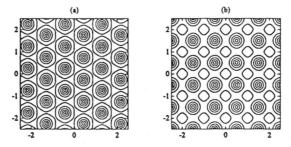

FIGURE 3. Absolute values: (a)hexagonal case, (b)square case

with

$$\theta_3(\frac{w}{r}; e^{i\theta}) = \sum_{n=-\infty}^{\infty} \exp(\frac{i\pi}{r}(2nw + n^2 re^{i\theta}))$$

the third Jacobi theta function of classical complex analysis (see [9]). In these formulas $r$ is the size of the lattice under consideration and $\theta$ is the angle between the translations that generate the lattice. See [1] for more details.

This complex function is a solution whose physical properties exhibit the symmetry of the lattice. It is amusing to see how this works, so to close we present contour plots of the real and imaginary parts of $v_0(w)$ and its absolute value in Figure 1 through Figure 3 for the important cases of hexagonal and square symmetry.

## Acknowledgment

The author is grateful to Marty Golubitsky and Jacek Turski who are coauthors of the original paper and also to Roland Glowinski and Helena Lopes for their help in proving the PDE result that was needed to complete the proof of the Gauge fixing theorem (Theorem 4.1). Their work appears in an appendix to [1]

## REFERENCES

1. E. Barany, M. Golubitsky and J. Turski. *Bifurcations with local gauge symmetries in the Ginzburg-Landau equations.* Physica D **56** (1992) 36–56
2. V. L. Ginzburg and L. D. Landau. *On the theory of superconductivity.* Zh. Eksp. Teor. Fiz. **20** (1950) 1064–1082 (Russian) .
3. M. Golubitsky and D. G. Schaeffer. *Singularities and Groups in Bifurcation Theory: Vol. I* Appl. Math. Sci. Ser. **51**, Springer-Verlag, New York, 1984.
4. M. Golubitsky, I. N. Stewart and D. G. Schaeffer. *Singularities and Groups in Bifurcation Theory: Vol. II* Appl. Math. Sci. Ser. **69**, Springer-Verlag, New York, 1988.
5. G. Lasher. *Series Solution of the Ginzburg-Landau equations for the Abrikosov mixed state.* Phys. Rev. **140**, 2A (1965) A523–A528.
6. F. Odeh. *Existence and bifurcation theorems for the Ginzburg-Landau equations.* J. Math. Phys. **8**, No. 12 (1967) 2351–2356.
7. T. Poston and I. N. Stewart. *Catastrophe Theory and its Applications*, Pitman, London, 1978.
8. D. R. Tilley and J. Tilley. *Superfluidity and Superconductivity*, Adam Hilger Ltd., Bristol, 1986.
9. E. T. Whittaker and G. N. Watson. *A Course of Modern Analysis*, MacMillan, New York, 1948.

DEPARTMENT OF MATHEMATICS, UNIVERSITY OF HOUSTON, HOUSTON, TX 77204

*Current address*: Department of Mathematical Sciences, New Mexico State University, Las Cruces, NM 88003

*E-mail address*: ebarany@nmsu.edu

Lectures in Applied Mathematics
Volume **29**, 1993

# An Overview of Potential Symmetries

## GEORGE BLUMAN

ABSTRACT. A given system of PDEs is embedded in an auxiliary system obtained by replacement of a PDE of the system by an equivalent conservation law. Point symmetries of the auxiliary system could yield nonlocal symmetries (potential symmetries) of the given system. The nonlinear diffusion equation is a prototypical example.

## 1. Introduction

Local symmetries admitted by a system of partial differential equations (PDEs) are useful for finding invariant solutions obtained by using group invariants to reduce the number of independent variables;[1-7] to accordingly simplify a well-posed boundary value problem provided that both its boundary and the conditions imposed on the boundary are invariant;[3,6] for finding conservation laws through Noether's theorem and its extensions;[4-6,8-11] to discover whether or not the system can be linearized by an invertible mapping and construct an explicit linearization when one exists;[6,12,13] to discover classes of PDEs equivalent to each other.[4,14,15]

Lie[1-7] gave an algorithm which yields "determining equations" to find infinitesimal generators of point symmetries admitted by a given differential equation. Such local symmetries are point transformations acting on the space of independent and dependent variables of the differential equation. Various symbolic manipulation programs now exist to implement Lie's algorithm: setting up determining equations,[16-21] finding the dimension of their solution space without explicitly solving them,[22,23] and/or yielding explicit generators.

For a given system of PDEs one could also find useful nonlocal symmetries through embedding it in an auxiliary "covering" system with auxiliary dependent variables.[24-26] A point symmetry of the auxiliary system, acting on the space consisting of the independent and dependent variables of the given system as well as the auxiliary variables, yields a nonlocal symmetry of the given

1991 Mathematics Subject Classification. Primary 58F37, 58G35.

Supported by the Natural Sciences and Engineering Research Council of Canada.

This paper is in final form and no version of it will be submitted for publication elsewhere.

system if it does not project onto a point symmetry acting on its space of independent and dependent variables. Such an auxiliary system is obtained by replacement of a PDE of the given system by an equivalent conservation law.[27,28] Corresponding nonlocal symmetries are called **potential symmetries**[6] of the given system. Consequently one could obtain more symmetry reductions of the given system,[6,27−29] its linearization by a noninvertible mapping,[6−30] find further equivalence transformations,[15] etc.

A local symmetry of a given system of PDEs always yields a symmetry of any of its auxiliary potential systems. However such a symmetry could be "lost" when determining local symmetries of some auxiliary system. The problem of finding "grand" potential systems to recover "lost" symmetries is considered. Two conjectures are presented.

## 2. Point symmetries

Consider a system of $m$ PDEs $R\{x, u\}$ given by

$$(1) \qquad G^\sigma(x, u, \underset{1}{u}, \underset{2}{u}, \ldots, \underset{k}{u}) = 0, \quad \sigma = 1, 2, \ldots, m$$

with $n$ independent variables $x = (x_1, x_2, \ldots, x_n)$, $m$ dependent variables $u = (u^1, u^2, \ldots, u^m)$; $\underset{j}{u}$ denotes the set of coordinates corresponding to all $j$th order partial derivatives of $u$ with respect to $x$ (a coordinate in $\underset{j}{u}$ is denoted by $u^\gamma_{i_1 i_2 \cdots i_j} = \frac{\partial^j u^\gamma}{\partial x_{i_1} \partial x_{i_2} \cdots \partial x_{i_j}}$ with $\gamma = 1, 2, \ldots, m$, $i_j = 1, 2, \ldots, n$, and $j = 1, 2, \ldots, k$).

DEFINITION 2.1. *A **symmetry** admitted by $R\{x, u\}$ is a transformation mapping any solution of $R\{x, u\}$ into another solution of $R\{x, u\}$.*

Note that a symmetry transformation admitted by a differential equation is defined topologically; there is no restriction to a particular set of coordinates. However to be of use a symmetry admitted by $R\{x, u\}$ must be defined in terms of some set of coordinates. Most useful is a symmetry corresponding to a Lie group of point transformations (point symmetry) acting on a space of variables including $(x, u)$ such that invariant solutions in the coordinate frame project onto solutions of $R\{x, u\}$.

DEFINITION 2.2 *A (Lie) **point symmetry** admitted by $R\{x, u\}$ is a symmetry characterized by an infinitesimal generator of the form*

$$(2) \qquad \mathbf{X} = \sum_{\mu=1}^m \eta^\mu(x, u, \underset{1}{u}) \frac{\partial}{\partial u^\mu}$$

*with $\eta = (\eta^1, \eta^2, \ldots \eta^m)$ linear in the components of $\underset{1}{u}$:*

$$\eta^\mu = \alpha^\mu(x, u) - \sum_{i=1}^{n} \xi_i(x, u) u_i^\mu, \quad \mu = 1, 2, \ldots, m.$$

Under the action of **X** a solution $u = \Theta(x)$ of (1) formally maps into a solution

(3) $$u = \Theta(x; \varepsilon) = e^{\varepsilon U} u|_{u=\Theta(x)}$$

where in (3) the extended (prolonged) infinitesimal generator $U = \eta^\gamma \frac{\partial}{\partial u^\gamma} + (D_j \eta^\gamma) \frac{\partial}{\partial u_j^\gamma} + \cdots + (D_{i_1} D_{i_2} \cdots D_{i_l} \eta^\gamma) \frac{\partial}{\partial u_{i_1 i_2 \cdots i_l}^\gamma} + \cdots$, in terms of the total derivative operators $D_i = \frac{\partial}{\partial x_i} + u_i^\gamma \frac{\partial}{\partial u^\gamma} + u_{ij}^\gamma \frac{\partial}{\partial u_j^\gamma} + \cdots + u_{i i_1 i_2 \cdots i_l}^\gamma \frac{\partial}{\partial u_{i_1 i_2 \cdots i_l}^\gamma} + \cdots$, $i = 1, 2, \ldots, n$. (We assume summation over a repeated index.) The infinitesimal generator (2) corresponds to the one-parameter ($\varepsilon$) Lie group of point transformations

$$x_i^* = x_i + \varepsilon \xi_i(x, u) + O(\varepsilon^2), \quad i = 1, 2, \ldots, n,$$
$$(u^\mu)^* = u^\mu + \varepsilon \alpha^\mu(x, u) + O(\varepsilon^2), \quad \mu = 1, 2, \ldots, m.$$

**2.1. How to find point symmetries.** If $(u^\mu)^* = e^{\varepsilon U} u^\mu = u^\mu + \varepsilon \eta^\mu(x, u, \underset{1}{u}) + O(\varepsilon^2)$, $\mu = 1, 2, \ldots, m$, then $G^\sigma(x, u^*, \underset{1}{u^*}, \ldots, \underset{k}{u^*}) = G^\sigma(x, u, \underset{1}{u}, \ldots \underset{k}{u}) + \varepsilon \mathcal{L}_\rho^\sigma[u] \eta^\rho + O(\varepsilon^2)$, where the linear operator

(4) $$\mathcal{L}_\rho^\sigma[u] = \frac{\partial G^\sigma}{\partial u^\rho} + \frac{\partial G^\sigma}{\partial u_j^\rho} D_j + \ldots + \frac{\partial G^\sigma}{\partial u_{i_1 i_2 \ldots i_k}^\rho} D_{i_1} D_{i_2} \cdots D_{i_k}, \sigma, \rho = 1, 2, \ldots, m,$$

is the Fréchet derivative[5] associated with $R\{x, u\}$.

THEOREM 2.3. **X** *is admitted by* $R\{x, u\}$ *if and only if*

(5) $$\mathcal{L}_\rho^\sigma[u] \eta^\rho = 0, \quad \sigma = 1, 2, \ldots, m,$$

*for any solution* $u = \Theta(x)$ *of* $R\{x, u\}$.[4-6]

The set of $m$ equations (5) reduces to an overdetermined system of linear PDEs, called the **determining equations**, whose unknowns are the coefficients (infinitesimals) $\alpha = (\alpha^1, \alpha^2, \ldots, \alpha^m)$, $\xi = (\xi_1, \xi_2, \ldots, \xi_n)$ of an infinitesimal generator **X**. Various symbolic manipulations programs[16-23], used effectively, often reduce the tedium of calculating both the dimension of the admitted Lie algebra of infinitesimal generators **X** and explicit infinitesimal generators for a particular system $R\{x, u\}$.

**2.2 Uses of point symmetries.** Infinitesimal generators admitted by a given system of PDEs $R\{x, u\}$ can be applied in various ways including the following:

(I) Construction of a one-parameter ($\varepsilon$) family of solutions from any known solution (eq. (3)).

(II) Construction of invariant solutions[1-7] when $\xi \not\equiv 0$. If the equations

(6) $$\eta^{\mu}(x, u, \underset{1}{u}) = 0, \quad \mu = 1, 2, \ldots, m,$$

are substituted into $R\{x, u\}$ one obtains a reduced system of $m$ PDEs with $n - 1$ independent variables. The $m + n - 1$ variables of this system are a set of $m + n - 1$ independent invariants of $\mathbf{X}$. This construction can be generalized to the wider class of "nonclassical" solutions obtained by determining when eqs. (6) and system $R\{x, u\}$ are compatible.[31-34]

(III) Determination of whether or not $R\{x, u\}$ can be linearized by an invertible point transformation; construction of an explicit linearization when one exists.[12,13]

(IV) Construction of conservation laws. If the Fréchet derivative (4) of $R\{x, u\}$ is self-adjoint, then system $R\{x, u\}$ is the set of Euler-Lagrange equations for some variational principle associated with $R\{x, u\}$.[5] If $\mathbf{X}$ is admitted by a corresponding action functional to within a divergence, then $\mathbf{X}$ yields an explicit conservation law of $R\{x, u\}$.[4-11] Moreover if the Fréchet derivative (4) is self-adjoint one can show that if $\sum_{\sigma=1}^{m} \lambda^{\sigma}(x, u, \underset{1}{u})G^{\sigma} = \sum_{i=1}^{n} D_i f^i$ for some $\{f^i(x, u, \underset{1}{u}, \ldots, \underset{k-1}{u})\}$ then $\mathcal{L}_{\rho}^{\sigma}[u]\lambda^{\rho} = 0$, $\sigma = 1, 2, \ldots, m$, for any solution $u = \Theta(x)$ of $R\{x, u\}$. Hence if a set of factors (multipliers) $\lambda = (\lambda^1(x, u, \underset{1}{u}), \lambda^2(x, u, \underset{1}{u}), \ldots, \lambda^m(x, u, \underset{1}{u}))$ yields a conservation law of $R\{x, u\}$ then the set must yield a symmetry $\mathbf{X} = \sum_{\mu=1}^{m} \lambda^{\mu}(x, u, \underset{1}{u}) \frac{\partial}{\partial u^{\mu}}$ admitted by $R\{x, u\}$.

**2.3 Prototypical example.** As an example let $R\{x, u\}$ be the nonlinear diffusion equation

(7) $$u_t = (K(u)u_x)_x = (L(u)_{xx}$$

with diffusivity $K(u) = L'(u) \neq$ const. The associated Fréchet derivative of (7) is
$$\mathcal{L}[u] = K(u)D_x^2 + 2K'(u)u_x D_x - D_t + K''(u)u_x^2 + K'(u)u_x$$
$$= D_x^2 \cdot K(u) - D_t$$

whose adjoint is $\mathcal{L}^*[u] = K(u)D_x^2 + D_t$. To within a scaling and translation of $u$ one can show that the point symmetries of (7) are:[1-3,5,6]

(I) $K(u)$ *arbitrary:*

$$\mathbf{X}_1 = u_x \frac{\partial}{\partial u}, \mathbf{X}_2 = u_t \frac{\partial}{\partial u}, \mathbf{X}_3 = (xu_x + 2tu_t)\frac{\partial}{\partial u}.$$

(II) $K(u) = u^{\lambda}$:

$$\mathbf{X}_1, \mathbf{X}_2, \mathbf{X}_3, \mathbf{X}_4 = (2u - \lambda xu_x)\frac{\partial}{\partial u}.$$

(III) $K(u) = u^{-4/3}$:

$$\mathbf{X}_1, \mathbf{X}_2, \mathbf{X}_3, \mathbf{X}_4, \mathbf{X}_5 = (3xu + x^2 u_x)\frac{\partial}{\partial u}.$$

Note that only $\mathbf{X}_5$ is not obvious by inspection.

### 3. Potential symmetries

DEFINITION 3.1. *A **local symmetry** admitted by $R\{x, u\}$ is a symmetry admitted by $R\{x, u\}$ characterized by an infinitesimal generator of the form*

$$(8) \qquad \mathbf{X} = \sum_{\mu=1}^{m} \eta^{\mu}(x, u, \underset{1}{u}, \ldots, \underset{p}{u})\frac{\partial}{\partial u^{\mu}};$$

$\mathbf{X}$ *is admitted by $R\{x, u\}$ if and only if $\mathcal{L}_{\rho}^{\sigma}[u]\eta^{\rho} = 0$, $\sigma = 1, 2, \ldots, m$, for any solution $u = \Theta(x)$ of $R\{x, u\}$.*

$\mathbf{X}$ of the form (8) defines a (nontrivial) **contact symmetry** if $m = 1$, $p = 1$ and $\eta = (\eta^1, \eta^2, \ldots, \eta^m)$ is not linear in the components of $\underset{1}{u}$; $\mathbf{X}$ of the form (8) defines a (nontrivial) **Lie-Bäcklund symmetry (higher symmetry, higher order symmetry, generalized symmetry, Noether symmetry)** if not of point or contact type.

DEFINITION 3.2. *A **nonlocal symmetry** admitted by $R\{x, u\}$ is a continuous symmetry admitted by $R\{x, u\}$ which is not of the form (8).*

One is naturally led to the following questions:
(I) What are proper conceptual frameworks for determining nonlocal symmetries?
(I) How can one find nonlocal symmetries algorithmically?
(III) What types of nonlocal symmetries could be useful?
(IV) Do nontrivial examples of nonlocal symmetries exist?

There have been various heuristic approaches to obtain nonlocal symmetries.[35-37] Krasilsh'chik and Vinogradov[24-26] have developed a conceptual framework for considering nonlocal symmetries. Their basic notion is to obtain a "covering" system for a given system $R\{x, u\}$; local symmetries of a covering system may yield nonlocal symmetries of $R\{x, u\}$ as well as its local symmetries. Nontrivial examples are not exhibited. A conceptual framework for obtaining nonlocal symmetries, related to that of Krasilsch'chik and Vinogradov, which is algorithmic and lends itself to applications, is now outlined.[6,27,28]

Suppose a PDE of $R\{x, u\}$, without loss generality, $G^m = 0$, is a conservation law $\sum_{i=1}^{n} D_i f^i(x, u, \underset{1}{u}, \ldots \underset{k-1}{u}) = 0$. Then $R\{x, u\}$ is the system

$$(9a) \qquad G^{\sigma}(x, u, \underset{1}{u}, \ldots, \underset{k}{u}) = 0, \qquad \sigma = 1, 2, \ldots, m - 1$$

(9b)
$$\sum_{i=1}^{m} D_i f^i(x, u, \underset{1}{u}, \dots, \underset{k-1}{u}) = 0.$$

Through (9b) one can introduce $n-1$ auxiliary dependent variables $v = (v^1, v^2, \dots, v^{n-1})$ and form an **auxiliary system** $S\{x, u, v\}$ of $m + n - 1$ PDEs given by:

$$f^1(x, u, \underset{1}{u}, \dots, \underset{k-1}{u}) = \frac{\partial v^1}{\partial x_2},$$

$$f^\ell(x, u, \underset{1}{u}, \dots, \underset{k-1}{u}) = (-1)^{\ell-1} \left[ \frac{\partial v^\ell}{\partial x_{\ell+1}} + \frac{\partial v^{\ell-1}}{\partial x_{\ell-1}} \right], \quad 1 < \ell < n,$$

$$f^n(x, u, \underset{1}{u}, \dots, \underset{k-1}{u}) = (-1)^{n-1} \frac{\partial v^{n-1}}{\partial x_{n-1}},$$

$$G^\sigma(x, u, \underset{1}{u}, \dots, \underset{k}{u}) = 0, \quad \sigma = 1, 2, \dots, m-1.$$

**3.1. Relationship between** $R\{x, u\}, S\{x, u, v\}$. If $(u(x), v(x))$ solves $S\{x, u, v\}$ then $u(x)$ solves $R\{x, u\}$; if $u(x)$ solves $R\{x, u\}$ then there is some $v(x)$ (**not unique**) such that $(u(x), v(x))$ solves $S\{x, u, v\}$. Consequently the relationship between $R\{x, u\}$ and $S\{x, u, v\}$ is noninvertible. However a boundary value problem (BVP) problem posed for $R\{x, u\}$ can be embedded in a BVP posed for $S\{x, u, v\}$; conversely a BVP posed for $S\{x, u, v\}$ can be solved by projection to a BVP posed for $R\{x, u\}$. This relationship is important without regard to symmetry considerations.

From this relationship between solutions of $R\{x, u\}$ and those of $S\{x, u, v\}$ it follows that a symmetry of $S\{x, u, v\}$ yields a symmetry of $R\{x, u\}$ and, conversely, a symmetry of $R\{x, u\}$ yields a symmetry of $S\{x, u, v\}$. Moreover a point symmetry of $S\{x, u, v\}$ could yield a nonlocal symmetry (**potential symmetry**) of $R\{x, u\}$; conversely, a point symmetry of $R\{x, u\}$ could yield a nonlocal symmetry of $S\{x, u, v\}$.

DEFINITION 3.3. *Suppose* $\mathbf{X}^S = (\alpha^\mu(x, u, v) - \xi_i^S(x, u, v) u_i^\mu) \frac{\partial}{\partial u^\mu} + (\beta^\nu(x, u, v) - \xi_i^S(x, u, v) v_i^\nu) \frac{\partial}{\partial v^\nu}$ *is a point symmetry admitted by* $S\{x, u, v\}$. $\mathbf{X}^S$ *induces a* (*nonlocal*) **potential symmetry** *of* $R\{x, u\}$ *if and only if* $(\alpha, \xi^S)$ *depends essentially on* $v$.

If $\mathbf{X}^S$ does not induce a potential symmetry of $R\{x, u\}$ then $\mathbf{X}^S$ projects onto a point symmetry of $R\{x, u\}$, namely $\mathbf{X} = (\alpha^\mu(x, u) - \xi_i^S(x, u,) u_i^\mu) \frac{\partial}{\partial u^\mu}$. Conversely a point symmetry admitted by $R\{x, u\}$, namely $\mathbf{X}^R = (\alpha^\mu(x, u) - \xi_i^R(x, u) u_i^\mu) \frac{\partial}{\partial u^\mu}$ yields a symmetry of $S\{x, u, v\}$ which is not a point symmetry if and only if $\mathbf{X}^S = \mathbf{X}^R + \zeta^\mu(x, u, v, \underset{1}{u}, \underset{1}{v}) \frac{\partial}{\partial v^\mu}$ defines no point symmetry of $S\{x, u, v\}$ for any choice of $\zeta = (\zeta^1, \zeta^2, \dots, \zeta^{n-1})$.

Since potential symmetries of $R\{x, u\}$ arise as point symmetries of $S\{x, u, v\}$ it follows that:

(I) The calculation of potential symmetries through a particular potential system $S\{x, u, v\}$ can be effected through existing symbolic manipulation programs.[16-23]

(II) The exponentiation action of infinitesimal generators of potential symmetries on solutions of $S\{x, u, v\}$ leads to families of solutions of $R\{x, u\}$ from known solutions of $R\{x, u\}$.

(III) Invariant solutions of $S\{x, u, v\}$ yield solutions of $R\{x, u\}$.[27,29,40] In principle this can lead to a further generalization of the ansatz yielding "nonclassical" solutions.

(IV) If $S\{x, u, v\}$ can be derived from a variational principle then a potential symmetry admitted by the corresponding action functional yields a nonlocal conservation law of $R\{x, u\}$.

(V) If $R\{x, u\}$ admits a potential symmetry leading to the linearization of $S\{x, u, v\}$ then $R\{x, u\}$ is linearized by a noninvertible mapping.[6,30].

3.2. **Prototypical example.** Let $R\{x, u\}$ be the nonlinear diffusion equation (7) which is already a conservation law. Correspondingly $S\{x, u, v\}$ is given by

(10a)
$$v_x = u,$$

(10b)
$$v_t = K(u)u_x = (L(u))_x.$$

One can show that through (10a,b) $R\{x, u\}$ has the following potential symmetries:[6,28]

(I) $K(u) = \frac{1}{1+u^2}e^{a\,\arctan u}$, $a = $ const:

$$\mathbf{X}^S = (u^2 + 1 + vv_x + atu_t)\frac{\partial}{\partial u} + (x + vv_x + atv_t)\frac{\partial}{\partial v}.$$

(II) $K(u) = u^{-2}$:

$$\mathbf{X}^S = [F^1(v, t)u_x + u^2 F^2(v, t)]\frac{\partial}{\partial u} + F^1(v, t)v_x\frac{\partial}{\partial v}$$

where $(F^1, F^2)$ are arbitrary solutions of the linear system $\frac{\partial F^1}{\partial t} = \frac{\partial F^2}{\partial v}, \frac{\partial F^1}{\partial v} = F^2$. This potential symmetry leads to the linearization of

(11)
$$u_t = (u^{-2}u_x)_x$$

by an explicit noninvertible mapping.[6,30]

However when $K(u) = u^{-4/3}$, the point symmetry $\mathbf{X}_5 = (3xu + x^2u_x)\frac{\partial}{\partial u}$ admitted by $R\{x, u\}$ is "lost" when one computes the point symmetries of potential system (10a,b) in the sense that (10a,b) admits no point symmetries projecting onto $\mathbf{X}_5$. There are many other examples of PDEs where point symmetries of $R\{x, u\}$ do not induce point symmetries of $S\{x, u, v\}$[6,27,28,38,41,42] and where potential symmetries exist, i.e. point symmetries of $S\{x, u, v\}$ do not project onto local symmetries of $R\{x, u\}$.[6,15,27,28,30,38-42] In the former case the Lie algebra of infinitesimal generators of point symmetries of $R\{x, u\}$ does not induce a Lie algebra of point symmetries of $S\{x, u, v\}$; in the latter case the Lie algebra of

infinitesimal generators of point symmetries of $S\{x, u, v\}$ does not project onto
a Lie algebra of infinitesimal generators of local symmetries of $R\{x, u\}$.

## 4. More potential symmetries

Let $v^{(1)} = v, S^{(1)} = S\{x, u, v^{(1)}\}$. Suppose a PDE of $S^{(1)}$ is a conservation
law. Then $n - 1$ more potential variables $v^{(2)}$ can be introduced. This leads to
another auxiliary potential system $S^{(2)} = S^{(2)}\{x, u, v^{(1)}, v^{(2)}\}$ of $m + 2(n - 1)$
PDEs with $m + 2(n - 1)$ dependent variables. Consequently one could obtain
additional nonlocal symmetries of $R\{x, u\}$. In particular a point symmetry of
$S^{(2)}$ is both a potential symmetry of $S^{(1)}$ and a nonlocal symmetry of $R\{x, u\}$
if the $(x, u)$ coefficients of its infinitesimal generator depend essentially on $v^{(2)}$.

In general if a PDE of $S^{(J-1)} = S^{(J-1)}\{x, u, v^{(1)}, v^{(2)}, ..., v^{(J-1)}\}, J = 1, 2, ...,$
is a conservation law then $n-1$ further potential variables $v^{(J)}$ can be introduced,
leading to another auxiliary potential system $S^{(J)} = S^{(J)}\{x, u, v^{(1)}, v^{(2)}, ..., v^{(J)}\}$
of $m + J(n - 1)$ PDEs with $m + J(n - 1)$ dependent variables. At any step
$J, S^{(J)}$ may not "cover" $R\{x, u\}$ , $S^{(1)}, S^{(2)}, ..., S^{(J-1)}$ in the sense that the
Lie algebra of infinitesimal generators of local symmetries of $S^{(J)}$ does not
project onto the Lie algebras of infinitesimal generators of local symmetries of
$R\{x, u\}, S^{(1)}, S^{(2)}, ..., S^{(J-1)}$, respectively.

4.1. **Prototypical example.** Consider again the nonlinear diffusion equation (7). Equation (10b) of potential system $S^{(1)} = S\{x, u, v\}$ is already a conservation law. Correspondingly one obtains potential system $S^{(2)} = T\{x, u, v, w\}$
given by

$$(12) \qquad \begin{aligned} v_x &= u, \\ w_x &= v, \\ w_t &= L(u). \end{aligned}$$

From previously exhibited calculations[38] one can show that the point symmetries
of (12) project onto **all** point symmetries of $R\{x, u\}$ and $S\{x, u, v\}$ for any choice
of $K(u)$, i.e. $T\{x, u, v, w\}$ "covers" both $R\{x, u\}$ and $S\{x, u, v\}$ for any choice
of $K(u)$. In particular $T\{x, u, v\}$ admits the following point symmetries not
obvious by inspection:

(I) $K(u) = \frac{1}{1+u^2} e^{a \arctan u}, a = \text{const}$:

$$\mathbf{X}^T = \mathbf{X}^S + \left( \frac{1}{2}(x^2 - v^2) + vw_x + atw_t \right) \frac{\partial}{\partial w}.$$

(II) $K(u) = u^{-2}$:

$$\mathbf{X}^T = \mathbf{X}^S + (F^3(v, t) - vF^1(v, t)) \frac{\partial}{\partial w} \quad \text{where } (F^1, F^2, F^3)$$

are arbitrary solutions of the linear system $\frac{\partial F^2}{\partial v} = F^1, \frac{\partial F^3}{\partial t} = F^2, \frac{\partial F^1}{\partial v} = F^2$,
which in turn leads to the linearization of PDE (11) by an explicit noninvertible
mapping.[6,30]

(III) $K(u) = u^{-4/3}$:

$$\mathbf{X}^T = (3xu + x^2 u_x)\frac{\partial}{\partial u} + (xu - w + x^2 v_x)\frac{\partial}{\partial u} + (x^2 w_x - xw)\frac{\partial}{\partial w}.$$

(Note that the "lost" symmetry $\mathbf{X}_5$ admitted by $R\{x, u\}$ is now "found". This
point symmetry of $T\{x, u, v, w\}$ is a potential symmetry of $S\{x, u, v\}$ but projects
onto a point symmetry of $R\{x, u\}$.)

(IV) $K(u) = u^{-2/3}$:

$$\mathbf{X}^T = (3uv + wu_x)\frac{\partial}{\partial u} + (v^2 + wv_x)\frac{\partial}{\partial v} + ww_x\frac{\partial}{\partial w}.$$

(This point symmetry of $T\{x, u, v, w\}$ is a potential symmetry of $S\{x, u, v\}$ which
also yields a new nonlocal symmetry of $R\{x, u\}$.)

4.2 **Two Conjectures.** Observations from numerous calculations of nonlocal
symmetries [38,41,42] lead one to the following conjectures:

CONJECTURE 4.1. *The outlined procedure for obtaining auxiliary systems*
$S^{(1)}, S^{(2)}, ..., S^{(N)}$ *continues to some finite* $N$ *when either*
*(I)* $S^{(N)}$ *has no conservation laws* **equivalent** *to one of its* $m + N(n-1)$ *PDEs,*
*or*
*(II)* $X^{(N)}$ *has an infinite number of equivalent conservation laws.*

CONJECTURE 4.2. *In Case (I) of Conjecture 4.1 the Lie algebra of infinites-*
*imal generators of point symmetries admitted by* $S^{(N)}$ *projects onto the Lie alge-*
*bra of infinitesimal generators of point symmetries of each of* $R\{x, u\}, S^{(1)}, S^{(2)},$
*...,* $S^{(N-1)}$ *as well as any subsystems[42] of these systems satisfied by subsets of*
*dependent variables.*

It should be noted that:
(I) Any auxiliary system $S^{(1)}, S^{(2)}, ..., S^{(N)}$ (or any subsystem of PDEs found
from a subset of dependent variables) could be of use to extract additional qualitative or quantitative information about solutions of $R\{x, u\}, S^{(1)}, S^{(2)}, ..., S^{(N)}$
or any subsystem. This could happen due to the noninvertible relationships
between these systems.
(II) A given system $R\{x, u\}$ could have more than one sequence of auxiliary
potential systems even when Case (I) of Conjecture 4.1 holds. In this case it is
an open problem whether or not $N$ is the same for all sequences and whether or
not for each sequence $S^{(N)}$ has the same Lie algebra to within equivalence.
(III) The problem of defining and constructing equivalent conservation laws for
a given system $R\{x, u\}$ is examined in another paper.[43] In general for a conservation law of $R\{x, u\}$ to be an equivalent conservation law it is necesssary that

multipliers $\lambda = (\lambda^1, \lambda^2, ..., \lambda^m)$ be independent of $\underset{1}{u}$. Hence in the case when
the Fréchet derivative of $R\{x, u\}$ is self-adjoint, a conservation law arising from
Noether's theorem can be an equivalent conservation law only if it arises from a
point symmetry with $\xi(x, u) \equiv 0$.

(IV) A linear system of PDEs $L_\rho^\sigma[x]u^\rho = 0, \sigma = 1, 2, ..., m$, has an equivalent
conservation law for any multipliers $\lambda(x) = (\lambda^1(x), \lambda^2(x), ..., \lambda^m(x))$ satisfying
$L_\rho^{*\sigma}[x]\lambda^\rho(x) = 0, \sigma = 1, 2, ..., m$, where $L^*[x]$ is the adjoint linear operator of
the linear operator $L[x]$. Consequently if a nonlinear system $R\{x, u\}$ can be
linearized by an invertible mapping then it must have an infinite number of
equivalent conservation laws. Moreover the determining equations satisfied by
the multipliers yield the coordinates and the adjoint of the linearizing system.[44]

## REFERENCES

1. S. Lie, "Über die Integration durch bestimmte Integrale von einer Klasse
   linearer partieller Differentialgleichungen," Arch. for Math. **6** (1881), 328-
   368; also *Gesammelte Abhandlungen*, Vol III, B.G. Teubner, Leipzig, 1922,
   492-523.

2. L.V. Ovsiannikov, *Group Properties of Differential Equations*, Novosibirsk
   (in Russian), 1962.

3. G.W. Bluman and J.D. Cole, *Similarity Methods for Differential Equations*,
   Springer, New York, 1974.

4. L.V. Ovsiannikov, *Group Analysis of Differential Equations*, Academic Press,
   New York, 1982.

5. P.J. Olver, *Applications of Lie Groups to Differential Equations*, Springer,
   New York, 1986.

6. G.W. Bluman and S. Kumei, *Symmetries and Differential Equations*, Springer,
   New York, 1989.

7. H. Stephani, *Differential Equations: Their Solution Using Symmetries*, Cambridge University Press, Cambridge, 1989.

8. E. Noether, "Invariante Variationsprobleme," Nachr. König. Gesell. Wissen,
   Göttingen, Math.-Phys. Kl. (1918), 235-257.

9. E. Bessel-Hagen, "Über die Erhaltungssätze der Elektrodynamik," Math.
   Ann. **84** (1921), 258-276.

10. T.H. Boyer, "Continuous symmetries and conserved currents," Ann. Phys.
    **42** (1967), 445-466.

11. N.H. Ibragimov, *Transformation Groups Applied to Mathematical Physics*,

Reidel, Boston, 1985.

12. S. Kumei and G.W. Bluman, "When nonlinear differential equations are equivalent to linear differential equations," SIAM J. Appl. Math. **42** (1982), 1157-1173.

13. G.W. Bluman and S. Kumei, "Symmetry-based algorithms to relate partial differential equations: I. Local symmetries," Euro. J. Appl. Math. **1** (1990), 189-216.

14. N.H. Ibragimov, M. Torrisi, and A. Valenti, "Preliminary group classification of equations $v_{tt} = f(x, v_x)v_{xx} + g(x, v_x)$," J. Math. Phys. **32** (1991), 2988-2995.

15. I. Lisle, Equivalence Transformations for Classes of Differential Equations, Ph.D. Thesis, University of British Columbia, 1992.

16. F. Schwarz, "Automatically determining symmetries of partial differential equations," Computing **34** (1985), 91-106.

17. P.H.M. Kersten, *Infinitesimal Symmetries: a Computational Approach*, Centrum voor Wiskunde en Informatica, Amsterdam, 1987.

18. A.K. Head, LIE: A muMATH Program for the calculation of the Lie Algebra of Differential Equations, CSIRO Division of Material Sciences, Clayton, Australia, 1990, 1992.

19. B. Champagne, W. Hereman, and P. Winternitz, "The computer calculation of Lie point symmetries of large systems of differential equations," Comp. Phys. Comm. **66** (1991), 319-340.

20. T. Wolf and A. Brand, "The computer algebra package CRACK for investigating PDEs," in *ERCIM Advanced Course in Partial Differential Equations and Group Theory*, Bonn, 1992.

21. G. Baumann, "Lie symmetries of differential equations: A *Mathematica* program to determine Lie symmetries," Wolfram Research Inc., Champaign, Ill., Math-Source 0202-622, 1992.

22. V.L. Topunov, "Reducing systems of linear differential equations to passive form," Acta Applic. Math. **16** (1989), 191-206.

23. G.J. Reid, "Algorithms for reducing a system of PDEs to standard form, determining the dimension of its solution space and calculating its Taylor series solution," Euro. J. Appl. Math. **2** (1991), 293-318.

24. I.S. Krasil'shchik and A.M. Vinogradov, "Nonlocal symmetries and the theory of coverings: an addendum to A.M. Vinogradov's Local symmetries and

conservation laws," Acta Applic. Math. **2** (1984), 79-96.

25. A.M. Vinogradov and I.S. Krasil'shchik," On the theory of nonlocal symmetries of nonlinear partial differential equations," Sov. Math. Dokl. **29** (1984), 337-341.

26. I.S. Krasil'shchik and A.M. Vinogradov, "Nonlocal trends in the geometry of differential equations: symmetries, conservation laws, and Bäcklund transformations," Acta Applic. Math. **15** (1989), 161-209.

27. G.W. Bluman and S. Kumei, "On invariance properties of the wave equation," J. Math. Phys. **28** (1987), 307-318.

28. G.W. Bluman, S. Kumei, and G.J. Reid, "New classes of symmetries for partial differential equations," J. Math. Phys. **29** (1988), 806–811; Erratum, J. Math. Phys. **29**, (1988), 2320.

29. G.W. Bluman and S.Kumei, "Exact solutions for wave equations of two-layered media with smooth transition,"J. Math. Phys. **29** (1989), 86-96.

30. G.W. Bluman and S. Kumei, "Symmetry-based algorithms to relate partial differential equations: II. Linearization by nonlocal symmetries," Euro. J. Appl. Math. **1** (1990), 217-223.

31. G.W. Bluman and J.D. Cole, "The general similarity solution of the heat equation," J. Math. Mech. **18** (1969), 1025-1042.

32. P.J. Olver and P. Rosenau, "On the "non-classical" method for group-invariant solutions of differential equations," SIAM J. Appl. Math. **47** (1987), 263-278.

33. M.C. Nucci and P.A. Clarkson, "The nonclassical method is more general than the direct method for symmetry reductions. An example of the Fitzhugh-Nagumo equation," Phys. Lett. A **164** (1992), 49-56.

34. E. Pucci, "Similarity reductions of partial differential equations," J. Phys.A **25** (1992), 2631-2640.

35. B.G. Konopelchenko and V.G. Mokhnachev, "On the group theoretical analysis of differential equations," J. Phys. A. **13** (1980), 3113-3124.

36. O.V. Kapcov, "Extension of the symmetry of evolution equations," Sov. Math. Dokl. **25** (1982), 173-176.

37. V.V. Pukhnachev, "Equivalence transformations and hidden symmetry of evolution equations," Sov. Math. Dokl. **35** (1987), 555-558.

38. I.S. Akhatov, R.K. Gazizov, and N.K. Ibragimov, "Nonlocal symmetries. Heuristic approach," J. Sov. Math. **55** (1991), 1401-1450.

39. M. Przanowski and A. Maciolek-Niedzwiecki, "Duality rotation as a Lie-Bäcklund transformation," J. Math. Phys. **33** (1992), 3978–3982.

40. E. Pucci and G. Saccomandi, "Potential symmetries and solutions by reduction of partial differential equations," Istituto di Energetica, U. di Perugia, Perugia, Italy, preprint, 1992.

41. A.Ma, Potential Symmetries for the Wave Equation, M. Sc.Thesis, University of British Columbia, 1990.

42. G. Bluman, "Use and construction of potential symmetries," to appear in J. Math. Comp. Modelling.

43. G. Bluman, "Potential symmetries and equivalent conservation laws," submitted for Proceedings of Workshop "Modern Group Analysis: advanced analytical and computational methods in mathematical physics", Acireale (Catania), Italy, Oct. 27-31, 1992, Kluwer.

44. G. Bluman, "Potential symmetries and linearization," to appear in Proceedings of NATO Advanced Research Workshop "Applications of analytic and geometric methods to nonlinear differential equations," Exeter, U.K., July 14-19, 1992, Kluwer.

DEPARTMENT OF MATHEMATICS, UNIVERSITY OF BRITISH COLUMBIA, VANCOUVER, B.C., CANADA V6T 1Z2

*E-mail address*: bluman@math.ubc.ca

Lectures in Applied Mathematics
Volume 29, 1993

# On The Computation of Strains and Stresses in Symmetrical Articulated Structures

## A. BOSSAVIT

ABSTRACT. Consider a linear system $Ax = b$. If $A$ commutes with all the matrices $U_g$ of a matrix group $\{U_g : g \in G\}$, there is a special basis, that can be built from the knowledge of the irreducible representations of group $G$, in which $A$ assumes a block-diagonal form (the number of blocks being correlated with the size of $G$), hence a splitting of the original problem into a family of smaller ones, hence substantial economies in computer time, etc. We purport to apply this general idea to structural computations, as done in civil engineering. Then $A$ is what is called the stiffness matrix of the structure, and $x$ is the (generalized) displacements vector. However, the algebraic structure of the problem is richer than that: instead of a *single* given square linear system, and a single group action, we have here two complementary rectangular systems of equations, and different actions of the same abstract group. Some algebraic manipulations (elimination) do lead to $Ax = b$, but this is not the mandatory course: there are other possibilities, and a more encompassing theory of symmetry reduction is needed in order to cover these as well. We attempt to outline such a theory.

## 1. Introduction

Consider the structure of Fig. 1, as loaded with a system of forces that is equivalent to 0 (null resultant, null moment). We wish to compute the forces in all members (compressions in bars, tensions in tendons). The problem is to take advantage of the symmetry in doing this, without however assuming that the loads are symmetrically applied.

Let us count. Five nodes provide 15 equilibrium equations, of which only nine are independent (because of the six constraints imposed to the system of loads), and there are 10 unknowns (the forces, or equivalently the strains, in the 10 members). We are short of one equation: the structure is hyperstatic. Algebraically, this means that there is one relation between the strains (an affine relation, if one assumes small displacements once and for all), which is the missing equation. If one denotes by $\epsilon$ and $\sigma$ the vectors (of dimension 10) of the strains and of the forces in the members, the system to be solved can be written

1991 Mathematics Subject Classification. Primary 65T, 73K99, Secondary 73B10, 58D19.
This paper is in final form and no version of it will be submitted for publication elsewhere.

as

(1)                                  $C\epsilon = h, \sigma = K\epsilon, E\sigma = f.$

Matrices $C$ (1 row, 10 columns), $K$ (10 ×10, diagonal), and $E$ (15 × 10) respectively express *compatibility* relations, the *constitutive law*, and the *equilibrium* relations relevant to this system. Right-hand sides $f$ and $h$ are the vectors of *nodal forces* and *incompatibilities*. Note that $h$ (here, a single parameter) is not necessarily 0: for instance, if one builds the structure of Fig. 1 with a mast a bit longer than is allowed by the rest-lengths of the horizontal bars and of the tendons, then $h \neq 0$. As a rule, each redundant member (i.e., each of those one would have to remove in order to render the structure isostatic) contributes one compatibility relation. These relations may not be independent, and the vector $h$ itself is constrained in that case (it has to lie in the range of $C$).

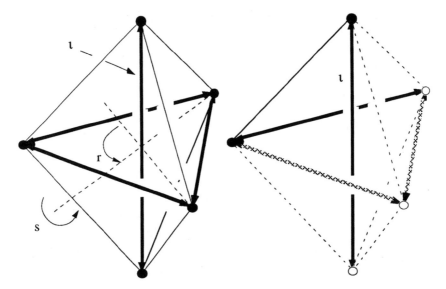

FIGURE 1. On the left, the simplest of all self-stressed structures (a kind of bumpy bicycle wheel): four bars and six tendons. The central bar (or mast) lies on the symmetry axis of an equilateral triangle, formed by the three horizontal bars. These as well as the six tendons (or cables) are identical: same lengths at rest, same stiffness. The structure is invariant under the action of the point group $D_3$. (Indeed, its symmetry group is $D_{3h}$, twice as big, but we ignore the horizontal mirror symmetry.) On the right, "the symmetry cell", as defined below (two bars, one tendon, two nodes, and the "redundant" bar $\iota$). The length of the mast is determined by the lengths of the other members, hence (by linearization) one affine relation between the strains. The right-hand side in this relation may be non-zero, since the rest lengths may not be compatible.

REMARK. Whereas matrix $E$ is easily obtained (just write equilibrium relations at all nodes), finding an explicit form for $C$ may be difficult. Because of that, structural programs do not expect more from the user than a description of $E$ (which contains all the relevant information: connectivity and lengths of the members). From this data, they construct a $C$ such that (2) hold. See, e.g., [16]. For an algorithm doing this, cf. [1]. ◊

One may easily see that (1) has here a unique solution: Uniqueness stems from the orthogonal decomposition

(2) $$X = \ker(C) \oplus \ker(E),$$

where $X = I\!R^{10}$ is the linear space common to $\epsilon$ and $\sigma$, and existence is warranted by the conditions imposed to $f$ and $h$, which amount to

(3) $$f \in \mathrm{cod}(E), h \in \mathrm{cod}(C)$$

(the images of $X$ by $E$ and by $C$), or equivalently after (2), $f \perp \ker(E^t)$ and $h \perp \ker(C^t)$, where $t$ denotes the transposition. In particular, if $f = 0$, there is one solution, that yields the forces in the bars in the absence of any loading. These are the forces that would be required in order to expand the tendons and to compress the bars until their lengths become compatible. That these forces be non-zero if $h \neq 0$ justifies the received terminology: such structures are called "self-stressed" or "self-constrained" (and "tensegrities" [17, 9], when the number of members is minimal).

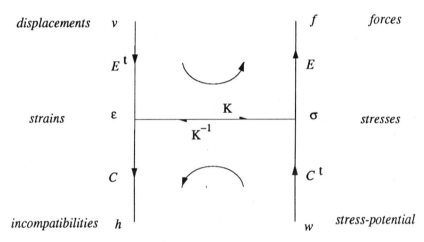

FIGURE 2. Tonti's diagram of problem (1). It suggests two ways to reduce pb. (1) to a standard linear system: in the upper part, $EK(E^t v + \epsilon_h) = f$ (the today well-accepted "displacement method") and below, $CK^{-1}(C^t w + \sigma_f) = h$ (the old "force method", just as good, in spite of fashion), where $\epsilon_h$ and $\sigma_f$ satisfy $C\epsilon_h = h$ and $E\sigma_f = f$ respectively. There are also "mixed" methods, in which one looks for the critical point $\{v, \sigma\}$ or $\{w, \epsilon\}$ of some Lagrangian ("Reissner's energy"), expressed in terms of the pair $\{v, \sigma\}$, or of its analogue $\{w, \epsilon\}$.

All this is general: all articulated structures, in the small displacements case, satisfy equations (1) with relations (2) and (3), so these will function for us as postulates. Hypothesis (2) automatically implies $\mathrm{cod}(E^t) = \ker(C)$ and $\mathrm{cod}(C^t) = \ker(E)$, so that one may draw the diagram of Fig. 2, called the *Tonti diagram* of the problem [21, 22]. Besides the linear spaces where $\epsilon$ and $\sigma$ live, this diagram displays the space of forces and its dual (the nodal displacements $v$), on top, and at the bottom the space of incompatibilities and its dual (the space of the $w$'s). What we denote by $w$ may be, in this particular case, the vector of forces in the members that one would have to discard in order to make the structure isostatic. (Having done that, one may then find a particular solution of $E\sigma = f$, hence all solutions, parameterized by $w$. This is the principle of the time-honored "force" method [16], now a bit out of fashion. See the caption of Fig. 2.) The same diagram, with the same notations, can serve for all problems in linear elasticity, $C$ and $E$ then being differential operators with appropriate boundary conditions. After finite-element discretization, such a problem takes the form (1), so any advantage there may be in the exploitation of symmetry as we shall propose here carries over to much more general situations.

Actually, as pointed out by Tonti, the structure of the diagram is common to an even wider array of physical problems. This remark is essential to understanding the real nature of the so-called "mixed" numerical methods and "mixed" finite elements. Though this is not our present concern (cf. [5]), the motivation for the present paper lies there. We wish to present methods of exploitation of symmetry in such a way that they apply to all the variant numerical treatments of (1), as based on various equivalent algebraic formulations, that are recalled in the caption of Fig. 2.

From this viewpoint, the form (1) of the problem ("symmetrical", too, although in a different way) is the most convenient one. Our aim will be achieved as follows: thanks to a kind of block-diagonalization that representation theory is able to provide, pb. (1) will split into a family of *smaller, independent* subproblems of *similar* form. Each of these may then be solved by whatever method seems preferable.

The method of exploitation of symmetry we wish to present relies on linear representation theory, so we shall briefly recall this standard material. Then, we shall show how the knowledge of the symmetry group and of its representations helps substitute for pb. (1) a family of subproblems. For example, in the case of Fig. 1, we shall be able to substitute for (1), which is a problem with 10 independent unknowns, four problems, with $3, 1, 3$ and $3$ unknowns respectively ($3 + 1 + 3 + 3 = 10$, and this is general). The number of subproblems is linked, in a non-obvious way, with the number of elements in the symmetry group of the structure, and thus the larger the group, the more economy one may expect, in many respects (computing time, data input, etc.)

There is no noticeable addition, in this paper, to the body of knowledge that crystallographers, chemists, quantum physicists, etc., have accumulated around symmetry [11, 23, 24]. The difficulty we mean to address (after many others, [6, 8, 20, 25], etc.) is the transfer of these ideas to another field of endeavor: structural analysis, as practiced in engineering. This seems to require some effort, if only to sort out what is essential to structural analysts from the lot of things that are important to chemists or solid physicists ... Even when the problems look quite similar, for instance when what is investigated is the structure of molecules quite comparable in their complexity and in their type of symmetry to that of Fig. 1 ($NH_3$ in [10, p. 60], $CH_3Cl$ in [12, p. 212], $H_2O$ in [14, p. 147], etc.), the orientation of chemists towards the problem of *vibrations* of molecules and or crystals does not help to make the transition to *static* loaded structures, a problem that classic or modern authors in this area ([3, 13, 15, 18, etc.], and those already quoted) do not seem to have been interested in. For all these reasons, this literature may not be the place where to look for the essential information. This can be obtained more quickly (is it a surprise?) from mathematicians: it is contained, for instance, in the very first pages of [19]. The problem is to take advantage of this information, in what we think is the relevant mathematical framework (Fig. 2), and this is the point of what follows.

## 2. Irreducible representations, projectors

The same "abstract" group $G$ may act on miscellaneous geometric objets in different ways, thereby generating as many "concrete" groups, which are in practice transformation groups, matrix groups, operator groups, etc. The structure of Fig. 1, for instance, has the "concrete" symmetry group[1] $D_3 = \{1, r, r^2, s, sr, sr^2\}$, a point group in $\mathbb{R}^3$, which is isomorphic to the "abstract" group $G = S(3)$ of all permutations on three objects. One says that $G$ acts on points, this action being described by $D_3$, and one will denote by $gx$ the image of point $x$ via the geometrical transformation associated with element $g$ of the abstract group. The same group also acts, in an obvious way, on three-dimensional vectors, and we shall denote by $g_*v$ the result of the corresponding action on a vector $v$. Thus, on Fig. 1, $G$ acts (by permutations) on the nodes of the structure, and also on the vectors that represent the nodal forces: if $f_n$ is the force at node $n$ (also denoted by $f(n)$ when necessary), $r_*f_n$ is the transform of vector $f_n$ by the rotation $r$, a vector that we shall later on consider as originating from node $rn$.

These elementary actions induce more complex ones. Consider for example the vector $f = \{f_n : n \in \mathcal{N}\}$ of nodal forces ($f$ has five 3-dimensional vector components, and thus spans a space of dimension 15), where $\mathcal{N}$ is the set of nodes. This $f$ is the loading of the structure. The group $G$ acts on the vector space of all loadings as follows: each group element $g$ maps $f$ to the new loading, that we shall denote as $V_g f$, obtained by applying the force $g_*f_n$ at node $gn$, and therefore defined by

$$(4) \qquad (V_g f)(gn) = g_* f(n).$$

This way, one gets a family of operators $\{V_g : g \in G\}$, that again constitute a group isomomorphic to $G$. (Note that $g_*f(n)$ should be parsed as $g_*(f(n))$.)

The group also acts on the members $m \in \mathcal{M}$ of the structure (bars or tendons), and there is a definite relationship between this action and the one on the nodes. Let us first assign to each member $m$ a unitary vector $u_m$, parallel to $m$, with an arbitrary but definite orientation. Thus, if $n$ and $n'$ are the end-nodes of $m$, $n' = n + l_m u_m$, where $l_m$ is the "natural" (unstrained) length of member $m$, and one will say that $n$ is its origin and $n'$ its extremity. One then sets $e_{nm} = -1, e_{n'm} = 1$, and $e_{n''m} = 0$ for all other nodes $n''$. Equilibrium

---

[1] Actually, the symmetry group is $D_{3h}$, which is too big for our expository purpose (while $D_3$, which is at the same time non-commutative and small, is all right). So we ignore the horizontal symmetry plane.

equations at node $n$ are then expressed as

$$(5) \qquad \sum_{m \in \mathcal{M}} e_{nm} u_m \sigma_m = f_n$$

(three scalar equations), and the collection of all such equations is the rectangular system $E\sigma = f$. If $g$ is a symmetry of the structure, there exists a member that links $gn_m$ and $gn'_m$, that will be denoted by $gm$ (take good note that $gm = m$ is possible!), and its unitary vector can be either $g_* u_m$ or $-g_* u_m$. Denoting from now on this sign by $o(g;m)$, we shall thus have

$$(6) \qquad u_{gm} = o(g;m)\, g_* u_m.$$

Then the relation

$$(7) \qquad \forall m \in \mathcal{M},\ \forall n \in \mathcal{N},\ \forall g \in G,\ e_{gn\,gm} = o(g;m)e_{nm}$$

holds, and this asserts that the structure is "symmetrically articulated". (Something is left to be said about the equal rigidity of similar members; cf. (14) below.)

The group also acts on the incompatibilities, as follows. The compatibility relations, one for each row of $C$, are of the form $\sum_m c_{\iota m}\epsilon_m = h_\iota$, where $\iota$ spans an abstract set $\mathcal{J}$. (Though $\mathcal{J}$ is identifiable with a subset of $\mathcal{M}$, namely the subset of "redundant" members, it seems better to view it as a distinct object.) By using (6) and (7), one may show that for each $\iota$ and each $g$, there is a $g\iota \in \mathcal{J}$ such that

$$(8) \qquad c_{g\iota\, gm} = c_{\iota m} \ \forall m \in \mathcal{M},$$

hence an action of $G$ on $\mathcal{J}$.

Last, there is an action on the strains (or on the stresses, just substitute $\sigma$ for $\epsilon$), defined by

$$(9) \qquad (U_g \sigma)(gm) = \sigma(m) \ \forall m \in \mathcal{M},$$

and an action $W$ on incompatibility vectors $h = \{h_\iota : \iota \in \mathcal{J}\}$, or on their dual elements, defined by $(W_g h)(g\iota) = h(\iota) \ \forall \iota \in \mathcal{J}$. Hence two new groups of operators, $U$ and $W$, both isomorphic to $G$.

As an exercise, one may see how the relations

$$(10) \qquad V_g f = f \ \forall g \in G, \quad U_g \sigma = \sigma \ \forall g \in G, \quad W_g h = h \ \forall g \in G,$$

do account, with such definitions, for the intuitive idea of "symmetrically loaded" or "symmetrically stressed" structure, as well as for the less familiar one of a structure with symmetrical incompatibilities.

One may now introduce the hypotheses that algebraically express the symmetry of the structure. One refers to them globally as *equivariance* (of the system (1))*with respect to the actions of group G.* They are as follows:

$$(11) \qquad V_g E = E U_g \ \forall g \in G,$$

$$(12) \qquad U_g K = K U_g \ \forall g \in G,$$

$$(13) \qquad W_g C = C W_g \ \forall g \in G.$$

Actually, (11) results from (6) and (7), as easily verified. Similarly, (8) implies (13). The only new hypothesis is thus (12). Since $K$ is diagonal ($K_{mm} = k_m$, the stiffness of member $m$, with $k_m > 0$ if the convention is that $\sigma > 0$ for tendons), (12) amounts to

$$(14) \qquad\qquad \forall m \in \mathcal{M}, \ \forall g \in G, \ k_{gm} = k_m,$$

the information that was missing: symmetric images of a given member all have the same stiffness.

REMARK. What characterizes tendons, as opposed to bars, is that $k_m = 0$ when $\epsilon_m < 0$. This introduces in the global problem a nonlinearity that does not fit with the present linear theory. Let's pretend we are only interested in situations where all $\epsilon_m$ are positive (no slackened cables). ◊

The symmetry of the structure is thus expressed either by (6)(7)(8) or by (11)(12)(13). One will remark that nothing has been said about the symmetry of the loading $f$: the first relation (10) is *not* part of the hypotheses. Same thing for $h$. Let us now unfold the theory.

DEFINITION 1. *Problem (1) is* equivariant *with respect to the actions of group $G$, the symmetry group of the structure, if (11)(12)(13) hold.*

DEFINITION 2. *Let $F$ be a finite-dimensional vector space on the complex field. One calls* linear representation *of group $G$ any map $\rho$ of $G$ into $GL(F)$ (the vector space of linear mappings from $F$ into itself) that satisfies*

$$\rho(gh) = \rho(g)\rho(h) \ \forall g, h \in G, \rho(1) = 1.$$

*The dimension $d$ of $F$ is called the* degree *of the representation.*

If in addition $F$ is equipped with a scalar product, one says that the representation is *unitary* if all the $\rho(g)$ are unitary, i.e., if $\rho(g^{-1}) = [\rho(g)]^{-1} = \rho(g)^*$, the adjoint of $\rho(g)$, for all $g$. As one may always find an equivalent scalar product for which this is true, it will be understood from now on that all representations mentioned are unitary.

According to the definition, the image of $G$ by $\rho$ is a matrix group (a subgroup of $GL(F)$), homomorphic to $G$, but not necessarily isomorphic, in contrast to the above actions of $G$. The latter, by the way, are all unitary with respect to the natural scalar products (such as $(f, f') = \sum_{n \in \mathcal{N}} f_n \cdot f'_n$, etc.).

DEFINITION 3. *A representation is* irreducible *if there is no subspace $L$ in $F$, other than $\{0\}$ and $F$ itself, such that $\rho(g)L \subset L \ \forall g \in G$.*

One will say "irrep" for shortness. For further use, let us recall the three irreps of group $D_3$, two of degree 1 and one of degree 2, as follows (Table 1):

| | $1$ | $r$ | $r^2$ | $s$ | $sr$ | $sr^2$ |
|---|---|---|---|---|---|---|
| $\rho_1:$ | $1$ | $1$ | $1$ | $1$ | $1$ | $1$ |
| $\rho_2:$ | $1$ | $1$ | $1$ | $-1$ | $-1$ | $-1$ |
| $\rho_3:$ | $\begin{pmatrix} 1 & 0 \\ 0 & 1 \end{pmatrix}$ | $\frac{1}{2}\begin{pmatrix} -1 & -\sqrt{3} \\ \sqrt{3} & -1 \end{pmatrix}$ | $\frac{1}{2}\begin{pmatrix} -1 & \sqrt{3} \\ -\sqrt{3} & -1 \end{pmatrix}$ | $\begin{pmatrix} 1 & 0 \\ 0 & -1 \end{pmatrix}$ | $\frac{1}{2}\begin{pmatrix} -1 & -\sqrt{3} \\ -\sqrt{3} & 1 \end{pmatrix}$ | $\frac{1}{2}\begin{pmatrix} -1 & \sqrt{3} \\ \sqrt{3} & 1 \end{pmatrix}$ |

TABLE 1. Irreducible representations of $D_3$. (The irreps of degree $> 1$ are defined up to equivalence, by a change of basis, and one has here selected a basis in which $\rho_3$ has real entries.)

The following *orthogonality relation*, one of the first results, with Schur's lemma, of representation theory, is the corner stone of the theory:

PROPOSITION 1. *If $\rho_\mu$ and $\rho_\nu$ are two irreps of the same group $G$, of degrees $d_\mu$ and $d_\nu$, one has*

$$(15) \qquad d_\nu \sum_{g \in G} \rho_\mu^{ij}(g)\, \rho_\nu^{kl}(g^{-1}) = |G|\, \delta_{il}\, \delta_{jk}\, \delta_{\mu\nu}$$

*where $|G|$ denotes the order of $G$* (or, in more compact form,

$$d_\nu \int_G \rho_\mu(g)\, \rho_\nu(g^{-1})\, dg = \delta_{\mu\nu},$$

if one denotes as an integration, $\int_G dg$, the summation $n^{-1} \sum_{g \in G}$, like in [3]).

From this, one may derive, in particular,

$$(16) \qquad \sum_\nu d_\nu^2 = |G|,$$

where the summation is with respect to all irreps, as indexed by $\nu$. For $D_3$, indeed, $1^2 + 1^2 + 2^2 = 6$. All the irreps of a *commutative* group (which $D_3$ is not) are of degree 1, and their number is thus equal to $|G|$.

Let now $U$ be an action of the group on some Hilbert space $X$, each $U_g$ unitary (for instance, the above $U_g$s). One introduces the following family of operators on $X$:

$$(17) \qquad P_\nu^{ij} = \frac{d_\nu}{|G|} \sum_{g \in G} \rho_\nu^{ij}(g) U_g$$

(or, in more compact form, $P_\nu = d_\nu \int_G \rho_\nu(g)\, U_g\, dg$, $P_\nu$ then being a matrix whose *blocks* are the $P_\nu^{ij}$, and which operates on the Cartesian product $X^{d_\nu}$). Then,

THEOREM 1. *The $P_\nu^{ii}$ defined in (17) are orthogonal projectors, their images $X_\nu^{ii} = P_\nu^{ii} X$ are mutually orthogonal closed subspaces of $X$, and one has the orthogonal decomposition*

$$(18) \qquad X = \oplus_{\nu,i} X_\nu^{ii},$$

*the summation being done with respect to all irreps $\rho_\nu$ and all indices $i = 1, ..., d_\nu$ (the degree of $\rho_\nu$). Moreover,*

$$(19) \qquad \sum_{k=1,...,d_\nu} \rho_\nu^{ik}(g)\, U_g P_\nu^{kj} = P_\nu^{ij}$$

(or in more compact form, $\rho_\nu(g)\, U_g P_\nu = P_\nu$).

*Proof.* It's a simple computation, which consists in using the orthogonality relation (15) in order to get

$$(20) \qquad P_\mu^{ij} P_\nu^{kl} = \delta_{\mu\nu}\, \delta_{jk}\, P_\nu^{il}$$

(compact form: $P_\mu P_\nu = \delta_{\mu\nu} P_\nu$), then (19). ◊

From the above $V_g$ and $W_g$, one may similarly construct operators $Q_\nu^{ij}$ and $R_\nu^{ij}$, which act on the spaces of forces and incompatibilities respectively, with similar properties.

The orthogonal decomposition (18) is an important step towards our objective. First, the conjugation relations (11)(12)(13), that are the basis of equivariance, extend to the new operators by linearity, so that

$$Q_\nu^{ij} E = E P_\nu^{ij} \ \forall \nu, \ \forall i, j = 1, ..., d_\nu,$$

$$P_\nu^{ij} K = K P_\nu^{ij} \ \forall \nu, \ \forall i, j = 1, ..., d_\nu,$$

$$R_\nu^{ij} C = C P_\nu^{ij} \ \forall \nu, \ \forall i, j = 1, ..., d_\nu.$$

Now, let us combine these new conjugation relations with the equations of system (1). By left-multiplying with the appropriate operators, we obtain

$$C P_\nu^{ij} \epsilon = R_\nu^{ij} h, \ P_\nu^{ij} \sigma = K P_\nu^{ij} \epsilon, \ E P_\nu^{ij} \sigma = Q_\nu^{ij} f,$$

that is, if we denote the images by $\epsilon_\nu^{ij}$, etc.,

$$(21) \qquad C \epsilon_\nu^{ij} = h_\nu^{ij}, \ \sigma_\nu^{ij} = K \epsilon_\nu^{ij}, \ E \sigma_\nu^{ij} = f_\nu^{ij}.$$

In particular, as regards the projections on subspaces, that is, the *Fourier components*[2] $\epsilon_\nu^{ii}$ and $\sigma_\nu^{ii}$ of the unknowns $\epsilon$ and $\sigma$, one has

$$(22) \qquad C \epsilon_\nu^{ii} = h_\nu^{ii}, \ \sigma_\nu^{ii} = K \epsilon_\nu^{ii}, \ E \sigma_\nu^{ii} = f_\nu^{ii},$$

a family of *independent* problems, as anticipated. Once these have been solved, one may proceed with the *Fourier synthesis* of the solution:

$$\epsilon = \sum_{\nu,i} \epsilon_\nu^{ii}, \ \sigma = \sum_{\nu,i} \sigma_\nu^{ii}$$

(and on the face of these formulas, it looks as if the off-diagonal terms $\epsilon_\nu^{ij}$ and $\sigma_\nu^{ij}$ were not needed, but this will prove wrong in a moment). The right-hand sides in (21) come from a *Fourier analysis* of $f$ and $h$. For $f$, after (17) and the definition of the $V_g$ (cf. (4)), this amounts to setting

$$(23) \qquad f_\nu^{ij}(n) = \frac{d_\nu}{|G|} \sum_{g \in G} \rho_\nu^{ij}(g^{-1})(g_*)^{-1} f(gn),$$

and for $h$, similarly,

$$(24) \qquad h_\nu^{ij}(\iota) = \frac{d_\nu}{|G|} \sum_{g \in G} \rho_\nu^{ij}(g^{-1}) h(g\iota).$$

Thus a *block-diagonalization* of the original problem (1) has been obtained, each block corresponding to one of the problems (22), with $d_\nu$ blocks for each irrep $\rho_\nu$ of degree $d_\nu$. In all, therefore, $\sum_\nu d_\nu$ subproblems, of the same structure as (1).

---

[2] We are not speaking of Fourier *coefficients*, that is, the components in some basis. The word "component" may generate some confusion here, but there is no alternative.

Beware, however, that Fourier components $\epsilon_\nu^{ii}$ and $\sigma_\nu^{ii}$ are here vectors of the same dimension as $\epsilon$ or $\sigma$. So the question of how exactly each of the problems (22) is "smaller" than pb. (1) is still pending. This comes from $X_\nu^{ii}$, a *strict* subspace of $X$, being isomorphic to a vector space of *reduced* dimension, that will be denoted as $\check{X}_\nu^{ii}$, and explicitly built in what follows.

REMARK. The method amounts to looking for a convenient basis for each $X_\nu^{ii}$. By putting together all the basis vectors, one then obtains a new basis for $X$, that is dubbed a *symmetry-adapted* basis. The components of a vector $x$ in this basis[3] are called *symmetrical components*, a terminology that is especially popular among electrical engineers. (Because of the use of three phases, machines and networks have a $C_3$ symmetry. The three symmetrical components are called *homopolar*, *direct* and *inverse* components respectively. The seminal paper in this direction was [7].)

## 3. Analysis of the subproblems

We need a few new notions.

DEFINITION 4. *If a group $G$ acts on a set $X$, one calls the* orbit *of the element $x$ the set $\{gx : g \in G\}$ of all images of $x$, and the* isotropy group *(or little group, or stabilizer) of $x$ the subgroup $G_x = \{g \in G : gx = x\}$.*

The set of orbits can be identified with a part of $X$, as obtained by picking one element $x$ in each orbit. For example, as regards the action on the members, one may extract from $\mathcal{M}$ a part $\mathcal{M}_G$, minimal, such that each member of the structure be the image of a member of $\mathcal{M}_G$ by some transform $g$. Same thing for the nodes, hence a subset $\mathcal{N}_G$ of $\mathcal{N}$, containing one node per orbit, and for the incompatibilities, hence $\mathcal{J}_G \subset \mathcal{J}$. One may always make the sampling in such a way that every node in $\mathcal{N}_G$ is either the origin or the extremity of one of the elements of $\mathcal{M}_G$, and that $\mathcal{J}_G$ corresponds to a part of $\mathcal{M}_G$.

DEFINITION 5. *The triple $\{\mathcal{N}_G, \mathcal{M}_G, \mathcal{J}_G\}$ is called a* symmetry cell *of the structure (cf. Fig. 1 for an example).*

DEFINITION 6. *A family $\{x^i : i = 1, ..., d_\nu\}$ of $d_\nu$ elements of $X$ is called $\nu$-symmetric when*

$$x^i = P_\nu^{ij} x^j \ \forall i, j \in \{1, ..., d_\nu\}.$$

(In that case, after (20), $x^i \in X_\nu^{ii}$, $i = 1, ..., d_\nu$. In particular, if $d_\nu = 1$, the vector $x$ is $\nu$-symmetric if it belongs to the subspace $X_\nu$.)

Then, after (20) and (19),

PROPOSITION 2. *The family $\{x^i : i = 1, ..., d_\nu\}$ is $\nu$-symmetric if and only if*

$$(25) \qquad \sum_{k=1,...,d\nu} \rho_\nu^{ik}(g) \, U_g x^k = x^i \ \forall g \in G.$$

Let us see what this implies, first in the simple case where $d_\nu = 1$. Let $\sigma \in X_\nu^{ii}$ be a set of stresses in members. Then, after (25) and the definition of $U_g$ in (9),

$$(26) \qquad \sigma(gm) = \rho_\nu(g) \, \sigma(m) \ \forall g \in G, \ \forall m \in \mathcal{M}.$$

---

[3] or else (the same confusion again) their products by the corresponding basis vectors,

So stresses in all members can be derived from a reduced number of them, those of the $m \in \mathcal{M}_G$. Same formula for the strains:

$$(27) \qquad \epsilon(gm) = \rho_\nu(g)\,\epsilon(m) \ \forall g \in G, \ \forall m \in \mathcal{M}.$$

So one selects as independent unknowns the families $\{\sigma(m) : m \in \mathcal{M}_G\}$ and $\{\epsilon(m) : m \in \mathcal{M}_G\}$. As for equilibrium equations, one may as well write only those relative to the nodes of the symmetry cell (three equations per node), for other equations are linear combinations of these. (To check this point, start from (5) expressed at node $gn$:

$$f_{gn} = \sum_{m \in \mathcal{M}} e_{gn\,m}\, u_m \sigma_m = \sum_{m \in \mathcal{M}} e_{gn\,gm}\, u_{gm} \sigma_{gm}$$

$$= \sum_{m \in \mathcal{M}} o(g;m)\, e_{nm}\, o(g;m)\, g_* u_m\, \rho_\nu(g)\sigma_m,$$

after (6) and (7), so $f_{gn} = \rho_\nu(g) f_n$.) Same thing for the incompatibilities, one for each $\iota \in \mathcal{J}_G$.

So each problem (22) corresponding to an irrep $\rho_\nu$ of degree 1 can be obtained by writing the equilibrium equations at the nodes of the symmetry cell, as loaded by the $f_\nu$ of formula (23), while imposing the incompatibilities (24). Each of these equations invokes some $\sigma_m$ and some $\epsilon_m$ for indices $m$ *not* included in $\mathcal{M}_G$, but the values of which are known nevertheless in terms of the $\{\sigma_m, \epsilon_m : m \in \mathcal{M}_G\}$, thanks to relations (27). Problem (22) is "small" because the genuine unknowns are the components of the *restrictions* of $\sigma_\nu$ and $\epsilon_\nu$ to the symmetry cell, and enough equations are obtained by only considering nodes in $\mathcal{N}_G$ and incompatibilities in $\mathcal{J}_G$.

One is thus tempted to conclude that the space $\tilde{X}_\nu^{ii}$ alluded to before (here, $\tilde{X}_\nu$, since $d_\nu = 1$), isomorphic to $X_\nu$, is spanned by the vectors $\tilde{x} = \{x_m : m \in \mathcal{M}_G\}$ which are restrictions of vectors of $X_\nu$ to the symmetry cell. But the latter space (let us call it $Y_\nu$) may contain $\tilde{X}_\nu$ as a *strict* subspace, because of relations (26)(27).

Indeed, (26)(27) imposes $\sigma(m) = \epsilon(m) = 0$ for every member such that $\rho_\nu(g) \neq 1$ for at least one element $g$ of its isotropy group. Thus, if the isotropy group of member $m$ is not trivial (which is the case for all the compressive members in our example), the corresponding strain and stress may be eliminated from the equations, for some of the irreps $\rho_\nu$.

For the trivial irrep $\rho_1$, this does not happen, since $\rho_1(g) = 1$ for all $g$, and $\tilde{X}_\nu$ is therefore of dimension 3 in our example. For the irrep $\rho_2$ (cf. Table 1), there is only one non zero $\sigma_m$, the one relative to the cable, for all other members are invariant by $s$, and $\rho_2(s) = -1$. The corresponding problem is thus of dimension 1, and reduces to a simple division.

When $d_\nu > 1$, things are messier. It happens in this example, since we are left with the two problems of the form (22) that correspond to $\rho_3$, for $i = 1$ and $i = 2$. Let $i = 1$ first, and let us denote by $\sigma^{11}$ the set of constraints looked for. As $\sigma^{11} \in X_\nu^{11}$, it is only natural, in view of the success obtained with $\rho_1$ and $\rho_2$, to try and characterize $\sigma^{11}$ by the values of its components $\sigma_m^{11}$ for all $m$ of the symmetry cell. For this, one has formula (25). But since this time $d_\nu = 2$, (25) is a relation between the values not only of $\sigma^{11}$, but also of its sibling (let us call it $\sigma^{12}$) in the $\nu$-symmetric family it belongs to. So the restriction of $\sigma^{11}$ to $\mathcal{M}_G$ is not enough: one will need the restrictions of $\sigma^{11}$ and of $\sigma^{12}$ to the symmetry cell, and hence $\tilde{X}_\nu^{11}$ is a priori a subspace of the Cartesian product $Y_\nu \times Y_\nu$, or said differently (in order to make the connection with the "tensor-product" notation of [2]), the unknowns are part of a set comprised of $d_\nu = 2$ values of $\sigma$ per member that belongs to $\mathcal{M}_G$. They form only a part of this set, because some of these values are a priori 0.

Indeed, by taking (9) into account, and according to Table 1, we may unfold relation (25) as

$$\begin{pmatrix} \sigma^{11} \\ \sigma^{12} \end{pmatrix}(sm) = \begin{pmatrix} 1 & 0 \\ 0 & -1 \end{pmatrix} \begin{pmatrix} \sigma^{11} \\ \sigma^{12} \end{pmatrix}(m)$$

and

$$\begin{pmatrix} \sigma^{11} \\ \sigma^{12} \end{pmatrix}(rm) = \begin{pmatrix} -1/2 & -\sqrt{3}/2 \\ \sqrt{3}/2 & -1/2 \end{pmatrix} \begin{pmatrix} \sigma^{11} \\ \sigma^{12} \end{pmatrix}(m)$$

So, as regards the horizontal bar of the symmetry cell (little group: $\{1, s\}$), $\sigma^{12} = 0$, and for the mast (little group: $G$ itself), $\sigma^{11} = \sigma^{12} = 0$. With the two unknowns borne by the tendon (whose little group is trivial), the number of genuine unknowns adds up to three, which is thus the dimension of $\tilde{X}_\nu^{11}$.

The structure of $\tilde{X}_\nu^{22}$ is obtained in the same manner: just replace $\sigma^{11}$ and $\sigma^{12}$ by $\sigma^{21}$ and $\sigma^{22}$ in the above relations. So each of the two subproblems (22) relative to $\rho_3$ has three unknowns, as announced.

In the general case ($d_\nu \geq 1$), the characterization of $\tilde{X}_\nu^{ii}$ is as follows: if $\tilde{\mathbf{x}}$ (boldface) denotes the vector $\{\tilde{x}^1, ..., \tilde{x}^{d_\nu}\}$ of restrictions to $\mathcal{M}_G$ of the $\nu$-symmetric family $\{x^1, ..., x^{d_\nu}\}$, one has

$$\tilde{X}_\nu^{ii} = \{\tilde{\mathbf{x}} \in Y_\nu^{d_\nu} : \gamma_\nu(\tilde{\mathbf{x}}) = 0\},$$

where $\gamma_\nu$ stands for the left-hand sides of relations (25). Now, to solve (22) subject to these constraints amounts to looking for $\nu$-symmetric families $\{\sigma^{ij}, j = 1, ..., d_\nu\}$ and $\{\epsilon^{ij}, j = 1, ..., d_\nu\}$ that satisfy (21) [4]. So, as anticipated, all problems (21) have to be solved (i.e., $\sum_\nu d_\nu^2 \equiv |G|$ subproblems in all, according to relation (16)). Each of them is "small", for the same reasons as above. Note that subproblems (21-$ij$), for $j = 1, ..., d_\nu$ and a fixed $i$, are coupled together.

## Conclusion

Thanks to representation theory, the computation of strains and stresses in a symmetrical structure splits into the series of subproblems (22). Each of these is a coupled problem, with $d_\nu$ components. Each of these components is similar to the original problem, but concerns a reduced structure (the "symmetry cell"). Members (bars and cables) that have a nontrivial isotropy group have been seen to be responsible for the coupling. As the isotropy group is trivial for most members in large structures, each problem (22) has a quasi block-diagonal structure, with $d_\nu$ weakly coupled blocks. From the point of view of the reduction in complexity, the situation is therefore roughly the same in both the commutative and the non-commutative case: instead of solving a single problem of size $M$ (the number of members in the structure), one deals with $|G|$ problems of approximate size $M/|G|$, some of them, possibly, weakly coupled.

Since, as a rule, the cost of solving a problem rises much faster than its size, the potential benefits are obvious.

## REFERENCES

1. P. Alfeld, D.J. Eyre: The Exact Analysis of Sparse Rectangular Linear Systems, ACM TOMS, **17**, 4 (1991), pp. 502-18.

2. E.L. Allgower, K. Georg, J. Walker: Exploiting Symmetry in $3D$ Boundary Element Methods", in *Contributions in Numerical Mathematics* (R.P. Agarwal, ed.), World Scientific (Singapore), 1993.

3. J. Barriol: *Applications de la théorie des groupes à l'étude des vibrations moléculaires et cristallines*, Masson (Paris), 1947.

4. A. Bossavit: Symmetry, Groups, and Boundary Value Problems. A Progressive Introduction to Noncommutative Harmonic Analysis of Partial Differential Equations in Domains with Geometrical Symmetry", Comp. Meth. Appl. Mech. Engng., **56** (1986), pp. 167-215.

5. A. Bossavit: A new viewpoint on mixed elements", Meccanica, **27** (1992), pp. 3-11.

6. D.A. Evensen: Vibration Analysis of Multi-Symmetric Structures", AIAA Journal, **14**, 4 (1976), pp. 446-53.

7. C.L. Fortescue: Method of Symmetrical Co-ordinates Applied to the Solution of Polyphase Networks", AIEE Trans., **37** (1918), pp. 1027-1140.

8. P.G. Glockner: Symmetry in Structural Mechanics", J. Struct. Div. ASCE, **98**, ST1 (1973), pp. 71-89.

9. N.S. Goel, V. Prakash: Tension icosahedron as a structure for use in terrestrial and outer space environments", Comp. & Structures, **41**, 2 (1991), pp. 189-196.

10. G.G. Hall: *Applied Group Theory*, Longmans (London), 1967.

11. T. Kahan et al.: *Théorie des groupes en physique classique et quantique, t. 1, 2, 3*, Dunod (Paris), 1960, 1971, 1972.

12. J.W. Leech, D.J. Newman: *How to use Groups*, Methuen (London), 1969.

13. J.S. Lomont: *Application of Finite Groups*, Academic Press (New York, London), 1959.

14. W. Ludwig, C. Falter: *Symmetries in Physics (Group Theory Applied to Physical Problems)*, Springer-Verlag (Berlin), 1988.

15. G.Ya. Lyubarskii: *The Application of Group Theory in Physics*, Pergamon Press (New York), 1960.

16. E.C. Pestel, F.A. Leckie: *Matrix methods in elastomechanics*, Mc Graw Hill (New York), 1963.

17. A. Pugh: *An introduction to tensegrity*, University of California Press (Berkeley), 1976.

18. D.S. Schonland: *La symétrie moléculaire: Introduction à la théorie des groupes et à ses applications à la Chimie* (transl. of *Molecular Symmetry: An Introduction to group theory and its uses in chemistry*, 1965), Gauthier-Villars (Paris), 1971.

19. J.P. Serre: *Représentations linéaires des groupes finis*, Hermann (Paris), 1978.

20. E. Stiefel, A. Fässler: *Gruppentheoretische Methode und Ihre Anwendung*, Teubner (Stuttgart), 1979. (New English edition: *Group Theoretical Methods and Their Applications*, Birkhaüser (Boston), 1992.)

21. E. Tonti: On the mathematical structure of a large class of physical theories", Rend. Acc. Lincei, **52** (1972), pp. 48-56.

22. E. Tonti: *La Struttura Formale delle Teorie Fisiche*, CLUP (Milano), 1976.

23. H. Weyl: *The Theory of Groups and Quantum Mechanics*, Dover (New York), 1950 (first German edition, 1928).

24. E.P. Wigner: *Gruppen Theorie und Ihre Anwendung auf die Quantenmechanik der Atomspektren*, Friedr. Vieweg, 1931 (English transl.: Ac. Press, 1959).

25. W. Zhong, C. Qiu: Analysis of Symmetric or Partially Symmetric Structures", Comp. Meth. Appl. Mech. & Engng., **38**, 1 (1983), pp. 1-18.

ÉLECTRICITÉ DE FRANCE, 1 AV. DU GAL DE GAULLE, 92141 CLAMART, FRANCE
*E-mail Address:* Alain.Bossavit@der.edf.fr

Lectures in Applied Mathematics
Volume **29**, 1993

# Symmetry Considerations in the Numerical Analysis of Bifurcation Sequences

## F.H. BUSSE AND R.M. CLEVER

ABSTRACT. The Galerkin method for the investigation of transitions in fluid flow is outlined and the symmetry considerations in the numerical analysis are discussed. The Galerkin method permits the exploitation of all symmetries of the fluid dynamical problem and exhibits the loss of symmetry in transitions from secondary to tertiary and then on to quarternary states of fluid flow. Some new results are described for knot convection and oscillatory bimodal convection.

## 1. Introduction

One of the major applications of bifurcation theory and symmetry is the area of transition in fluid flows. Nearly all fluid systems that are not in hydrostatic equilibrium experience changes from the primary laminar state to complex turbulent states as some control parameter such as the Reynolds number is increased. In cases when the external conditions of the system exhibit a high degree of symmetry it is often possible to distinguish discrete transitions in the evolution from simple to complex states of fluid flow. One or more symmetries are typically broken as the more complex state bifurcates from the simpler one. Rayleigh-Bénard convection in a layer heated from below and the Taylor-Couette system are the best known, but by no means the only examples.

For many fluid dynamical problems it is possible to find an idealized configuration in which the basic state depends only on a single coordinate while it is homogeneous with respect to the two other coordinates of a orthogonal system. Circular Couette flow between two co-axial, differentially rotating cylinders is a typical example. From the theoretical point of view these configurations of maximum symmetry have several advantages:

---

1991 *Mathematics Subject Classification.* Primary 76E15, 76E38; Secondary 76E05, 76M15.
The research was supported by NSF grant ATM-8913715 and by a NATO travel fund.
This paper is in final form and no version of it will be submitted for publication elsewhere.

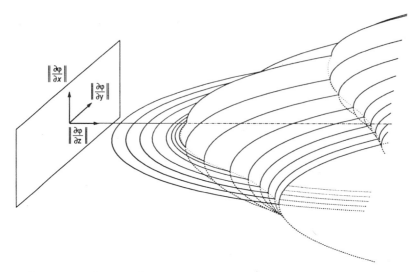

FIGURE 1. Sketch of subsequent bifurcations in the phase space
of solutions. The thick line indicates the primary solution depending on one coordinate only.

- The physical mechanisms of instability are realized in their simplest
  form.
- Transitions from the basic state and from higher order states occur
  through bifurcations, instead of imperfect bifurcations in inhomogeneous
  configurations.
- Symmetries broken by bifurcations permit an easy identification of transitions.
- High degrees of symmetry facilitate the numerical analysis.

In figures 1 and 2 some typical examples of the fluid systems of interest have
been sketched. In the following we shall go a step further by restricting the
attention to a planar configuration in which case a Cartesian system can be
used with $x, y$-coordinates in the homogeneous direction. But this restriction
is chosen only for mathematical convenience since all our considerations can
easily be generalized to other configurations described by systems of orthogonal
coordinates.

In the numerical analysis the Galerkin method will be used in which the dependent variables are expanded in complete systems of functions which satisfy all
boundary conditions. Trigonometric functions characterize the $x, y$-dependences
in the homogeneous dimensions. While it usually is not possible to obtain exact
solutions in this way, good approximations can be obtained by truncating the
infinite sums and by solving the equations for the expansion coefficients numerically. The quality of the approximation can be checked through changes of the
truncation parameter.

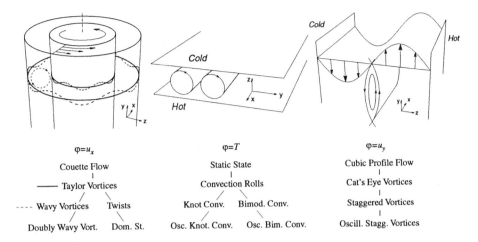

$\varphi = u_x$          $\varphi = T$          $\varphi = u_y$

Couette Flow      Static State      Cubic Profile Flow
|  |  |
—— Taylor Vortices     Convection Rolls     Cat's Eye Vortices
/ \     / \     |
- - - Wavy Vortices   Twists    Knot Conv.   Bimod. Conv.    Staggered Vortices
/ \     / \     |
Doubly Wavy Vort.   Dom. St.    Osc. Knot. Conv.   Osc. Bim. Conv.    Oscill. Stagg. Vortices

FIGURE 2. Examples of subsequent bifurcations in the Taylor-Couette system (left), the Rayleigh-Bénard layer (middle) and the vertical fluid layer heated from the side (right). Different sequences of bifurcation occur depending on the parameters of the problem. Dom. St. refers to domain states which occur as quarternary states in a certain parameter region of the Taylor-Couette system [1].

## 2. Secondary Solutions

The basic equations for the type of problems shown in figure 1 can be written in the form

$$(2.1) \qquad L\varphi - RB\varphi + V\frac{\partial}{\partial t}\varphi = N(\varphi, \varphi)$$

where in general $\varphi$ is a vector function and $L, B, N, V$ are matrix operators, while $R$ is the control parameter. But for simplicity we may just think of $\varphi$ as the $z$-component of the velocity field. The nonlinear operator $N$ is usually quadratic in the variable $\varphi$. The linearized problem in which the right hand side of (2.1) vanishes yields solutions of the form

$$(2.2) \qquad \varphi_0 \propto \exp\{il \cdot \mathbf{x}\} + \sigma t\}$$

where $l$ is an arbitrary vector in the $x, y$-plane. The lowest value of $R$ for which a growthrate with non-negative real part can be obtained is called the critical value $R_c$. The corresponding wavevector $l_c$ determines orientation and wavelength of growing roll-like solutions as the instability grows. Only in degenerate problems there may exist two or more different wavevectors $l$ which correspond to $R_c$. A special case are problems exhibiting the property of isotropy in the $x, y$-plane, as for example Rayleigh-Bénard convection in a horizontal layer heated from below. But even in this case roll-like solutions become the preferred secondary solution as the control parameter $R$ exceeds the critical value by a finite amount [2], [3].

Restricting the attention to the case of steady rolls corresponding to a monotone bifurcation we rotate the coordinate system such that the $y$-direction coincides with the vector $\mathbf{l}_c$ and write the secondary solution in the form

$$(2.3) \quad \varphi = \sum_{m,n \geq 1} a_{mn} \exp\{im\alpha y\} f_n(z) \quad \text{with} \quad a_{-mn} = a_{mn}^+, \quad \alpha \equiv |\mathbf{l}_c|$$

where the superscript$^+$ indicates the complex conjugate. In addition to the
symmetries exhibited by solution (2.3),

$$(2.4) \qquad \frac{\partial}{\partial t}\varphi = 0 \quad , \quad \frac{\partial}{\partial x}\varphi = 0 \quad , \quad \varphi(y + \frac{2\pi}{\alpha}, z) = \varphi(y, z)$$

rolls often exhibit additional symmetries, as for instance the existence of symmetry planes $y =$const. or an inversion symmetry about their axis,

$$(2.5a) \qquad \varphi(-y, z) = \varphi(y, z) \quad \text{corresponding to} \quad a_{-mn} = a_{mn}$$

(2.5b)

$$\varphi(\frac{\pi}{\alpha} - y, -z) = -\varphi(y, z) \quad \text{corresponding to} \quad a_{mn} = 0 \quad \text{for} \quad m + n = \text{odd}$$

These symmetries can simplify considerably the numerical computation of solutions of the form (2.3). For the numerical computations the infinite sums in
expression (2.3) must be truncated. A convenient way is the neglection of all
coefficients $a_{mn}$ satisfying

$$(2.6) \qquad\qquad m + n \geq N_T$$

The numerical effort is thus reduced to the solution of a finite system of nonlinear algebraic equations which are obtained when the original partial differential equations are projected onto the space of the expansion functions used
in expression (2.3). By increasing the truncation parameter $N_T$ the numerical
convergence of the procedure can be checked. A sufficiently high value $N_T$ must
be adopted such that a reasonably good approximation of the exact solution is
obtained. Using Floquet theory we can analyze the instabilities of rolls leading
to transitions to tertiary solutions by superimposing infinitesimal disturbances
onto solution (2.3),

$$(2.7) \qquad \tilde{\varphi} = \left\{ \sum_{m,n \geq 1} \tilde{a}_{mn} \exp\{im\alpha y\} f_n(z) \right\} \exp\{ibx + idy + \sigma t\}$$

For given values $R, \alpha$ the growthrate $\sigma$ with maximum real part $\sigma_r$ must be
determined as function of $b$ and $d$. When there exists a growing disturbance
with $\sigma_r > 0$, the steady solution (2.3) is unstable; otherwise it is stable.

### 3. Tertiary Solutions

In contrast to the generic secondary solution in the form of two-dimensional rolls, there is large variety of tertiary solutions which are three-dimensional in general. In the case of a monotone bifurcation from the roll solution with $\sigma_i = 0$ or when the traveling wave mode is preferred in the case of a Hopf-bifurcation, $\sigma_i \neq 0$, the tertiary solution can be written in the form

$$(3.1) \quad \varphi = \sum_{lmn} a_{lmn} \exp\{il\alpha_x x + im\alpha_y y\} f_n(z) \quad \text{with} \quad \alpha_{-l-mn} = a_{lmn}^{+}$$

where $\alpha_x$ represents the value $b$ of the strongest growing disturbance of the form (2.7) and where $\alpha_y$ usually corresponds to $\alpha$ or $\alpha/2$. Other values of $\alpha_y$ can not be excluded; but we shall disregard in the following the possibility of an aperiodic dependence on $y$. In the case of wave solutions traveling in the $x$-direction, $x$ must be replaced by $x \pm ct$ in expression (3.1).

Probably the simplest non-trivial solution of the form (3.1) describes bimodal convection in a layer heated from below [11]. In this case there exist symmetry planes $x =$const. as well as $y =$const. and an inversion symmetry is also realized when the properties of the fluid layer are symmetric about its midplane. In terms of the coefficients $a_{lmn}$ these symmetries are described by

(3.2)
$$a_{-lmn} = a_{l-mn} = a_{-l-mn} = a_{lmn} \quad , \quad a_{lmn} \equiv 0 \quad \text{for} \quad l + m + n = \text{odd}$$

Bimodal convection cells are typically observed when the ratio $P$ between viscous and thermal diffusivities of the fluid, $P = \nu/\kappa$, is rather high, as it is for oils and glycerol. For an experimental study see [6].

Knot convection is another form of convection which is also described by a tertiary solution with the symmetry properties (3.2). In contrast to bimodal convection which is characterized by a high value of the second wavenumber $\alpha_x$, $\alpha_x > \alpha_y$, knot convection exhibits low values of $\alpha_x$, $\alpha_x < \alpha_y$. In experiments knot convection is typically observed for values of the Prandtl number $P$ in the range $2 \lesssim P \lesssim 15$. In figure 3 a typical solution describing knot convection is shown. For more details on the numerical analysis we refer to [9]. While the view of the midplane of the layer, $z = 0$, shows the symmetry between ascending and descending motion in the convection pattern, the physically more interesting properties of the solution can be seen in a plane closer to the boundaries of the fluid layer. In particular the deviations from the horizontal mean of the normal derivative of the temperature at the boundary shown in the lower left plot of the figure demonstrate the mechanisms of heat transfer from fluid to solid. The star-shaped areas of positive $\partial(\Theta - \overline{\Theta})/\partial z$ indicate the emergence of hot fluid from the lower thermal boundary layer most of which then moves in the form of a plume towards the opposite boundary. The acceleration owing to the buoyancy of the plume produces a relative high velocity as the plume impinges on the opposite boundary, causing a compression of the thermal boundary layer and

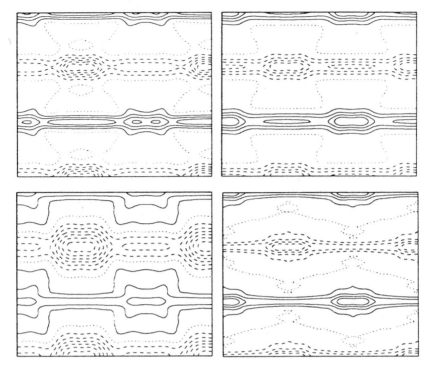

FIGURE 3. Knot convection in a fluid layer with Rayleigh number $R = 4.10^4$, Prandtl number $P = 7$. Lines of constant vertical velocity in the planes $z = -0.3$ (upper left) and $z = 0$ (upper right) and lines of constant $\partial(\Theta - \overline{\Theta})/\partial z$ at $z = -0.5$ (lower left) and isotherms in the plane $z = 0$ (lower right) are shown. $\Theta - \overline{\Theta}$ denotes the deviation of the temperature from its horizontally averaged value. The wavenumbers $\alpha_x = 1.6$ and $\alpha_y = 2.0$ have been used. Solid (dashed) lines indicate positive (negative) values; the dotted lines correspond to zero.

thus an effective heat transport. Since the plots of figure 3 for negative values of $z$ are the same as those for positive values of $z$ if positive and negative lines are reversed, the impingement process can be seen in the form of the broken lines in the lower left plot.

For the study of the stability of tertiary solutions of the form (3.1) a representation for the infinitesimal disturbances in analogy to expression (2.7) can be assumed,

$$(3.3) \qquad \tilde{\varphi} = \sum_{lmn} \tilde{a}_{lmn} \exp\{i(l\alpha_x + d)x + i(m\alpha_y + b)y + \sigma t\} f_n(z)$$

As before, the maximum of the real part of the growthrate $\sigma$ as a function of the parameter $d$ and $b$ determines the stability of the tertiary solution (3.1). Because of the considerable computational expense associated with the general eigenvalue problem posed by the superposition of the disturbance onto the stationary solution (3.1), the stability analysis has often been reduced to the case of vanishing Floquet wavenumbers $b$ and $d$. Experimental observations support this restriction of the stability analysis. They indicate that many instabilities do not change the two-dimensional periodicity lattice in the $x, y$-plane of the solution (3.1).

The assumption of vanishing $b$ and $d$ simplifies the stability analysis considerably since symmetries of the tertiary solution in the $x, y$-plane can be used. For example, in the cases of knot convection or bimodal convection it is possible to separate the disturbances into eight classes,

(3.4)    $ECC, \quad ECS, \quad ESC, \quad ESS, \quad OCC, \quad OCS, \quad OSC, \quad OSS$

depending on whether they are symmetric or antisymmetric with respect to the symmetry planes $x = 0$ and $y = 0$ of the stationary solution. Symmetry (antisymmetry) with respect to $x = 0$ is indicated by the second letter $C(S)$ in the above symbols; the third letter describes the symmetry in the $y$-direction. The first letter $E(O)$ indicates that coefficients $\tilde{a}_{lmn}$ with $l+m+n =$ odd (even) vanish. This latter symmetry also holds in the case of non-vanishing $b, d$. The symmetry (3.2) of the tertiary solution is given by $ECC$ in this notation.

## 4. Quarternary Solutions

When the growing disturbances fit the $x, y$-periodicity interval established by the tertiary solution, usually some of the symmetries of the form (3.2) or similar ones are broken in the transition to quarternary states of fluid flow. The representation (3.1) can still be used if the growthrate is purely real. But often the symmetry in time in broken and quarternary solutions become periodic time. In that case it is convenient to assume that the coefficients $a_{lmn}$ are functions of time. Instead of nonlinear algebraic equations we obtain a system of ordinary differential equations in time which can be solved numerically by Runge-Kutta-methods or by a Crank-Nicolson scheme or by other numerical techniques. In the following we restrict the attention to two examples of time-periodic quarternary states. The first example shown in figure 4 describes oscillatory knot convection arising from the growth of a disturbance of $OCC$ symmetry. The descending cold blob and the ascending hot blob are half a period and half a wavelength in the $x$-direction out a phase. The Nusselt numbers at the top and bottom boundaries are thus also half a period out of phase. Similarly oscillatory blob instabilities of steady knot convection occur in the form of disturbances with the other symmetries listed in (3.4) except for the case $ECC$. It thus appears that the break in the translational symmetry in time is typically connected with the breaking of a spatial symmetry. Another example of a time-periodic quarternary solution is presented in figure 5 where oscillatory bimodal convection pattern is shown. This solution is also described by the superposition of $ECC$- and $OCC$-

FIGURE 4. Oscillatory knot convection with $R = 4.10^4, P = 4, \alpha_x = 1.6, \alpha_y = 2.0$. The left, middle and right columns show lines of constant vertical velocity in the planes $z = 0$, $z = -0.3$, and isotherms in the plane $z = 0$, respectively. Equally distant time steps $\pi/3\omega$ are used for the plots from top to bottom such that the seventh plot would be identical to the first one.

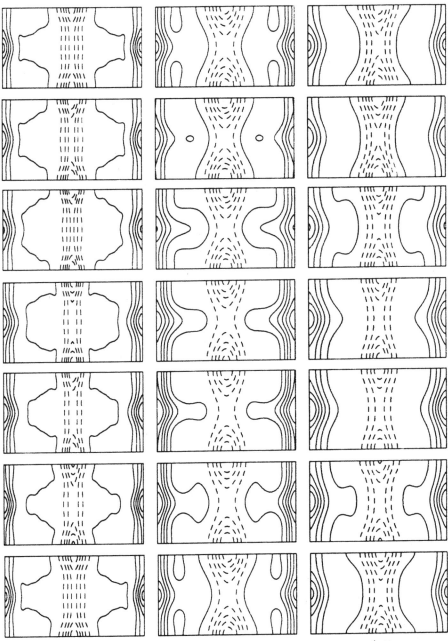

FIGURE 5. Oscillatory bimodal convection with $R = 4.10^4, P = 16, \alpha_x = 4.5, \alpha_y = 2.5$. Left, middle and right columns show lines of constant vertical velocity in the planes $z = 0, z = -0.3$ and isotherms in the plane $z = 0$. Equal time steps of $\pi/3\omega$ are used from top to bottom. As in figure 3 and 4 the solid line adjacent to the dashed lines indicates zero.

modes such that planes of symmetry given by $y = n\pi/\alpha_y$ or $x = m\pi/\alpha_x$ with integer values $n, m$ are preserved. Oscillatory bimodal cells are often observed in convection experiments with fluids with moderate to large Prandtl number [13]. As the Rayleigh number is increased further, the periodic oscillations tend to be replaced by an aperiodic time dependence. But this transition has not yet been investigated in a systematic fashion.

## 5. Concluding Remarks

In this short review only a brief outline of methods has been given that permit the computation of tertiary and quarternary states of fluid flow. Because of the homogeneity in the $x, y$-dimensions and in time, solutions that are periodic in these coordinates can be assumed, which includes the case when they are independent of one of the coordinates. These solutions describe experimentally observed flows in a variety of systems; those shown in figure 1 are only the most familiar ones. For the Taylor-Couette system we refer to [14], [15], [18]. For the Rayleigh-Bénard case the references [4], [5], [8], [10], [11] are of interest in addition to those already mentioned, and higher order bifurcations in the cubic profile case have been treated in [16], [17], [7]. Of special interest are those type of solutions which occur under different physical circumstances such as wavy rolls in an inclined layer heated from below which are essentially identical to wavy Taylor vortices between differentially rotating co-axial cylinders. A systematic analysis of all possible periodic tertiary solutions has not yet been attempted, but appears to be feasible.

## REFERENCES

1. C.D. Andereck, S.S. Lin and H.L. Swinney, *Flow regions in a circular Couette system with independently rotating cylinders*, J. Fluid Mech. **164**, 155–183 (1986)
2. F.H. Busse, *The stability of finite amplitude cellular convection and its relation to an extremum principle*, J. Fluid Mech. **30**, 625–649 (1967)
3. F.H. Busse, *Nonlinear Properties of Convection*, Rep. Progress in Physics **41**, 1929–1967 (1978)
4. F.H. Busse and R.M. Clever, *Transitions to more complex patterns in thermal convection*, pp. 37–45 in "New Trends in Nonlinear Dynamics and Pattern Forming Phenomena: The Geometry of Nonequilibrium", P. Coullet and P. Huerre, eds., NATO ASI Series, Plenum Press, (1990)
5. F.H. Busse and R.M. Clever, *Three-dimensional convection in an inclined layer heated from below*, J. Engin. Math. **26**, 1–19 (1992)
6. F.H. Busse and J.A. Whitehead, *Instabilities of convection rolls in a high Prandtl number fluid*, J. Fluid Mech. **47**, 305–320 (1971)
7. A. Chait and S.A. Korpela, *The secondary flow and its stability for natural convection in a tall vertical enclosure*, J. Fluid Mech. **200**, 189–216 (1989)
8. R.M. Clever and F.H. Busse, *Nonlinear oscillatory convection*, J. Fluid Mech. **176**, 403–417 (1987)
9. R.M. Clever and F.H. Busse, *Three-dimensional knot convection in a layer heated from below*, J. Fluid Mech. **198**, 345–363 (1989)
10. R.M. Clever and F.H. Busse, *Nonlinear oscillatory convection in the presence of a vertical magnetic field*, J. Fluid Mech. **201**, 507–523 (1989)

11. R.M. Clever and F.H. Busse, *Convection at very low Prandtl numbers* , Physics of Fluids **A2**, 334–339 (1990)

12. H. Frick, F.H. Busse and R.M. Clever, *Steady three-dimensional convection at high Prandtl number*, J. Fluid Mech. **127**, 141–153 (1983)

13. R. Krishnamurti, *On the transition to turbulent convection, Part 2: The transition to time-dependent flow*, J. Fluid Mech. **42**, 309–320 (1970)

14. P.S. Marcus, *Simulation of Taylor-Couette flow Part 2: Numerical Results for wavy-vortex flow with one travelling wave*, J. Fluid Mech. **146**, 65–113 (1984)

15. M. Nagata, *Three-dimensional finite-amplitude solutions in plane Couette-flow: bifurcation from infinity*, J. Fluid Mech. **217**, 519–527 (1990)

16. M. Nagata and F.H. Busse, *Three-dimensional tertiary motions in a plane shear layer*, J. Fluid Mech. **135**, 1–26 (1983)

17. M. Nagata and F.H. Busse, *Secondary and tertiary bifurcations in a plane parallel shear flow*, pp. 65–71 in "Turbulence and Chaotic Phenomena in Fluids", T. Tatsumi (editor), Elsevier Science Publ. (1984)

18. E. Weisshaar, F.H. Busse and M. Nagata, *Twist vortices and their instabilities in the Taylor-Couette system*, J. Fluid Mech. **226**, 549–564 (1991)

INSTITUTE OF PHYSICS, UNIVERSITY OF BAYREUTH, 8580 BAYREUTH
*E-mail address*: Busse@Uni-bayreuth.de

INSTITUTE OF GEOPHYSICS AND PLANETARY PHYSICS, UNIVERSITY OF CALIFORNIA AT LOS ANGELES, LOS ANGELES, CA 90024

Lectures in Applied Mathematics
Volume **29**, 1993

# On the existence of rotating waves in a steady-state bifurcation problem with $O(3)$ symmetry

## P. CHOSSAT AND E. PROTTE

ABSTRACT. This report refutes a previous claim about the existence of rotating waves in one-parameter steady-state bifurcation problems which are equivariant by the natural irreducible representation of O(3) of dimension 11. A computer-assisted study corrects the value of a hand-computed coefficient in a bifurcation equation which in fact appears to be completely degenerate, as a numerical simulation confirms.

## 1. Introduction

In steady-state bifurcation problems for one-parameter families of O.D.E.'s, it is generic that the linearized vector field at criticality has a simple eigenvalue at 0. Then the only possibility is that a branch of equilibria bifurcates from the basic state and the local dynamics is very easily described, unless there are some higher order degeneracies in the Taylor expansion of the vector field at the bifurcation point. The situation becomes more complicated if the vector field is $G$-equivariant, where $G$ is some compact group acting in the phase space. Then 0 need not be simple, but instead its eigenspace is generically associated to an absolutely irreducible representation $\Gamma$ of $G$, i.e. the only linear operators which commute with $\Gamma$ are scalar multiples of the identity. In this case one can ask whether other solutions than equilibria could bifurcate in the absence of other degeneracies. Examples of more complicated dynamics have been found in [**FS**] for certain classes of subgroups of reflection groups.

In a previous paper Chossat [**C**] considered the case $G = O(3)$, a compact Lie group, and $\Gamma$ being the natural irreducible representation of dimension 11, a basis of which is given by spherical harmonics of order 5 $Y_5^m(\theta, \phi)$, $m = -5, ..., 5...$

---

1991 *Mathematics Subject Classification.* 58F14.
*Key words and phrases.* bifurcation with symmetry, spherical symmetry, rotating waves.
This paper is in final form and no version will be published elsewhere

The difficulty with $G = O(3)$ comes from the fact that $\Gamma$-equivariant polynomial vector fields of degree 3 are always gradients of invariant polynomials, hence do not possess non stationary bounded solutions. Therefore a local bifurcation analysis taking into account only the Taylor expansion of the vector field up to order three does not allow one to find non-stationnary solutions. Moreover it has been proved in [**CLM**] that if all the equilibria which bifurcate for the truncated vector field are $G$-hyperbolic, i.e. have hyperbolic group orbits, then this bifurcation diagram does persist when the rest of the vector field is taken into account. This is precisely what happens when $\Gamma$ is an irreducible representation of $O(3)$ of dimension less than 11 [**CLM**]. What makes the problem interesting in the 11-dimensional case is that there exist at cubic order non-hyperbolic, but generic $G$-orbits of bifurcated equilibria. The aim of the method developed in [**C**] was to analyse the dynamics near the non-hyperbolic branches of equilibria, and to show that adding the rest to the truncated vector field removes the degeneracy by the creation of *relative equilibria*, i.e. flow-invariant group orbits which, instead of consisting of pure equilibria, are foliated by periodic orbits (also called rotating waves). In other words, time acts in such solutions as a one-parameter subgroup of $SO(3)$. Unfortunately, as the paper was being printed, the author realized that an argument was incorrect in the proof, precluding the result. This was indicated in a note added at the end of the paper. A new, computer assisted proof has been performed since then. The use of a computer was made necessary because of the heavy algebraic computations involved in a Lyapunov-Schmidt reduction. A first calculation concluded to the existence of the rotating waves. However, being aware of how easy it is to make a mistake in these calculations, we also made an attempt tocheck directly, by a numerical simulation, the existence of the rotating waves. The difficulty is that fifth order terms have to be introduced in the equations in order to remove the degeneracy, and that means hundreds of degree 5 monomials... Instead of this, we considered the truncated, cubic equations, and tried to check numerically that the $G$-orbits of the non-hyperbolic equilibria are isolated, which is a sufficient condition for the existence of nearby relative equilibria when higher order terms are introduced. It turned out that the numerical simulations did not fit with the analytical prediction... We finally realised that a few terms in the cubic vector field introduced in the program had wrong coefficients. After correction, both calculations agreed that these rotating waves don't exist. In section 2 we recall the general setting of the problem and the method of [**C**]. In section 3 we describe the application of this method and the numerical simulations, and we conclude with a conjecture.

## 2. A description of the problem

We do not provide here a complete description of known results about the stationary bifurcation with $\Gamma$ being from now on the natural irreducible representation of $O(3)$ of dimension 11 ("natural" means that the symmetry through the origin in $O(3)$ acts as $-I$). All relevant information concerning this problem can be found in [**CLM**]. We consider an ODE

(1) $$\dot{x} = F(x, \lambda)$$

where $F : R^n \times R \mapsto R^n$ is smooth and $\Gamma$-equivariant. Since $\Gamma$ is absolutely irreducible, $F(0, \lambda) = 0$ and $D_x F(0, \lambda) = \sigma(\lambda)I$, with $\sigma$ a smooth real function. We assume that: (i) $\sigma(0) = 0$, (ii) $\sigma'(0) \neq 0$. Hence $(x, \lambda) = (0, 0)$ is a bifurcation point from the trivial solution. By a suitable change of variables, we can assume that $\sigma(\lambda) \equiv \lambda$.

The $\Gamma$-equivariance of $F$ has the following strong consequence on the form of this vector field near the origin:

(2) $$F(x, \lambda) = \lambda x + \alpha C(x) - ||x||^2 x + R(x, \lambda)$$

where $\alpha \in R$, $C(x)$ is a homogeneous cubic equivariant, the form of which is given in [**CLM**], and $R$ is a rest containing higher order terms in $(x, \lambda)$. In fact, the Taylor expansion of $F$ with respect to $x$ only contains odd order terms. The coefficient of the cubic radial term has been normalized to $-1$, so as to give supercritical instability of the trivial solution. The study of this bifurcation problem was made in [**CLM**] by considering the truncated equation (2) when $R$ is set equal to 0.

Before we proceed further, let us recall the following basic facts. Given a point $x \in R^{11}$, the isotropy group of $x$ is the subgroup of $G$ defined by $G_x = \{g \in G/gx = x\}$, where $G$ is abusively identified with its action $\Gamma$ in $R^{11}$. The space $Fix(G_x) = \{y \in R^{11}/G_x y = y\}$ is *flow-invariant* for the equation (1). An isotropy type is a conjugacy class of isotropy groups. The classification of isotropy types is the main tool in bifurcation theory with symmetry. The isotropy groups for the irreducible representations of $O(3)$ are known. We use the following notations for subgroups of $O(3)$. Let us denote by $R_m$ the rotation by $\frac{2\pi}{m}$ around e.g. the vertical axis, and by $S$ the reflection through the plane $yOz$ in $R^3$. Then $D_m^z$ is the group generated by $R_m$ and $S$ ($m = 4, 5$) and $Z_2^- = \{id, S\}$. These three groups define isotropy types and clearly $Fix(D_m^z) \subset Fix(Z_2^-)$. Moreover $dim Fix(D_m^z) = 2$ and $dim Fix(Z_2^-) = 6$.

The following facts have been proved for $\Gamma$-equivariant bifurcation problems:

**1.** There generically exist bifurcated branches of equilibria with types $D_4^z$ and $D_5^z$ . There exists a parametrization $(x_m(\epsilon), \epsilon^2)$ of these branches, which satisfies $x_m(\epsilon) \sim \epsilon(\zeta_m + O(\epsilon^2))$ near 0. The form of $\zeta_m$ will be precised in the next section.

**2.** The eigenvalues of $D_x F(x_m(\epsilon), \epsilon^2)$ are listed below:

*a.* $\sigma_0^m \equiv 0$, of multiplicity 3 and eigenspace tangent to the $SO(3)$-orbit of $x_m(\epsilon)$ at this point (degeneracy due to the group invariance of the problem);

*b.* $\sigma_1^m \sim -c_1^m \epsilon^2$ (simple eigenvalue), $c_1^m$ a positive constant for $m = 4, 5$;

*c.* $\sigma_j^m \sim -c_j^m \alpha \epsilon^2$, $j = 2 - 4$ (one simple and two double eigenvalues), $c_j$ positive constants for $m = 4, 5$;

*d.* $\sigma_5^m \sim c_5^m \epsilon^4$ (double eigenvalue), $c_5^m$ coefficients depending on terms $O(||x||^5)$ in $F$ and such that $c_5^4 = -c_5^5$.

It follows that if $\alpha > 0$, then precisely one of the branches of type $D_4^z$ or $D_5^z$ is stable, depending on the sign of the coefficient $c_5^m$. However, for the truncated equation

$$(3) \qquad \dot{x} = \lambda x + \alpha C(x) + \beta ||x||^2 x$$

the eigenvalue $\sigma_5^m$ is 0, hence the equilibria $x_m$ are not $G$-hyperbolic. The idea of [C] was to analyse the situation near the branch $x_m(\epsilon)$ as a bifurcation at $\epsilon = 0$ in the eigendirection of the degenerate eigenvalue $\sigma_5^m$. For this, we first blow-up the singularity by setting: $x = \epsilon u$, $\lambda = \epsilon^2$, $t = s/\epsilon^2$. After dividing by $\epsilon^2$ ($\epsilon \neq 0$), the equation for $u$ reads

$$(4) \qquad \dot{u} = P(u) + \epsilon^2 \tilde{R}(u, \epsilon^2)$$

where $P(u) = u + \alpha C(u) + \beta ||u||^2 u$ and $\tilde{R}(u, \epsilon^2) = \epsilon^{-2} R(\epsilon u, \epsilon^2)$ is smooth. Then $u_m(\epsilon) = \epsilon^{-1} x_m(\epsilon)$ is solution of (4), and $u_m(0) = \zeta_m$ is solution of (4) with $\epsilon = 0$. Moreover the linearized operator in (4) near $u_m(\epsilon)$ has eigenvalues $\tilde{\sigma}_j^m = \epsilon^{-2} \sigma_j^m$, $j = 0, ..., 5$. Notice that $\tilde{\sigma}_5^m(\epsilon) \sim \epsilon^2 c_5^m$, hence $\tilde{\sigma}_5^m(0) = 0$ while the other eigenvalues are off the imaginary axis, except of course $\tilde{\sigma}_0^m = 0$ due to the group invariance. We are therefore led to a bifurcation problem for eq. (4), from the branch $u_m(\epsilon)$ at $\epsilon = 0$.

This bifurcation problem has two important features: (i) it is a bifurcation from a *group orbit* of equilibria (with the consequence that an eigenvalue $\tilde{\sigma}_0^m$ is forced to be 0); (ii) since $u_m$ has isotropy $D_m^z$, the bifurcation equations are $D_m^z$-invariant. We can in principle study this bifurcation by the following method (about this methodology see [K]). Let $J_1, J_2, J_3$ be a basis of the Lie algebra of $\Gamma$ and define $\theta_k = J_k u_m(0)$, which span the null space associated with $\tilde{\sigma}_0^m$. We set

$$(5) \qquad u(s) = e^{(\omega_1 J_1 + \omega_2 J_2 + \omega_3 J_3)s}[u_m(\epsilon) + w]$$

where $\omega_j$ are real and $w \perp \theta_j$, $j = 1, 2, 3$. If some $\omega_j$ are not zero, then the resulting solution is periodic in time, assuming the form of a rotating wave. This is because one-parameter connected subgroups of $O(3)$ are isomorphic to $SO(2)$. Let $\Pi$ be the orthogonal projection onto the space spanned by the $\theta_j$'s. If the $\theta_k$'s are normalized, then

$$\Pi u = \sum_{i=1}^{3} < u, \theta_i > \theta_i.$$

By replacing $u$ by (5) in (4) and decomposing along $\Pi$ and $1 - \Pi$, eq. (4) is transformed into the two systems

$$(6) \quad \omega_k + \sum_{j=1}^{3} \omega_j < J_j w, \theta_k^* > = < P(u_m + w) + \epsilon^2 \tilde{R}(u_m + w, \epsilon^2), \theta_k^* > \quad (k = 1, 2, 3)$$

$$(7) \qquad 0 = (1 - \Pi)[P(u_m + w) + \epsilon^2 \tilde{R}(u_m + w, \epsilon^2) - \sum_{i=1}^{3} \omega_i J_i w]$$

The system (6) is readily solved for $\omega_k$'s by the implicit function theorem. Then (7) is an equation for $w$ with parameter $\epsilon$, and moreover the linearized operator on the r.h.s. at $\epsilon = 0$ has only a double 0 eigenvalue. A Lyapunov-Schmidt decomposition on (7) will provide the bifurcation equations (in $R^2$). Hence one is led to a standard steady-state bifurcation problem with $D_m$ symmetry (indeed the 2-d actions od $D_m^z$ and $D_m$ are equivalent). By identifying the plane with $C$, these equations have the general form (see [GSS])

$$(8) \qquad 0 = g(|z|^2, Re(z^m), \epsilon)z + h(|z|^2, Re(z^m), \epsilon)\bar{z}^{m-1} \quad (m = 4, 5)$$

where $g, h$ are rel functions. What remains to do is to check that these equations are not degenerate and that corresponding solutions of (6)-(7) have non-zero frequency. This will be the topic of the next section.

## 3. Computations

The basic remark here is that the computations can be made in the 6-d subspace $V = Fix(Z_2^-)$, which reduces substantially the difficulty. Indeed, not only $Fix(D_m^z) \subset V$, but also the eigenspace associated with the degenerate eigenvalue $\sigma_m^5$ intersects $Fix(Z_2^-)$ one-dimensionaly. Restricting to $V$ does in fact correspond to restrict to one symmetry axis for the 2-d $D_m$ -equivariant bifurcation equation (8). The general form of the bifurcation equation is

$$(9) \qquad 0 = x(a\epsilon^2 + cx^2 + O[(\epsilon^2 + x^2)^2])$$

We already know that $a = c_m^5$ (leading part of the degenerate eigenvalue $\sigma_m^5$), as a standard consequence of the Lyapunov-Schmidt reduction. Therefore it is enough to compute $c$ in order to finish the proof, i.e. to consider the system (6)-(7) with $\epsilon = 0$.

Notice that $N(Z_2^-) = O(2) \times \{I, -I\}$. With the definition of $Z_2^-$ given in section 2, the $SO(2)$ part is the group of rotations around the axis $Oy$ in $R^3$. Then

$$\frac{N(Z_2^-)}{Z_2^-} \simeq O(2)$$

which implies that we can look for the solutions in $V$ in the form $u(s) = e^{\omega J s}[u_m(\epsilon) + w]$, and generically $\omega \neq 0$ [K]. Now $J$ is the infinitesimal generator of the action of $SO(2)$ in $V$, $w \perp \theta = \frac{J\zeta_m}{\|J\zeta_m\|}$ and $\Pi u = < u, \theta > \theta$. The cubic vector field $C$ restricted to $V$ is easily deduced from the expressions given in [**CLM**] (table B3) for $C$ in $R^{11}$. However, in using this reference, be aware of the fact that, in the component along $Y_5^0$, coefficients of the monomials which are not symmetric under complex conjugation have to be multiplied by 2 because the conjugate terms were not listed.

We now enter into the computational part of the proof.

**Computation of the coefficient** $c$. In terms of orthonormal spherical harmonics $Y_5^k$, $k = -5, ..., 5$, the elements of $V$ have the form

$$\sum_{k=0}^{5} x_k (Y_5^k + (-1)^k Y_5^{-k})$$

where $x_k \in R$. The branches of solutions of (4) with isotropy $D_m^z$, $m = 4, 5$, are defined at the leading order by $\zeta_m$, which is given by the following relations (only non-zero coordinates are indicated):

*case* $m = 4$. $x_4^2 = 5/6 x_0^2$, $\epsilon^2 = 8/3 (75\alpha - \beta) x_0^2$,

*case* $m = 5$. $x_5^2 = 1/3 x_0^2$, $\epsilon^2 = 5/3 (75\alpha - \beta) x_0^2$.

The subgroup $SO(2)$ acts in $V$ as the rotations around the axis $Oy$ in $R^3$. Its infinitesimal generator acts on spherical harmonics as

$$JY_l^k = -\frac{i}{2}(c_k Y_l^{k+1} + c_{-k} Y_l^{k-1}), \quad c_k = \sqrt{(l-k)(l+k+1)}.$$

From this we compute $\theta$. Let $\xi$ be an eigenvector of $DP(\zeta_m)$ in $V$. After calculation we get :

*case* $m = 4$. $\theta = \frac{1}{28}(0, \sqrt{18}, 0, \sqrt{9}, 0, 1)$, $\xi = (0, \sqrt{10}, 0, \sqrt{5}, 0, -3)$

*case* $m = 5$. $\theta = \frac{1}{10}(-1, 0, 0, 0, 3, 0)$, $\xi = (0, 0, 1, -\sqrt{14}, 0, 0)$.

We are now ready to perform a Lyapunov-Schmidt decomposition on $(7)_V$ [**IJ**]. We define a projection on the eigenspace spanned by $\xi$:

$$Qw = <w, \xi^*> \xi$$

where $\xi^*$ is the vector in $ker|_V(DP(\zeta)^*)$ such that $< \xi, \xi^* >= 1$ and $< \theta, \xi^* >= 0$. Then we set

$$w = x\xi + y, \quad y = (I - Q)w$$

and decompose $(7)_V$ by applying $Q$ and $I - Q$. The resulting equations are (with $\epsilon = 0$)

$$0 = < (1 - \Pi)[P(u_m + x\xi + y) + \epsilon^2 \tilde{R}(u_m + x\xi + y, \epsilon^2) - \omega J(x\xi + y)], \xi^* >$$

$$0 = (I - Q)(1 - \Pi)[P(u_m + x\xi + y) + \epsilon^2 \tilde{R}(u_m + x\xi + y, \epsilon^2) - \omega J(x\xi + y)].$$

The second equation can be rewritten

$$(11) \qquad\qquad (I - Q)(1 - \Pi)DP(\zeta).y = N(x, y, \epsilon^2)$$

where $N(0,0,0) = D_y N(0,0,0) = 0$. We know from classical theory that (11) is solved for $y$ by the implicit function theorem. Hence we get a unique and smooth $y(x,\epsilon)$, and replacing $y$ by this expression in the first equation of the decomposition we get the bifurcation equation (9). Before we proceed to the computations, notice that $\omega = 0$ when $\epsilon = 0$. Therefore, since we are just interested in computing $c$ in (9), we can also set $\omega = 0$, which simplifies the expressions. The computation consists in expending $y(x,0) = x\xi + x^2 y^{(2)} + ...$ and identifying terms of same order in the equations. A few algebra shows that

$$c = < (I - \Pi)[1/6 D^3 P(\zeta_m).\xi.\xi.\xi + 1/2 D^2 P(\zeta_m).\xi.y^{(2)}], \xi^* >$$

and

$$DP(\zeta_m).y^{(2)} = 1/2(I - Q)(1 - \Pi)D^2 P(\zeta_m).\xi.\xi.$$

It remains to compute $D^i P(\zeta_m)$, $i = 1,2,3$, then to solve the equation for $y^{(2)}$, and finally to compute $c$. Details can be found in [**P**]. The intermediate expressions are very heavy but the result is simple: in both cases, $c = 0$. Hence (9) is degenerate at order $O(\|x\|^3)$.

**A numerical simulation.** We have tried to simulate the dynamics near the $D_m^z$ equilibria by numerical integration of the vector field in $V$, using the software package $DSTOOL$ of Kim and Guckenheimer. However it turned out to be very difficult, because terms of order at least 5 are needed, and there are hundreds of these to introduce in the code. Therefore we have proceeded indirectly, by exploiting the following remark. Consider the truncated, cubic system in $V$

(12) $$\dot{u} = P(u)$$

and its group orbit of equilibria $\Gamma\zeta_m$. We know that there exists a center manifold to $\Gamma\zeta_m$ of dimension 2 and tangent to $\Gamma\xi$ on the orbit. If the local dynamics on this center manifold is not degenerate, i.e. if it does not consist of a manifold of equilibria, then $\Gamma\xi$ is either attracting or repelling on this center manifold. In either case, adding higher order terms to the vector field may locally change the attractor to a repellor or vice-versa by removing the degeneracy, which implies the existence of a bounded, bifurcated solution nearby. In order to check dynamically that $\Gamma\zeta_m$ is an attractor or repellor, we need to identify the group orbit. For this we introduce an "artificial" dynamics on $\Gamma\zeta_m$ by looking for solutions of

(13) $$\dot{v} = P(v) + \eta J v$$

where $\eta \in R$ is fixed. Solutions of this equation have the form $e^{\eta J t} u(t)$, where $u(t)$ is solution of (12). In particular $\Gamma\zeta_m$ is now a relative equilibrium. Taking initial conditions on the cross section $\{\zeta_m + \mu\xi, \ \mu \in R\}$ near $\Gamma\zeta_m$, we have observed the presence of relative equilibria for (13) with isotropy $Z_2^-$ as near as we wish from $\Gamma\zeta_m$. This indicates that $\Gamma\zeta_m$ is not isolated as an orbit of equilibria of (12), therefore that no rotating wave of (4) can bifurcate from $\Gamma\zeta_m$ as $\epsilon \neq 0$.

**Discussion.** Both the analytical resolution of (4) and the numerical simulation indicate that the presumed rotating waves do not exist. We conjecture that (12) has a manifold of equilibria connecting $\Gamma\zeta_4$ and $\Gamma\zeta_5$. For eq. (4) (and hence eq. (1)), this means a heteroclinic connection between the orbits of equilibria of types $D_4^z$ and $D_5^z$. This idea is supported by the fact that the degenerate eigenvalues $\sigma_5^m$ have opposite sign for both types of equilibria. If it were not the case, these solutions would be connected to something else with isotropy $Z_2^-$, i.e. to rotating waves.

**Acknoledgments.** P. Chossat is grateful to R. Lauterbach and I. Melbourne for their helpful comments.

## REFERENCES

[C]    P. Chossat, *Branching of rotating waves in a one-parameter problem of steady-state bifurcation with spherical symmetry*, Nonlinearity **4**, 1123-1129.

[CLM]  P. Chossat, R. Lauterbach, I. Melbourne, *Steady-State bifurcation with O(3) symmetry*, Arch. Rat. Mech. Anal. **113** (1990), 313-376.

[FS]   M. Field, J. Swift, *Stationary bifurcation to limit cycles and heteroclinic cycles*, Nonlinearity **4** (1991), 1001-1043.

[GSS]  M. Golubitsky, I. Stewart, D. Schaeffer, *Singularities and groups in Bifurcation Theory*, vol.2, Appl. Math. Sci. series 69, Springer Verlag, New-York, 1988.

[IJ]   G. Iooss, D. Joseph, *Elementary stability and bifurcation theory*, Undergraduate texts in mathematics, Springer Verlag, New-York, 1980.

[K]    M. Krupa, *Bifurcation of relative equilibria*, SIAM J. Math. Anal. **21** (1990), 1453-1486.

[P]    E. Protte, *Bifurcation d'ondes rotatives dans un problème de codimension un avec symétrie sphèrique*, Rapport de stage de DEA, Université de Nice (1992).

*Current address*: Institut Nonlinéaire de Nice, UMR CNRS 129, Parc Valrose, F-06034 Nice

*E-mail address*: chossat@ecu.unice.fr

Lectures in Applied Mathematics
Volume **29**, 1993

# Dynamics of Waves in Extended Systems

## G. DANGELMAYR, J.D. RODRIGUEZ, AND W. GÜTTINGER

ABSTRACT. Oscillatory instabilities in spatially extended systems are governed by coupled systems of Ginzburg-Landau equations. The presence of sidewalls breaks translational invariance and requires Robin boundary conditions for the Ginzburg-Landau system. Assuming weak symmetry breaking, we derive a Galerkin approximation based on the eigenfunctions of the weakly perturbed problem. We study the bifurcation sequence of waves and show that the appearance of confined traveling waves and "blinking" states occurs as predicted by previous analytical work. Moreover, these local analytical results are shown to be globally valid for real nonlinear coefficients. For complex coefficients the bifurcation sequence of the confined traveling wave state leads to quasi-periodic and chaotic dynamics as successive eigenmodes become unstable.

## 1. Introduction

In spatially extended convective fluid systems the invariance of the governing physical equations under reflection and translation define an action of the group $O(2)$. Translation invariance is lost, however, when the domain of the original physical system is of finite extent. The loss of translation invariance implies the breaking of the $SO(2)$ subgroup of $O(2)$ leaving only the reflective subgroup $Z(2)$. This loss of rotational symmetry has profound consequences on the spatio-temporal behavior of the convective system.

Oscillatory instabilities of uniform states in extended systems typically result in traveling waves[6]. When translation invariance is broken traveling waves cannot occur. Instead, the dynamical behavior in a finite geometry is dominated by spatially confined waves of various types including confined traveling waves and the "blinking" states. Confined traveling waves are characterized by the motion of a roll pattern through the fluid which is in turn generated and absorbed at the wall boundaries. The blinking wave states consist of a class of

1991 *Mathematics Subject Classification*. Primary 76E15, 35B32; Secondary 35A40, 58G28.
JDR acknowledges with thanks the support of the Alexander von Humboldt Stiftung
This paper is in final form and no version of it will be submitted for publication elsewhere

confined wave motions of particular interest which are characterized by periodic oscillations between right and left confined traveling waves. These states have been observed in convection experiments with binary fluids in boxes with large aspect ratio [7],[9]. Moreover, the roll patterns observed by Kolodner et. al. [9] in alcohol-water mixtures suggest that they may be interpreted as superimposed left and right traveling waves at the center of the box.

The bifurcation of wave patterns from a basic state is governed by amplitude equations which describe the spatio-temporal evolution of nonlinear waveforms. In the generic case these equations appear as two coupled Ginzburg-Landau equations which are derived from the basic hydrodynamic equations as amplitude equations for slowly varying envelopes of weakly nonlinear traveling wave patterns [2]. When the spatial extent of the system is infinite, these equations possess solutions corresponding to pure right and pure left traveling waves. The existence of such "pure" wave states is a consequence of the phase shift symmetry $SO(2)$ of the Ginzburg-Landau equations which in turn is induced by the translation invariance of the original equations. If the translation invariance is broken, which is always the case if side walls are present, the solutions of the coupled Ginzburg-Landau equations can be interpreted as superpositions of right and left traveling waves leading to the confined states described above.

A natural approach to investigating finite domains is to consider the case of "distant" sidewalls. In this case the symmetry breaking is weak and the dynamics may be investigated more readily and compared to the ideal geometry. For the weakly perturbed case the problem may be approached analytically using appropriate small amplitude perturbation expansions. This method is particularly facilitated by the development of a general theory of the Hopf bifurcation under broken $O(2)$ symmetry [3]. This approach was used in [4] to study spatio-temporal wave states which are in qualitative agreement with those observed in experiments [7]. In the relevant Ginzburg-Landau equation the symmetry breaking is reflected in the appearance of general Robin boundary conditions[1].

The goal of this paper is to obtain a global bifurcation picture of the dynamical states occurring in the Ginzburg-Landau equations. Our main results are that the local bifurcation scenarios derived in [4] are also globally valid if the coefficients of the nonlinear terms of the Ginzburg-Landau system are real. However, we find that complicated dynamical states are created after the primary bifurcation when the coefficients are complex. In §2 we review the results of local perturbative analysis for symmetry breaking Hopf bifurcation in $O(2)$ symmetric systems. In §3 we present global numerical results for the perfect symmetric problem and for weak symmetry breaking perturbations. For the perturbed problem we derive a reduced system of O.D.E.s via Galerkin projection with an ansatz based on a perturbation of the eigenfunctions of the perfectly symmetric system. We show that the predictions of the local analysis are qualitatively well reproduced in the full system near the primary bifurcation. For larger values of the bifurcation parameter we investigate the fate of confined traveling wave states as additional eigenmodes become unstable and detail the transition to chaotic dynamics. In §4 we briefly conclude and summarize our results.

## 2. Local behavior

We consider a system of two coupled Ginzburg-Landau equations governing the spatio-temporal evolution of traveling wave amplitudes,

$$(2.1) \qquad A_t = A_{xx} + 2sA_x + \Lambda A + a|B|^2 A + b(|A|^2 + |B|^2)A$$

$$(2.2) \qquad B_t = B_{xx} - 2sB_x + \Lambda B + a|A|^2 B + b(|A|^2 + |B|^2)B,$$

with the boundary conditions,

$$(2.3) \quad x = -L: \qquad A + \epsilon(\bar{\mu} A_x + \bar{\nu} B_x) = 0, \qquad B + \epsilon(\mu B_x + \nu A_x) = 0$$

$$(2.4) \quad x = L: \qquad A - \epsilon(\mu A_x + \nu B_x) = 0, \qquad B - \epsilon(\bar{\mu} B_x + \bar{\nu} A_x) = 0,$$

where $\epsilon$ is a small parameter and complex conjugation is denoted by an overbar. This P.D.E. system can be derived [1] from the original fluid dynamical equations in the vicinity of an oscillatory instability. The amplitudes $A$ and $B$ represent complex envelopes of left and right traveling waves which vary on the scales $t = \epsilon^2 T$ and $x = \epsilon X$ of the original, physical time $T$ and the horizontal coordinate $X$ with the horizontal side of the box given by $-L/\epsilon \leq X \leq L/\epsilon$, i.e., the idealized case of an infinitely extended system corresponds to the limit $\epsilon \to 0$.

The stream function $\Phi(T, X, Z)$ associated to a solution $((A(t, x), B(t, x))$ of (2.1)-(2.4) is of the form,

$$(2.5)$$
$$\Phi(T, X, Z) = \epsilon\big(A(t, x)e^{i(\omega_c T + k_c X)} + B(t, x)e^{i(\omega_c T - k_c X)} + c.c.\big)\Psi(Z) + O(\epsilon^2),$$

where $\Psi(Z)$ describes variations in the bounded, vertical direction and $\omega_c$ and $k_c$ are the critical frequency and wave number, respectively. Because $L$ is considered as a quantity of $O(1)$ we can assume, by suitable rescalings, that $L = 1$. The parameter $\Lambda$ is related to the Rayleigh number $R$ by $R - R_c = \epsilon^2\Lambda$ with $R_c$ being the critical Rayleigh number at which the original fluid dynamical equations encounter the oscillatory instability in the infinitely extended limit. In our analysis $\Lambda$ is chosen as the basic, global ($\Lambda = O(1)$) bifurcation parameter. The values of the complex coefficients $\mu, \nu$ and $a$ are selected following [1]: $\mu = 3.053e^{2.725i}, \nu = -4.254e^{.6346i}$ and $a = -3$. The wave speed $s \in \mathbf{R}$ can be arbitrary but is assumed to be small in the numerical calculations: $s = 0.1$. The nonlinear coefficient $b$ is allowed to vary and we present results in §3 for both real ($b = -1$) and complex ($b = -1 + 4i$) cases.

For $\epsilon = 0$ the equations (2.1)-(2.4) are invariant under the symmetry operations,

$$(2.6) \qquad\qquad S^1 : (A, B) \to (e^{i\varphi} A, e^{i\varphi} B),$$

(2.7)                    $$Z(2) : (A, B, x) \to (B, A, -x),$$

(2.8)                    $$SO(2) : (A, B) \to (e^{i\varphi} A, e^{-i\varphi} B).$$

These symmetries correspond, respectively, to temporal translation $T \to T + \tau$, spatial reflection $X \to -X$, and spatial translation $X \to X + \xi$ of the wave pattern (2.5). For $\epsilon \neq 0$ the $SO(2)$ invariance is broken by virtue of the non-trivial boundary conditions (2.3),(2.4) which in turn follow from the presence of sidewalls, *i.e.*, from the broken translational invariance of the original fluid dynamical equations.

If $\epsilon = 0$, a bifurcation from the trivial solution occurs at

(2.9)              $$\Lambda = \Lambda_n \equiv s^2 + (n\pi/2)^2 \qquad (n = 1, 2, \dots).$$

The center eigenspace is four-dimensional and spanned over $\mathbf{C}$ by $(e^{-sx} u_n(x), 0)$ and $(0, e^{sx} u_n(x))$ where $u_n(x) = \sin(n\pi(x - 1)/2)$. When $\Lambda$ is close to $\Lambda_n$ and $\epsilon$ is small the system (2.1)-(2.4) can be reduced to a system of O.D.E.s for the center manifold variables $(v, w) \in \mathbf{C}^2$ [4],[5],

(2.10)       $$\dot{v} = \left(\lambda + K_n a|w|^2 + K_n b(|v|^2 + |w|^2)\right)v + \epsilon d_n w + \dots,$$

(2.11)       $$\dot{w} = \left(\lambda + K_n a|v|^2 + K_n b(|v|^2 + |w|^2)\right)w + \epsilon d_n v + \dots,$$

where

$$K_n = \frac{3n^4\pi^4 \sinh(2s)}{8s(4s^2 + n^2\pi^2)(s^2 + n^2\pi^2)}, \qquad d_n = -\left(\frac{n\pi}{2}\right)^2 \left(\nu e^{2s} + \bar{\nu} e^{-2s}\right)$$

and $\lambda = \Lambda - \Lambda_n - \left(\frac{n\pi}{2}\right)^2 (\mu + \bar{\mu})\epsilon$ is the local bifurcation parameter. The O.D.E. system (2.10), (2.11) stays also invariant under the symmetry operations $S^1$ and $Z(2)$, where $S^1$ acts as before and $Z(2)$ acts according to $(v, w) \to (w, v)$. Moreover, in the limit $\epsilon \to 0$ of the infinitely extended case they possess in addition to $S^1$ and $Z(2)$ the translation invariance $SO(2)$ in the same way as the underlying coupled Ginzburg-Landau equations. Thus (2.10),(2.11) coincide in this limit with the normal form for a Hopf bifurcation with $O(2)$ symmetry [8] $\left(O(2) = SO(2) \times Z(2)\right)$ in a rotating frame of reference.

In the "perfect" case $\epsilon = 0$ the equations (2.10), (2.11) possess two types of stationary solutions which we shall refer to as standing waves $SW, (|v| = |w|)$, and traveling waves $TW$ which are either left-moving ($v \neq 0, w = 0$) or right-moving ($v = 0, w \neq 0$) [8]. The bifurcation of these wave patterns from the basic state is shown in Figure 1a for the choice of parameters $a = -3$, $Re(b) = -1$ and $n = 1$.

This figure shows that the first bifurcation at $\Lambda = \Lambda_1$ leads to stable traveling and unstable standing waves. In the "perturbed" case $\epsilon \neq 0$ the O.D.E.s (2.10), (2.11) take the form of a Hopf bifurcation normal form with broken circular symmetry which is analyzed in detail in [3]. Because the system is no longer $SO(2)$

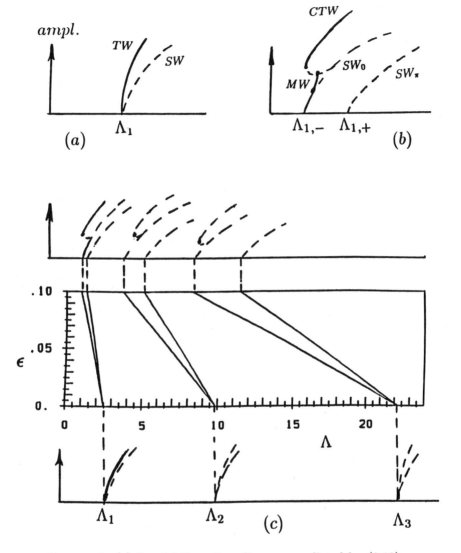

FIGURE 1. (a) Local bifurcation diagram predicted by (2.10), (2.11) for $\epsilon = 0$ near $\Lambda = \Lambda_1$. Solid (dashed) lines correspond to stable (unstable) branches. (b) Same as (a) for $\epsilon \neq 0$. (c) Splitting of the bifurcation values $\Lambda_n$ into $\Lambda_{n\pm}(\epsilon)$ according to (2.12) and the local bifurcation branches predicted by (2.10), (2.11) for $n = 1, 2, 3$.

invariant pure traveling waves cannot exist. Moreover, the four-fold degeneracy for $\Lambda = \Lambda_n$ splits into two bifurcations at

$$(2.12) \qquad \Lambda = \Lambda_{n\pm} \equiv \Lambda_n + \left(\frac{n\pi}{2}\right)^2 \left(\mu + \bar{\mu} \pm (e^{2s}\nu + e^{-2s}\bar{\nu})\right)\epsilon + O(\epsilon^2),$$

from which two types of standing waves, denoted $SW_0$ and $SW_\pi$ in [3], bifurcate in succession from the basic state. The imperfect bifurcation diagram to which Figure 1a evolves when $\epsilon > 0$ is shown in Figure 1b for the selection of the values of $\mu, \nu$ chosen in this paper. The branch denoted $CTW$ in this figure approaches the $TW$ solution of Figure 1a in the limit $\epsilon \to 0$ and corresponds to confined traveling wave envelopes of the Ginzburg-Landau equations. The additional bounded branch $MW$ disappears in the limit $\epsilon \to 0$. It corresponds to a modulated (two-frequency) wave state and is referred to as the "blinking" state. Further details on analytical results concerning the local bifurcation near $\Lambda = \Lambda_1$, including an investigation of the phenomena that occur when the $S^1$ symmetry is broken as well can be found in [4] and [5].

For $n > 1$ the center manifold reduced system (2.10), (2.11) has the same form as for $n = 1$. This means that also near the bifurcation values $\Lambda_n$ with $n > 1$ the bifurcation diagram of Figure 1a and Figure 1b occur, however, all branches predicted by the local analysis are unstable because at $(\Lambda, \epsilon) = (\Lambda_n, 0)$ the dimension of the unstable eigenspace is $4(n-1)$. In Figure 1c we show the splitting of the eigenvalues and the corresponding local bifurcating diagrams for $n = 1, 2, 3$. Although the bifurcating branches near $\Lambda_{n>1}$ are unstable, their existence should have a profound influence on the global attractors of (2.1)-(2.4) when $\Lambda$ increases. In the next section we describe results of a numerical analysis of the coupled Ginzburg-Landau equations for larger values of the bifurcation parameter $\Lambda$.

## 3. Global behavior

We seek to determine the global behavior of the coupled Ginzburg-Landau system as $\Lambda$ increases beyond the instability of successive eigenvalues of the linearized problem. As each additional eigenvalue becomes unstable the dimension of the center-unstable manifold increases by four and analytical techniques become impractical. To proceed further we must employ numerical methods. The problem divides naturally into the perfect problem for $\epsilon = 0$ and the perturbed problem for $\epsilon > 0$. A summary of our results is shown in Figure 2. for $\epsilon = 0$ and $\epsilon = 0.05$. The results for periodic states were obtained by continuation. Quasi-periodic and chaotic states were determined by evolving specific initial conditions as a function of $\Lambda$ at intervals of $\Lambda = 0.1$. Near bifurcation points this sampling interval was decreased to a minimum of $\Lambda = 0.001$.

An inspection of Figure 2 will reveal that the qualitative behavior is highly complex with regions of multiple stability. In the following sections we describe the results as displayed in Figure 2 in greater detail. We first present results

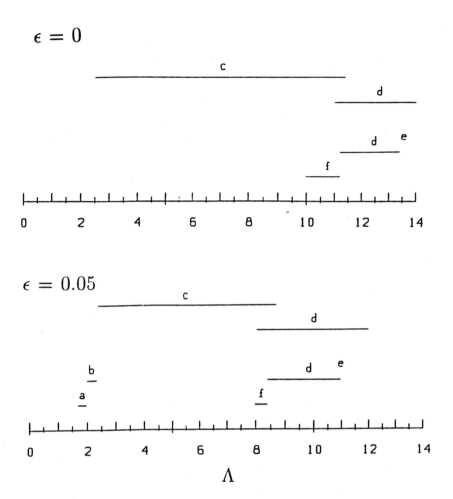

FIGURE 2. Global attracting states observed numerically in different $\Lambda$ regimes for $\epsilon = 0$ and $\epsilon = 0.05$. a: standing wave, b: two-torus (blinking state), c: single-mode confined traveling wave, d: two torus, e: period doubling accumulation point, f: two-mode confined traveling wave.

for the bifurcation of standing and confined traveling waves for $\epsilon = 0$. We next discuss the perturbed system again paying particular attention to confined wave states. We finally discuss the transition to chaos, these results apply to both perfect and perturbed systems.

**3.1. Standing and traveling wave states in the perfect system.** For an infinite domain the fluid dynamical equations are translationally invariant implying $\epsilon = 0$ in the boundary conditions (2.3),(2.4). The coupled Ginzburg-Landau equations are then invariant under $O(2) \times S^1$ symmetry and the boundary conditions are of Dirichlet type. The decoupling at the boundaries implies that in the linearized system the amplitude of one wave may be set to zero. For this case, we have employed a pseudo-spectral Fourier expansion [10] to obtain numerical results. Formally, we expand;

$$(3.1) \qquad A(x,t) = e^{-sx} \sum_{n=1}^{N} a_n(t) \sin\left(\frac{n\pi(x-1)}{2}\right)$$

$$(3.2) \qquad B(x,t) = e^{sx} \sum_{n=1}^{N} b_n(t) \sin\left(\frac{n\pi(x-1)}{2}\right)$$

The bifurcation structure of traveling and standing wave states which results from numerical calculations based on (3.1), (3.2) ($N = 9$) is shown in Figure 3. We obtained the results shown in Figure 3 by first exploiting the $S^1$ invariance of the system (2.1)-(2.4) to reduce the bifurcation to an algebraic system via the ansatz, $A(x,t) = e^{i\omega t}f(x), B(x,t) = e^{i\omega t}g(x)$. The numerical calculations then proceeded by continuation of solutions of the now steady-state system.

Both traveling and standing waves bifurcate directly from the trivial solution, with an additional pair of standing and traveling waves appearing as the bifurcation parameter $\Lambda$ exceeds each bifurcation value ($\Lambda_n$). Near the instability of the first eigenvalue, the qualitative behavior coincides with that predicted by the local bifurcation analysis.

To obtain global results we increased $\Lambda$ beyond the stability threshold of succeeding eigenmodes. For real $b$ we find the surprising result that the traveling wave state is stable for large regions of the parameter space. The frequency of the traveling wave decreases as an inverse power of $\Lambda$ and is $O(10^{-4})$ at $\Lambda = 100$. This result holds for the pure traveling wave $TW$ in the perfect case as well as for the confined traveling wave $CTW$ in the perturbed case which is described in the next subsection.

For complex $b$ the system displays a rich variety of dynamical behavior. In our initial investigation, we have extensively studied the bifurcation of the unperturbed problem ($\epsilon = 0$) for $b = -1 + 4i$, i.e., the real part of $b$ is fixed to preserve the qualitative bifurcation structure near the primary instability $\Lambda = \Lambda_1$ as shown in Figure 1a.

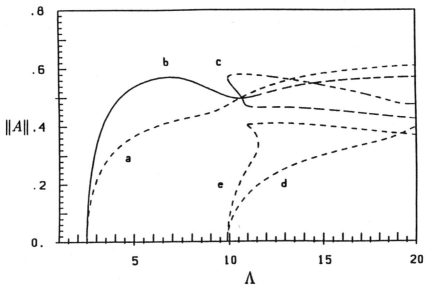

FIGURE 3. Global bifurcation diagram $\|A, B\| \equiv \left(\sum_i |p_i|^2 + |q_i|^2\right)^{1/2}$ vs. $\Lambda$ for $\epsilon = 0.05$ and $b = -1$. For the quasi-periodic $MW$ we plot the time average of $\|A, B\|$ vs. $\Lambda$.

From the global bifurcation diagram of Figure 3 we infer that the primary traveling wave branch ($b$) loses stability at $\Lambda = 11.4$. By a local analysis we can identify this loss of stability with a subcritical Hopf bifurcation. In addition to the primary bifurcation of traveling and standing waves we also find the appearance of two-mode traveling waves, ($c$), so called because the first two eigenmodes are of comparable magnitude. This state appears from a saddle node bifurcation at $\Lambda = 9.95$ and eventually loses stability through a Hopf bifurcation to an invariant torus at $\Lambda = 11.2$.

The second primary traveling wave state, ($e$), bifurcates at $\Lambda = \Lambda_2$ and has two saddle node points. Bistabilty of tori can occur when a second stable torus is generated in a bifurcation from the second traveling wave at the upper saddle node point at $\Lambda = 10.98$. In addition, the first primary traveling wave, ($a$), is also stable for values of $\Lambda$ which overlap the range where stable tori exist. There are then up to three attractors at ranges of $\Lambda$ between the instability of the second and third eigenvalues. Beyond $\Lambda = 11.2$ no stable pure wave states exist and the dynamics become more complex. Since for larger values of $\Lambda$ the qualitative behaviors for $\epsilon = 0$ and $\epsilon = 0.05$ are similar, we confine the discussion of the more complicated dynamical phenomena to the perturbed case.

**3.2. Standing and traveling wave states in the perturbed system.** In the case of a finite system with distant sidewalls, we have the full Robin boundary conditions (2.3), (2.4) with $0 < \epsilon \ll 1$. For the numerical calculations we derive a modal system with a diagonal linear part based on the eigenfunctions of the linearized operator of the r.h.s. of (2.1),(2.2) with Robin boundary conditions. Since $\epsilon$ is small these are calculated using a perturbation analysis. Substituting the ansatz

$$(3.3) \qquad A(x) = e^{-sx}(a_+ e^{\theta x} + a_- e^{-\theta x})$$

$$(3.4) \qquad B(x) = e^{sx}(b_+ e^{\theta x} + b_- e^{-\theta x})$$

into the eigenvalue equation and expanding $\theta = \theta_0 + \epsilon \theta_1 + O(\epsilon^2)$ yields at leading order $\theta_0 = \frac{in\pi}{2}$ for $n = 1, 2, \ldots$. The perturbed eigenfunctions are then calculated by requiring that the boundary conditions are satisfied to $O(\epsilon)$. The first order perturbation $\theta_1$ is obtained using a solvability condition for $a_-$ and $b_-$,

$$(3.5) \qquad \theta_1 = \frac{in\pi}{4}(\mu + \bar{\mu} \pm (e^{2s}\nu + e^{-2s}\bar{\nu})),$$

leading to the eigenvalues $\lambda_{n\pm} = \Lambda - \Lambda_{n\pm}(\epsilon)$. By solving the linear equations for $(a_\pm, b_\pm)$ to order $\epsilon$ we finally obtain the desired eigenfunctions as

$$(3.6) \quad A_{n\pm} = e^{-sx} \cos\left(\frac{n\pi x}{2}\right)$$
$$+ \frac{\epsilon n\pi}{4} e^{-sx}\left((\bar{\mu} \pm e^{-2s}\bar{\nu})(1-x) - (\mu \pm e^{2s}\nu)(1+x)\right)\sin\left(\frac{n\pi x}{2}\right)$$

$$(3.7) \quad B_{n\pm} = \pm e^{sx} \cos\left(\frac{n\pi x}{2}\right)$$
$$\pm \frac{\epsilon n\pi}{4} e^{sx}\left((\mu \pm e^{2s}\nu)(1-x) - (\bar{\mu} \pm e^{-2s}\bar{\nu})(1+x)\right)\sin\left(\frac{n\pi x}{2}\right)$$

for $n$ odd, and,

$$(3.8) \quad A_{n\pm} = e^{-sx} \sin\left(\frac{n\pi x}{2}\right)$$
$$+ \frac{\epsilon n\pi}{4} e^{-sx}\left((\mu \pm e^{2s}\nu)(1+x) - (\bar{\mu} \pm e^{-2s}\bar{\nu})(1-x)\right)\cos\left(\frac{n\pi x}{2}\right)$$

$$(3.9) \quad B_{n\pm} = \pm e^{sx} \sin\left(\frac{n\pi x}{2}\right)$$
$$\pm \frac{\epsilon n\pi}{4} e^{sx}\left((\bar{\mu} \pm e^{-2s}\bar{\nu})(1-x) - (\mu \pm e^{2s}\bar{\nu})(1-x)\right)\cos\left(\frac{n\pi x}{2}\right)$$

for $n$ even. Similar expressions are also obtained for the adjoint eigenfunctions.

We now make an expansion in terms of the $A_{n\pm}(x), B_{n\pm}(x)$,

$$(3.10) \qquad A(x,t) = \sum_n p_n(t) A_{n+}(x) + q_n(t) A_{n-}(x)$$

$$(3.11) \qquad B(x,t) = \sum_n p_n(t) B_{n+}(x) + q_n(t) B_{n-}(x),$$

where the $p_n, q_n$ are complex variables which may be interpreted as amplitudes of standing waves. In contrast, the $a_n, b_n$ introduced in the expansion (3.1), (3.2)

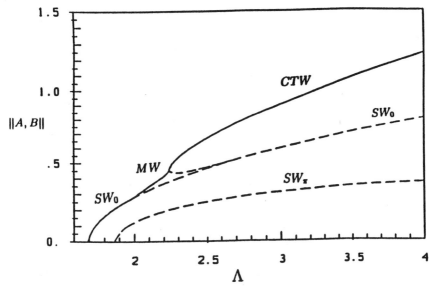

FIGURE 4. Global bifurcation diagram $\|A, B\| \equiv \left( \sum_i |p_i|^2 + |q_i|^2 \right)^{1/2}$ vs. $\Lambda$ for $\epsilon = 0.05$ and $b = -1$. For the quasi-periodic $MW$ we plot the time average of $\|A, B\|$ vs. $\Lambda$.

for $\epsilon = 0$ correspond to amplitudes of traveling waves. A Galerkin projection then yields a reduced system of ordinary differential equations:

$$(3.12) \quad \dot{p}_m = \lambda_{m+}p_m + \sum_{j,k,l} c^1_{mjkl}p_jp_k\bar{p}_l + c^2_{mjkl}q_jp_k\bar{p}_l + c^3_{mjkl}p_jp_k\bar{q}_l$$
$$+ c^4_{mjkl}q_jq_k\bar{p}_l + c^3_{mjkl}p_jq_k\bar{q}_l + c^5_{mjkl}q_jq_k\bar{q}_l$$

$$(3.13) \quad \dot{q}_m = \lambda_{m-}q_m + \sum_{j,k,l} d^1_{mjkl}p_jp_k\bar{p}_l + d^2_{mjkl}q_jp_k\bar{p}_l + d^3_{mjkl}p_jp_k\bar{q}_l$$
$$+ d^4_{mjkl}q_jq_k\bar{p}_l + d^3_{mjkl}p_jq_k\bar{q}_l + d^5_{mjkl}q_jq_k\bar{q}_l,$$

with complex constants $d^n_{mjkl}$ and $c^n_{mjkl}$.

For small amplitudes and near the primary instability $\Lambda = \Lambda_1$ we have confirmed that the qualitative transition to the confined traveling wave state occurs as predicted by the local analysis. The effect of symmetry breaking is to split each bifurcation value $\Lambda_n$ of the perfect system into a pair $\Lambda_{n\pm}$. Standing wave states bifurcate from the trivial solution as $\Lambda$ exceeds each of these split bifurcation values. Unlike the perfect symmetric system, confined traveling wave states now bifurcate from standing waves rather than the trivial solution. A global bifurcation diagram is shown in Figure 4.

Both confined traveling waves ($CTW$) and standing waves ($SW$) are shown together with the blinking state ($MW$). The blinking state is a two-torus in the $\{p_n, q_n\}$ space but appears as a periodic orbit in the $\{|p_1|, |q_1|\}$ plane. The bifurcation structure of the traveling and standing waves was calculated as for Figure 3. The blinking state was calculated by direct time evolution of (3.24), (3.25) at intervals of $\Lambda = 0.02$ for $1.98 \leq \Lambda \leq 2.3$ and at intervals of $\Lambda = 0.001$ $2.3 \leq \Lambda \leq 2.35$. The spatial structure of the confined traveling wave state with $b = -1$ for increasing values of $\Lambda$ is shown in Figure 5a, 5b.

The first eigenmode remains dominant and thus we refer to this state as a single-mode confined traveling wave. In the limit of large $\Lambda$ the confined traveling wave state has an asymptotic boundary layer character. The magnitude of one envelope of the confined traveling wave approaches zero and the second approaches unity excepting a boundary layer correction.

As in the perfect system for complex $b$ $(b = -1 + 4i)$, the confined traveling wave state loses stability in a subcritical Hopf bifurcation for $\epsilon = 0.05$ at $\Lambda = 8.7$. Two-mode confined traveling waves are also present for this case appearing at $\Lambda = 7.98$. Additional tori are generated via bifurcation from confined traveling wave states which differ from the blinking state in that the wave vector oscillates about that of a single confined traveling wave state. The blinking state is created in a Hopf bifurcation of the standing wave state and terminates in a global heteroclinic bifurcation linking the right and left moving confined traveling wave states. Since this transition is continuous the qualitative dynamics of the blinking state may be interpreted as oscillation between confined traveling wave states. A temporal trace of the blinking state torus $(MW)$ in the plane of the real parts of the two lowest eigenmodes for $A$ and $B$ is shown in Figure 6.

In this plane, confined traveling wave states are simple closed curves with symmetry about the lines $Re(p_1) = \pm Re(q_1)$. The additional tori generated from confined traveling waves are destroyed in period doubling cascades and are not constricted to linking confined traveling waves. The corresponding trace for the torus bifurcating from the second single-mode confined traveling wave $(p_2, q_2$ dominant) is shown in Figure 7.

### 3.3. Transition to chaotic dynamics.
The emergence of chaotic confined traveling wave states can occur by period-doubling cascade or by fractalization of basin boundaries in instances of bistable periodic orbits and tori. In Figure 8 we show three stable attractors at $\Lambda = 8.64$ and $\epsilon = 0.05$.

The plot is a temporal trace of the moduli of the first two eigenmode amplitudes. In these variables the dimension of the attractors is decreased by one. Tori are closed curves and periodic orbits are fixed points. The tori are generated from the two-mode, and from the second single-mode confined traveling waves, respectively. The fixed point is the first single-mode confined traveling wave. As $\Lambda$ is increased, the proximity of the torus, $(c)$, to the stable periodic orbit, $(a)$, in Figure 8 results in a fractalization of their respective basins of attraction and leads to the appearance of chaotic transients. Initially transient chaotic trajectories appear which typically are drawn into the basin of attraction of the stable

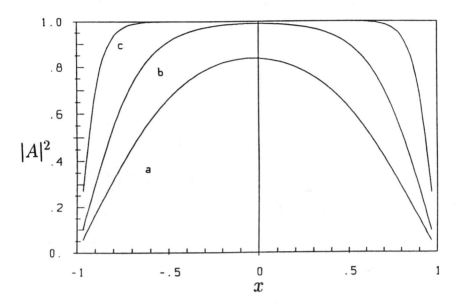

FIGURE 5a. Spatial structure of the primary confined left traveling wave branch for the wave amplitude $A(x)$, $\epsilon = 0.05$, $b = -1$. a: $\Lambda = 4$, b: $\Lambda = 14$, c: $\Lambda = 104$.

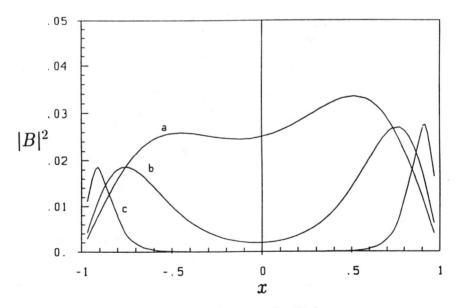

FIGURE 5b. As above for $B(x)$.

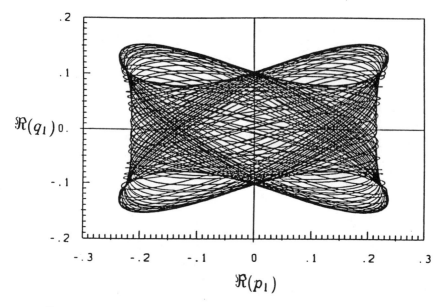

FIGURE 6. The blinking state at $\Lambda = 2.15$, for $\epsilon = 0.05$, $b = -1 + 4i$. Trace of the real parts of the first eigenmodes $p_1$ and $q_1$.

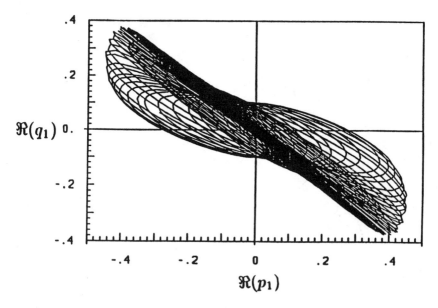

FIGURE 7. Same as Figure 6 for a two-torus generated from the second primary traveling wave, $\Lambda = 8.65, \epsilon = 0.05$.

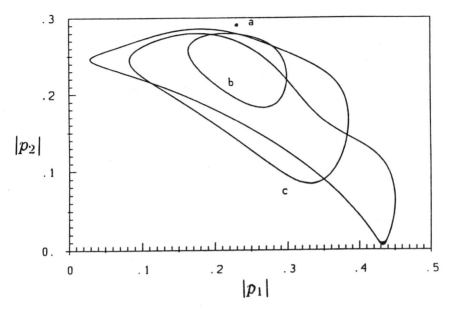

FIGURE 8. Multiple stable attractors at $\Lambda = 8.65, \epsilon = 0.05$, $b = -1 + 4i$. a: primary traveling wave, b: two torus from two-mode traveling wave, c) two-torus from second single-mode traveling wave.

confined traveling wave state. As $\Lambda$ is increased chaotic trajectories may persist for extended time intervals. Figure 9 shows a chaotic trajectory for the same parameter values as those for which the attractors in Figure 8 were calculated.

Small perturbations of the initial conditions may result in chaotic or quasi-periodic states. In addition, the flow near the observed tori is on a Möbius band indicating the presence of period-doubling at nearby points of the parameter space. For higher values of $\Lambda$ the tori undergo period-doubling cascades and create chaotic attractors. The accumulation point of the period doubling cascade for the torus $b$ of Figure 8, generated from the two-mode confined traveling wave state, is $\Lambda = 10.99$ for $\epsilon = 0.05$ and $\Lambda = 13.4$ for $\epsilon = 0$. The period-doubling cascade is shown in Figure 10 with a sequence of traces of $|p_1|$ vs. $|p_2|$ immediately following the first three period-doubling bifurcations as well as the resultant chaotic attractor.

## 4. Conclusion

The results of the numerical computations show that the qualitative predictions of the local analysis are effective in predicting quantitative behavior. In particular for $b = -1$ the confined traveling wave state is stable throughout the range of parameter space and the local results are globally valid.

Contrary to that, for $b = -1 + 4i$ the dynamics are extremely complicated beyond the validity regime of the local analysis. Both chaotic attractors and

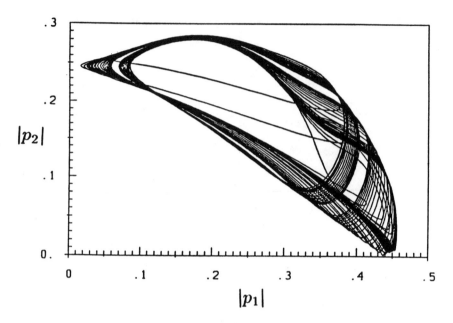

FIGURE 9. Chaotic trajectory at $\Lambda = 8.65, \epsilon = 0.05, b = -1 + 4i$.

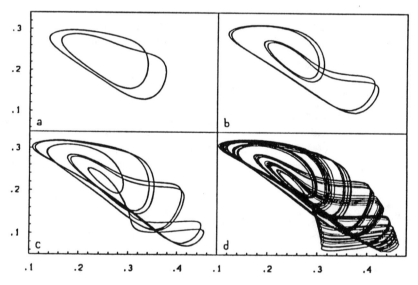

FIGURE 10. Period-doubling cascade, $\epsilon = 0.05$, $b = -1 + 4i$, a: $\Lambda = 9.5$, b: $\Lambda = 10.1$, c: $\Lambda = 10.7$, d: $\Lambda = 11.0$.

fractalized basins of attraction have been observed already for $\Lambda_2 < \Lambda < \Lambda_3$. Moreover, we find that the qualitative global dynamics are similar in both the perfect and unperturbed system except for a shift of bifurcation values. Thus the global behavior of the system is not essentially changed by the symmetry breaking. Nevertheless, much of the dynamical phenomena we observe results from interaction of multiple attractors and thus relative shifts in the parameter values resulting in instability of particular attractors may eliminate regions of multiple stability and induce changes in the qualitative dynamics.

When $a$ and $b$ are both real the perfect system possesses in addition to $O(2) \times S^1$ the $Z(2)$ symmetry $(A, B) \rightarrow (\bar{A}, \bar{B})$ which probably accounts for for the relatively simple behavior observed for $b = -1$. An interesting task for future investigations is therefore to analyze the behavior of the coupled Ginzburg-Landau equations when the imaginary part of $b$ decreases. In this case we expect that the bifurcation values at which the more complicated attractors are created are shifted to increasingly higher values of $\Lambda$, but then the attractors are also influenced by an increasing number of unstable modes. This might lead to high dimensional attractors, $i.e.$, to spatio-temporal chaos in contrast to the still low dimensional chaotic states observed here for $b = -1 + 4i$.

## References

1. M.C. Cross, P.G. Daniels, P.C. Hohenburg, and E.D. Siggia, J. Fluid Mech. **127**, 155 (1983).
2. M.C. Cross, Phys. Rev. Lett. **57**, 2935 (1986); M.C. Cross, Phys. Rev. A. **38**, 3593 (1988).
3. G. Dangelmayr, E. Knobloch, Nonlinearity **4**, (1991).
4. G. Dangelmayr, E. Knobloch, M. Wegelin, Europhys. Lett., **16** (8), (1991).
5. G. Dangelmayr, E. Knobloch, in *Nonlinear Evolution of Spatio-Temporal Structures in Dissipative Continuous Systems*, F.H. Busse and L. Kramer (eds.), Plenum Press, 399 (1990).
6. P.G. Drazin and W.H. Reid, *Hydrodynamic Stability*, Cambridge Univ. Press (1981).
7. J. Fineberg, E. Moses, V. Steinberg, Phys. Rev. Lett. 61, 838 (1988); P. Kolodner, C.M. Surko, Phys. Rev. Lett. 61, 842 (1988).
8. M. Golubitsky, I. Stewart and D.G. Schaeffer, *Singularities and Groups in Bifurcation Theory*, Vol.II, (Springer Verlag, New York, 1988).
9. P. Kolodner, A. Passner, C.M. Surko, and R.W. Walden Phys. Rev. Lett. 56, (1986).
10. L. Sirovich, J.D. Rodriguez and B.W. Knight, Physica D, Vol. 43 No. 1 (1990).

INSTITUT FÜR INFORMATIONSVERARBEITUNG, UNIVERSITÄT TÜBINGEN, KÖSTLINSTR. 6, D-7400 TÜBINGEN, GERMANY

*E-mail address*: piida01@convex.zdv.uni-tuebingen.de

Lectures in Applied Mathematics
Volume **29**, 1993

# THE EQUIVARIANT DARBOUX THEOREM

MICHAEL DELLNITZ AND IAN MELBOURNE

ABSTRACT. The classical Darboux Theorem states that symplectic forms are locally constant up to isomorphism, or equivalently that any two symplectic forms are locally isomorphic. We consider the corresponding results for symplectic forms that are invariant under the action of a compact Lie group. In this context, it is still true that symplectic forms are locally constant up to isomorphism but it is not true that any two symplectic forms are locally isomorphic.

## 1. INTRODUCTION

The Darboux theorem plays a fundamental role in the theory of Hamiltonian systems. Roughly speaking, the theorem states that locally all finite-dimensional symplectic manifolds of the same dimension look the same. The equivariant Darboux theorem plays an analogous role for Hamiltonian systems that are equivariant with respect to a symplectic action of a compact Lie group of symmetries, see for example [8], [2].

It will be convenient to divide the Darboux theorem and its equivariant counterpart into two parts. Suppose that $X$ is a finite-dimensional manifold with symplectic form $\omega$. In the absence of symmetry, the following statements are valid:

(a) Locally, there is a change of coordinates so that the transformed symplectic form is constant.

(b) There is a further change of coordinates yielding the 'canonical' symplectic form

$$\sum_{i=1}^{n} dq_i \wedge dp_i.$$

Together statements (a) and (b) imply

(c) Any two symplectic forms on symplectic manifolds of the same dimension are locally isomorphic.

In the literature, either (a) or (c) is called the Darboux Theorem. We shall distinguish these statements by referring to (a) as the 'locally constant' result and (c) as the 'locally isomorphic' result. We note that statement (b) follows from a standard result in linear algebra.

1991 *Mathematics Subject Classification.* 58F05, 58F14.

The first author was supported in part by the Deutsche Forschungsgemeinschaft

The second author was supported in part by NSF Grant DMS-9101836 and by the Texas Advanced Research Program (003652037)

This paper is in final form and no version of it will be submitted for publication elsewhere

Our aim in this paper is to clarify the corresponding results in the equivariant context where there is a compact group of symmetries present. Although results corresponding to statements (a) and (b) can be found in the literature, nowhere are they stated together correctly. We note first that the 'locally constant' result still holds. This is an easy consequence of the Darboux-Weinstein Theorem, see Guillemin and Sternberg [5, Theorem 22.1]. However, the analog of the 'locally isomorphic' result ([5, Theorem 22.2]) is false as is implicit in the work of Montaldi, Roberts and Stewart [8]. In short, statement (a) holds but statements (b) and (c) are invalid. On the other hand it is possible to classify the nonisomorphic symplectic forms that fill the role of the single canonical symplectic form in (b), see [8].

The existence of nonisomorphic symplectic forms is intimately related to the representation of the group of symmetries $\Gamma$. Indeed, statement (c) is valid if and only if none of the irreducible representations that appear in the representation of $\Gamma$ are of complex type (that is, each irreducible representation that occurs is of real or quaternionic type). For example, this is the case for the groups 1 (no symmetry), $\mathbf{O}(2)$, $\mathbf{SO}(3)$ and $\mathbf{O}(3)$ which have only real representations, and $\mathbf{SU}(2)$ which has only real and quaternionic representations.

However the nontrivial representations of the circle group $\mathbf{SO}(2)$ are of complex type and so uniqueness fails for nontrivial actions of $\mathbf{SO}(2)$. For example, suppose that $\mathbf{SO}(2)$ is acting in the standard way on $\mathbb{R}^2$ which we identify with $\mathbb{C}$. Then the (real) $\mathbf{SO}(2)$-invariant symplectic forms $\frac{1}{2}(dz \wedge id\bar{z})$ and $-\frac{1}{2}(dz \wedge id\bar{z})$ are not isomorphic (see Section 2 for an explicit verification of this fact).

The existence of nonisomorphic symplectic forms (in particular symplectic forms that are not isomorphic to the usual canonical symplectic form) is of some significance in the local bifurcation theory for equivariant Hamiltonian vector fields. For example, certain symplectic forms force spectral stability of equilibria (cyclospectrality in [8]) and the existence of Liapunov centers (weak cyclospectrality in [7]). Also, the expectation that certain collisions of eigenvalues will be dangerous in the sense of Krein is dependent on the symplectic structure present (see [2]).

In Section 2 we show by direct computation that the locally isomorphic result cannot be valid when there is symmetry. In Section 3 we state the Darboux-Weinstein theorem and deduce from this the locally constant result. Then in Section 4 we describe the nonisomorphic symplectic forms to which a symplectic form may locally be transformed by an equivariant change of coordinates. We illustrate our results by listing the nonisomorphic symplectic forms for an action of $\mathbf{SO}(2)$ on $\mathbb{R}^{10}$.

## 2. Nonisomorphic symplectic forms on $\mathbb{R}^2$

In this section, we illustrate by explicit calculation the failure of the 'locally isomorphic' result (statement (c)) in the equivariant context. The simplest example is the standard action of the circle group $\mathbf{SO}(2)$ on $\mathbb{R}^2$.

It is convenient to identify $\mathbb{R}^2$ with $\mathbb{C}$. In these coordinates the standard action of $\mathbf{SO}(2)$ is given by

$$z \to e^{i\theta}z, \; \theta \in \mathbf{SO}(2).$$

A symplectic form $\omega$ is $\mathbf{SO}(2)$-invariant if

$$\omega(e^{i\theta}z, e^{i\theta}w) = \omega(z, w),$$

for all $\theta \in \mathbf{SO}(2)$ and $z, w \in \mathbb{R}^2$. It is readily shown that

$$\Im(i\bar{z}w), \quad -\Im(i\bar{z}w),$$

are real $\mathbf{SO}(2)$-invariant symplectic forms on $\mathbb{R}^2$. Moreover they correspond to the putative nonisomorphic symplectic forms mentioned in Section 1.

Two symplectic forms $\omega_1$ and $\omega_2$ on $\mathbb{R}^2$ are $\mathbf{SO}(2)$-isomorphic if there is an invertible linear map $P : \mathbb{R}^2 \to \mathbb{R}^2$ commuting with the action of $\mathbf{SO}(2)$ such that $\omega_1(Pz, Pw) = \omega_2(z, w)$ for all $z, w \in \mathbb{R}^2$. Again, it is an easy computation to show that $P$ is given by

$$Pz = ke^{i\theta}z,$$

for some fixed $\theta \in \mathbf{SO}(2)$ and $k > 0$. Suppose now that $\omega(z, w) = \Im(i\bar{z}w)$. Then

$$
\begin{aligned}
\omega(Pz, Pw) &= \Im(i\overline{Pz}Pw) \\
&= \Im(i\overline{ke^{i\theta}z}ke^{i\theta}w) \\
&= \Im(ik^2 e^{-i\theta}\bar{z}e^{i\theta}w) \\
&= k^2\omega(z, w).
\end{aligned}
$$

In particular, $\omega$ and $-\omega$ are nonisomorphic as required.

## 3. The Darboux-Weinstein Theorem

In this section we consider the equivariant analogue to the 'locally constant' result (statement (a)) in the classical Darboux theorem.

**Definition 1.** Suppose that $X$ is a finite-dimensional manifold. A *symplectic form* on $X$ is a closed two-form $\omega$ such that for each $x \in X$, $\omega_x : T_xX \times T_xX \to \mathbb{R}$ is nondegenerate, that is if $\omega_x(v, w) = 0$ for all $v \in T_xX$ and some $w \in T_xX$ then $w = 0$.

Suppose that $\Gamma$ is a compact Lie group acting smoothly on a manifold $X$ and that $\omega$ is a symplectic form on $X$. If $x \in X$ and $\gamma \in \Gamma$ there is an induced linear map $T_x\gamma : T_xX \to T_{\gamma x}X$. We shall abuse notation and refer also to this induced map as $\gamma$. Then we say that $\omega$ is $\Gamma$-*invariant* or that the group action is *symplectic* if $\gamma^*\omega = \omega$ for all $x \in X$, that is

$$\omega_{\gamma x}(\gamma v, \gamma w) = \omega_x(v, w),$$

for all $x \in X$, $v, w \in T_xX$.

If the group $\Gamma$ acts on manifolds $X$ and $Y$, then a mapping $f : X \to Y$ is $\Gamma$-*equivariant* if $f(\gamma x) = \gamma f(x)$ for all $\gamma \in \Gamma$, $x \in X$. It is clear that if $\omega$ is a $\Gamma$-invariant symplectic form on $Y$ and that $f : X \to Y$ is a $\Gamma$-equivariant diffeomorphism, then $f^*\omega$ is a $\Gamma$-symplectic form on $X$.

We are now in a position to state the Darboux-Weinstein Theorem.

**Theorem 1.** *Suppose that $\Gamma$ is a compact Lie group acting on a finite-dimensional manifold $X$ and let $\omega_0, \omega_1$ be two $\Gamma$-invariant symplectic forms on $X$. Suppose that $Y$ is a $\Gamma$-invariant submanifold of $X$ and that $\omega_0|_Y = \omega_1|_Y$. Then there exists an open $\Gamma$-invariant neighborhood $U$ of $Y$ and a $\Gamma$-equivariant diffeomorphism $f : U \to X$ such that $f|_Y = Id_Y$ and $f^*\omega_1 = \omega_0$.*

If we take the submanifold $Y$ to consist of a single point $x$ we obtain the 'locally constant' theorem (cf [1, Theorem 8.1.2] in the nonequivariant context).

**Corollary 2.** *Suppose that $\Gamma$ is a compact Lie group acting on a finite-dimensional manifold $X$ and let $\omega$ be a $\Gamma$-invariant symplectic form on $X$. Let $x \in X$ be a $\Gamma$-invariant point and let $\omega'$ denote the constant symplectic form that agrees with $\omega$ at $x$. Then there is an open $\Gamma$-invariant neighborhood $U$ of $x$ and a $\Gamma$-equivariant diffeomorphism $f : U \to X$ such that $f(x) = x$ and $f^*\omega = \omega'$.*

Observe that since $x \in X$ is $\Gamma$-invariant, the action of the group $\Gamma$ on $X$ induces a linear group action of $\Gamma$ on $T_x X$.

## 4. CANONICAL SYMPLECTIC FORMS

Suppose that $\omega$ is a symplectic form on an $m$-dimensional manifold $X$, and let $x \in X$. Then $\omega$ is locally constant and isomorphic to the *canonical* (and constant) symplectic form

$$\sum_{i=1}^{n} dq_i \wedge dp_i$$

where $n = m/2$.

In this section we describe the canonical symplectic forms in the presence of a compact Lie group $\Gamma$. As mentioned in the introduction, it is not necessarily the case that there is a unique canonical symplectic form. However, by Corollary 2 we may assume that the symplectic form is locally constant, and thus reduce the problem to one of listing the possible 'canonical' symplectic forms for the action of $\Gamma$ on a finite-dimensional vector space.

The classification of $\Gamma$-invariant symplectic forms was first stated in Montaldi, Roberts and Stewart [8] and follows from Lie-theoretic results in [6]. The result is also an immediate consequence of the linear-algebraic results in Melbourne and Dellnitz [7]. Of course the Lie-theoretic proof is more direct and intrinsic. Here we simply state the results and refer to [8] and [6], or alternatively [7], for the proofs.

In Subsection 4.1 we recall some basic representation theory, see for example [4]. This allows us to reduce to working with symplectic forms over a real division ring. There are three nonisomorphic division rings: the reals, complexes, and quaternions. Then in Subsection 4.2 we list the canonical symplectic forms over each division ring. It is the complexes that lead to nonisomorphic symplectic forms.

**4.1. Some representation theory.** Suppose that $\Gamma$ is a compact Lie group acting on a vector space $V$. Define $\mathrm{Hom}_\Gamma(V)$ to be the vector space of $\Gamma$-equivariant real matrices

$$\mathrm{Hom}_\Gamma(V) = \{L : V \to V \text{ linear}; \gamma L = L\gamma \text{ for all } \gamma \in \Gamma\}.$$

A subspace $U$ is said to be $\Gamma$-*irreducible* if it is invariant under $\Gamma$ and has no proper invariant subspaces. If $U$ is an irreducible subspace, then $\mathrm{Hom}_\Gamma(U)$ is a real division ring and hence is isomorphic to $\mathbb{R}$, $\mathbb{C}$ or $\mathbb{H}$.

The space $V$ may be written as a direct sum of irreducible subspaces

$$V = U_1 \oplus \cdots \oplus U_k.$$

Group together those $U_i$ on which $\Gamma$ acts isomorphically to obtain the *isotypic decomposition*

$$V = W_1 \oplus \cdots \oplus W_\ell,$$

where each *isotypic component* $W_j$ is the sum of isomorphic irreducible subspaces. The isotypic decomposition is unique, and moreover each isotypic component is left invariant by matrices in $\mathrm{Hom}_\Gamma(V)$. It follows that

$$\mathrm{Hom}_\Gamma(V) = \mathrm{Hom}_\Gamma(W_1) \oplus \cdots \oplus \mathrm{Hom}_\Gamma(W_\ell).$$

Next suppose that $W$ is an isotypic component. We may write $W = U \oplus \cdots \oplus U = \bigoplus_{i=1}^m U$ where $U$ is irreducible and $\mathrm{Hom}_\Gamma(U) \cong \mathcal{D} = \mathbb{R}, \mathbb{C}$ or $\mathbb{H}$. Let $A \in \mathrm{Hom}_\Gamma(W)$. Then $A = \{A_{jk}\}_{1 \leq j,k \leq m}$ where $A_{jk} : U \to U$. It is easy to check that $A_{jk} \in \mathrm{Hom}_\Gamma(U)$. Since $\mathrm{Hom}_\Gamma(U) \cong \mathcal{D}$ we have shown that

$$\mathrm{Hom}_\Gamma(W) \cong \mathrm{Hom}(\mathcal{D}^m),$$

where $\mathrm{Hom}(\mathcal{D}^m)$ denotes the space of $m \times m$ matrices with entries in $\mathcal{D}$. Often it will be convenient to denote the isotypic component $W$ by $\mathcal{D}^m$. We say that an isotypic component $\mathcal{D}^m$ is *real, complex* or *quaternionic* depending on $\mathcal{D}$. Also we define the *dimension* of the isotypic component $\mathcal{D}^m$ to be the integer $m$. Note that the dimension of the corresponding (real) subspace $W$ is a multiple of $m$ but is in general not equal to $m$.

A symplectic form on $\mathcal{D}^m$ is a nondegenerate anti-symmetric bilinear map $\omega : \mathcal{D}^m \times \mathcal{D}^m \to \mathbb{R}$. Two symplectic forms $\omega$ and $\omega'$ are isomorphic (over $\mathcal{D}$) if there is a $\mathcal{D}$-linear map $P : \mathcal{D}^m \to \mathcal{D}^m$ such that $\omega(Pv, Pw) = \omega'(v, w)$ for $v, w \in \mathcal{D}^m$. We have the following result (see [3], [8] and [7]).

**Proposition 3.**     (a) *Suppose that $\omega$ is a $\Gamma$-invariant symplectic form on $V$. Let $\omega_i = \omega|_{W_i}$. Then $\omega_i$ is a $\Gamma$-invariant symplectic form on $W_i$. Moreover two $\Gamma$-symplectic forms $\omega$ and $\omega'$ on $V$ are $\Gamma$-isomorphic if and only if the corresponding summands $\omega_i$ and $\omega_i'$ are $\Gamma$-isomorphic for each $i$.*

    (b) *Suppose that $W$ is an isotypic component for $\Gamma$ so that $\mathrm{Hom}_\Gamma(W) \cong \mathcal{D}^m$ where $\mathcal{D}$ is a real division ring. Then there is a one-to-one correspondence between (real) $\Gamma$-invariant symplectic forms on $W$ and symplectic forms on $\mathcal{D}^m$. Moreover, two $\Gamma$-invariant symplectic forms on $W$ are isomorphic if and only if the corresponding symplectic forms on $\mathcal{D}^m$ are isomorphic (over $\mathcal{D}$).*

**4.2. Canonical symplectic forms over real division rings.** Let $W = \mathcal{D}^m$ be an isotypic component of dimension $m$. In this subsection we list the nonisomorphic symplectic forms on $\mathcal{D}^m$. Using Proposition 3 we can then construct the nonisomorphic $\Gamma$-invariant symplectic forms on $V$ and hence the canonical locally constant symplectic forms on a $\Gamma$-invariant manifold.

We choose coordinates $x_1, \ldots, x_m$ on $\mathbb{R}^m$, $z_1, \ldots, z_m$ on $\mathbb{C}^m$, and $w_1, \ldots, w_m$ on $\mathbb{H}^m$.

**Theorem 4.** *Suppose that $\omega$ is a symplectic form on $\mathcal{D}^m$. Then $\omega$ is isomorphic to precisely one of the following canonical symplectic forms.*

$$\mathcal{D} = \mathbb{R}: \quad \sum_{j=1}^{n} dx_j \wedge dx_{j+n}, \quad m = 2n \text{ even.}$$

$$\mathcal{D} = \mathbb{C}: \quad \Re \sum_{j=1}^{n} dz_j \wedge dz_{j+n} + \tfrac{1}{2}\rho \sum_{k=2n+1}^{m} dz_k \wedge id\bar{z}_k, \quad 0 \le n \le m/2, \rho = \pm 1.$$

$$\mathcal{D} = \mathbb{H}: \quad \Re \sum_{j=1}^{n} dw_j \wedge dw_{j+n}, \quad m = 2n \text{ even,}$$

$$\Re \sum_{j=1}^{n} dw_j \wedge dw_{j+n} + \tfrac{1}{2}dw_m \wedge id\bar{w}_m, \quad m = 2n + 1 \text{ odd.}$$

*Remark 1.*     (a) It follows from Theorem 4 that the equivariant version of Darboux's theorem described in [5] is incorrect whenever there are complex isotypic components in the representation of $\Gamma$. In particular there are $m + 1$ nonisomorphic symplectic forms on each complex isotypic component of dimension $m$.

(b) Our choices of canonical symplectic forms are somewhat different from those in [8]. It turns out that the analysis of linear Hamiltonian vector fields is slightly simplified when working with the symplectic forms listed here (see [7]).

As an example we consider an action of the group $\mathbf{SO}(2)$ on $\mathbb{R}^{10}$. Identify $\mathbb{R}^{10}$ with $\mathbb{R}^2 \times \mathbb{C}^4$ and choose coordinates $v = (x_1, x_2, z_1, z_2, z_3, z_4)$. The action of $\theta \in \mathbf{SO}(2)$ is given by

$$\theta v = (x_1, x_2, e^{i\theta}z_1, e^{i\theta}z_2, e^{i\theta}z_3, e^{i\theta}z_4).$$

In this case there are two isotypic components, $\mathbb{R}^2$ corresponding to two trivial representations of $\mathbf{SO}(2)$ and $\mathbb{C}^4$ which corresponds to four copies of the standard representation of $\mathbf{SO}(2)$.

Applying the results of Subsection 4.1 we can build the canonical symplectic forms on $\mathbb{R}^{10}$ out of the canonical symplectic forms on $\mathbb{R}^2$ and $\mathbb{C}^4$. There are five canonical symplectic forms on $\mathbb{C}^4$:

$$\pm\tfrac{1}{2}(dz_1 \wedge id\bar{z}_1 + dz_2 \wedge id\bar{z}_2 + dz_3 \wedge id\bar{z}_3 + dz_4 \wedge id\bar{z}_4),$$
$$\Re\, dz_1 \wedge dz_2 \pm \tfrac{1}{2}(dz_3 \wedge id\bar{z}_3 + dz_4 \wedge id\bar{z}_4),$$
$$\Re(dz_1 \wedge dz_3 + dz_2 \wedge dz_4).$$

Hence there are five canonical symplectic forms on $\mathbb{R}^{10}$ given by the direct sum of the symplectic form $dx_1 \wedge dx_2$ on $\mathbb{R}^2$ and one of the symplectic forms on $\mathbb{C}^4$.

## Acknowledgment

We are grateful to Ernest Barany for helpful discussions.

## REFERENCES

1. R. Abraham, J. E. Marsden and T. Ratiu. *Manifolds, tensor analysis, and applications*, Addison-Wesley, New York, 1983.
2. M. Dellnitz, I. Melbourne and J. E. Marsden. *Generic bifurcation of Hamiltonian vector fields with symmetry*, Nonlinearity **5** (1992), pp. 979–996.
3. M. Golubitsky and I. Stewart. *Generic bifurcation of Hamiltonian systems with symmetry*, Physica D **24** (1987), pp. 391–405.

4. M. Golubitsky, I. Stewart and D. Schaeffer. *Singularities and groups in bifurcation theory*. II, Springer-Verlag, New York, 1988.

5. V. Guillemin and S. Sternberg. *Symplectic techniques in physics*, Camb. Univ. Press, Cambridge, 1984.

6. I. MacDonald. *Algebraic structure of Lie groups*, Representation Theory of Lie Groups (M. F. Atiyah, ed.), London Math. Soc. Lecture Notes, vol. 34, Camb. Univ. Press, Cambridge, 1979.

7. I. Melbourne and M. Dellnitz. *An equivariant version of the Williamson normal form theorem*, Math. Proc. Camb. Phil. Soc. (to appear).

8. J. Montaldi, M. Roberts and I. Stewart. *Periodic solutions near equilibria of symmetric Hamiltonian systems*, Phil. Trans. R. Soc. **325** (1988), pp. 237–293.

DEPARTMENT OF MATHEMATICS, UNIVERSITY OF HOUSTON, HOUSTON, TEXAS 77204-3476, USA

*Current address*: Institut für Angewandte Mathematik, Universität Hamburg, D-2000 Hamburg 13, Germany

*E-mail address*: am70010@dhhuni4.bitnet


DEPARTMENT OF MATHEMATICS, UNIVERSITY OF HOUSTON, HOUSTON, TEXAS 77204-3476, USA

*E-mail address*: ism@math.uh.edu

Lectures in Applied Mathematics
Volume **29**, 1993

# Invariant Boundary Conditions for the Generalized Diffusion Equations

## M.J. ENGLEFIELD

### Abstract

Requiring invariance of boundary conditions given with an equation leads to boundary conditions on the determining equations for the symmetry algebra. Some examples of these conditions are given for an arbitrary scalar equation in two variables. For any particular equation with known symmetry algebra the form of all admissible boundary conditions may be calculated. Examples are given for the diffusion convection equation and for the hyperbolic diffusion equation.

## 1. Introduction

The symmetry algebra of a partial differential equation is a Lie algebra of first-order differential operators $V$ which are generators of coordinate transformations that leave the equation invariant, and transform solutions into solutions. We call $V$ a symmetry, and say the equation admits $V$. By writing the equation in terms of new variables $\zeta$ that are invariants of some chosen symmetry ($V\zeta = 0$) the number of variables is reduced by one; an equation in two variables is thereby reduced to an ordinary differential equation. A boundary condition that is also invariant under $V$ may also be expressed in terms of $\zeta$, and hence applied to the reduced equation. It is thus important, if possible, to make the reduction using a symmetry that also leaves the boundary condition invariant.

Let $x$ and $t$ denote the independent variables, and $u$ the dependent variable. A general symmetry has the form

$$(1) \qquad V = \xi(x,t,u)\partial/\partial x + \tau(x,t,u)\partial/\partial t + \eta(x,t,u)\partial/\partial u$$

where $\xi$, $\tau$, and $\eta$ satisfy the determining equations. Invariance of a given type of boundary condition on $u(x,t)$ gives linear boundary conditions on the determining equations. For example, $u = M(x)$ when $t = 0$ is invariant if

$$(2) \qquad \tau(x,0,M(x)) = 0 \text{ and } \xi(x,0,M(x))M'(x) = \eta(x,0,M(x)).$$

In the next section various other results of this type are given. Such requirements restrict the symmetry algebra to some subalgebra.

Unfortunately this subalgebra is usually dimension zero, meaning that no group-invariant solution can satisfy the required boundary condition. The sec-

---

1991 *Mathematics Subject Classifications.* 17B99, 58G35, 35G30

This paper is in final form and no version of it will be submitted for publication elsewhere.

ond part of (2) has the form of a differential equation for $M$. Given the symmetry algebra, such differential equations may be solved to determine what boundary conditions are invariant. The results of these calculations are given in section 3 for the diffusion-convection equation

$$(3) \qquad \frac{\partial u}{\partial t} = \frac{\partial}{\partial x}\left[F(u)\frac{\partial u}{\partial x}\right] + G(u)\frac{\partial u}{\partial x}$$

and in section 4 for the hyperbolic diffusion equation

$$(4) \qquad \frac{\partial^2 u}{\partial t^2} + \frac{\partial u}{\partial t} = \frac{\partial}{\partial x}\left[H(u)\frac{\partial u}{\partial x}\right],$$

taking cases of $F(u)$, $G(u)$ and $H(u)$ which allow sufficient symmetry for nontrivial results.

The calculation of conditions that are invariant also gives the subalgebras that leave the conditions invariant. These subalgebras, to be used for reduction, are also given. Any statement below that a boundary condition is invariant means invariant under some (usually stated) subalgebra of the symmetry algebra of equation (3) or (4).

## 2. Invariance Conditions

First consider the derivation of equation (2). The given condition $u = M(x)$ when $t = 0$ is invariant under $V$ if

$$(5) \qquad V(t - 0) = 0 \quad \text{and} \quad V(u - M(x)) = 0$$

when $t = 0$ and $u = M(x)$. Using (1) immediately gives (2).

A Neumann condition involves partial derivatives, and therefore requires a first prolongation, for example

$$(6) \qquad V + \left[\frac{\partial \eta}{\partial x} + \left(\frac{\partial \eta}{\partial u} - \frac{\partial \xi}{\partial x}\right)u_x - \frac{\partial \tau}{\partial x}u_t - \frac{\partial \xi}{\partial u}u_x^2 - \frac{\partial \tau}{\partial u}u_x u_t\right]\frac{\partial}{\partial u_x}$$

where $u_x$ and $u_t$ are regarded as variables independent of $x$, $t$ and $u$.

The following are examples of invariance conditions that may be obtained:

$$(7) \qquad u(x, t_0) = M(x) \text{ admits (1) if}$$

$$(8) \qquad \tau(x, t_0, M(x)) = 0 \text{ and } \xi(x, t_0, M(x))M'(x) = \eta(x, t_0, M(x));$$

$$(9) \qquad u = u_0 \text{ on the boundary } x = f(t) \text{ admits (1) if}$$

$$(10) \qquad \eta(f(t), t, u_0) = 0 \text{ and } \tau(f(t), t, u_0)f'(t) = \xi(f(t), t, u_0);$$

$$(11) \qquad \partial u/\partial x = M(x) \text{ when } t = t_0 \text{ admits (1) if, when } t = t_0,$$

(12)                    then $\tau = D(M)\tau = 0$ and $D(M)(\eta - \xi M) = 0$,

(13)                    where $D(M) = \partial/\partial x + M(x)\partial/\partial u$;

(14)        $\partial u/\partial x = M(t)$ when $x = x_0$ admits (1) if, when $x = x_0$,

(15)        then $\xi = D(M)\tau = 0$ and $D(M)(\eta - \xi M) = \tau M'(t)$;

(16)  $\partial u/\partial x = V_0$ on the boundary $x = f(t)$ admits (1) if, when $x = f(t)$,

(17)        then $\xi = \tau f'(t), D(V_0)\tau = 0$ and $D(V_0)(\eta - V_0\xi) = 0$.

Equations (12), (15) and (17) have been obtained assuming that $u$ is not also specified on the same boundary; if $u$ is specified the given conditions are sufficient but may not be necessary. In particular the derivative (11), which is tangential to the boundary, is determined by specifying $u$.

Further examples may be obtained from (7) – (17) by interchanging variables and functions, for example take (11) and (12), interchange $x$ and $t$, $\xi$ and $\tau$, $x_0$ and $t_0$, to get conditions for invariance of $u_t(x_0, t) = M(t)$.

## 3. The Diffusion Convection Equation

(i) For (3) with $F(u) = au^p$ and $G(u) = cu^b$, the general symmetry (1) has

(18)            $\xi = \alpha + (p - b)x, \tau = \beta + (p - 2b)t, \eta = u$

where $\alpha$ and $\beta$ are arbitrary, i.e. a 3-dimensional algebra. Symmetries not containing $\partial/\partial u$ generate only translations in $x$ and $t$, and are not considered here as the corresponding invariant solution can be written down for arbitrary $F$ and $G$. Substituting (18) into (7) – (17), and solving the resulting differential equations for $M$ (or for $f$), yields possible invariant boundary conditions.

To state these results it is convenient to first give the invariant curves that are the possible boundaries:

(19)        $x = f(t) = B + K(t + A)^q, q = (p - b)/(p - 2b) \neq 0$;

(20)            $x = g(t) = A + Ke^{Bt}$ when $p = 2b$;

(21)        $x = h(t) = K + B\log(t + A)$ when $p = b$.

In these equations, the constants $A$ and $B$ determine the subalgebra under which the boundary curve is invariant, but this subalgebra does not depend on $K$. On these curves, (9), (10), (16) and (17) show that the conditions $u = 0$ or $u_x = 0$ or $u_t = 0$ are invariant, under the following subalgebras:

(22)        for $x = f(t), \alpha = B(b - p) \neq 0, \beta = A(p - 2b) \neq 0$;

(23)            for $x = g(t), \alpha = -Ab, \beta = b/B, (p = 2b)$;

(24)                 for $x = h(t), \alpha = -Bb, \beta = -Ab, (p = b)$.

For (22) or (23) with $p = b + 1$, $u_x = V_0$ is invariant, and for (22) or (24) with $p = 2b + 1$, $u_t = V_0$ is invariant.

For the special cases where the boundaries are $x = x_0$ or $t = t_0$, (7), (8), (11), (12), (14) and (15) give the following conditions invariant under the indicated subalgebras:

(25)    $u(x, t_0) = K(x + A)^r, r = 1/(p - b), \alpha = A(p - b), \beta = (2b - p)t_0$;

(26)             $u(x, t_0) = Ke^{Ax}, \alpha = 1/A, \beta = bt_0(= pt_0)$;

(27)    $u(x_0, t) = K(t + A)^s, s = 1/(p - 2b), \alpha = (b - p)x_0, \beta = (p - 2b)A$;

(28)        $u(x_0, t) = Ke^{At}, \alpha = -bx_0 \left( = -\frac{1}{2}px_0 \right), \beta = 1/A$;

(29) $u_x(x, t_0) = rK(x + A)^{r-1}, r = 1/(p - b), \alpha = A(p - b), \beta = (2b - p)t_0$;

(30)             $u_x(x, t_0) = AKe^{Ax}, \alpha = 1/A, \beta = bt_0(= pt_0)$;

(31)             $u_x(x_0, t) = K(t + A)^s, s = (1 + b - p)/(p - 2b)$,
                 $\alpha = (b - p)x_0, \beta = A(p - 2b)$;

(32)        $u_x(x_0, t) = Ke^{At}, p = 2b, \alpha = -bx_0, \beta = (1 - b)/A$.

In (29) with $r = 1$, $\alpha$ is arbitrary, giving a 2-dimensional subalgebra. Similarly $\beta$ is arbitrary in (32) when $p = -1 (b = -2)$ and $A = 0$.

(ii) For (3) with $F(u) = ae^{pu}$ and $G(u) = ce^{bu}$, the general symmetry (1) has $\eta = 1$, and $\xi, \tau$ as in (18). Hence the boundaries (19) – (21) remain invariant, under the same subalgebras given in (22) – (24). The invariant conditions now include $u_x = 0$ for (22), and $u_x = V_0$ for (24).

The special cases where the boundaries are $x = x_0$ or $t = t_0$ are:

(33)    $u(x, t_0) = K + \frac{1}{p - b} \log(x + A), \beta = (2b - p)t_0, \alpha = (p - b)A \neq 0$;

(34)             $u(x, t_0) = Ax + K, \beta = pt_0(= bt_0), \alpha = 1/A$;

(35)    $u(x_0, t) = K + \frac{1}{p - 2b} \log(t + A), \alpha = (b - p)x_0, \beta = (p - 2b)A \neq 0$;

(36)             $u_x(x, t_0) = A, \beta = (2b - p)t_0, \alpha = 1/A, b = p$;

(37)    $u_x(x, t_0) = (x + A)^{-1}/(p - b), \beta = (2b - p)t_0, \alpha = (p - b)A \neq 0$;

(38)             $u_x(x_0, t) = K(t + A)^q, q = (b - p)/(p - 2b)$,
                 $\alpha = (b - p)x_0, \beta = (p - 2b)A$;

(39) $\qquad u_x(x_0, t) = Ke^{At}, p = 2b, \alpha = -bx_0, \beta = -b/A;$

(40) $\qquad u_x(x_0, t) = V_0 \neq 0, p = b, \alpha = (b-p)x_0;$

(41) $\qquad u_x(x_0, t) = 0, \alpha = (b-p)x_0.$

The subalgebras in (36), (40) and (41) are 2-dimensional ($\alpha$ or $\beta$ arbitrary).

The tangential derivatives (29), (30), (36) or (37) can be satisfied by an invariant solution only if the corresponding conditions (25), (26), (34) or (33) also hold.

## 4. The Hyperbolic Diffusion Equation

(i) For (4) with $H(u) = ae^{pu}$, the general symmetry is (Oron and Rosenau, 1986)

(42) $\qquad (\alpha + px)\partial/\partial x + \beta\partial/\partial t + 2\partial/\partial u.$

Invariant boundary conditions are:

(43) $\qquad u(x, t_0) = K + (2/p)\log(x + A), \alpha = Ap, \beta = 0;$

(44) $\qquad u(x_0, t) = K + At, \alpha = -px_0, \beta = 2/A;$

(45) $\qquad$ either $u_t = V_0$ or $u_x = 0$ on $x = B + Ke^{At}, \alpha = -Bp, \beta = p/A;$

(46) $\qquad u_x(x, t_0) = K(x + A)^{-1}, \alpha = Ap, \beta = 0;$

(47) $\qquad u_x(x_0, t) = Ke^{At}, \alpha = -x_0 p, \beta = -p/A;$

(48) $\qquad u_t(x_0, t) = A, \alpha = -px_0, \beta = 2/A;$

(49) $\qquad u_t(x, t_0) = V_0, \beta = 0, \alpha$ arbitrary.

If $K = 0$ in (45) or (47), then $\beta$ is arbitrary, and the subalgebra is 2-dimensional.

(ii) When $H(u) = a(u + \ell)^p$ in (4), then the general symmetry is (Oron and Rosenau 1986)

(50) $\qquad (\alpha + px)\partial/\partial x + \beta\partial/\partial t + 2(u + \ell)\partial/\partial u$

The following boundary conditions are invariant for arbitrary $p$:

(51) $\qquad u_x = 0$ on $x = B + Ke^{At}, \alpha = -Bp, \beta = p/A;$

(52) $\qquad u(x, t_0) = -\ell + K(x + A)^{2/p}, \alpha = Ap, \beta = 0;$

(53) $\qquad u(x_0, t) = -\ell + Ke^{At}, \alpha = -px_0, \beta = 2/A;$

(54) $\qquad u_x(x, t_0) = K(x + A)^q, q = -1 + 2/p, \alpha = Ap, \beta = 0;$

(55) $\qquad u_x(x_0, t) = Ke^{At}, \alpha = -px_0, \beta = (2-p)/A;$

(56) $\qquad u_t(x, t_0) = K(x + A)^{2/p}, \alpha = Ap, \beta = 0;$

(57) $\qquad u_t(x_0, t) = Ke^{At}, \alpha = -px_0, \beta = 2/A.$

If $K = 0$ in (55), then $\beta$ is arbitrary, and the subalgebra is 2-dimensional. If $p = 2$, then $u_x = V_0$ is invariant in (51), also on $t = t_0$ with $\beta = 0$ and $\alpha$ arbitrary, and also on $x = x_0$ with $\alpha = -2x_0$ and $\beta = 0$.

(iii) If $p = -4/3$ in (ii), then the general symmetry is (cf. (50))

$$(58) \qquad (\alpha + \gamma p x + \delta x^2)\frac{\partial}{\partial x} + \beta\frac{\partial}{\partial t} + (2\gamma - 3\delta x)(u + \ell)\frac{\partial}{\partial u}$$

where $\alpha$, $\beta$, $\gamma$, $\delta$ are arbitrary so that the algebra is 4-dimensional. Then there are the following extra invariant conditions:

$$(59) \quad u(x, t_0) = -\ell + K(x^2 + Ax + B)^{-3/2}, \alpha = B, \beta = 0, \gamma = A/p, \delta = 1;$$

$$(60) \qquad u(x_0, t) = -\ell + Ke^{At}, \alpha = \frac{4}{3}\gamma x_0 - \delta x_0^2, \beta = (2\gamma - 3\delta x_0)/A;$$

$$(61) \qquad\qquad u(x_0, t) = V_0, \gamma = 3\delta x_0/2, \alpha = \delta x_0^2.$$

$$(62) \quad u_x(x, t_0) = K(x^2 + Ax + B)^{-5/2}, \alpha = B, \beta = 0, \gamma = -3A/4, \delta = 1;$$

$$(63) \qquad u_x(x_0, t) = Ke^{At}, \alpha = \frac{4}{3}\gamma x_0 - \delta x_0^2, \beta = \left(\frac{10}{3}\gamma - 5\delta x_0\right)/A;$$

$$(64) \qquad\qquad u_x(x_0, t) = V_0, \gamma = 3\delta x_0/2, \alpha = \delta x_0^2;$$

$$(65) \quad u_t(x, t_0) = K(x^2 + Ax + B)^{-3/2}, \alpha = B, \beta = 0, \gamma = -3A/4, \delta = 1;$$

$$(66) \qquad u_t(x_0, t) = Ke^{At}, \alpha = \frac{4}{3}\gamma x_0 - \delta x_0^2, \beta = (2\gamma - 3\delta x_0)/A;$$

$$(67) \qquad\qquad u_t(x_0, t) = V_0, \gamma = 3\delta x_0/2, \alpha = \delta x_0^2.$$

The following pairs of conditions (the derivative being tangential) have related constants if both conditions are applied: (43) and (46); (44) and (48); (52) and (54); (53) and (57); (59) and (62); (60) and (66); (61) and (67).

(iv) If $p = -2$ in (ii), then the symmetry algebra is infinite-dimensional, and equation (4) may be linearized.

## 5. Conclusion

Requiring invariance of a boundary condition gives conditions on the generators of the symmetry algebra that are boundary conditions on the determining equations for the generators. Because the number of realizable boundary conditions is quite restricted, it is feasible (for a given equation) to list the possible invariant conditions, as illustrated in the previous two sections. In general the Dirichlet-type boundary conditions (which specify values of $u$) correspond to invariant curves of the point transformations generated by the symmetry algebra. Similarly Neumann-type boundary conditions (specifying $u_x$ and/or $u_t$) correspond to invariant curves of the transformations generated by the first prolongation.

An explicit listing of the invariant conditions may be useful for modelling, in the same way as the explicit solution of the classification problem which determines the cases of the equation having extra symmetries. For example a model using the hyperbolic diffusion equation (4) may have sufficient flexibility to allow $H(u)$ to be chosen to be one of the functions considered in section 4, and might also allow one of the given boundary conditions to be used. The corresponding solution may then be found analytically by using the invariants of the given subalgebra. The possible boundaries may include a family of curves invariant under the same subalgebra, for example by varying $K$ in (19), (20) or (21). Then more than one value of $K$ may be used if required by the physical boundary.

The suggestion here that subalgebras for the reduction should be chosen according to required boundary conditions is an alternative to the standard method (Ovsiannikov, 1982; Olver, 1986) using an optimal system of subalgebras. Most of the subalgebras given above are one-dimensional, in which case it would be easy (and more useful) to give the invariants. These invariants also provide a check, in that the boundary condition must be expressible in terms of them. This approach to the method of reduction has been used for radiation hydrodynamics equations (Coggleshall and Axford, 1986).

The methods above apply when the domain of the problem is unbounded. Similar progress for the case of a finite region appears possible only if the symmetries indicate the equation is linearizable.

## References

S.V. Coggleshall and R.V. Axford, "Lie group invariance properties of radiation hydrodynamics equations and their associated similarity solutions", *Phys. Fluids* **29** (1986) 2398.

O. Oron and P. Rosenau, "Some symmetries of the nonlinear heat and wave equations", *Phys. Lett.* **A118** (1986) 172.

L.V. Ovsiannikov, "Group analysis of differential equations" (Academic Press, N.Y., 1982).

P.J. Olver, "Applications of Lie groups to differential equations" (Springer, N.Y., 1986).

Dr M.J. Englefield
Department of Mathematics
Monash University
Clayton, Vic. 3168
Australia.

Lectures in Applied Mathematics
Volume **29**, 1993

# THE POWER OF THE GENERALIZED SCHUR'S LEMMA

## ALBERT FÄSSLER

ABSTRACT. If a linear operator $M$ has the symmetry of a group $G$, that is, if $M$ commutes with a representation of $G$, then the operator $M$ can be block-diagonalized, if a symmetry adapted basis is known. Hence, for instance, solving the eigenvalue problem of $M$ will be considerably simplified. This technique, based on a generalization of Schur's lemma, will be applied to a network problem, thus representing problems from various fields.

### 1. SYMMETRY ADAPTED BASIS, BLOCK DIAGONALIZATION

Let us consider a representation $\vartheta$ of a finite group $G$ and let the complex representation space $V$ be of finite dimension $n$. For a given basis in $V$ the representation $\vartheta$ assigns to each element $s$ of the group a unique linear transformation described by a non-singular $n \times n$ matrix $D(s)$ such that

$$D(s) \cdot D(t) = D(st) \quad \text{for all} \quad s, t \in G$$

As every representation $\vartheta$ of a finite group is completely reducible, the representation $\vartheta$ decomposes into

(1) $$\vartheta = c_1 \vartheta_1 \oplus c_2 \vartheta_2 \oplus \ldots \oplus c_N \vartheta_N$$

where the $\vartheta_j$ are irreducible and mutually inequivalent representations of $G$. The number $c_j \in \{1, 2, 3, \ldots\}$ indicates the multiplicity, $n_j$ the dimension of $\vartheta_j$. Accordingly, the representation space $V$ of $\vartheta$ decomposes into

(2) $$V = V_1 \oplus V_2 \oplus \ldots \oplus V_N$$

Here $V_j$ consists of $c_j$ invariant subspaces $V_j^1, V_j^2, \ldots, V_j^{c_j}$ each of which has dimension $n_j$ and transforms after the manner of $\vartheta_j$. More precisely, for each $\rho$ the representation obtained by restricting $\vartheta$ to the space $V_j^\rho$ is equivalent to $\vartheta_j$. The $(c_j n_j)$-dimensional subspaces $V_j$ of $V$ are called the isotypic components (or the conglomerates ) of type $\vartheta_j$ for $\vartheta$.

1991 Mathematics Subject Classification. Primary 20C30, 15A18.

Supported by the Ingenieurschule Biel/Bienne and therewith by the Canton of Bern, Switzerland, which enabled me to spend my sabbatical leave at Colorado State University in Fort Collins from October 1, 1992 until March 31, 1993. I am obliged to the CSU for the excellent working environment.

This paper is in final form and no version of it will be submitted for publication elsewhere.

**Remark**: By construction, the isotypic decomposition (2) is unique. However, the decomposition of an isotypic component $V_j$ into the $n_j$-dimensional irreducible subspaces is in general not unique.

The $n_j c_j$ basis vectors of an isotypic component $V_j$ shall be arranged in an array as follows:

$$
(3) \qquad
\left.
\begin{array}{llllll}
V_j^1: & b_1^1 & b_2^1 & b_3^1 & \ldots & b_{n_j}^1 \\
V_j^2: & b_1^2 & b_2^2 & b_3^2 & \ldots & b_{n_j}^2 \\
\vdots & \vdots & \vdots & \vdots & & \vdots \\
V_j^{c_j}: & b_1^{c_j} & b_2^{c_j} & b_3^{c_j} & \ldots & b_{n_j}^{c_j}
\end{array}
\right\}
$$

The $\rho$th row contains the basis vectors of the subspace $V_j^\rho$. These subspaces all are transformed equivalently.

**Definition**: If the basis vectors of $V$ are chosen in a way such that the transformation behavior of irreducible subspaces transformed equivalently is described by one and only one matrix, then the basis of $V$ is said to be **symmetry adapted**.

**Remark**: From the theoretical point of view one would value the definition as inessential. But it will prove that for applications the assumption of having a symmetry adapted basis will be important.

With respect to a symmetry adapted basis the representing matrices $D(s)$ have the structure

$$
(4) \qquad D(s) = c_1 D_1(s) \oplus c_2 D_2(s) \oplus \ldots \oplus c_N D_N(s)
$$

where $\dim D_j(s) = n_j$ and $c_j \in \{1, 2, \ldots\}$ are the multiplicities. Now we are ready for a

**Generalization of Schur's Lemma.** *Assume that a linear operator $M$ has the symmetry of a representation $\vartheta : s \to D(s)$, that is, that*

$$
(5) \qquad M D(s) = D(s) M \quad \text{for all} \quad s \in G
$$

*Then relative to a favorably labeled symmetry adapted basis, the operator $M$ has the following block diagonal form:*

$$
(6) \qquad M = n_1 M_1 \oplus n_2 M_2 \oplus \ldots \oplus n_N M_N
$$

*where $\dim M_j(s) = c_j$ for all $j$. In comparing $\vartheta$ with $M$, we notice an interchange of roles for multiplicities and dimensions. Here, for every isotypic component, each of the $n_j$ subspaces invariant under the operator $M$ and of dimension $c_j$ is spanned by the vectors in the **columns** of (3). Hence in (3), for every isotypic component, labeling favorably means labeling along columns.*

This result is already mentioned in [4]. A more detailed discussion, complete with proof, is found in [1].

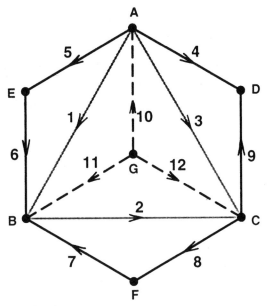

## 2. An application to networks

Let us consider a network of resistors or pipes arranged as in the figure below:

The underlying structure has the **symmetry** of the dihedral group $D_3$, that is, it is carried into itself by rotations through the angles $k \cdot 2\pi/3$, $k = 0, 1, 2$ and the reflections in the three axes through the nodes A, B, C. The dihedral group $D_3$ is isomorphic with the symmetric group $S_3$ on three objects; namely, rotations correspond to cyclic permutations, reflections correspond to transpositions.

The $7 \times 12$ incidence matrix has the form

$$
N_0 = \begin{bmatrix}
-1 & +1 & 0 & 0 & 0 & 0 & 0 \\
0 & -1 & +1 & 0 & 0 & 0 & 0 \\
-1 & 0 & +1 & 0 & 0 & 0 & 0 \\
-1 & 0 & 0 & +1 & 0 & 0 & 0 \\
-1 & 0 & 0 & 0 & +1 & 0 & 0 \\
0 & +1 & 0 & 0 & -1 & 0 & 0 \\
0 & +1 & 0 & 0 & 0 & -1 & 0 \\
0 & 0 & -1 & 0 & 0 & +1 & 0 \\
0 & 0 & -1 & +1 & 0 & 0 & 0 \\
+1 & 0 & 0 & 0 & 0 & 0 & -1 \\
0 & +1 & 0 & 0 & 0 & 0 & -1 \\
0 & 0 & +1 & 0 & 0 & 0 & -1
\end{bmatrix}
\begin{matrix}
1 \\ 2 \\ 3 \\ 4 \\ 5 \\ 6 \\ 7 \\ 8 \\ 9 \\ 10 \\ 11 \\ 12
\end{matrix}
$$

$$\text{node} \quad A \quad B \quad C \quad D \quad E \quad F \quad G$$

Thus, for instance, the last row means that edge no. 12 connects the $(-1)$ node $G$ with the $(+1)$ node $C$. The matrix $N_0$ describes the flow within the network.

Let us denote by $u_A$, $u_B$, ..., $u_G$ the potentials in the nodes $A$, $B$, ..., $G$. In an electrical network node $G$ is grounded, and in a network of pipes node $G$ is a reservoir. Therefore, the potential or the pressure $u_G$ equals zero. Consequently, we need to take care only of the remaining potentials $\vec{u} = (u_A, u_B, \ldots, u_F)^T$ and of the $6 \times 12$ incidence matrix $N$, the latter being obtained from $N_0$ by removing the last column. $N$ enters in the two Kirchhoff rules as follows:

(i) The algebraic sum of the changes in potential encountered in a complete traversal of a circuit must be zero; correspondingly for pipes, the sum of pressure differences along a loop must be zero; i. e.,

(7) $$\vec{d} = N\vec{u}$$

where $\vec{d} = (d_1, d_2, \ldots, d_{12})^T$ denotes the potential or the pressure differences, respectively.

(ii) At any junction the algebraic sum of the currents must be zero; or for pipes, the total flow approaching a node equals the total flow leaving that node:

(8) $$\vec{e} = N^T \cdot \vec{c}$$

where $\vec{e} = (e_A, e_B, \ldots, e_F)^T$ denotes the external sources and $\vec{c} = (c_1, c_2, \ldots, c_{12})^T$ denotes the currents.

The relation between the potential differences $\vec{d}$ and the currents $\vec{c}$ is governed by Ohm's law, or, in the case of pipes, by Poiseuille's law. We assume that not only the geometry, but also the physical structure respects the symmetry of the dihedral group $D_3$, that is (compare figure),

(9) $$\vec{c} = R \cdot \vec{d}$$

where $R$ is a $12 \times 12$ diagonal matrix with the main diagonal entries $\{r, r, r, s, s, s, s, s, s, t, t, t\}$, which themselves describe the resistances. Thus (7), (8), and (9) give us a direct relation between the external sources $\vec{e}$ and the potential $\vec{u}$; namely,

$$\vec{e} = N^T R N \vec{u}$$

Thus the above equation, known as the Euler-Lagrange equation in calculus of variations, is described by the symmetric matrix $M = N^T R N$. It is of the form

(10) $$M = \begin{bmatrix} 2(r+s)+t & -r & -r & -s & -s & 0 \\ -r & 2(r+s)+t & -r & 0 & -s & -s \\ -r & -r & 2(r+s)+t & -s & 0 & -s \\ -s & 0 & -s & 2s & 0 & 0 \\ -s & -s & 0 & 0 & 2s & 0 \\ 0 & -s & -s & 0 & 0 & 2s \end{bmatrix}$$

Let $p$ be the permutation of nodes that is obtained by reflecting the figure in the vertical axis, that is,

$$p = (A)(ED)(BC)(F)$$

and let $P$ be the corresponding permutation matrix, where $A$ is assigned to the first, $B$ to the second, ..., and $F$ to the sixth basis vector. Hence

$$P = \begin{bmatrix} 1 & 0 & 0 & 0 & 0 & 0 \\ 0 & 0 & 1 & 0 & 0 & 0 \\ 0 & 1 & 0 & 0 & 0 & 0 \\ 0 & 0 & 0 & 0 & 1 & 0 \\ 0 & 0 & 0 & 1 & 0 & 0 \\ 0 & 0 & 0 & 0 & 0 & 1 \end{bmatrix}$$

Correspondingly, let us consider the reflection in the axis through the nodes $C$ and $E$:

$$Q = \begin{bmatrix} 0 & 1 & 0 & 0 & 0 & 0 \\ 1 & 0 & 0 & 0 & 0 & 0 \\ 0 & 0 & 1 & 0 & 0 & 0 \\ 0 & 0 & 0 & 0 & 0 & 1 \\ 0 & 0 & 0 & 0 & 1 & 0 \\ 0 & 0 & 0 & 1 & 0 & 0 \end{bmatrix}$$

The linear 3-parameter operator $M$ commutes with $P$ and $Q$, that is,

$$MP = PM \qquad \text{and} \qquad MQ = QM$$

Hence $M$ commutes with the entire group of all 6 permutation matrices. This group is known to be generated by $P$ and $Q$, which form a representation $\vartheta$ of $D_3$. Thus the hypotheses of the generalized Schur's lemma are satisfied. The following is true:

$$\vartheta = 2\vartheta_1 \oplus 2\vartheta_2$$

Here $\vartheta_1$ is the unit representation assigning to each group element the number 1. The representation $\vartheta_2$ is the 2-dimensional irreducible representation of $D_3$ obtained by describing the reflections and the rotations of the plane. Relating to the geometry of the regular hexagon $A$, $B$, $C$, $D$, $E$, $F$ shown in the figure, one immediately finds the following symmetry adapted basis vectors, which transform under $\vartheta$

(i) **like $\vartheta_1$, that is, identically.**

The components from top to bottom correspond to $A$, $B$, $C$, $D$, $E$, $F$ in alphabetical order. There are two irreducible subspaces of dimension 1:

(11) $$\vec{x}_1 = (1, 1, 1, 0, 0, 0)^T$$

(12) $$\vec{x}_2 = (0, 0, 0, 1, 1, 1)^T$$

In fact, we have

$$P\vec{x}_i = Q\vec{x}_i = \vec{x}_i \qquad i = 1, 2$$

(ii) **like $\vartheta_2$.**

The two irreducible and invariant subspaces of dimension 2 are spanned by $(\vec{v}_1, \vec{w}_1)$ or $(\vec{v}_2, \vec{w}_2)$, respectively.

(13) $$\vec{v}_1 = (1, \bar{\omega}, \omega, 0, 0, 0)^T, \qquad \vec{w}_1 = (1, \omega, \bar{\omega}, 0, 0, 0)^T = \overline{\vec{v}_1}$$

(14)         $\vec{v}_2 = (0,0,0,\alpha,\bar{\alpha},-1)^T, \quad \vec{w}_2 = (0,0,0,\bar{\alpha},\alpha,-1)^T = \bar{\vec{v}_2}$

where $\omega = e^{i2\pi/3}$, $\alpha = e^{i\pi/3}$.
In fact,

(15)                     $P\vec{v}_i = \vec{w}_i, \qquad P\vec{w}_i = \vec{v}_i \qquad i = 1, 2$

(16)                     $Q\vec{v}_i = \bar{\omega}\vec{w}_i, \qquad Q\vec{w}_i = \omega\vec{v}_i \qquad i = 1, 2$

Hence each of the three pairs $[\vec{x}_1, \vec{x}_2]$, $[\vec{v}_1, \vec{v}_2]$, and $[\vec{w}_1, \vec{w}_2]$ from the symmetry adapted basis spans an invariant subspace of the operator $M$.

As a matter of fact, relative to the symmetry adapted basis $\vec{x}_1$, $\vec{x}_2$, $\vec{v}_1$, $\vec{v}_2$, $\vec{w}_1$, $\vec{w}_2$, the operator $M$ is described by the block diagonal matrix

(17)
$$\begin{bmatrix} 2s+t & -2s & 0 & 0 & 0 & 0 \\ -2s & 2s & 0 & 0 & 0 & 0 \\ 0 & 0 & 3r+2s+t & -s & 0 & 0 \\ 0 & 0 & -s & 2s & 0 & 0 \\ 0 & 0 & 0 & 0 & 3r+2s+t & -s \\ 0 & 0 & 0 & 0 & -s & 2s \end{bmatrix}$$

with the following eigenvalues $\lambda_i$ and associated eigenspaces $E_{\lambda_i}$ :

$$\lambda_1 = \frac{4s+t-\sqrt{16s^2+t^2}}{2} \qquad \text{of multiplicity 1}$$

$$\lambda_2 = \frac{4s+t+\sqrt{16s^2+t^2}}{2} \qquad \text{of multiplicity 1}$$

$$\lambda_3 = \frac{3r+4s+t-\sqrt{9r^2+4s^2+t^2+6rt}}{2} \qquad \text{of multiplicity 2}$$

$$\lambda_4 = \frac{3r+4s+t+\sqrt{9r^2+4s^2+t^2+6rt}}{2} \qquad \text{of multiplicity 2}$$

The 1-dimensional eigenspace $E_{\lambda_1}$ is spanned by the vector

(18)                     $a \cdot \vec{x}_1 + \vec{x}_2 = (a,a,a,1,1,1)^T$

with

$$a = \frac{4s}{t+\sqrt{16s^2+t^2}}$$

The 1-dimensional eigenspace $E_{\lambda_2}$ is spanned by the vector (18) with

$$a = \frac{4s}{t-\sqrt{16s^2+t^2}} \qquad .$$

The 2-dimensional eigenspace $E_{\lambda_3}$ is spanned by the vector

$$(19) \qquad b \cdot \vec{v_1} + \vec{v_2} = (b, b\frac{-1-\sqrt{3}\,i}{2}, b\frac{-1+\sqrt{3}\,i}{2}, \frac{1+\sqrt{3}\,i}{2}, \frac{1-\sqrt{3}\,i}{2}, -1)^T$$

with

$$b = \frac{2s}{3r + t + \sqrt{9r^2 + 4s^2 + t^2 + 6rt}}$$

and their complex conjugate.

The 2-dimensional eigenspace $E_{\lambda_4}$ is spanned by the vector (19) with

$$b = \frac{2s}{3r + t - \sqrt{9r^2 + 4s^2 + t^2 + 6rt}}$$

and their complex conjugate.

The obtained results can be verified symbolically or numerically with a software system such as, for instance, $Mathematica^{TM}$ (see [5]). Furthermore, such programs are able to give the symbolic eigenvalues, but in general not the symbolic eigenvectors, directly from the full matrix [10].

**Remarks:**

- In [3] a simpler case of network is discussed where the eigenvalue problem can be solved without group theoretical methods.
- Also in case of a high-dimensional problem (such as a network with many nodes and edges) whose physical and topological structure still respects the symmetry of the dihedral group $D_3$, the technique of using a symmetry adapted basis to get a block diagonal matrix works as well. For more information see [1].
- For problems with other symmetries, see [2], where tables of symmetry adapted bases for permutation representations of **all** finite point groups of dimensions 2 and 3 are given.

### REFERENCES

[1] A. Fässler and E. Stiefel, *Group Theroretical Methods and Their Applications*, Birkhäuser Boston, 1992.

[2] A. Fässler and R. Mäder, *Symmetriegerechte Basisvektoren für Permutationsdarstellungen aller endlichen Punktgruppen der Dimensionen 3 und 2*, ZAMP **31** (1980), 277-292.

[3] G. Strang, *Pattern in Linear Algebra*, Amer. Math. Monthly **96** (1989), 105-117.

[4] H. Weyl, *The Classical Groups*, Princeton University Press, 2nd ed., 1946.

[5] S. Wolfram, *Mathematica, A System for Doing Mathematics by Computer*, Addison Wesley, 2nd ed., 1991.

**Acknowledgement.** I would like to thank Gilbert Strang for his stimulating paper mentioned above and Baoswan Dzung Wong for editing the TEX-files.

INGENIEURSCHULE BIEL, POSTFACH 1180,
CH-2500 BIEL/BIENNE, SWITZERLAND

E-MAIL ADDRESS: FAESSLER@ODIS.ISBIEL.CH

CURRENT ADDRESS: DEPARTMENT OF MATHEMATICS,
COLORADO STATE UNIVERSITY, FORT COLLINS, CO 80523

Lectures in Applied Mathematics
Volume **29**, 1993

# COMPUTATION OF BIFURCATION GRAPHS

KARIN GATERMANN

ABSTRACT. The numerical treatment of equivariant parameter-dependent nonlinear equation systems, and even more its automation with Computer Algebra, requires the intensive use of group theory. The bifurcation subgroups and isotropy groups are determined in the symbolic part. Depending on the symmetry of a point a transformation of the Jacobian has block diagonal structure which is used by the numerical part. The innerrelationship between irreducible representations of the group and the subgroups explains the change from one block diagonal structure to another one. The principle of conjugacy is used everywhere to make symbolic and numerical computations even more efficient. Finally, the fact that the given representation is a quasi-permutation representation is exploited automatically.

## 1. INTRODUCTION

Figure 1 shows some stationary solutions of a parameter-dependent nonlinear system in the form

$$\dot{x} = F(x, \lambda), \quad F : \mathbb{R}^n \times \mathbb{R} \to \mathbb{R}^n, \quad \text{where}$$

$$\vartheta(t) F(x, \lambda) = F(\vartheta(t)x, \lambda), \quad \forall t \in G,$$

with a linear representation $\vartheta$, holds. Such *equivariant* systems have branches of stationary solutions having the same symmetry, i.e. are invariant with respect to *isotropy groups* $H$. As different branches may have different isotropy groups, all subgroups of $G$ which occur as isotropy groups have to be taken into account. There may also exist *conjugate* solutions $\vartheta(t)x$ which are given by a group operation $t \in G$. At symmetry breaking bifurcation points branches with different isotropy groups intersect such that one group is a common supergroup of the others. The subgroups fulfill the conditions of a *bifurcation subgroup*, which are arranged in the *bifurcation graph* (DELLNITZ, WERNER [1]).

The analysis of one bifurcation point may be found in GOLUBITSKY, STEWART, SCHAEFFER [8], VANDERBAUWHEDE [17] while the numerical treatment was first considered by DELLNITZ, WERNER [1], and HEALEY [10].

1991 Mathematics Subject Classification: 20C15, 20C40, 65H10, 58F14, 90C99
This paper is in final form and no version of it will be submitted for publication elsewhere.

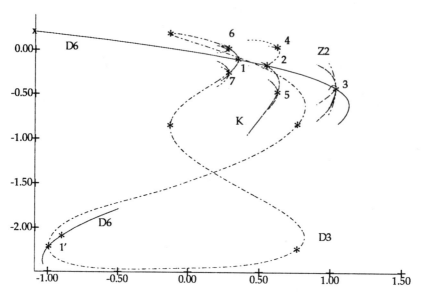

FIGURE 1. Part of a bifurcation diagram, $x_1$ versus parameter $\lambda$. The isotropy groups of branches and different numbers for different types of symmetry breaking bifurcation points are given.

The basic concept of numerical treatment is to apply the numerical pathfollowing procedure to the *symmetry reduced systems* only for one of the conjugate branches. The program SYMCON for example includes the numerical pathfollowing algorithm ALCON (DEUFLHARD, FIEDLER, KUNKEL [2]). While performing the numerical pathfollowing the bifurcation points leading to higher symmetry are detected by sign change of *symmetry monitor functions* and the points leading to smaller isotropy by sign check of determinants ([4]). The bifurcation points are computed by Newton's method applied to an augmented system ([4], [18]). These techniques exploite the block diagonal structure of the Jacobian (see also [11], [12], [15], and [18]). When a bifurcation point is found, the numerical pathfollowing procedure is restarted and possibly applied to a different symmetry reduced system checking determinants of a different block diagonal structure.

While the numerical mathematician concentrates on the numerical methods for determination and computation of bifurcation points we are interested in the automated preparation of examples which are equivariant with respect to different finite groups $G$. Different tasks and questions arise from those proposed by numerical mathematicians.

(1) How are the bifurcation subgroups computed automatically?
(2) How are the relevant subgroups for a given problem determined?
(3) How is the block diagonal structure wrt a subgroup obtained if the transformation to the structure corresponding to a supergroup is known?
(4) How does the group $G$ act on the bifurcation diagram?
(5) Which group theoretical computations may be avoided by conjugation?

(6) How to organize automatically the numerical evaluation of reduced systems and Jacobian blocks such that the numerical evaluation is as efficient as possible?

This paper continues the work in [4] and uses a database of irreducible representations (theory: STIEFEL, FÄSSLER [3], program: SYMMETRY [6]). The computation of the bifurcation graph and its relevant part exploits the interrelationship between irreducible representations of a group and its subgroups (Section 3).

A deeper understanding of innerconnectivity of irreducible representations of $G$ and its subgroups gives the transformation between the block diagonal structures corresponding to different groups (Section 4). In the literature this is only mentioned for the example of $D_6$ (IKEDA, MUROTA [12]). The handling of conjugate bifurcation points necessitates a deeper understanding of the principle of conjugacy (Section 5). Section 6 gives an overall view. In Section 7 is explained the exploitation of the special property of the quasi–permutation representation $\vartheta$ where Computer Algebra is used to generate the most efficient numerical algorithm.

## 2. ANALYSIS OF EQUIVARIANT SYSTEMS

The group theoretic investigations in the Sections 3-5 are derived, because we want to investigate the following problem automatically:

Let $F : \mathbb{R}^{n+1} \to \mathbb{R}^n$ be an explicitly given function which is $G$–equivariant, i.e.

$$(1) \qquad F(\vartheta(t)x, \lambda) = \vartheta(t)F(x, \lambda), \quad \forall t \in G,$$

where $\vartheta : G \to GL(\mathbb{R}^n)$ is a real orthogonal linear representation $(\vartheta(t)\vartheta(s) = \vartheta(ts) \forall t, s \in G)$. Throughout this paper we assume that $G$ is finite. The different types of stationary solutions $(x, \lambda)$, i.e.

$$(2) \qquad F(x, \lambda) = 0,$$

are of interest. Because the system (2) is parameter dependent the solutions appear in continua. In this section the analysis of this problem class is summarized (see [1], [4], and [8]).

The elements $x \in \mathbb{R}^n$ are distinguished by their symmetries. Mathematically speaking

$$(3) \qquad G_x := \{t \in G \mid \vartheta(t)x = x\}$$

denotes the *isotropy group* $H = G_x$ of $x$. In turn $x$ is called $H$–*invariant*, if $\vartheta(t)x = x, \forall t \in H$. The isotropy group $H = G_x$ is the maximal subgroup of $G$ with the property, that $x$ is $H$–invariant. A simple but fundamental fact is that equivariant systems have continua of solutions with the same isotropy group. Different solution paths with different isotropy groups may intersect in the so-called bifurcation points. Depending on whether different or equal groups interact they are called *symmetry breaking* or *symmetry preserving bifurcation points*. Since the Jacobian $DF(x^*, \lambda^*)$ is singular in a bifurcation point $(x^*, \lambda^*)$, the consequences of the equivariance (1) for the Jacobian have to be considered.

For $x \in \mathbb{R}^n$ with isotropy group $H = G_x$

$$(4) \qquad \begin{aligned} \vartheta(t) D_x F(x, \lambda) &= D_x F(x, \lambda) \vartheta(t), \ \forall \, t \in H \,, \\ \vartheta(t) D_\lambda F(x, \lambda) &= D_\lambda F(x, \lambda), \ \forall \, t \in H \,, \end{aligned}$$

hold. By the theory of symmetry adapted basis [3], the Jacobian $D_x F(x, \lambda)$ may be block diagonalized with an orthogonal transformation matrix $M_H \in \mathbb{R}^{(n,n)}$ which depends on $H$. For this we remind that an irreducible representation has the property that it does not split into subrepresentations. Up to isomorphy a finite group has a finite number $h_H$ of real irreducible representations $\vartheta_H^i : H \to Gl(\mathbb{R}^{n_i})$ each of dimension $n_i$. The supplement real stresses the fact that we restrict to real vector spaces. $\chi_H^i : H \to \mathbb{R}, i = 1, \dots, h_H$ are the corresponding characters. One distinguishes three different types (real, complex, quaternonian) of real irreducible representations [18]. The type depends on whether the algebra of matrices commuting with $\vartheta_H^i$ is isomorphic to the field of real, complex or quaternonian numbers. The real type is also called absolutely irreducible [8]. The type of a real irreducible representation of complex type consists of 2 complex irreducible representations which are complex conjugate to each other. We restrict to groups which have no irreducible representations of quaternonian type because groups appearing in applications have only irreducible representations of real and complex type.

For $\vartheta : H \to GL(\mathbb{R}^n)$, there exists *multiplicities* $c^i$ with $\vartheta = \sum_{i=1}^{h_H} c^i \vartheta_H^i$. For matrices $A$ with $\vartheta(t) A = A \vartheta(t) \, \forall \, t \in H$ the transformation matrix has the form

$$\begin{aligned} M_H &= (M^1, \dots, M^{h_H}), & M^i &\in \mathbb{R}^{n, c^i \cdot n_i}, & i &= 1, \dots, h_H \,, \\ M^i &= (M^{i1}, \dots, M^{i n_i}), & M^{ij} &\in \mathbb{R}^{n, c^i}, & j &= 1, \dots, n_i \,, \end{aligned}$$

such that $A_i \in \mathbb{R}^{c_i, c_i}$ with $A_i := (M^{ij})^T A M^{ij}$ are the blocks appearing $n_i$ times. In [3] a straightforward computation of $M^{ij}$ using projections and the Gram-Schmid process is described. But for this the irreducible representations $\vartheta_H^i$ and its characters $\chi_H^i, i = 1, \dots, h_H$ have to be known.

The matrix $M_H$ introduces the coordinate transformation $x = M_H u$.

**Definition 2.1.** [4]: *Let $F : \mathbb{R}^{n+1} \to \mathbb{R}^n$ be $G$-equivariant. Let $H$ be a subgroup of $G$ and $M_H$ the transformation matrix. Then the function $g : \mathbb{R}^{n+1} \to \mathbb{R}^n$ defined by*

$$g(u, \lambda) := (M_H)^T F(M_H u, \lambda)$$

*is called the* symmetrical normal form *of $F$ with respect to $H$.*

The name symmetrical normal form is justified by the fact that the Jacobian $D_u g(u, \lambda)$ has block diagonal form for every $H$-invariant point $x = M_H u$.

Let $\vartheta^1 : H \to \mathbb{R}, \ \vartheta^1(t) \equiv 1$. Then each $H$-invariant $x \in \mathbb{R}^n$ corresponds to one $u = (M_H)^T x = (\tilde{u}, 0)$ with $\tilde{u} \in \mathbb{R}^{c^1}$. Because $F$ is equivariant (1), the set of $H$-invariant solutions of (2) is equivalent to the solution set of the $H$-*reduced equations*

$$(5) \qquad \tilde{g}(\tilde{u}, \lambda) = (M^1)^T g(M^1 \tilde{u}, \lambda) = 0 \,,$$

where $\tilde{g} : \mathbb{R}^{c^1 + 1} \to \mathbb{R}^{c^1}$. This is a well known fact and is often used.

If a block $A_i$ ($i \neq 1$, $\vartheta_H^i$ of real type) becomes singular, a *symmetry breaking bifurcation point* may occur. Because the kernel of $DF$ has dimension $n_i$, multi-dimensional irreducible representations gives raise to a multiple bifurcation point. In the *equivariant branching lemma* of VANDERBAUWHEDE [17] and CICOGNA (see also [8], p. 82 and [1]) the multiple problem is reduced to a simple bifurcation phenomena. The following definition was introduced in [1]:

**Definition 2.2.** *Let $\vartheta_H^i : H \to GL(\mathbb{R}^{n_i})$ be a real irreducible representation of real type. $K$ is called a* bifurcation subgroup *of type $\vartheta_H^i$, if*

    a) *there exists $v \in \mathbb{R}^{n_i}$, $v \neq 0$, such that $K = H_v$,*
    b) *for every $K$–invariant $w \in \mathbb{R}^{n_i}$ exists $a \in \mathbb{R}$ with $w = av$.*

**Definition 2.3.** *A group $H$ is called a* bifurcation supergroup *of $K$, if $K$ is a bifurcation subgroup of $H$ of some type.*

If on a branch of solutions $(u(\lambda), \lambda) = (\tilde{u}(\lambda), 0, \lambda)$ with isotropy group $H$ one block $A_i, i \geq 2$ becomes singular at $(u^*, \lambda^*)$ then it follows from the equivariant branching lemma that generically branches of solutions emanate having the isotropy of bifurcation subgroups $K$ of type $\vartheta_H^i$. These branches are solutions of the $K$–reduced problem. A singular block $A_1$ indicates a turning point, a symmetry preserving bifurcation point or a symmetry breaking bifurcation point where $x^*$ has the isotropy of a bifurcation supergroup of $H$.

Definition 2.2 leads to the definition of a *bifurcation graph* showing all bifurcation subgroups and bifurcation supergroups (see Fig. 2). The first aim of automation is the computation of the bifurcation graph. But the numerical pathfollowing applied to the reduced systems and the evaluation of Jacobian blocks is needed only for the isotropy groups of the given system (2).

### 3. COMPUTATION OF RELEVANT BIFURCATION SUBGROUPS

In this and the next section the relation of irreducible representations of a group $H$ with the irreducible representations of its subgroups is fundamental. We start with a technical definition. The restriction of a representation $\vartheta : H \to GL(\mathbb{R}^n)$ to a subgroup $K$ of $H$ is denoted by $\vartheta \downarrow K$.

The real irreducible representations $\vartheta_H^i, i = 1, \ldots, h_H$ of $H$ may be restricted to representations $\vartheta_H^i \downarrow K$ of $K$. Then a canonical decomposition with respect to $K$ exists, i. e. integers $d_{i\underline{j}} \in \mathbb{N}$ exists with

$$(6) \qquad \vartheta_H^i \downarrow K = \sum_{\underline{j}=1}^{h_K} d_{i\underline{j}} \, \vartheta_K^{\underline{j}}, \quad i = 1, \ldots, h_H,$$

where $\vartheta_K^{\underline{j}}, \underline{j} = 1, \ldots, h_K$ are the real irreducible representations of $K$. For clarification indices corresponding to the subgroup $K$ are underlined.

**Definition 3.4.** *The integers $d_{i\underline{j}}, i = 1, \ldots, h_H, \; \underline{j} = 1, \ldots, h_K$, are called* inner-connectivity multiplicities.

The innerconnectivity multiplicities $d_{ij}$ are easily computed by a formula for multiplicities (see [3]). They have two applications. For the given representation $\vartheta : G \to GL(\mathbb{R}^n)$ in (1) the multiplicities $c_G^i$ of $\vartheta_i^G$ in the canonical decomposition are obtained by the formula mentioned above. ($c_G^1$ is the dimension of the $G$-reduced system and $c_G^i$ are the dimensions of the Jacobian blocks $A_i$.) Computing $H$-invariant solutions of (2) means consideration of $\vartheta \downarrow H$ where $\vartheta_H^j$ appear with multiplicity $c_H^j$, $j = 1, \ldots, h_H$. The multiplicities $c_H^j$ for subgroups $H$ may be obtained easily from $c_G^i$ and with the innerconnectivity multiplicities with respect to $G$ and $H$ (see [7]).

Second the innerconnectivity multiplicities enable the computation of bifurcation subgroups.

**Lemma 3.5.** *Let $\vartheta_H^i : H \to GL(\mathbb{R}^{n_i})$ be a non-trivial irreducible representation and $\vartheta_H^i \downarrow K$ its restriction to a subgroup $K$ of $H$. Then $K$ is a bifurcation subgroup of type $\vartheta_H^i$, iff*

$$\vartheta_H^i \downarrow K = \sum_{j=1}^{h_K} d_{i\underline{j}} \, \vartheta_{\underline{K}}^j ,$$

*with $d_{i1} = 1$ and if $K$ is maximal with this property.*

Bifurcation subgroups of type $\vartheta_H^i$ are isotropy groups of $\vartheta_H^i$. But by this definition it is not considered whether they are isotropy groups for the given representation $\vartheta : G \to GL(\mathbb{R}^n)$ in (1).

**Definition 3.6.** *A subgroup $H$ of $G$ is called a* relevant bifurcation subgroup *of $\vartheta : G \to GL(\mathbb{R}^n)$ of* level 1, *if*

    a) $c_G^1 \geq 1$, where $\vartheta = \sum_{i=1}^{h_G} c_G^i \, \vartheta_G^i$,
    b) $i \in \{2, \ldots, h_G\}$ exists, such that $H$ is a bifurcation subgroup of type $\vartheta_G^i$,
    c) and $c_G^i \geq 1$.

**Definition 3.7.** *A subgroup $K$ of $G$ is called a* relevant bifurcation subgroup *of $\vartheta : G \to GL(\mathbb{R}^n)$ of* level $\nu$ *($\nu > 1$), if*

    a) *a relevant bifurcation subgroup $H$ of $\vartheta$ of level $\nu - 1$ exists and*
    b) $i \in \{2, \ldots, h_H\}$ *exists, such that $K$ is a bifurcation subgroup of $H$ of type $\vartheta_H^i$ and*
    c) $c_H^i \geq 1$, *where $\vartheta \downarrow H = \sum_{i=1}^{h_H} c_H^i \, \vartheta_H^i$.*

For finite groups $G$ there is a level $\mu$ such that there are no relevant bifurcation subgroups of level $\nu > \mu$ for all linear representations of $G$.

**Definition 3.8.** *The* relevant bifurcation subgroups *of $\vartheta : G \to GL(\mathbb{R}^n)$ of all levels $\nu \geq 1$ are called* relevant subgroups *of $G$ with respect to $\vartheta$.*

**Lemma 3.9.** *The relevant subgroups of $G$ wrt $\vartheta$ are isotropy groups of $\vartheta$.*

The relevant subgroups of $G$ give the relevant part of the bifurcation graph (compare Fig. 2 and Fig. 3) which is needed in the numerical computation of (2).

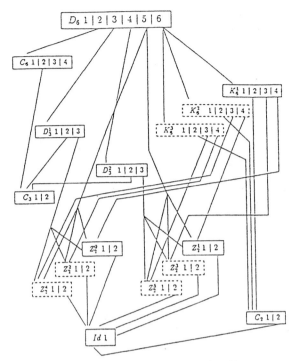

FIGURE 2. Bifurcation graph for $D_6$.

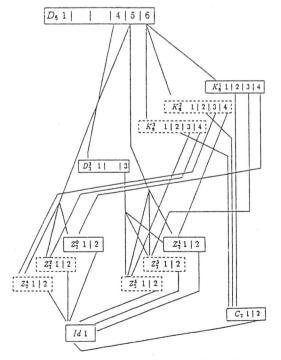

FIGURE 3. Relevant part of bifurcation graph for $D_6$–brusselator

## 4. Innerconnectivity

Once the transformation matrix $M_G$ for the supergroup $G$ is computed by means of projections, the transformation matrices $M_H$, where $H$ is a relevant subgroup of $G$, are easily obtained in the following way.

Recall that for $i = 1, \ldots, h_G$ the restricted irreducible representations $\vartheta^i_G \downarrow H$ have a canonical decomposition (see (6)). Furthermore there exist orthogonal coordinate transformations $C^i \in \mathbb{R}^{n_i, n_i}$ such that $(C^i)^T \vartheta^i_G(t) C^i$ are simultaneously block diagonal for all $t \in H$, where the blocks consist of $d_{ij}$ matrices $\vartheta^j_H(t), \underline{j} = 1, \ldots, h_H$:

$$(7) \qquad (C^i)^T \vartheta^i_G(t) C^i = \operatorname{diag}(\vartheta^j_H(t)), \quad \forall t \in H, \quad i = 1, \ldots, h_G.$$

These *innerconnectivity matrices* $C^i$ are computed as usual by means of projections applied to $\vartheta^i_G \downarrow H$. Note that the columns of $C^i$ form a symmetry adapted basis of $\mathbb{R}^{n_i}$.

Based on $d_{ij}$ and $C^i$ a *connection matrix* $C_{GH} \in \mathbb{R}^{n,n}$ is defined such that

$$(8) \qquad\qquad\qquad M_H = M_G C_{GH},$$

is a coordinate transformation matrix for $H$. The computation of $M_H$ or its parts $M^{ij}_H$ using the connection matrix needs less time than by application of the projections, but the definition of $C_{GH}$ is complicated to describe (for details see Fig. 4 in [7]). For this we introduce the notation

$$(9) \qquad\qquad\qquad D := \coprod_{i=1}^{m} D_i = (D_1, \ldots, D_m),$$

where $D_i$ are given $n \times j_i$ matrices and $D$ denotes the matrix with $n$ rows and $\sum_{i=1}^{m} j_i$ columns consisting of the collection of the matrices $D_i$. In this notation a decomposition exists

$$(10) \qquad C^i = \coprod_{\substack{j = 1 \\ d_{ij} \neq 0}}^{h_H} C^{i\underline{j}}, \qquad C^{i\underline{j}} = \coprod_{\underline{k}=1}^{n_{\underline{j}}} C^{i\underline{j}\underline{k}}, \quad i = 1, \ldots, h_G,$$

where $C^{i\underline{j}}$ are real $n_i \times (n_{\underline{j}} \cdot d_{ij})$ matrices and $C^{i\underline{j}\underline{k}}$ are $n_i \times d_{ij}$ matrices. $n_{\underline{j}}$ is the dimension of $\vartheta^j_H$. $C^{i\underline{j}}$ correspond to the irreducible representations $\vartheta^j_H$. Note that for $d_{ij} > 1$ the matrix $C^{i\underline{j}}$ is not unique. Recall the partitions of $M_G$ and $M_H$ with respect to the irreducible representations

$$M_G = \coprod_{i=1}^{h_G} M^i_G, \qquad M^i_G = \coprod_{l=1}^{n_i} M^{il}_G,$$

$$M_H = \coprod_{j=1}^{h_H} M^{\underline{j}}_H, \qquad M^{\underline{j}}_H = \coprod_{\underline{k}=1}^{n_{\underline{j}}} M^{\underline{j}\underline{k}}_H,$$

Then the innerconnectivity of irreducible representations of $G$ and $H$ (see (6), (7)) implies

$$(11) \qquad M_{\underline{H}}^{jk} = \coprod_{\substack{i=1 \\ d_{i\underline{j}} \neq 0}}^{h_G} \coprod_{\mu=1}^{d_{i\underline{j}}} \sum_{l=1}^{n_i} M_G^{il} C_{l\mu}^{ij\underline{k}},$$

which is written in (8) in compact form. If $\vartheta_{\underline{H}}^{\underline{j}}$ is of complex type the upper limit $d_{i\underline{j}}$ has to be replaced by $2d_{i\underline{j}}$.

If $H$ is a proper subgroup, the innerconnectivity matrices $C^i$ are given, and $M_G$ was already computed, then $M_H$ is uniquely defined by this procedure.

For a proper subgroup $K$ of $H$ which is a proper subgroup of $G$ the problem arises that $M_K$ may be computed by $M_G C_{GH} C_{HK}$ or $M_G C_{GK}$ which may happen to be different. In SYMCON the transformation matrices $M_K = M_G C_{GK}$ are used. In [15] it is stated that one can rearrange the order of irreducible representations and choose right bases for them and choose between conjugate groups such that for a maximal chain of subgroups the same transformation matrix $M = M_G = M_H = M_K \cdots$ is valid. But the relevant subgroups are not arrangeable in one chain in general. The counterexample of $D_6$ shows that it is not possible to determine one common matrix $M$ for all non-conjugate subgroups.

By $x = M_G u_G$ and $x = M_H u_H$ two coordinate transformations are defined. One may switch between $G$- and $H$-coordinates by

$$(12) \qquad u_G = C_{GH} u_H$$

The matrix-vector operations $C_{GH} u_H$, $C_{GH}^1 u_H^1$ as well as $C_{GH}^T u_G$ involves a lot of multiplications with 1 and addition of zero because of the special structure of $C_{GH}$. In [7] formulas for this operations are given using the special structure. For this, the following splittings are introduced.

$$(13) \qquad \begin{aligned} u_H &= (u_H^1, \dots, u_H^{h_H}), \\ u_{\underline{H}}^{j} &= (u_H^{j1}, \dots, u_H^{jn_{\underline{j}}}), \qquad \underline{j} = 1, \dots, h_H, \\ u_{\underline{H}}^{jk} &= (u_H^{jk1}, \dots, u_H^{jkh_G}), \qquad \underline{k} = 1, \dots, n_{\underline{j}}, \\ C_{GH} &= \coprod_{\underline{j}=1}^{h_H} C_{GH}^{\underline{j}} \end{aligned}$$

The formulas for the coordinate changes are needed to determine the directions of emanating branches which are used for offset of the numerical pathfollowing. From the numerical determination of a bifurcation point $(\tilde{u}_H^*, \lambda^*)$ with isotropy group $H$ and type $\vartheta_H^i$ corresponding to an irreducible representation $\vartheta_H^i$ the kernel of $Dg_H(\tilde{u}_H^*, 0, \lambda^*)$ is known to be spanned by vectors $t_1, \dots, t_{n_i}$ with

$$(14) \qquad \begin{aligned} t_{\mu H}^{\tilde{i}} &= 0 \qquad \forall \mu = 1, \dots, n_i, \forall \tilde{i} = 1, \dots, h_H, \tilde{i} \neq i, \\ t_{\mu H}^{il} &= 0 \qquad \forall l = 1, \dots, n_i, l \neq \mu, \ t_{\mu H}^{i\mu} = z, \quad z \in \mathbb{R}^{c_H^i}. \end{aligned}$$

Then a $K$–invariant vector in the kernel in reduced $K$–coordinates, where $K$ is a bifurcation subgroup of type $\vartheta_H^i$, is given by

$$(15) \qquad \hat{t} = t_K^1 = (C_{GK}^1)^T C_{GH} \left( \sum_{l=1}^{n_i} b_l \cdot t_i \right),$$

where $b_l$ are arbitrary numbers (for example $b_1 = 1, b_2 = 2, b_3 = 1$).

While numerical pathfollowing is done a branch of solutions with isotropy $K$ a symmetry breaking bifurcation point $(\tilde{u}^*, 0, \lambda^*)$ with the isotropy of a bifurcation supergroup $H$ has to be detected. This is done by a *symmetry monitor* sm $: \ \mathbb{R}^{c_K^1} \rightarrow \mathbb{R}^{c_K^1 - c_H^1}$. Let $u_K = (\tilde{u}, 0) = (u_K^1, 0)$ be a $K$–invariant point in $K$–coordinates which we have chosen. If $G = H$ then $u_K$ is decomposed as in (13). This is not ascertained for $H \neq G$. Then

$$(16) \qquad \hat{u}_K^1 = (C_{HK}^1)^T (C_{GH})^T C_{GK}^1 \, u_K^1,$$

is a reduced $K$–invariant vector in such coordinates that

$$(17) \qquad (\hat{u}_K^1)_l = 0, \quad \forall l = 1 + c_H^1, \dots, c_K^1,$$

if this vector is $H$–invariant. Thus a symmetry monitor is given by

$$(18) \qquad \mathrm{sm}(\tilde{u}) = (\hat{u}_K^{112}, \dots, \hat{u}_K^{11h_H}).$$

## 5. Action of $G$ on the bifurcation graph

The aim of this section is to show how unnecessary symbolic and numerical computations are avoided by the principle of conjugacy. Recall the linear representation $\vartheta : G \rightarrow GL(\mathbb{R}^n)$, and the system $F(x, \lambda) = 0$ in (2) which is equivariant with respect to $\vartheta$. $G$ acts on $\mathbb{R}^n$ by $\vartheta(t)$. If for a given $x \in \mathbb{R}^n$ the transformed $\vartheta(t)x \neq x$, then $\vartheta(t)x$ is called a *conjugate* vector to $x$. If $x$ is a solution of (2), then also its conjugates are solutions. $G$ acts on the set of subgroups of $G$ by $tHt^{-1}$ which in case $tHt^{-1} \neq H$ is called *conjugate subgroup* of $H$. Both fit together in the sense that the isotropy group of $\vartheta(t)x$ is $tG_x t^{-1} = G_{\vartheta(t)x}$.

The conjugate elements of $x$ form the *orbit* $O_x$. The order of $O_x$ is equal to the *index* $m$ of $H = G_x$ in $G$, which is the number of left cosets of $H$ in $G$. Once a solution $x$ is found numerically the conjugates are easily obtained by a cycle of group operations

$$(19) \qquad \begin{aligned} x_0 &:= x, \\ x_1 &:= \vartheta(r_i)x_{i-1}, \qquad i = 1, \dots, m-1, \end{aligned}$$

where $s_i := r_i \cdot \ldots \cdot r_1$ are representatives of the *left cosets* of $G/H$ and $r_m := s_{m-1}^{-1}$ gives the original vector $\vartheta(r_m)x_{m-1} = x$.

The group

$$N_G(H) := \{t \in G | \ tHt^{-1} = H\},$$

is called the normalizer of $H$ in $G$. For the isotropy group $H = G_x$ the conjugate element $\vartheta(t)x, t \in N_G(H) - H$ has the same isotropy group $G_x = G_{\vartheta(t)x}$.

If $K$ is a bifurcation subgroup of $H$ there are only two possibilities for the normalizer. Either $N_H(K)/K \cong \mathbb{Z}_2$ or $N_H(K) = K$. The first case $N_H(K)/K \cong \mathbb{Z}_2$ corresponds to a pitchfork bifurcation point and $K$ is thus called *symmetrical* [1].

While the action of $G$ on the subgroups is determined, the normalizer $N_G(K)$ is derived which easily gives $N_H(K) = N_G(K) \cap H$ for the subgroups $H$ as well.

The action of $G$ on its subgroups is much more sophisticated.

**Definition 5.10.** *(see [13]) : Let $H$ be a subgroup of $G$, $\rho : H \to GL(V)$ a linear representation, $t \in G$.*

$$\rho^t : tHt^{-1} \to GL(V), \quad \rho^t(tst^{-1}) := \rho(s), \quad \forall s \in H,$$

*is called the* conjugate representation *of $\rho$.*

Of course it may happen that $\rho^t$ is equivalent to $\rho$. The group operations with this property form the *inertia group* $H^\rho$ of $\rho$. For each irreducible representation $\vartheta^i_H$ the conjugate representation $(\vartheta^i_H)^t$ is an irreducible representation of the conjugate group $tHt^{-1}$ and thus equivalent to one $\vartheta^\iota_{tHt^{-1}}$. For $t \in N_G(H)$ this is a permutation of irreducible representations of the same dimensions. By

$$(20) \qquad t^I_H : \{1, \ldots, h_H\} \to \{1, \ldots, h_{tHt^{-1}}\}, \quad \left((\vartheta^i_H)^t \sim \vartheta^{t^I_H(i)}_{tHt^{-1}}\right)$$

we denote the induced mapping between indices of irreducible representations.

Because it is not convenient to handle with the equivalence of representations, the mapping $t^I_H$ is determined with the characters $\chi^i$.

**Proposition 5.11.** *Let $H$ be a proper subgroup of $G$ and $K$ a proper subgroup of $H$. Let $d_{ij}$ and $C^{ij}$ denote the innerconnectivity multiplicities and matrices with respect to $H$ and $K$. Let $c^i_H$ denote the multiplicity of $\vartheta^i_H$ of $\vartheta \downarrow H$ for a given representation $\vartheta : G \to GL(\mathbb{R}^n)$. Then*

a) $d_{ij} = d_{t^I_H(i) t^I_K(j)} \quad \forall t \in G \quad \forall i = 1, \ldots, h_H, j = 1, \ldots, h_K.$

b) $\vartheta^i_H(t) C^{i1}$ *is for all $t \in H$ the innerconnectivity matrix with respect to $H$ and $tKt^{-1}$.*

c) $c^i_H = c^{t^I_H(i)}_{tHt^{-1}} \quad \forall t \in G.$

**Proposition 5.12.** *If $K$ is a bifurcation subgroup of $H$ of type $\vartheta^i_H$, then $\forall t \in G$ the conjugate group $tKt^{-1}$ is a bifurcation subgroup of $tHt^{-1}$ of type $\vartheta^{t^I_H(i)}_{tHt^{-1}}$. If $H$ is a relevant subgroup with respect to $\vartheta : G \to GL(\mathbb{R}^n)$ then $tHt^{-1}, \forall t \in G$ are relevant subgroups.*

## Restriction to non-conjugate groups

This principle of conjugacy has consequences for the numerical computations as well as for the preparing group theoretic computations. The numerical pathfollowing procedure is applied to the reduced systems (5) with respect to non-conjugate relevant subgroups only.

Solutions of (2) including symmetry breaking bifurcation points are conjugated by cycle elements. Because the type and bifurcation subgroups are stored together with a bifurcation point, a sophisticated administration of these points is needed. This necessitates the knowledge of $t_H^I$ and the action of $t$ on the subgroups of $G$.

Once a symmetry breaking bifurcation point with isotropy of a non-conjugate $H$ is computed, formula (15) gives the directions of emanating branches. In SYM-CON the offset directions are computed for all bifurcation subgroups $K$ which are non-conjugate with respect to $H$. The other directions are given by conjugation. Computation of offset directions for subgroups $K$ which are non-conjugate in $G$ could have the disadvantage that the bifurcation supergroup is a conjugate group of $H$ and thus more of the innerconnectivity matrices $C^{i\underline{j}}$ are needed in the symbolic part.

Pathfollowing a branch with non-conjugate isotropy the symmetry monitor functions with respect to all bifurcation supergroups have to be considered.

The innerconnectivity multiplicities $d_{ij}$ are computed with respect to non-conjugate subgroups $H$ and non-conjugate subgroups $K$ of $H$ (conjugacy in $H$!). The others are given by conjugation. Especially for $t \in N_H(K)$ the relation $d_{i\underline{j}} = d_{i\,t_K^I(\underline{j})}$ holds where $\underline{j}$ and $t_K^I(\underline{j})$ are indices of conjugate irreducible representations.

As few as possible innerconnectivity matrices $C^{i\underline{j}}$ should be computed. They have to be computed for $G$ and all non-conjugate $H$. Additionally, some other $C^{i1}$ are needed. Because conjugate solutions with isotropy $\tilde{H}$ are computed numerically using $M_{\tilde{H}}^1$, in the symbolic preparation of $M_{\tilde{H}}^1$ the matrices $C^{i1}$ with respect to $G$ and conjugate groups of $H$ are needed. In the computation of these $C^{i1}$ itself Proposition 5.11 is applied. Secondly, these includes matrices $C^{i1}$ which are needed for the determination of offset directions. $K$ being a bifurcation subgroup of a non-conjugate subgroup $H$ and $K$ being non-conjugate in $H$, the matrices $C^{i1}$ with respect to $G$ and $K$ are needed. Thirdly, some $C^{i1}$ are needed for the symmetry monitor functions. To avoid the use of $C^{i\underline{j}}, \underline{j} \geq 2$ with respect to $G$ and conjugate subgroups $\tilde{H}$, the formulas (16) and (17) for the symmetry monitor functions have to be modified [7].

For this formula the matrices $C^{i1}$ with respect to non-conjugate $H$ and its bifurcation subgroups $K$ are necessary. Computing these $C^{i1}$ itself again Propostion 5.11 b.) may be used.

During the numerical computations the action $t_{\tilde{H}}^I$ on irreducible representations (20) is needed only for some group elements $t$ and subgroups $\tilde{H}$, if $H$ is a non-conjugate relevant subgroup and $t = r_i$ a member of its cycle with $s = s_{i-1} = r_{i-1} \cdot \ldots \cdot r_1$ and $\tilde{H} = sHs^{-1}$.

## 6. OVERVIEW OF THE ALGORITHM

The group theoretic computations in SYMCON simulate a mathematician who knows the analysis (equivariant braching lemma), looks up the irreducible representations, and then prepares and implements a given equivariant system (2).

The SYMMETRY Package [6] is used which contains functions for the computation of symmetry adapted bases and a database containing irreducible representations for the small dihedral groups, $S_4$ and others.

The first task of the algorithm is the determination of all subgroups which is implemented like the other grouptheoretic computations in RLISP. Starting with an arbitrary set of elements of fixed order including a known subgroup elements are eliminated and new are chosen with a weighting procedure until a subgroup is found. This works fine for the small groups from the SYMMETRY Package.

A group isomorphism between these computed subgroups and a stored abstract one from the Package are constructed based on a search of generators which fulfill the conditions stored in the abstract group. This enables the use of the stored irreducible representations for the subgroups.

While conjugacy is exploited the rest of the action of $G$ on the subgroup tree is determined giving the normalizers as well. The algorithm from representation theory gives the innerconnectivity multiplicities $d_{ij}$ and matrices $C^{ij}$ which determines with Lemma 3.5 the bifurcation graph. The computation of cycles is the last point in the problem independent part. Once this problem independent computations have been done for $G$, they are valid for all $G$-equivariant systems. There is no need to run it a second time.

Based on some stored information $(\vartheta_G^i, d_{ij}, C_{ij}, t_H^I$, bifurcation graph with conjugate and symmetric groups, cycles) the equivariant systems (2) are tackled starting with determination of multiplicities $c_G^i$, $c_H^i$, the relevant subgroups (see Def. 3.8) and parts $M_G^{ij}$ of the transformation matrix.

During these computations and the following C code generation the intensive exploitation of conjugacy was the important aim.

First the group operations $\vartheta(t)$ including its action on the tree of relevant subgroups and the actions on irreducible representations $t_H^I$ (see (20)) needed for bifurcation point administration are generated with GENTRAN. Then for each relevant subgroup coordinate transformation (11), symmetry monitors (18) and offset directions (15) are generated.

A clear distinction between a set of non-conjugate relevant subgroups and their conjugates are made as the reduced system (5) and Jacobian blocks $A_i$ are generated for the non-conjugates only. Their efficient evaluation uses a special property, see Section 7. In the numerical part the pathfollowing of non-conjugate solutions is done with ALCON (see [2], [4]) applied to the reduced systems. The conjugate solutions are computed by group operations in a cycle.

While pathfollowing the symmetry monitor functions and determinants of Jacobian blocks are evaluated detecting bifurcation points leading to higher or smaller symmetry. The Jacobian block $A_i$ appears in the augmented system which has the bifurcation point as a solution ([18]).

## 7. EXPLOITATION OF QUASI-PERMUTATION REPRESENTATION

**Definition 7.13.** *A linear representation $\vartheta$ is called a **permutation representation**, if all entries of the matrices $\vartheta(t)$ are zero or 1.*

In [9] this property is exploited. If most of the $\vartheta(t)$–entries are zeros and 1, but a few couples are different numbers, we call $\vartheta$ a **quasi–permutation representation**.

The improved version of SYMCON uses this property to make the numerical evaluation of the symmetry reduced systems and Jacobian blocks more efficient using Computer Algebra methods.

The equivariance, rewritten as

$$F(x, \lambda) \;=\; \vartheta(t)^T \; F\Big(\vartheta(t)x, \lambda\Big) \qquad \forall t \in G$$
$$D_x F(x, \lambda) \;=\; \vartheta(t)^T \; D_x F\Big(\vartheta(t)x, \lambda\Big)\vartheta(t)$$

implies the existence of sets of indices $I \subset \{1, 2, \ldots, n\}$ and $J \subset \{1, \ldots, n\}^2$ with the property

$$F_i(x, \lambda) = \sum_{k \in I}\sum_{t \in G} a_{ik}^t F_k\Big(\vartheta(t)x, \lambda\Big) \qquad\qquad \forall i \notin I\,,$$
$$(D_x F)_{ij}(x, \lambda) = \sum_{(\nu, u) \in J}\sum_{t \in G} b_{ij\nu u}^t (D_x F)_{\nu u}\Big(\vartheta(t)x, \lambda\Big) \quad \forall (i, j) \notin J\,,$$

where $a_{ik}^t$ and $b_{ij\nu u}^t$ are real numbers. In case of a permutation representation for each $i \notin I\Big((i, j) \notin J\Big)$ only one coefficient is unequal zero. An algorithm determining the sets $I$ and $J$ is easily implemented using symbolic computations. Then, in REDUCE, the matrix multiplications are performed for the reduced systems and Jacobian blocks

$$\tilde{g}(\tilde{u}, \lambda) \;=\; (M_H^1)^T \; F(M_H^1\tilde{u}, \lambda)\,,$$
$$A_i(\tilde{u}, \lambda) \;=\; (M_H^{i1})^T \; D_x F(M_H^1\tilde{u}, \lambda)M_H^{i1}\,,$$

substituting the expressions above. Some group elements $t$ may be eliminated by choosing one representative of each left coset of $H$ in $G$.

The functions $F_k$, $k \in I$ and $DF_{\nu u}$, $(\nu, u) \in J$ are generated as C–code as well as the reduced equations $\tilde{g}(\tilde{u}, \lambda)$ and blocks $A_i$ for each non–conjugate relevant subgroup $H$. Then the numerical evaluation of $\tilde{g}(\tilde{u}, \lambda)$ $(A_i(\tilde{u}, \lambda)$ resp.) consists of function calls as $F_k$, $k \in I$ $(DF_{\nu u}$, with $(\nu, u) \in J$ resp.) at $M_H^1\tilde{u}$ and some conjugate vectors. This is also the best way in view of minimization the amount of code.

| example | REDUCE | compile | link | numeric | file size |
|---------|--------|---------|------|---------|-----------|
| brussD3 | 20 sec | 9.2u sec | 6.5u sec | 2.3 sec | 16960 Bytes |
| brussD4 | 55 sec | 19.3u sec | 6.3u sec | 8.38 sec | 33211 Bytes |
| brussD6 | 350 sec | 73.7u sec | 6.7u sec | 16.8 sec | 101741 Bytes |
| brussS4 | 127 sec | 43.0u sec | 6.7u sec | 8.28 sec | 61168 Bytes |
| dome | 3120 sec | 333.2u sec | 8.4u sec | 1558.6 sec | 308723 Bytes |

TABLE 1. Performance of SYMCON on a Data General

| group | $K_4$ | $D_3$ | $D_4$ | $C_4$ | $C_5$ | $C_6$ | $D_6$ | $A_4$ | $S_4$ |
|-------|-------|-------|-------|-------|-------|-------|-------|-------|-------|
| cpu in ms | 330 | 820 | 2140 | 150 | 240 | 690 | 11530 | 8520 | 242 670 |

TABLE 2. Computation of bifurcation graphs on a Data General

## 8. Example: Hexagonal lattice dome

In [5] the example of an hexagonal lattice dome (Healey [10], see also [16]) is fully described and an overview of stable solutions is given. We found more than 600 symmetry breaking bifurcation points of different types which are connected by ca. 200 branches of solutions with non-conjugate isotropy groups. Table 1 gives the cpu for the hexagonal lattice dome and the brusselators (see e.g. [4]) with respect to different groups which have been often treated. The table shows that the computing time is dominated by the symbolic part which automates work done normally by pencil and paper. The exploitation of symmetry brings reliability and robustness to the numerical investigations.

**Acknowledgement.** I would like to thank J. Neubüser, A. Hohmann, B. Werner.

### References

1. M. Dellnitz, B. Werner, *Computational Methods for bifurcation problems with symmetries — with special attention to steady state and Hopf bifurcation points.* J. Comp. Appl. Math. **26**, 97–123 (1989).
2. P. Deuflhard, B. Fiedler, P. Kunkel, *Efficient Numerical Pathfollowing Beyond Critical Points.* SIAM J. Numer. Anal. **24**, 912–927 (1987).
3. A. Fässler. E. Stiefel, *Group Theoretical Methods and Their Applications.* Birkhäuser (1992).
4. K. Gatermann, A. Hohmann, *Symbolic Exploitation of Symmetry in Numerical Pathfollowing.* Impact of Computing in Science and Engineering **3**, 330–365 (1991).
5. K. Gatermann, A. Hohmann, *Hexagonal Lattice Dome - Illustration of a Nontrivial Bifurcation Problem.* Konrad–Zuse–Zentrum für Informationstechnik Berlin, Preprint SC 91-8 (1991).
6. K. Gatermann, *SYMMETRY.* Reduce network library, available at netlib@ rand.org, (1991).
7. K. Gatermann, *Computation of Bifurcation Graphs.* Konrad–Zuse–Zentrum für Informationstechnik Berlin, Preprint SC 92-13 (1992).
8. M. Golubitsky, I. Stewart, D. G. Schaeffer, *Singularities and Groups in Bifurcation Theory.* Vol. **I**, **II**, Springer, New York (1985), (1988).
9. K. Georg, R. Miranda, *Exploiting Symmetry in Solving Linear Equations.* In E.L. Allgower, K. Böhmer, M. Golubitsky (Eds.) Bifurcation and Symmetry, pp. 157–168. Birkhäuser, Basel (1992).
10. T. J. Healey, *A group–theoretic approach to computational bifurcation problems with symmetry.* Comp. Meth. Appl. Mech. Eng. **67**, 257–295 (1988).
11. T. J. Healey, *Numerical bifurcation with symmetry: diagnosis and computation of singular points.* In L. Kaitai, J. E. Marsden, M. Golubitsky, G. Iooss (Eds.), Int. Conf. Bifurcation Theory and its Num. Anal., pp. 218–227. Xian University Press (1989).
12. K. Ikeda, K. Murota, *Bifurcation analysis of symmetric structures using block-diagonalization.* Comp. Meth. Appl. Mech. Engr. **86**, 215–243 (1991).
13. G. James, A. Kerber, *Representation Theory of the Symmetric Group.* Addison-Wesley Publ. Comp., Reading, Massachusetts (1981).
14. A. C. Hearn, *REDUCE User's Manual, Version 3.3.* The RAND Corp., Santa Monica (1987).
15. K. Murota, K. Ikeda, *Computational use of group theory in bifurcation analysis of symmetric structures.* SIAM J. Sci. Stat. Comp. **12**, 273–297 (1991).
16. P. Stork, B. Werner, *Symmetry adapted block diagonalization in equivariant steady state bifurcation problems and its numerical application.* Advances in Math. **20**, 455–487 (1991).
17. A. Vanderbauwhede, *Local Bifurcation and Symmetry.* Pitman, Boston (1982).
18. B. Werner, *Eigenvalue Problems with the Symmetry of a Group and Bifurcations.* In D. Roose, B. de Dier, A. Spence (Eds.), Continuation and Bifurcation Numerical Techniques and Applications, NATO ASI SERIES C Vol. 313, pp. 71–88. Kluwer, Dordrecht (1990).

Konrad-Zuse-Zentrum Berlin, Heilbronner Str. 10, D–1000 Berlin 31, Federal Republic of Germany, Email address: Gatermann@sc.zib-berlin.de

Lectures in Applied Mathematics
Volume **29**, 1993

# Caustics in Optimal Control: An Example of Bifurcation When the Symmetry is Broken

Zhong Ge

## 1    Introduction

This is a report on a recent joint work with Ivan Kupka, which provides an interesting example of bifurcations of caustics when symmetries are broken.

Let $M$ be a three-dimensional manifold. We consider a Hamiltonian system on $T^*M$,

$$H(x,p) = < X_1(x), p >^2 + < X_2(x), p >^2, \quad (x,p) \in T^*M, \qquad (1)$$

where $< \cdot, \cdot >$ is the dual bracket between $TM$ and $T^*M$, $X_1, X_2$ are vector fields on $M$, satisfying the bracket generating condition

$$X_1, X_2, [X_1, X_2] \ span \ \mathbb{R}^3.$$

( This is satisfied almost everywhere for a pair of generic vector fields $X_1, X_2$. )

1991 Mathematics Subject Classification. Primary 58F05; Secondary 49J15.

Supported by the Ministry of Colleges and Universities of Ontario and the Natural Sciences and Engineering Research Council of Canada

This paper is in final version.

This kind of Hamiltonian systems has appeared in sub-Riemannian geometry ( [5], [6], [14], [16]), optimal control ([3], [13]) and the theory of sub-elliptic operators ([14]). More precisely, this is the Hamiltonian system for the optimal problem of minimizing

$$\int_0^1 (u_1^2 + u_2^2)dt \quad \rightarrow \quad min$$

over the curves $x$ subject to the constraint

$$\dot{x} = u_1 X_1 + u_2 X_2, \tag{2}$$

with fixed end point condition

$$x(0) = x_0, x(1) = x_1.$$

The bracket generating condition ensures that for any two points $x_0, x_1$, there is a path $x$ satisfying ( 2) and joining $x_0$ to $x_1$.

**Example.**    Consider a point mass joined to a flywheel by an elastic, massless rod. The flywheel is in turn joined to a fixed table in such a way that the flywheel can rotate without friction. Let $(x, y)$ be the coordinates of the point mass. We assume that the total angular momentum is zero,

$$\dot{\theta} = x\dot{y} - y\dot{x}.$$

We consider the problem of minimizing the kinetic energy

$$\int_0^1 ((\dot{x})^2 + (\dot{y})^2)dt \rightarrow min$$

over the space of curves $(x(t), y(t), \theta(t))$, subject to the constraint, with fixed end points.

This problem can also be reformulated as an isoperimetric problem on $\mathbf{R}^2$. Let $(x(t), y(t))$ be a curve on $\mathbf{R}^2$, and

$$\theta(t) = \int_0^t (x(t)dy(t) - y(t)dx(t)),$$

be the " area " of the path, so

$$\dot{\theta} = x(t)\dot{y}(t) - y(t)\dot{x}.$$

The isoperimetric problem is to minimize

$$\int_0^1 ((\dot{x})^2 + (\dot{y})^2)dt$$

subject to the boundary condition that $(x(0), y(0), \theta(0)), (x(1), y(1), \theta(1))$ are fixed.

By Pontryagin's Maximal principle, the Hamiltonian of this optimal problem is

$$H = (p_1 - yp_3)^2 + (p_2 + xp_3)^2. \tag{3}$$

More generally, the isoperimetric problem on any surface can be put in the form (1).

**Definition of caustics.**

Now we fix the starting point $x_0$ and let the end point $x_1$ vary. We consider the dependence of the extremals on $x_1$. Roughly speaking, the caustic of $x_0$ is the set of points where this dependence is not smooth; in other words, **the caustic is where the extremals bifurcate.**

We fix $x_0$ and consider the set of extremals from $x_0$, i.e. the solutions of the Hamiltonian system ( 1) with the initial condition

$$x(0) = x_0.$$

A solution ( at time 1 ) depends on the value of $p$ at time $t = 0$, $p(0) = p_0$. So we can write the solution as a function of $p_0$

$$x_1 = \Theta(p_0).$$

**Definition.** The caustic of $x_0$ is the set of $x_1 = \Theta(p_0)$, where the Jacobian

$$Det\frac{D\Theta(p_0)}{Dp_0} = 0.$$

In the classical calculus of variations ( e.g. optics, mechanics, Riemannian geometry), the caustic does not occur nearby, and hence is a global phenomenon. However, in our case, the situation is different, because of the presense of the constraint.

**Lemma 1** *([3]) The caustic of $x_0$ occurs arbitrarily close to $x_0$, and so is a local phenomenon.*

**Example.** As mentioned earlier, for the isoperimetric problem on $\mathbb{R}^2$, the Hamiltonian is

$$H = (p_1 - yp_3)^2 + (p_2 + xp_3)^2.$$

There is a rotation symmetry around the z-axis, and a rotation will permute the extremals. Thus there is an infinite number of extremals joining a point $(0, 0, z)$ to $(0, 0, 0)$. So the caustic of $(0, 0, 0)$ is the z-axis.

This is also true for a right-invariant control system on any three dimensional Lie group ([16]). In all these examples, there is a rotation symmetry, and the caustic is a straight line. And it has been conjectured that the caustic is always a straight line.

On the other hand, however, for arbitrary $X_1, X_2$, there is no such rotation symmetry. So a natural question to ask is how the caustics bifurcate when the symmetry is lost? Is it still a straight line ?

Our result shows that this bifurcation is rather complicated, and the caustic is in general not a straight line.

## 2    Main Results

To state our result, we need a normal form of the Hamiltonian function.

**Lemma 2** *We can choose a local coordinates at $x_0$ in which $H$ is written as*

$$H = (p + a_0xyq + (y/2 + b_1y^3/3 + b_2xy^2/2 + b_3x^2y + fzy)r)^2$$
$$+(q - (x/2 + cx^2/2)r)^2 + higher\ order$$

*Here $a_0, b_1, b_2, b_3, c, f$ are constants.*

**Theorem 3** *In the local coordinates above, if we introduce $y' = y - cz$, then the caustic of $0$ within the half space $z > 0$ is given by*

$$x = \pi\rho^3\{(\frac{7}{2}c^2 + 2b_1 - 2b_3)sin^3(\theta_0) +$$
$$+ b_2(-3cos(\theta_0) + 2cos^3(\theta_0))\} + O(\rho^4)$$
$$y' = \pi\rho^3\{-(\frac{7}{2}c^2 + 2b_1 - 2b_3)cos^3(\theta_0) +$$
$$+ b_2(-3sin(\theta_0) + 2sin^3(\theta_0))\} + O(\rho^4)$$
$$z = \pi\rho^2 + O(\rho^3)$$

*where $0 \leq \theta_0 \leq 2\pi$.*

Though the rotation symmetry is lost, the caustic still seems to exhibit some discrete symmetries. For example, the above normal form has a 4-fold symmetry, i.e. a $Z_4$-symmetry. These discrete symmetries in the normal form have not been well understood yet.

This theorem is proved by a lengthy and complicated computation. We expand a solution of the Hamiltonian system ( 1) in the energy $\rho = H^{1/2}$. Then the first term of the expansion is a solution of the Hamiltonian system ( 3) and the other terms can be determined inductively. Instead of reproducing the computation here, we will discuss topics related to this result.

# 3    Discussions

1. It follows that for a generic sub-Riemannian metric, the caustic of $x_0$ within a sufficiently small neighborhood of $x_0$ has two branches. Each one is homeomorphic to one of the following six surfaces:

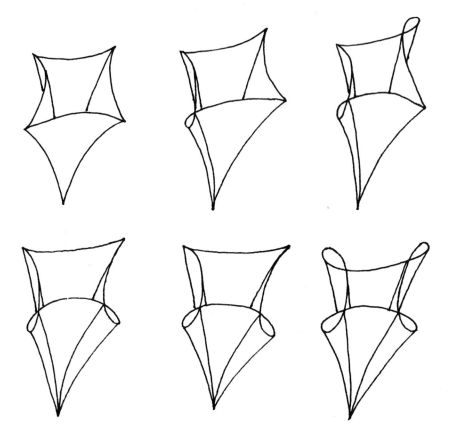

In these pictures, the total number of the cusps and the self-intersections of the caustic is less than or equal to 4.

2. It has been an open problem as how to define the curvature in sub-Riemannian geometry. Recently, Bryant-Hsu, Wilkens have computed the local invariants of this metric. However, the geometric meaning of their invariants is not clear yet. The above result about the caustics gives a geometric meaning of their local invariants, though the exact formulation has not been worked out yet. This formulation probably is similar to the well-known relation between curvature and conjugate locus in Riemannian geometry.

3. It would be interesting to know the structure of the caustics in the large as well. For example, when does every extremal from $x_0$ has a conjugate point in finite time? This is related to the problem of defining

the curvature in sub-Riemannian geometry. This has been studied in [8], which we briefly describe in the following. Recall that the structure equation for the sub-Riemannian metric is given by

$$
\begin{aligned}
d\omega_1 &= A\omega_2 \wedge \omega_3 + B\omega_1 \wedge \omega_2, \\
d\omega_2 &= D\omega_1 \wedge \omega_3 + E\omega_1 \wedge \omega_2, \\
d\omega_3 &= \omega_1 \wedge \omega_2
\end{aligned}
$$

where $\omega_1, \omega_2$ is a coframe for the sub-Riemannian metric, and $\omega_3$ is a 1-form which defines the distribution by $\omega_3 = 0$. Let $e_i, i = 1, 2, 3$, be vector fields dual to $\omega_i, i = 1, 2, 3$. A vector field $v = a_1 e_1 + a_2 e_2 + a_3 e_3$ is a Killing field if it is a contact transformation and preserves the sR-metric. We say $v$ is **transversal** if $a_3 \neq 0$ everywhere. The **Gaussian curvature with respect to the transversal Killing field** $v$ is defined to be

$$
\begin{aligned}
K_v &= e_2(B - \frac{a_1}{a_3}) - e_1(E - \frac{a_2}{a_3}) - (B - \frac{a_1}{a_3})^2 - (E - \frac{a_2}{a_3})^2 + \frac{A - D}{2} \\
&\quad + \frac{1}{2a_3}(e_1(a_2) - Ea_2 + Ba_1 - e_2(a_1)) + \frac{1}{4}(e_2(a_1) - e_1(a_2))^2.
\end{aligned}
$$

**Theorem 4** *Suppose $v = a_1 e_1 + a_2 e_2 + a_3 e_3$ is a transversal Killing field. If $K_v > c > 0$, then every sub-Riemannian geodesic has a conjugate point in finite time and hence $M$ is compact.*

4. The caustics we obtained do not seem to have appeared in the list of generic caustics in the book by Arnold et al. [1], [2]. This suggests that the caustics in the sub-Riemannian context are rather special.

5. Closely related to the caustics is the cut-locus, the set of points past which an extremal loses its global optimality. If the metric is analytic, the cut-locus of $x_0$ is the set of points at which the sub-Riemannian distance function $d(x, x_0)$ is not analytic. The cut-locus is studied in detail in [6].

Moreover, if the sub-Riemannian metric is analytic, then combining the result that the cut-locus is sub-analytic ( [6], [15]) and a structure theorem on the cut-locus ([6]), we obtain the following result

**Corollary 5** *If the sub-Riemannian metric is analytic and generic, then the cut-locus consists of four half-planes, the boundaries of which are its intersection with the caustic, and are either the cusps or the self-intersections of the caustic.*

It is interesting to compare the generic caustic in the Riemannian case: pictures of the caustics in that case are very different, cf. Buchner [4].

6. It should be pointed out that our result is only valid for the generic case. In fact, there is a sub-class of sub-Riemannian metrics for which the above result is not valid: those from the isoperimetric problems on surfaces. In view of the fact that the isoperimetric problem has an even older history than sub-Riemannian geometry, it would be interesting to study the generic caustics within this sub-class of metrics.

# References

[1] V. I. Arnold, *Singularities in the variational calculus,* J. of Sov. Math., 27(1984), 2679-2713.

[2] V. I. Arnold, S. M. Gusein-Zade, A. N. Varchenko, *Singularities of differentiable maps,* vol. 1, 1985, Birkhauser, Boston.

[3] R. Brockett, *Control theory and singular Riemannian geometry,* in New Directions in Applied Mathematics, Springer-Verlag, New York, 1981.

[4] M. Buchner, *The structure of the cut locus in dimension less than or equal to six,* Comp. Math., 37(1978), 103-119.

[5] Zhong Ge, *On a variational problem and the spaces of horizontal paths*, Pacific J. of Mathematics, 149(1991), 61-93.

[6] Zhong Ge, *Horizontal paths space and Carnot-Caratheodory metrics*, Pacific J. of Mathematics, 1993.

[7] Zhong Ge, *On the cut points and conjugate points in a constrained variational problem*, The Fields Institute Preprint, 1992.

[8] Zhong Ge, *On the global geometry of sub-Riemannian three-manifolds*, The Fields Institute Preprint, FI93-CT01, 1993.

[9] Zhong Ge, *Collapsing Riemannian metrics to sub-Riemannian metrics and Laplacians to sub-Laplacians*, Canadian J. Math., 1993.

[10] J. Kogan, *Bifurcation of extremals in optimal control*, Lect. Notes in Math., vol.1216, Springer, 1980.

[11] S. Kobyashi, *On conjugate and cut locus*, in Studies in Global Geometry and Analysis, S. S. Chern ed., MAA Studies in Mathematics, vol.4, 1967, 96-122.

[12] J. Marsden, *Geometric mechanics, stability and control*, preprint.

[13] R. Montgomery, *The isoholonomic problem and some applications*, Comm. Math. Phys, 128(1990), 565-592.

[14] R. S. Strichartz, *Sub-riemannian geometry*, J. Differential Geom., 24(1986), 221-263.

[15] H. Sussmann, *Subanalyticity of the distance functions for real analytic sub-Riemannian metrics on three dimensional manifolds*, Report SYCON-91-05a, Rutgers, 1991.

[16] A. M. Vershik, C. Ya. Gershkovich, *Non-holonomic manifolds and nilpotent analysis*, J. Geom. Phys., 5(1988), no 3.

[17] A. Weinstein, *The cut locus and conjugate locus of a Riemannian manifold,* Ann. of Math., 87(1968), 29-41.

Fields Institute for Research in Mathematical Sciences, 185 Columbia St. West, Waterloo, Ontario N2L 5Z5, Canada

Lectures in Applied Mathematics
Volume **29**, 1993

# Symmetry Aspects in Numerical Linear Algebra With Applications to Boundary Element Methods

KURT GEORG AND RICK MIRANDA

ABSTRACT. A linear operator equation $\mathcal{L}f = g$ is considered in the context of boundary element methods, where the operator $\mathcal{L}$ is equivariant under the action of a given finite symmetry group. By introducing a suitable discretization, it is seen that the discretization matrix is also equivariant under the given group which now acts as a permutation group on indices. It is hence possible to transform the matrix into a symmetry adapted block diagonal form in the sense of Stiefel and Fässler. In the present paper, an equivalent decomposition of the given linear problem is given via tensor products. The reduced linear equations can be efficiently solved without first determining bases for the corresponding subspaces. A small numerical example illustrates the approach.

## Introduction

Recent efforts have shown the efficacy of applying group theoretical methods to the numerical treatment of boundary value problems via finite differences and finite elements, see [1], [2], [3], [7], [8], [9], [10], [16]. The articles, e.g., [1], [3], [8], have demonstrated that the use of discretizations of partial differential equations which are suitably adapted to respect symmetry properties yield highly useful decompositions which can reduce the computational effort, improve the numerical conditioning of problems, and significantly facilitate the study of bifurcation behavior at singularities. The results in the present paper extend those in [1], [3] via a systematic exploitation of group representation theory.

1991 *Mathematics Subject Classification.* 65R20, 65F05, 20C15, 45L10, 31C20.

*Key words and phrases.* Equivariant linear operator equations, discretization methods, boundary element methods, representation theory, symmetry groups, linear equation solvers.

Partially supported by the National Science Foundation under grant number DMS-9104058

This paper is in final form and no version of it will be submitted elsewhere for publication.

July 1991, revised March 1993

Our attention is centered on solving the linear equations obtained via a suitable discretization. It is assumed that the underlying operator equation respects the symmetries of a given group action, i.e., we assume that the operator is equivariant. We are particularly interested in the case of boundary integral operators since they lead to full discretization matrices and since they typically come up in "classical" problems which display many symmetries. By employing only "suitable" discretization methods, it is seen that the discretization matrix is also equivariant under the given group which now acts as a permutation group on indices. It is hence possible to transform the matrix into a symmetry adapted block diagonal form, see [16] and also [13], [19].

Our method is based on the same diagonalization. We show how a tensor product approach leads to an efficient way of generating the reduced linear equations. In [11], we also argue that the given method is most efficient in the sense that a general problem cannot be further decomposed. A numerically equivalent approach introducing a generalization of the Fourier transform is given in [17].

The tensor products lead to a natural block structure suitable for numerical implementations. A small numerical example illustrates this point. It is possible to dismiss the tensor aspect altogether and perform a similar approach in terms of index manipulations. This has subsequently been done by the authors in [11].

## 1. Representations of Finite Groups

We begin our investigation by recalling some facts about representations of finite groups, see, e.g., the books [14], [15] for details. The main reason for doing this is to introduce the above mentioned tensor notation.

If $V$ is a complex Hilbert space, then we denote by $L(V)$ the algebra of bounded linear operators from $V$ into $V$, and by $U(V) \subset L(V)$ the multiplicative group of unitary operators.

Let $\Gamma$ be a finite group. We call a map $T : \Gamma \to L(V_T)$ a *unitary representation* if $T$ is a group homomorphism from $\Gamma$ into $U(V_T)$, where $V_T$ is a complex Hilbert space associated with $T$. The dimension of $T$ is by definition the dimension of the space $V_T$ on which $\Gamma$ acts via $T$, i.e., $\dim T = \dim V_T$.

Given two unitary representations $S, T$ of $\Gamma$, we call a map $F : V_T \to V_S$ *equivariant* if $S(a) F = F T(a)$ holds for all $a \in \Gamma$. $S, T$ are called *equivalent* if there is a linear isomorphism $F : V_T \to V_S$ which is equivariant.

A unitary representation $T$ is called *reducible* if there is a nonzero proper closed linear subspace $V_0 \subset V_T$ which is $T$-invariant. Otherwise $T$ is called *irreducible*. We will usually denote the irreducible unitary representations with lower case Greek letters. Given a finite group $\Gamma$, we denote by $\mathcal{U}_\Gamma$ a maximal set of irreducible non-equivalent unitary representations. For many groups $\Gamma$ such lists $\mathcal{U}_\Gamma$ can be looked up in literature.

In order to be able to conveniently express our method in a block structured way, we introduce the following definition: Let $T$ be a unitary representation with finite dimension. Recall that $L(V_T)$ is a Hilbert space under the scalar product $(A, B) \mapsto \text{trace}(AB^*)$. We define the map $\hat{T} : \Gamma \to L(L(V_T))$ via the composition $\hat{T}(a)\alpha := T(a)\alpha$ for $a \in \Gamma$ and $\alpha \in L(V_T)$. It follows immediately that $\hat{T}$ is again a unitary representation.

Given a unitary representation $T$ of $\Gamma$ and a function $f : \Gamma \to \mathbb{C}$, we introduce the operator

$$T(f) := \frac{1}{|\Gamma|} \sum_{a \in \Gamma} f(a) T(a)$$

where $|\Gamma|$ denotes the order of the group. If we endow the group algebra

$$\mathcal{F}(\Gamma) := \{f : \Gamma \to \mathbb{C}\}$$

with the convolution

$$(f * g)(a) = \frac{1}{|\Gamma|} \sum_{\substack{b,c \in \Gamma \\ bc=a}} f(b)\, g(c),$$

then it turns out that $f \in \mathcal{F}(\Gamma) \mapsto T(f) \in \mathrm{L}(V_T)$ is a *-homomorphism. In particular, $T(1)$ (for the function $1 \in \mathcal{F}(\Gamma)$) is an orthogonal projector, and it is easy to see that its range is the fixed point set of $T$, i.e.,

(1) $$\text{range}\, T(1) = \{x \in V_T : T(a)x = x \text{ for all } a \in \Gamma\}.$$

In particular, for an irreducible representation $\tau \in \mathcal{U}_\Gamma$, we obtain the orthogonal projectors

(2) $$P_{T,\tau} := T\big((\dim \tau)(\text{trace}\, \tau)\big) : V_T \to V_T$$

and

(3) $$Q_{T,\tau} := (T \otimes \hat{\tau})(1) : V_T \otimes L(V_\tau) \to V_T \otimes L(V_\tau).$$

The projectors $P_{T,\tau}$ are classical in group representation theory. A basic fact is the following orthogonal decomposition

(4) $$V_T = \bigoplus_{\tau \in \mathcal{U}_\Gamma} (\text{range}\, P_{T,\tau}).$$

The projectors $Q_{T,\tau}$ are introduced here as a convenient tool to describe our block structured approach to the above decomposition. The $P_{T,\tau}$ operators can be easily retrieved from the $Q_{T,\tau}$ operators via the following commutative diagram

(5)

$$
\begin{array}{ccc}
V_T \otimes L(V_\tau) & \xrightarrow{\ Q_{T,\tau}\ } & V_T \otimes L(V_\tau) \\
A \uparrow & & \downarrow B \\
V_T & \xrightarrow[\ P_{T,\tau}\ ]{} & V_T
\end{array}
$$

where

$$A : x \qquad \mapsto x \otimes \mathbb{I}_\tau$$
$$B : x \otimes \alpha \; \mapsto (\dim \tau)(\operatorname{trace} \alpha)x$$

and $\mathbb{I}_\tau$ denotes the identity on $V_\tau$.

## 2. Equivariant Linear Operators

Let us now consider the problem of solving the linear equation

$$(6) \qquad \mathcal{L}u = v, \quad u \in V_T, \; v \in V_S,$$

where $\mathcal{L} : V_T \to V_S$ is an equivariant linear operator with respect to two unitary representations $S$ and $T$ of the finite group $\Gamma$. We want to investigate how we can make use of the equivariance of $\mathcal{L}$ in order to solve (6) efficiently.

It is well-known that the decomposition (4) leads to a decomposition of the problem (6), see, e.g., [16] and also [13], [19]. Namely, the equivariance of $\mathcal{L}$ implies $S(f)\mathcal{L} = \mathcal{L}T(f)$ for all $f \in \mathcal{F}(\Gamma)$. In particular, we obtain the reduced problems by applying the projectors $P_{T,\tau}$ to both sides of (6), which gives

$$\mathcal{L}P_{T,\tau}u = P_{S,\tau}v.$$

Thus we treat the linear problem (6) as a direct sum of reduced linear subproblems

$$(7) \qquad \mathcal{L}_\tau u_\tau = v_\tau, \quad \tau \in \mathcal{U}_\Gamma,$$

where

$$u_\tau := P_{T,\tau}\, u, \quad v_\tau := P_{S,\tau}\, v$$

are the components of $u$ and $v$,

$$V_{T,\tau} := \operatorname{range} P_{T,\tau}, \quad V_{S,\tau} := \operatorname{range} P_{S,\tau}$$

are the orthogonal components of the Hilbert spaces $V_T$ and $V_S$ respectively, and the restricted operators

$$\mathcal{L}_\tau := \mathcal{L}|_{V_{T,\tau}} : V_{T,\tau} \to V_{S,\tau}$$

are the components of $\mathcal{L}$. Note that the solvability of the subproblems (7) is ensured by the solvability of the original system: $\mathcal{L}$ is bijective if and only if $\mathcal{L}_\tau$ are bijective for all $\tau \in \mathcal{U}_\Gamma$.

The question remains of how to solve the subproblems (7) efficiently. We propose to do this by considering the projectors $Q$ in (3) and retrieving the components of the direct sum via (5).

Let us denote by $\hat{\mathbb{I}}_\tau$ the identity map $L(V_\tau) \to L(V_\tau)$. Then $\mathcal{L} : V_T \to V_S$ has a natural extention to $\mathcal{L} \otimes \hat{\mathbb{I}}_\tau : V_T \otimes L(V_\tau) \to V_S \otimes L(V_\tau)$, and it is clear that this extension is again equivariant with respect to $T \otimes \hat{\tau}, S \otimes \hat{\tau}$. This allows us to extend the linear problem also, and in fact $\mathcal{L}u = v$ is equivalent to

$$(8) \qquad (\mathcal{L} \otimes \hat{\mathbb{I}}_\tau)(u \otimes \mathbb{I}_\tau) = (v \otimes \mathbb{I}_\tau).$$

The equivariance of $\mathcal{L} \otimes \hat{\mathbb{I}}_\tau$ implies in particular

$$Q_{S,\tau}(\mathcal{L} \otimes \hat{\mathbb{I}}_\tau) = (\mathcal{L} \otimes \hat{\mathbb{I}}_\tau)Q_{T,\tau},$$

which shows that the restricted operator

(9)  $\qquad (\mathcal{L} \otimes \hat{\mathbb{I}}_\tau)\big|_{\mathrm{range}\,Q_{T,\tau}} : \mathrm{range}\,Q_{T,\tau} \to \mathrm{range}Q_{S,\tau}$

exists, and is bijective if $\mathcal{L}$ is bijective.

Applying $Q_{S,\tau}$ to both sides of (8), and using the equivariance, gives

(10)  $\qquad (\mathcal{L} \otimes \hat{\mathbb{I}}_\tau)Q_{T,\tau}(u \otimes \mathbb{I}_\tau) = Q_{S,\tau}(v \otimes \mathbb{I}_\tau)$

These equations are the ones we propose to solve numerically. Note that by applying $(\dim \tau)(\mathbb{I}_S \otimes \mathrm{trace})$ to (10) and using (5) we obtain back our reduced linear equations (7).

The above discussions lead us to the following method which we propose for solving $\mathcal{L}u = v$:

*Method.*
(i) For each $\tau \in \mathcal{U}_\Gamma$, form $v \otimes \mathbb{I}_\tau \in V_S \otimes L(V_\tau)$.

If we view $v$ as a column vector with complex entries $v(i)$ with respect to some basis, then $v \otimes \mathbb{I}_\tau$ is a column vector with $L(V_\tau)$-entries $v(i)\mathbb{I}_\tau$. For example, if $N = \dim V_S$ is finite and $L(V_\tau)$ is identified with $\dim \tau$-by-$\dim \tau$ matrices, then $v \otimes \mathbb{I}_\tau$ can be implemented as a block-structure:

$$v \otimes \mathbb{I}_\tau = \begin{pmatrix} v(1)\mathbb{I}_{\dim \tau} \\ \vdots \\ v(N)\mathbb{I}_{\dim \tau} \end{pmatrix}.$$

(ii) For each $\tau$, generate $\beta_\tau = Q_{S,\tau}(v \otimes \mathbb{I}_\tau)$.

Again, using column vectors with respect to a basis, we have

$$\beta_\tau(i) = \frac{1}{|\Gamma|} \sum_{a \in \Gamma} \big(S(a)v\big)(i)\,\tau(a).$$

Note that $\beta_\tau$ is a block-structured column: the entry $\beta_\tau(i)$ is in fact a $\dim \tau$-by-$\dim \tau$ matrix since $\big(S(a)v\big)(i)$ is a complex number and $\tau(a)$ can be viewed as such a matrix.

(iii) For each $\tau$, solve the symmetric subproblem $(\mathcal{L} \otimes \hat{\mathbb{I}}_\tau)\alpha_\tau = \beta_\tau$.

Note that the solution is in fact $\alpha_\tau = Q_{T,\tau}(u \otimes \mathbb{I}_\tau)$ where $\mathcal{L}u = v$. In terms of indices, we solve a block-structured problem

$$\sum_j \mathcal{L}(i,j)\alpha_\tau(j) = \beta_\tau(i),$$

where the unknowns $\alpha_\tau(j)$ are of course $\dim \tau$-by-$\dim \tau$ matrices.

(iv) Calculate $u_\tau := \dim \tau\,(\mathbb{I}_T \otimes \mathrm{trace})\,\alpha_\tau = P_{T,\tau}\,u$, see (7) and (5). In terms of indices: $u_\tau(i) = (\dim \tau)(\mathrm{trace}\,\alpha_\tau(i))$.

(v) Now the solution of (6) is given by $u := \sum_{\tau \in \mathcal{U}_\Gamma} u_\tau$,   see (4).

It seems that in step 3 of the above methods, we are presenting a blown up version of the subproblems (7). However, this is mainly for mathematical exposition, our whole point is: we can utilize (5) in such a way that a computationally efficient method for solving (7) and hence (6) with a minimal amount of overhead costs results. This will enable us to reduce the problem dramatically, as will be seen, giving a considerable overall reduction to the original problem.

## 3. Equivariant Matrices

Let now $\mathcal{L}$ be a nonsingular square matrix over $\mathbb{C}$ of dimension $N$. An entry of $\mathcal{L}$ will be denoted by $\mathcal{L}(i,j)$. Here $i,j$ varies over an index set $J$ of $N$ elements.

In the applications that we envision, $\mathcal{L}$ is a system matrix corresponding to the discretization of an operator equation, usually a differential or integral equation. This operator equation often is defined over a geometric domain and hence may inherit an equivariance from the geometric symmetry with respect to a group $\Gamma$ of symmetries. The discretization typically is a Galerkin or collocation method, employing basis functions (and possibly collocation points). In the presence of symmetry of the domain, it is advantagous to choose the basis functions in a way that the group $\Gamma$ acting on the domain simply permutes these basis functions. In Section 5 we will explain this situation in more detail for the case of a boundary integral equation, and give a simple but significant example in Section 6.

Hence, if the operator equation is equivariant, then the system matrix $\mathcal{L}$ can be chosen to be equivariant in a sense which we will now explain. Since the index set $J$ corresponds in practice to basis functions (and possibly collocation points), which are permuted by the group $\Gamma$, the domain and range $\mathbb{C}^J$ of $\mathcal{L}$ carry a priori two unitary representations of $\Gamma$ which are induced by permutations. This works as follows:

Let $\Sigma_J$ be the multiplicative group of all permutations $J \to J$. The above-mentioned permutations of $J$ coming from the $\Gamma$-action can be viewed as homomorphisms $\tilde{S}, \tilde{T} : \Gamma \to \Sigma_J$ which may be equal or different. These group homomorphisms induce unitary representations $S, T : \Gamma \to U(\mathbb{C}^J)$ by setting $(S(a)x)(i) = x(\tilde{S}(a^{-1})i)$, (and similarly for $\tilde{T}$), where the components of a column $x \in \mathbb{C}^J$ are denoted by $x(i)$.

Since we now have unitary representations for both the domain and range of $\mathcal{L}$, the notion of equivariance is well defined as in Section 1: $S(a)\mathcal{L} = \mathcal{L}T(a)$ for all $a \in \Gamma$. The following characterization of equivariance in terms of the entries of the matrix and the permutations is easy to prove, see [1]:

LEMMA 11. *The matrix $\mathcal{L}$ is equivariant with respect to $T, S$ if and only if*

(12) $$\mathcal{L}(\tilde{S}(a)i, j) = \mathcal{L}(i, \tilde{T}(a^{-1})j)$$

*holds for all $i, j \in J$ and $a \in \Gamma$.*

With these representations in place we have that both $V_S$ and $V_T$ are equal to the column space $\mathbb{C}^J$.

In order to simplify our discussion, we make the following

*Assumption* 13. For all $i \in J$ and all $a \in \Gamma$, the condition $\tilde{S}(a)i = i$ implies that $a = e :=$ unit element of $\Gamma$. For $\tilde{T}$ an analogous condition is assumed to hold.

The above assumption exactly means that $\tilde{S}(a)$ and $\tilde{T}(a)$ are fixed point free permutations if $a \neq e$. It is possible to drop this assumption, see [**4**].

For $i \in J$ we introduce the orbit

$$\mathcal{O}_S(i) := \{\tilde{S}(a)i : a \in \Gamma\}.$$

It is clear that two orbits are either identical or disjoint. By Assumption 13, all orbits have the cardinality of $\Gamma$, and the map $a \in \Gamma \mapsto \tilde{S}(a)i$ is a one-to-one correspondence between $\Gamma$ and $\mathcal{O}_S(i)$.

By a *selection* from $\tilde{S}$-orbits we mean a minimal subset $J' \subset J$ such that $\bigcup_{i \in J'} \mathcal{O}_S(i) = J$; this means that $J'$ contains exactly one element from every orbit. The following is an immediate consequence of Assumption 13:

LEMMA 14. *Let $J'$ be a selection from $\tilde{S}$-orbits. Then for every $j \in J$ there is a unique $a \in \Gamma$ and a unique $k \in J'$ such that $j = \tilde{S}(a)k$.*

The orbits and selections are defined analogously for $T$.

Let us now consider the tensor products for a fixed $\tau \in \mathcal{U}_\Gamma$. It is clear that we can view the product $\mathbb{C}^J \otimes L(V_\tau) = L(V_\tau)^J$ as the space of $J$-columns $\alpha : J \to L(V_\tau)$ which have entries $\alpha(i)$ in $L(V_\tau)$. Under this identification, the unitary representation $S \otimes \hat{\tau}$ acts in the following way:

$$\Big(\big((S \otimes \hat{\tau})a\big)\alpha\Big)(i) = \tau(a)\,\alpha\big(\tilde{S}(a^{-1})i\big) \text{ for } a \in \Gamma \text{ and } i \in J.$$

Hence, by (1) and (3), it can be seen that the following symmetry condition holds:

$$(15) \quad \alpha \in \text{range}\,Q_{S,\tau} \iff \alpha\big(\tilde{S}(a)i\big) = \tau(a)\alpha(i) \text{ for all } a \in \Gamma \text{ and } i \in J.$$

We need to identify the space $\text{range}\,Q_{S,\tau}$ with a suitable column space, and this is possible by using a selection of indices:

LEMMA 16. *Let $J'$ be a selection from $\tilde{S}$-orbits. Then the restriction map* $\text{Res}_\tau : \alpha \in L(V_\tau)^J \mapsto \alpha|_{J'} \in L(V_\tau)^{J'}$ *induces an isomorphism $\text{Res}_\tau$ from* $\text{range}\,Q_{S,\tau}$ *onto* $L(V_\tau)^{J'}$.

PROOF. Clearly, $\text{Res}_\tau$ is linear. Let us assume that $\alpha \in \text{range}\,Q_{S,\tau}$ is given so that $\text{Res}_\tau \alpha = 0$. Fix $j \in J$. According to Lemma 14 we have a unique representation $j = \tilde{S}(a)i$. From $\text{Res}_\tau \alpha = 0$ we obtain $\alpha(i) = 0$, and (15) implies that $\alpha(j) = \tau(a)\alpha(i) = 0$ for all $j \in \mathcal{O}_S(i)$. Hence $\alpha = 0$, and the injectivity of $\text{Res}_\tau$ is shown.

To prove surjectivity, let $\beta \in L(V_\tau)^{J'}$ be given. By using again $j = \tilde{S}(a)i$, we define $\alpha(j) := \tau(a)\beta(i)$. From (15) it follows that $\alpha \in \text{range } Q_{S,\tau}$, and $\text{Res}_\tau \alpha = \beta$ is obvious. $\square$

Let us now discuss the method for solving the equation $\mathcal{L}u = v$. Therefore, we assume that two representations $S, T$ are given as above, and that the matrix $\mathcal{L}$ is equivariant.

Furthermore, we assume that $v \in \mathbb{C}^J$ is given. We fix $\tau \in \mathcal{U}_\Gamma$. Recall that $\mathbb{I}_\tau$ denotes the identity map $V_\tau \to V_\tau$. We calculate

$$\beta_\tau := Q_{S,\tau}(v \otimes \mathbb{I}_\tau),$$

$$\text{i.e.,} \quad \beta_\tau(i) = \frac{1}{|\Gamma|} \sum_{a \in \Gamma} v\big(\tilde{S}(a^{-1}i)\big)\tau(a) \quad \text{for } i \in J.$$

Hence $\beta_\tau \in \text{range } Q_{S,\tau}$. Therefore the linear equation

$$(17) \qquad \sum_{j \in J} \mathcal{L}(i,j)\,\alpha_\tau(j) = \beta_\tau(i), \quad i \in J,$$

has a unique solution $\alpha_\tau \in L(V_\tau)^J$, and furthermore $\alpha_\tau \in \text{range } Q_{T,\tau}$. Note that by (15), $\alpha_\tau$ and $\beta_\tau$ are determined by their co-ordinates in a selection of indices.

We now collect the terms of the sum (17) in a different way. Let $J'$ be a selection from $\tilde{S}$-orbits and $J''$ a selection from $\tilde{T}$-orbits. We split the running index $j$ in (17) according to Lemma 14 and apply (15):

$$\sum_{k \in J''} \sum_{a \in \Gamma} \mathcal{L}\big(i, \tilde{T}(a)k\big)\,\alpha_\tau\big(\tilde{T}(a)k\big)$$

$$(18) \qquad = \sum_{k \in J''} \left(\sum_{a \in \Gamma} \mathcal{L}\big(i, \tilde{T}(a)k\big)\tau(a)\right)\alpha_\tau(k) = \beta_\tau(i), \quad i \in J'.$$

Hence, by setting

$$\mathcal{A}_\tau(i,k) := \sum_{a \in \Gamma} \mathcal{L}\big(i, \tilde{T}(a)k\big)\tau(a)$$

$$(19) \qquad = \sum_{a \in \Gamma} \mathcal{L}\big(\tilde{S}(a^{-1})i, k\big)\tau(a) \quad i \in J', \ k \in J'',$$

we obtain the linear equations

$$(20) \qquad \sum_{k \in J''} \mathcal{A}_\tau(i,k)\,\alpha_\tau(k) = \beta_\tau(i), \quad i \in J'.$$

Note that the coefficients of this equation are in $L(V_\tau)$. The matrix $\mathcal{A}_\tau$ corresponds to the restricted operator (9), and in view of Lemma 16 it is clear that $\mathcal{A}_\tau$ is nonsingular.

Once the solution $\alpha_\tau(k)$ is calculated for $k \in J''$, we extend $\alpha_\tau \in$ range $Q_{T,\tau}$ in the usual way by setting $\alpha_\tau(\tilde{T}(a)k) := \tau(a)\alpha_\tau(k)$. Now we conclude the program outlined our method by calculating

$$(21) \qquad u_\tau(j) := \dim \tau \, \operatorname{trace} \alpha_\tau(j), \quad j \in J$$

and summing up

$$(22) \qquad u := \sum_{\tau \in \mathcal{U}_\Gamma} u_\tau$$

to obtain the solution $u$.

Let us discuss the computational cost of the method. At least for large problems, the essential effort of the method consists in solving the subproblems (20). Note that $\mathcal{A}_\tau$ is a block-structured matrix, where the number of blocks in one row is the number of orbits, which is $N/|\Gamma|$ by Assumption 13; each block is a $\dim \tau$-by-$\dim \tau$ matrix. Hence, we see that $\mathcal{A}_\tau$ is a square matrix of dimension

$$N_\tau := \frac{N \dim \tau}{|\Gamma|}.$$

Assuming that the linear systems involved are full, we see that the essential part of the computational effort is described by $\sum_{\tau \in \mathcal{U}_\Gamma} C N_\tau^3$ flops, as opposed to $C N^3$ flops when not making use of the group symmetries. Here $C$ is a constant defined by the linear equation solver which is used. The notion "flop" describes a standard way of counting arithmetic machine operations, see [12] for details.

Taking the quotient, we obtain the following reduction factor

$$(23) \qquad \rho := \frac{1}{|\Gamma|^3} \sum_{\tau \in \mathcal{U}_\Gamma} (\dim \tau)^3.$$

This factor can be interpreted in the following way: If one unit of work describes the computational effort for solving the linear system $\mathcal{L}u = v$ without making use of the symmetry group, then $\rho$ units of work essentially describe the computational effort of our method. For simplicity, we did not distinguish here between real and complex arithmetic.

An additional and important saving of computational expense can be obtained when evaluating the entries of the matrix $\mathcal{L}$. Namely, the entries $\mathcal{L}(i,j)$ only need to be evaluated for all $i \in J', j \in J$ or for all $i \in J, j \in J''$. The remaining entries are obtained from the equivariance condition (12), and in fact are not needed for generating the reduced matrix $\mathcal{A}_\tau$ in (19).

To illustrate the potential savings of the method, let us calculate the reduction factor $\rho$ for several important groups. We refer the reader to Serre's book [15] for the descriptions of all irreducible representations of the groups discussed here.

We begin with the cyclic group of order $n$, which has exactly $n$ irreducible representations, all of dimension 1. Therefore, the reduction factor is $\rho = n^{-2}$. The reader should note that the method outlined in this paper for cyclic groups is equivalent to using the discrete Fourier transform.

More generally, if $\Gamma$ is any finite abelian group of order $n$, $\Gamma$ has exactly $n$ irreducible representations of dimension 1. Hence we obtain the same reduction factor $\rho = n^{-2}$.

Next we consider the dihedral group $D_n$ of order $2n$, which is the isometry group of a regular $n$-gon in $\mathbb{R}^2$. If $n$ is even, say $n = 2m$, then there are 4 representations of dimension 1, and $m-1$ irreducible representations of dimension 2. Hence, the reduction factor is $\rho = (n - 1)/(2n^3)$. If $n$ is odd, then there are only two representations of dimension 1, and $(n-1)/2$ irreducible representations of dimension 2. Therefore, the reduction factor in this case is $\rho = (2n-1)/(4n^3)$. In either case, this is asymtotically $\rho \approx \frac{1}{2}n^{-2}$. This is about half as good as the reduction for a cyclic group of the same order. But it is twice as good as using only the cyclic subgroup of order $n$ in $D_n$.

Let us next consider the group $A_4$ of even permutations on 4 symbols; it arises as the group of rotations of the regular tetrahedron in $\mathbb{R}^3$ and has order 12. We have 3 irreducible representations of dimension 1 and one of dimension 3. Hence the reduction factor is $30/12^3 \approx 0.01736$.

Related to this group is the group $\Sigma_4$ of all permutations on 4 symbols, which arises as the group of all isometries of the regular tetrahedron in $\mathbb{R}^3$ (including reflections), and as the group of rotations of the cube: it has order 24. There are 2 irreducible representations of dimension 1, one of dimension 2, and two of dimension 3. The reduction factor is $64/24^3 \approx 0.00463$.

Finally we discuss the full symmetry group of the cube (including reflections). This group is a direct product of $\Sigma_4$ and the cyclic group of order 2, hence it has 48 elements. There are 4 irreducible representations of dimension 1, two of dimension 2, and four of dimension 3. The reduction factor is $128/48^3 \approx 0.00116$.

## 4. Simplifications for Real Numbers

Typically, linear problems occuring in applications are formulated over the field of real numbers. However, the natural field for the representation theory as outlined above is the field of complex numbers. Of course, from a point of view of computational efficiency, it is usually preferable to handle real instead of complex numbers. Here we want to point out a few ideas of how to obtain savings in the computational expense under the assumption that the original problem (6) is formulated over the field of real numbers.

In particular, the above assumption implies that the matrix $\mathcal{L}$ discussed in the last section has real entries $\mathcal{L}(i,j)$, and that the data entries $v(i)$ are also real. On the other hand, the linear system (20) which we finally want to solve, involves complex arithmetic. We consider three cases for a fixed $\tau \in \mathcal{U}_\Gamma$.

First, if $\tau$ is one-dimensional and real valued, then nothing needs to be done since all arithmetic in (20) is real.

Second, if $\tau$ is one-dimensional and has at least one value which is complex and not real, then the complex conjugate $\bar{\tau}$ is also an irreducible representation, and $\tau$ and $\bar{\tau}$ are disjoint. It follows that both the $\alpha_\tau(j)$ and the $\alpha_{\bar{\tau}}(j)$ have to be calculated according to a system like (20), and from

$$\overline{\beta_\tau(i)} = \beta_{\bar{\tau}}(i),$$

and (17) we obtain that

$$\overline{\alpha_\tau(i)} = \alpha_{\bar{\tau}}(i)$$

holds. Hence, only the solution $\alpha_\tau$ has to be calculated, and $\alpha_{\bar{\tau}}$ is obtained for free. In fact, in the sum (22), the summand $u_\tau + u_{\bar{\tau}}$ can be replaced by $2\mathcal{R}(u_\tau)$, where $\mathcal{R}$ indicates the real part.

Third, if $\dim \tau > 1$, then it may be possible to choose the representative $\tau$ of the class $\tau$ such that it consists of matrices with real entries (which are hence orthogonal). This is often possible when the group $\Gamma$ is an isometry group of a body in $\mathbb{R}^n$, an important case for applications. Then again we have real arithmetic in (20). The numerical example given in Section 6 is of this type.

## 5. Collocation Methods for Boundary Integral Equations

Let us briefly indicate how linear systems admitting an equivariant matrix as described in Section 3 may be obtained. A typical case is a boundary element method, but finite element methods may be regarded as well.

As an illustration, we discuss a specific example. This example will be explicitly calculated in the next section for a particular case. It will be clear from the discussion below how other discretizations for other problems may be viewed.

Let $\mathcal{D} \subset \mathbb{R}^n$ be a bounded open nonempty domain. We consider Laplace's equation:

$$\Delta u = 0 \text{ in } \mathcal{D},$$
(24)
$$u = g \text{ on } \partial\mathcal{D}.$$

The boundary $\partial\mathcal{D}$ is supposed to be sufficiently regular.

Let us introduce the fundamental solution via

$$s(x,y) := \begin{cases} \dfrac{1}{2\pi} \log(|x-y|) & \text{for } n = 2, \\[2mm] \dfrac{1}{4\pi} \dfrac{1}{|x-y|} & \text{for } n = 3, \\[2mm] \dfrac{1}{\omega_n(n-2)} \dfrac{1}{|x-y|^{n-2}} & \text{for } n > 3, \end{cases}$$

where $\omega_n$ denotes the area of the $(n-1)$-dimensional unit sphere and $|\cdot|$ the Euclidean norm in $\mathbb{R}^n$.

There are several ways to associate a boundary integral equation to (24). Let us briefly describe one possibility, see, e.g., [6]: We assume a solution in the form (ansatz)

$$u(x) = \int_{\partial\mathcal{D}} \frac{\partial s(x,y)}{\partial\nu(y)} \sigma(y) \, d\mu(y),$$

where $\sigma$ is an unknown function on $\partial\mathcal{D}$, $\nu(y)$ indicates the outer normal of $\partial\mathcal{D}$ at $y \in \partial\mathcal{D}$, and $d\mu$ is the standard surface element of integration. Letting $x$ go to the boundary gives

$$(25) \qquad \int_{\partial\mathcal{D}} \frac{\partial s(x,y)}{\partial\nu(y)}\sigma(y)\,d\mu(y) - \frac{\theta(x)}{\omega_n}\,\sigma(x) = g(x), \quad x \in \partial\mathcal{D},$$

where $\theta(x)$ denotes the solid angle generated by $\partial\mathcal{D}$ at $x \in \partial\mathcal{D}$. In particular $\theta(x) = \frac{1}{2}\omega_n$ if $\partial\mathcal{D}$ is smooth at $x$. Equation (25) is a Fredholm integral equation of the second kind for the unknown function $\sigma$ defined on $\partial\mathcal{D}$.

In order to solve (25) approximately, we view it as an operator equation

$$E\sigma = g,$$

where $E$ is a bounded linear operator on a suitable space of (generalized) functions on $\partial\mathcal{D}$, and discretize it. A typical discretization method is a collocation method. One introduces basis functions $\{\phi_j\}_{j\in J}$ on $\partial\mathcal{D}$ and corresponding collocation points $\{x_i\}_{i\in J}$ in $\partial\mathcal{D}$ to obtain an interpolation of $\sigma$:

$$\sigma(x_i) = \sum_{j\in J} u(j)\,\phi_j(x_i), \quad i \in J.$$

The following collocation equation determines the coefficients $u(j)$ of this interpolation:

$$(26) \qquad \mathcal{L}u = v, \quad \text{where} \quad \mathcal{L}(i,j) := (E\phi_j)(x_i) \quad \text{and} \quad v(i) := g(x_i).$$

Let us now assume that the domain $\mathcal{D}$ admits a symmetry group $\Gamma$ of orthogonal transformations $\mathbb{R}^n \to \mathbb{R}^n$. Because the kernel function and the surface element used to define $E$ involve essentially only distances, and every element of $\Gamma$ is distance preserving, this symmetry is respected by the integral equation:

$$(27) \qquad\qquad\qquad (E\sigma) \circ a = E(\sigma \circ a)$$

for $a \in \Gamma$.

In order to make use of this symmetry, we have to choose the basis functions $\{\phi_j\}_{j\in J}$ and collocation points $\{x_i\}_{i\in J}$ in such a way that this symmetry is respected. Under the action of $a \in \Gamma$, the collocation point $x_i$ is sent to $ax_i$ and the basis function $\phi_j$ is transformed into $\phi_j \circ a^{-1}$. For our purposes we require that both the set of collocation points and the set of basis functions are preserved under these transformations. We do not require that the individual elements of these sets be left fixed, but only that they be permuted by each $a \in \Gamma$. This then leads to the following assumption which is easy to obtain in cases of practical interest, see also [1]. The numerical example in the next section will further illustrate this.

*Assumption* 28. There are two group homomorphisms $\tilde{S}, \tilde{T} : \Gamma \to \Sigma_J$ as described in Section 3, such that

$$ax_i = x_{\tilde{S}(a)i} \text{ and } \phi_j \circ a^{-1} = \phi_{\tilde{T}(a)j} \text{ for } a \in \Gamma \text{ and } i, j \in J.$$

In particular, we also assume that Assumption 13 holds, i.e., that there are no fixed points for any element of either action.

Let us show that the equivariance (27) of the operator $E$ implies the equivariance of the system matrix $\mathcal{L}$ defined in (26). In fact, the following calculation verifies the equivariance condition (12).

$$
\begin{aligned}
\mathcal{L}\big(\tilde{S}(a)i, j\big) &= (E\phi_j)(x_{\tilde{S}(a)i}) \\
&= (E\phi_j)(ax_i) \\
&= (E\phi_j \circ a)(x_i) \\
&= (E\phi_{\tilde{T}(a^{-1})j})(x_i) \\
&= \mathcal{L}\big(i, \tilde{T}(a^{-1})j\big).
\end{aligned}
$$

Hence, the collocation methods for boundary integral equations are an example where our method outlined in Section 3 may be applied. Galerkin methods may be treated similarly.

Finite element methods may also be treated similarly, only the system matrix $\mathcal{L}$ obtained in this case is sparse, and it is desirable to make use of an induced sparseness for the reduced matrices $\mathcal{A}_\tau$. This induced sparseness is related to the domain decomposition techniques which have been investigated in [2], [3], [7], [9], [10]. We will further pursue this topic elsewhere.

## 6. A Simple Example Involving the Isometry Group of an Equilateral Triangle

Let us continue the discussion of the last section by giving a specific and very simple example which, however, already displays the essential features of our approach. More complex examples are presented in [5] and [18].

We consider Laplace's equation in the plane and assume that $\mathcal{D}$ is an equilateral triangle with the nodes

$$v^0 := \begin{pmatrix} 0 \\ \frac{2}{\sqrt{3}} \end{pmatrix}, \ v^{2m} := \begin{pmatrix} 1 \\ -\frac{1}{\sqrt{3}} \end{pmatrix}, \ v^{4m} := \begin{pmatrix} -1 \\ -\frac{1}{\sqrt{3}} \end{pmatrix}.$$

It has an inscribed circle about the origin with radius $\frac{1}{\sqrt{3}}$.

The isometry group of $\mathcal{D}$ is the so-called dihedral group of six elements

$$\Gamma = \{e, r, r^2, s, sr, sr^2\}$$

where $e$ is the identity, $r$ a rotation with $\frac{2}{3}\pi$ and $s$ a reflection at the $y$-axis. It permits three irreducible representations: $\tau$, $\sigma$ have dimension 1 and $\nu$ has dimension 2. They are described in the following table.

|        | $e$ | $r$ | $r^2$ | $s$ | $sr$ | $sr^2$ |
|--------|-----|-----|-------|-----|------|--------|
| $\tau$ | 1 | 1 | 1 | 1 | 1 | 1 |
| $\sigma$ | 1 | 1 | 1 | $-1$ | $-1$ | $-1$ |

$$\mu \quad \begin{pmatrix} 1 & 0 \\ 0 & 1 \end{pmatrix} \begin{pmatrix} -\frac{1}{2} & -\frac{\sqrt{3}}{2} \\ \frac{\sqrt{3}}{2} & -\frac{1}{2} \end{pmatrix} \begin{pmatrix} -\frac{1}{2} & \frac{\sqrt{3}}{2} \\ -\frac{\sqrt{3}}{2} & -\frac{1}{2} \end{pmatrix} \begin{pmatrix} -1 & 0 \\ 0 & 1 \end{pmatrix} \begin{pmatrix} \frac{1}{2} & \frac{\sqrt{3}}{2} \\ \frac{\sqrt{3}}{2} & -\frac{1}{2} \end{pmatrix} \begin{pmatrix} \frac{1}{2} & -\frac{\sqrt{3}}{2} \\ -\frac{\sqrt{3}}{2} & -\frac{1}{2} \end{pmatrix}$$

For convenience, we apply a cyclic notation by using the cyclic group of natural numbers modulo $n := 4m$ as our index set $J$. Hence, we have defined $v^k$ for $k = 0, 2m, 4m$. We subdivide each of the three sides of the triangle into $2m$ equidistant line segments and thus obtain the other nodes $v^k$ for $k \in J$, counting them in a counterclockwise fashion.

We use the simplest collocation method available, namely constant elements conforming to this subdivision. Hence, the $n$ basis functions $\phi_k$ as the characteristic functions of the line segments from $v^k$ to $v^{k+1}$, and the collocation points $x_k$ as the centers of these line segments. The two homomorphisms $\tilde{S}, \tilde{T} : \Gamma \to \Sigma_J$ coincide and are described in the following table:

| $e$ | $r$ | $r^2$ | $s$ | $sr$ | $sr^2$ |
|-----|-----|-------|-----|------|--------|
| $k \mapsto k$ | $k \mapsto k + 2m$ | $k \mapsto k + 4m$ | $k \mapsto 4m + 1 - k$ | $k \mapsto 2m + 1 - k$ | $k \mapsto 1 - k$ |

A selection $J'$ of indices is readily obtained: $J' := \{0, 1, \ldots, m - 1\}$. We calculated the example for the very coarse grid $m = 2$, and for the input function $g$ being the characteristic function of the line segment vom $v^{2m-1}$ to $v^{2m}$.

The following table represents the discretization matrix $\mathcal{L}$. Note that the equivariance condition (12) holds.

| 1.0000 | .0000 | .0000 | .0000 | .0894 | .1034 | .0983 | .0787 | .0131 | .0274 | .0894 | .5000 |
|--------|-------|-------|-------|-------|-------|-------|-------|-------|-------|-------|-------|
| .0000 | 1.0000 | .0000 | .0000 | .1300 | .1427 | .1102 | .0713 | .0456 | .0894 | .1833 | .2271 |
| .0000 | .0000 | 1.0000 | .0000 | .2271 | .1833 | .0894 | .0456 | .0713 | .1102 | .1427 | .1300 |
| .0000 | .0000 | .0000 | 1.0000 | .5000 | .0894 | .0274 | .0131 | .0787 | .0983 | .1034 | .0894 |
| .0131 | .0274 | .0894 | .5000 | 1.0000 | .0000 | .0000 | .0000 | .0894 | .1034 | .0983 | .0787 |
| .0456 | .0894 | .1833 | .2271 | .0000 | 1.0000 | .0000 | .0000 | .1300 | .1427 | .1102 | .0713 |
| .0713 | .1102 | .1427 | .1300 | .0000 | .0000 | 1.0000 | .0000 | .2271 | .1833 | .0894 | .0456 |
| .0787 | .0983 | .1034 | .0894 | .0000 | .0000 | .0000 | 1.0000 | .5000 | .0894 | .0274 | .0131 |
| .0894 | .1034 | .0983 | .0787 | .0131 | .0274 | .0894 | .5000 | 1.0000 | .0000 | .0000 | .0000 |
| .1300 | .1427 | .1102 | .0713 | .0456 | .0894 | .1833 | .2271 | .0000 | 1.0000 | .0000 | .0000 |
| .2271 | .1833 | .0894 | .0456 | .0713 | .1102 | .1427 | .1300 | .0000 | .0000 | 1.0000 | .0000 |
| .5000 | .0894 | .0274 | .0131 | .0787 | .0983 | .1034 | .0894 | .0000 | .0000 | .0000 | 1.0000 |

Next we list the linear subsystems and their solution, corresponding to the three irreducible representations $\tau$, $\sigma$, $\mu$ described above.

$$\begin{pmatrix} 1.681330 & .318670 \\ .474213 & 1.525787 \end{pmatrix} \begin{pmatrix} -.210665 \\ .065475 \end{pmatrix} = \begin{pmatrix} -.333333 \\ .000000 \end{pmatrix}$$

$$\begin{pmatrix} .523821 & -.056864 \\ -.122808 & .938607 \end{pmatrix} \begin{pmatrix} -.645519 \\ .084460 \end{pmatrix} = \begin{pmatrix} .333333 \\ .000000 \end{pmatrix}$$

$$\begin{pmatrix} .488090 & .002094 & -.105752 & .019402 \\ .134313 & 1.409335 & .150877 & -.025151 \\ -.279374 & -.011362 & .755659 & .049317 \\ .134900 & .103672 & .141609 & 1.012144 \end{pmatrix} \begin{pmatrix} -.373998 & .650191 \\ .255928 & .027429 \\ -.137218 & .248906 \\ .042831 & -.124293 \end{pmatrix} = \begin{pmatrix} -.166667 & .288675 \\ .288675 & .166667 \\ .000000 & .000000 \\ .000000 & .000000 \end{pmatrix}$$

We note that the last equation displays a block structure of 2-by-2 matrices since the third irreducible representation $\mu$ has dimension 2. From the three subsolutions above we now recover the solution via (21-22).

$$u = \begin{pmatrix} -.258285 \\ -.373087 \\ -.537215 \\ -2.827057 \\ 1.464306 \\ .768378 \\ .473393 \\ .311833 \\ .098538 \\ .054513 \\ .006865 \\ -.053328 \end{pmatrix}$$

## Acknowledgement

We thank John Walker for carrying out the numerical computations.

## REFERENCES

1. E. L. Allgower, K. Böhmer, K. Georg, and R. Miranda, *Exploiting symmetry in boundary element methods*, SIAM J. Numer. Anal. **29** (1992), 534–552.

2. E. L. Allgower, K. Böhmer, and Z. Mei, *A generalized equibranching lemma with applications to $D_4 \times Z_2$ symmetric elliptic problems. Part I*, Preprint, University of Marburg, Fed. Rep. Germany, 1990.

3. _____, *On a problem decomposition for semi-linear nearly symmetric elliptic problems*, Parallel Algorithms for PDE's (Braunschweig, Fed. Rep. Germany) (W. Hackbusch, ed.), Notes on Numerical Fluid Mechanics, vol. 31, Vieweg Verlag, 1991, pp. 1–17.

4. E. L. Allgower, K. Georg, and R. Miranda, *Exploiting permutation symmetry with fixed points in linear equations*, Exploiting Symmetry in Applied and Numerical Analysis (Providence, RI) (E. L. Allgower, K. Georg, and R. Miranda, eds.), Lectures in Applied Mathematics, vol. 28, American Mathematical Society, 1993.

5. E. L. Allgower, K. Georg, and J. Walker, *Exploiting symmetry in 3D boundary element methods*, Contributions in Numerical Mathematics (Singapore) (R. P. Agarwal, ed.), World Scientific Series in Applicable Analysis, vol. 2, World Scientific Publ. Comp., Singapore, 1992, to appear.

6. K. E. Atkinson, *The numerical solution of Laplace's equation in three dimensions*, SIAM J. Numer. Anal. **19** (1982), 263–274.

7. A. Bossavit, *Symmetry, groups, and boundary value problems. A progressive introduction to noncommutative harmonic analysis of partial differential equations in domains with geometric symmetry*, Computer Methods in Applied Mechanics and Engineering **56** (1986), 165–215.

8. M. Dellnitz and B. Werner, *Computational methods for bifurcation problems with symmetries — with special attention to steady state and Hopf bifurcation points*, J. Comput. Appl. Math. **26** (1989), 97–123.

9. C. C. Douglas and J. Mandel, *The domain reduction method: High way reduction in three dimensions and convergence with inexact solvers*, (Philadelphia), Fourth Copper Mountain Conference on Multigrid Methods, SIAM, 1989, pp. 149–160.

10. _____ , *A group theoretic approach to the domain reduction method: The commutative case*, Computing **47** (1990).

11. K. Georg and R. Miranda, *Exploiting symmetry in solving linear equations*, Bifurcation and Symmetry (Basel, Switzerland) (E. L. Allgower, K. Böhmer, and M. Golubitsky, eds.), ISNM, vol. 104, Birkhäuser Verlag, 1992, pp. 157–168.

12. G. H. Golub and C. F. van Loan, *Matrix computations*, second ed., J. Hopkins University Press, Baltimore, London, 1989.

13. T. J. Healey, *Numerical bifurcation with symmetry: Diagnosis and computation of singular points*, Bifurcation Theory and its Numerical Analysis (Xi'an, P. R. China) (L. Kaitai, J. Marsden, M. Golubitsky, and G. Iooss, eds.), Xi'an Jiaotong University, Xi'an Jiaotong University Press, 1989, pp. 218–227.

14. A. A. Kirillov, *Elements of the theory of representations*, Springer, 1976.

15. J.-P. Serre, *Linear representations of finite groups*, Graduate Texts in Mathematics, vol. 42, Springer Verlag, Berlin, Heidelberg, New York, 1977.

16. E. Stiefel and A. Fässler, *Gruppentheorethische Methoden und ihre Anwendung*, Teubner, Stuttgart, Fed. Rep. Germany, 1979.

17. J. Tausch, *A generalization of the discrete Fourier transformation*, Exploiting Symmetry in Applied and Numerical Analysis (Providence, RI) (E. L. Allgower, K. Georg, and R. Miranda, eds.), Lectures in Applied Mathematics, vol. 28, American Mathematical Society, 1993.

18. J. Walker, *Numerical experience with exploiting symmetry groups for bem*, Exploiting Symmetry in Applied and Numerical Analysis (Providence, RI) (E. L. Allgower, K. Georg, and R. Miranda, eds.), Lectures in Applied Mathematics, vol. 28, American Mathematical Society, 1993.

19. B. Werner, *Eigenvalue problems with the symmetry of a group and bifurcations*, Continuation and Bifurcations: Numerical Techniques and Applications (Dordrecht, The Netherlands) (D. Rose, B. de Dier, and A. Spence, eds.), NATO ASI Series C, vol. 313, Kluwer Academic Publishers, 1990, pp. 71–88.

DEPARTMENT OF MATHEMATICS, COLORADO STATE UNIVERSITY, FT. COLLINS, COLORADO 80523

*E-mail address*: Georg@Math.ColoState.edu, Miranda@Riemann.Math.ColoState.edu

Lectures in Applied Mathematics
Volume **29**, 1993

# Numerical Results on the Zeros of Faber Polynomials For $m$-fold Symmetric Domains

## Matthew He

**Abstract**

Faber polynomials, generated by a conformal mapping $\Phi(z)$ of the exterior of a set $E$ onto the exterior of a circle, have well-known classical applications in numerical analysis as basis sets for polynomial and rational approximations in the complex plane. The structure of the Faber polynomials of a given set $E$ is essential for such applications. In this paper we study the Faber polynomials associated with $m$-fold symmetric domains. A new determinant representation which relates the zeros of Faber polynomials to the eigenvalues of a certain matrix is derived and numerical computations on the zeros of Faber polynomials associated with symmetric lunes and $m$-gons are illustrated.

# 1   Introduction

Let $E$ be any closed continuum (not a single point) in the complex plane $\mathbf{C}$ and let $\overline{\mathbf{C}}$ denote the extended complex plane. The Riemann mapping theorem asserts that there exists a conformal mapping $w = \Phi(z)$ of $\overline{\mathbf{C}} \setminus E$ onto the exterior of a circle $|w| = \rho_E$ in the $w$-plane. For a unique choice of $\rho_E$, we insist that

$$\Phi(\infty) = \infty, \ \Phi'(\infty) = 1$$

so that, in a neighborhood of infinity,

$$\Phi(z) = z + a_0 + \frac{a_1}{z} + \frac{a_2}{z^2} + \cdots. \tag{1.1}$$

1991 Mathematics Subject Classifications: 30C15, 41A58.

This paper is in final form and no version of it will be submitted for publication elsewhere.

With this normalization, the constant $\rho_E$ is the logarithmic capacity or transfinite diameter of the set $E$. The polynomial part of $\Phi(z)^n$, denoted by $F_n(z) := F(z; E) = z^n + \cdots$, is called the **Faber polynomial** of degree $n$ generated by the set $E$. (For a survey of the theory of Faber polynomials see [2]).

Let

$$\Psi(w) = w + b_0 + \frac{b_1}{w} + \frac{b_2}{w^2} + \cdots \qquad (1.2)$$

be the inverse function of $w = \Phi(z)$. Then $\Psi(w)$ maps the domain $|w| > \rho_E$ conformally onto $\overline{\mathbf{C}} \setminus E$ and Faber [5] proved that

$$\frac{\Psi'(w)}{\Psi(w) - z} = \sum_{n=0}^{\infty} \frac{F_n(z)}{w^{n+1}}, \quad |w| > \rho_E, \ z \in E. \qquad (1.3)$$

For the unit disk the Faber polynomial of degree $n$ is $z^n$ and the corresponding Faber series for an analytic function is its Taylor series about the origin. If $E = [-1, 1]$, then the multiples of the Chebyshev polynomials are the Faber polynomials.

In recent years there has been a growing interest in studying the Faber polynomials for specific regions. Elliott [4] computed the coefficients of some Faber polynomials for the semi-disk $|z| \leq 1$, $\mathrm{Re}\, z \geq 0$ and for the square $|\mathrm{Re}\, z| \leq 1$, $|\mathrm{Im}\, z| \leq 1$. Coleman and Smith [3] as well as Gatermann, Hoffman and Opfer [6] have studied the coefficients of the Faber polynomials on circular sectors.

Many of the results contain substantial contributions to the available information about the structure of the Faber polynomials. However, only a few results are known on the zeros of Faber polynomials associated with the set $E$(see [10]). In this report, we consider the following problem.

**Problem. The Distribution of Zeros of Faber Polynomials.** *Let $F_n(z; E)$ denote the Faber polynomials associated with the set $E$ and let*

$$A_n = \{z_{n,1}, z_{n,2}, \cdots, z_{n,n}\} \qquad (1.4)$$

*be the set of zeros of $F_n(z; E)$ and*

$$B = \{z : \lim_{n \to \infty} z_n = z, \ z_n \in A_{i_n}, \ i_n \to \infty\}. \qquad (1.5)$$

*Determine the location of the set $A_n$ and properties of the set $B$.*

The answer to this problem is far from known except for some special important cases. In the case when $E$ has empty interior we actually know the asymptotic distribution of the zeros of $F_n(z; E)$, namely it coincides with equilibrium measure of the set $E$ (see [7]). This result is no longer true if $E$ has nonempty interior, as the above mentioned example of the unit disk shows. It seems to be a very difficult problem to determine the limiting distribution of zeros (if it exists at all) for general $E's$. In this report, we explore possible zero distributions of the Faber polynomials $F_n(z)$ associated with $m$-fold symmetric domains. In addition, we derive a new representation which relates the zeros of Faber polynomials to the eigenvalues of a certain matrix. Furthermore, we compute the zeros of Faber polynomials associated with symmetric lunes, regular $m$-polygons.

# 2 Faber Polynomials For m-Fold Symmetric Domains

As we have seen that sequences of Faber polynomials $F_n(E; z)$ depend on the set $E$ or the mapping function $\Phi(z)$. In this section, we shall explore some algebraic properties of Faber polynomials associated with a set satisfying the following symmetry.

**Definition.** *Let $m$ be a positive integer. A domain $E$ is said to be m-fold symmetric if a rotation of $E$ about the origin through an angle $\frac{2\pi}{m}$ carries $E$ on itself.*

It follows that if $E$ is $m$-fold symmetric, then

$$\Phi(e^{\frac{2\pi i}{m}} w) = e^{\frac{2\pi i}{m}} \Phi(z). \tag{2.1}$$

It follows also from (1.1) and (1.2), that the Laurent expansion near $\infty$ of the mapping function $\Phi(z)$ and $\Psi(w)$ are respectively of the form

$$\Phi(z) = z \left( 1 + \sum_{k=1}^{\infty} \frac{a_{mk-1}}{z^{mk}} \right). \tag{2.2}$$

and

$$\Psi(w) = w \left( 1 + \sum_{k=1}^{\infty} \frac{b_{mk-1}}{w^{mk}} \right). \tag{2.3}$$

Alternatively, $m$-fold symmetric domain can also be reformulated in terms of the theory of linear representations.

Let

$$C_m = \{1, r, \ldots, r^{m-1}\}$$

denote the cyclic group and

$$\chi^k(r^j) = e^{\frac{2\pi i(k-1)j}{m}} \quad j = 0, 1, \ldots, m-1, \quad k = 1, \ldots, m$$

be its irreducible representations. Then a m-fold symmetric domain may also be called $C_m$-**invariant**. The Faber polynomials $F_{mn+k}(z)$ span a $C_m$-**invariant space** corresponding to $\chi^{k+1}$. Each polynomial corresponding to $\chi^k$ is the product of $z^{k-1}$ and a polynomial in the unknown $z^m$.

For some special $C_m$-invariant domains, we have the following:

**Example 1.** Let $E = S_m$ denote the regular $m$-Star

$$S_m = \{x\omega^k; \ 0 \le x \le 4^{\frac{1}{m}}, \ k = 0, 1, \ldots, m-1, \ \omega^m = 1, \ m = 2, 3, \ldots\}.$$

It was proved in [1] that $B = S_m$.

**Example 2.** Let $E = H_m$ be the closed region bounded by $m$-Cusped Hypocycloid, with parametric equation

$$z = \exp(i\theta) + \frac{1}{(m-1)} \exp(-(m-1)i\theta), \quad 0 \le \theta < 2\pi, \quad m = 2, 3, 4, \ldots.$$

The location and asymptotic behavior of the zeros of Faber polynomials associated with $H_m$ were determined by using the $m$-th order three term recurrence relation for $F_n(z)$ and showed that $B = S_m$ in [8].

We remark that the domains $S_m$ and $H_m$ from Examples 1 and 2 are additionally invariant with respect to conjugation $z \to \bar{z}$ and are invariant with respect to the dihedral group $D_m$. The sets $A_n$ and B corresponding to $S_m$ and $H_m$ are $C_m$-invariant, respectively.

In general, for the zeros of Faber polynomials associated with a $C_m$ invariant domain we have the following:

**Theorem 2.1.**    *If the domain   $E$   is $C_m$-invariant then the set of zeros of $F(z; E)$ is $C_m$-invariant.*

**Proof.**    From (2.2),

$$\Phi(z) = z + \sum_{k=1}^{\infty} \frac{a_{mk-1}}{z^{mk-1}} = z \left( 1 + \sum_{k=1}^{\infty} \frac{a_{mk-1}}{z^{mk}} \right),$$

$$\Phi(z)^n = z^n \left( 1 + \sum_{k=1}^{\infty} \frac{a_{mk-1}}{z^{mk}} \right)^n$$

and

$$\Phi(e^{\frac{2\pi i}{m}} z)^n = e^{\frac{2\pi n i}{m}} \Phi(z)^n.$$

Then, for $n = 0, 1, 2, ...$,

$$F_n(e^{\frac{2\pi i}{m}} z) = e^{\frac{2\pi n i}{m}} F_n(z). \tag{2.4}$$

Consequently, for $k = 0, 1, 2, ..., m - 1$,

$$F_{mn+k}(z) = z^k \prod_{j=1}^{n} (z^m - z_j^m), \tag{2.5}$$

where the $z_j's$ $(j = 1, 2, ..., n)$ are the nonvanishing zeros of $F_{nm+k}(z)$.  ∎

Little is known about the exact location and asymptotic distribution of zeros $z_j's$. In the following section, we shall present a new representation of Faber polynomials in terms of determinant of certain matrix, which provide an efficient algorithm to compute the zeros of Faber polynomials.

# 3   A Determinant Representation for Faber Polynomials

Faber polynomials have various representations such as integral formula, recurrence relation and generating function equation (cf. [9, p. 106]). Those representations give little information on the zeros of Faber polynomials. The following new determinant representation establishes the relation between the zeros of Faber polynomials and the eigenvalues of a well-structured matrix.

**Theorem 3.1.** *Let $F_n(z)$ be the Faber polynomials associated with the mapping function* $\Psi(w) = w + b_0 + \dfrac{b_1}{w} + \cdots$. *For $n = 1, 2, \ldots$, we have the following representation:*

$$F_n(z) = \det(zI_n - H_n), \qquad (3.1)$$

*where*

$$H_n = \begin{bmatrix} b_0 & 1 & 0 & \ldots & 0 & 2b_1 \\ b_1 & b_0 & 1 & \ldots & 0 & 3b_2 \\ b_2 & b_1 & b_0 & \ldots & 0 & 4b_3 \\ \vdots & \vdots & \vdots & & \vdots & \\ b_{n-2} & b_{n-3} & b_{n-4} & \ldots & b_0 & nb_{n-1} \\ 1 & 0 & 0 & \ldots & 0 & b_0 \end{bmatrix} \qquad (3.2)$$

*is an $n \times n$ matrix, $I_n$ is an $n \times n$ identity matrix. Consequently, the zeros of $F_n(z)$ are the eigenvalues of $H_n$ and*

$$F_n(z) = z^n + c_{n-1}z^{n-1} + \cdots + c_{n-j}z^{n-j} + \cdots + c_1 z + c_0, \qquad (3.3)$$

*where*

$$c_{n-j} = (-1)^{n-j} \sum_{1 \leq i_1 < i_2 < \cdots < i_j \leq n} \det[b_{i_p i_q}],$$

*where $j = 1, 2, \ldots, n$, sum over all positive integers on principle submatrices of $H_n$. In particular, if the domain $E$ is $C_m$-invariant then for $k = 0, 1, 2, \ldots, m-1$,*

$$F_{mn+k}(z) = z^{mn+k} + c_{mn+k-m}z^{mn+k-m} + \cdots + c_{mn+k-jm}z^{mn+k-jm} + \cdots + c_k z^k.$$

**Proof.** The Faber polynomials $F_n(z)$ satisfy the well-known recurrence relation

$$F_{n+1}(z) = (z - b_0)F_n(z) - \sum_{k=1}^{n-1} b_k F_{n-k}(z) - (n+1)b_n,$$

with initial value $F_0(z) = 1$ (see (1.3)). This gives the following determinant representation:

$$F_n(z) = \begin{vmatrix} 1 & 0 & 0 & \cdots & 0 & -(b_0 - z) \\ b_0 - z & 1 & 0 & \cdots & 0 & -2b_1 \\ b_1 & b_0 - z & 1 & \cdots & 0 & -3b_2 \\ \vdots & \vdots & \vdots & & \vdots & \vdots \\ b_{n-3} & b_{n-4} & b_{n-5} & \cdots & 1 & -(n-1)b_{n-2} \\ b_{n-2} & b_{n-3} & b_{n-4} & \cdots & b_0 - z & -nb_{n-1} \end{vmatrix}. \qquad (3.4)$$

Now we interchange the first row with the second row and the second row with the third row and so on until the first row gets down to the last row. We get

$$F_n(z) = (-1)^n \det(H_n - zI_n),$$

where $H_n$ is as in (3.2) and $I_n$ is an $n \times n$ identity matrix. Using the identity

$$\det(H_n - zI_n) = (-1)^n \det(zI_n - H_n),$$

we have (3.1). ∎

# 4    Computations of the Zeros of $F_n(z)$

We now use (3.1) to compute zeros of Faber polynomials associated with certain $C_m$-invariant domains such as symmetric lunes $L_\alpha$ $(0 < \alpha \le 2)$ (see 4.1) and $m$-gons $G_m$ $(m = 3, 4, ...)$ (see 4.2). As we shall see from the definitions of $L_\alpha$ and $G_m$ that $L_\alpha$ is $C_2$ and $C_m$-invariant, respectively. In addition, $L_\alpha$ is also invariant with respect to the Kleinian group $Z_2 \times Z_2$ and $G_m$ is invariant with respect to the dihedral group $D_m$.

The computational procedure for the zeros of $F_n(z)$ consists of three steps:

1. Finding the coefficients $b_k$ of the mapping function $\Psi(w)$.

2. Restoring $b_k$'s into the matrix $H_n$ (see (3.2)).

3. Computing and plotting the eigenvalues of $H_n$ in the complex plane.

Our numerical results on the zeros of $F_n(z; E)$ when $E$ is $C_m$-invariant are very stable and lead us to make the following conjecture.

**Conjecture:**    *If the domain $E$ is $C_m$-invariant then the set of limit points of zeros of $F(z; E)$ B is $C_m$-invariant.*

## 4.1    Zeros of $F_n(z)$ for a Lune

Let $L_\alpha$ be the closed region bounded by a circular lune symmetric about both axes with vertices at the points $\pm a$ and exterior angles $\alpha\pi$ $(0 < \alpha \le 2)$. The normalized mapping function from $|w| > 1$ onto $\overline{C} \setminus L_\alpha$ is given by

$$z = \Psi(w) = \alpha \frac{1 + (\frac{w-1}{w+1})^\alpha}{1 - (\frac{w-1}{w+1})^\alpha}, \quad 0 < \alpha \le 2.$$

The branch of the root is selected such that $a^\alpha > 0$ when $a > 0$.

If $\alpha = 1$, then $L_\alpha$ becomes an unit disk and $\Psi(w) = w$. The Faber polynomials of degree $n$ is $z^n$. For $\alpha = 2$, $L_2$ is a segment [-2,2], the Faber polynomials are the first kind Chebyshev polynomials $T_n(x)$ for the segment $[-2, 2]$. It is known from [11] that the zeros of $T_n(x)$ are located on $(-2, 2)$ for every $n \ge 1$ and the asymptotic behavior of the zeros of $T_n(x)$ is given by the arcsine distribution

$$d\mu(t) = \frac{1}{\pi} \frac{1}{\sqrt{4 - t^2}} dt.$$

What can be said about the zeros of $F_n(z; L_\alpha)$ when $0 < \alpha < 1$ or $1 < \alpha < 2$? When $\alpha = \frac{1}{2}$, we have

$$\begin{aligned}
\Psi(w) &= \frac{1}{2} \frac{\sqrt{w+1} + \sqrt{w-1}}{\sqrt{w+1} - \sqrt{w-1}} \\
&= w + \sum_{k=1}^{\infty} \frac{b_{2k-1}}{w^{2k-1}},
\end{aligned}$$

where

$$b_{2k-1} = -\frac{(2k-3)!!}{2^{k+1}k!}, \quad k = 1, 2, ....$$

When $\alpha = \frac{3}{2}$, we have

$$\Psi(w) = \frac{3}{2}\frac{(w+1)^{3/2}+(w-1)^{3/2}}{(w+1)^{3/2}-(w-1)^{3/2}}$$

$$= w + \frac{5}{12}\frac{1}{w} + \sum_{k=2}^{\infty}\frac{b_{2k-1}}{w^{2k-1}},$$

where

$$b_{2k-1} = \frac{7}{16}\frac{(-1)^k}{3^k} + \frac{27}{8}\sum_{j=1}^{k-2}\frac{(-1)^{k+j}}{3^{k-j}}\frac{(2j-1)!!}{2^j(j+2)!}, \quad k = 2, 3, ....$$

In the following, we shall give our computations for the zeros of the Faber polynomials. Our numerical results show that both vertices of $L_{1/2}$ and $L_{3/2}$ are the limits of the zeros of the Faber polynomials and zeros of $F_n(z; L_{1/2})$ and $F_n(z; L_{3/2})$ stay away from the analytic portions of the boundaries of $L_{1/2}$ and $L_{3/2}$, respectively.

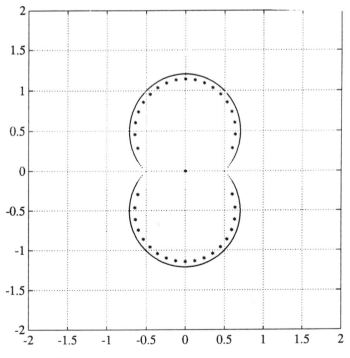

Figure 1. Zeros of $F_n(z)$ of $L_{\frac{1}{2}}$ when $n = 39$

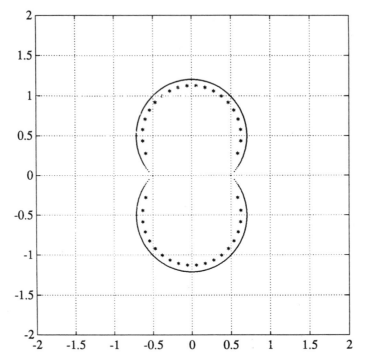

Figure 2. Zeros of $F_n(z)$ of $L_{\frac{1}{2}}$ when $n = 40$

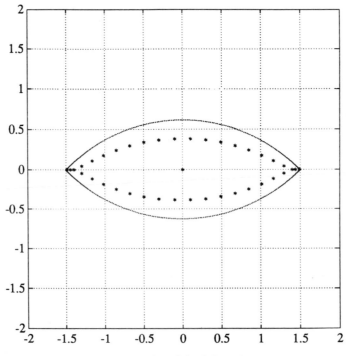

Figure 3. Zeros of $F_n(z)$ of $L_{\frac{3}{2}}$ when $n = 39$

## 4.2   Zeros of $F_n(z)$ for a Polygon

Let $G_m$ be the closed set bounded by an $m$-gon. By using Schwarz-Christoffel formula for conformal mapping of a polygon, we find that the normalized inverse mapping function $\Psi(w)$ for $G_m$ is given by

$$\Psi(w) \;=\; \int \left(1 - \frac{1}{w^m}\right)^{2/m} dw$$

$$=\; w + \sum_{k=1}^{\infty} \frac{b_{mk-1}}{w^{mk-1}},$$

where

$$b_{m-1} = \frac{1}{m-1},$$

$$b_{mk-1} = \frac{2(m-2)(2m-2)\cdots(m(k-1)-2)}{(mk-1)m^k k!}, \quad k = 2, 3, \dots.$$

It follows from theorem 2.1 that the Faber polynomials for $G_m$ has the following form,

$$F_n(z; G_m) = \sum_{k=0}^{\left[\frac{n}{m}\right]} c_{n-mk} z^{n-mk},$$

where $\left[\frac{n}{m}\right] = \begin{cases} \frac{n}{m} & \text{if } n \equiv 0(\mathrm{mod}\ m), \\ \frac{n-k}{m} & \text{if } n \equiv k(\mathrm{mod}\ m), \quad k = 0,1,2,\dots,m-1. \end{cases}$

We find that, In particular,

$$F_k(z) = z^k, \; 1 \le k \le m-1,$$

$$F_m(z) = z^m - \frac{m}{m-1},$$

$$F_{m+k}(z) = z^k\left(z^m - \frac{m+k}{m-1}\right), \; 1 \le k \le m-1,$$

$$F_{2m}(z) = z^{2m} - \frac{2m}{m-1}z^m + \frac{3m^2 - 5m + 2}{m(2m-1)(m-1)^2}.$$

As we shall see from our computations for the zeros of $F_n(z; G_m)$ when $m = 3, 4, 5, 6$ that every vertex of $G_m$ is the limit of the zeros of $F_n(z; G_m)$ and no limit points of the zeros of $F_n(z; G_m)$ on the analytic portions of the boundaries of $G_m$.

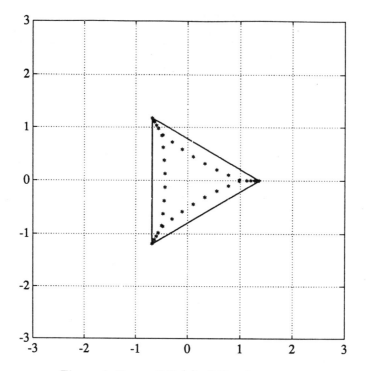

Figure 4. Zeros of $F_n(z)$ of $G_3$ when $n = 39$

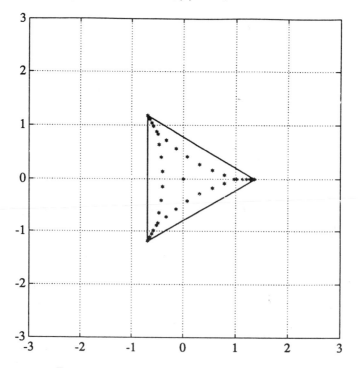

Figure 5. Zeros of $F_n(z)$ of $G_3$ when $n = 40$

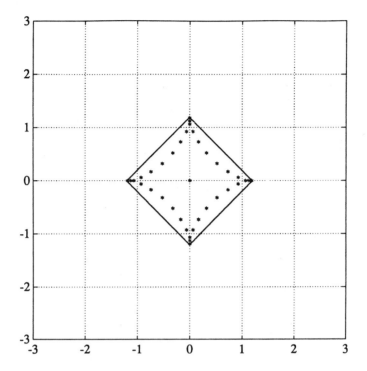

Figure 6. Zeros of $F_n(z)$ of $G_4$ when $n = 39$

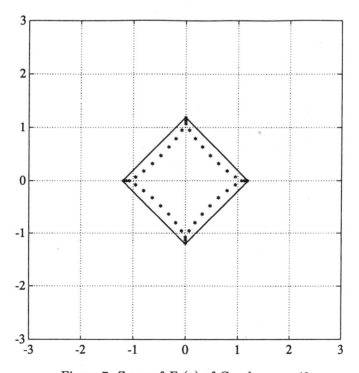

Figure 7. Zeros of $F_n(z)$ of $G_4$ when $n = 40$

# References

[1]  J. Bartalomeo and M. He, *On Faber Polynomials Generated by an m-Star*, to appear in Math. Comp.

[2]  J. H. Curtiss, *Faber Polynomials and the Faber Series*, Amer. Math. Monthly, 78 (1971), 577-596.

[3]  J. P. Coleman and R. A. Smith, *The Faber Polynomials for Circular Sector*, Math. Comp., 49 (1987), 231-241.

[4]  S. W. Elloiott, *Computation of Faber Series with Application to Numerical Polynomials Approximation in the Complex Plane*, Math. Comp., 40 (1983), 575-587.

[5]  G. Faber, *Über Polynomische Entwickelungen*, Math. Ann., 57 (1903), 389-408.

[6]  K. Gatermann, C. Hoffmann and G. Opfer, *Explicit Faber Polynomials on Circular Sector*, Math. Comp., Vol. 59(1992), 241-253.

[7]  M. He, *Weighted Polynomial Approximation and Zeros of Faber Polynomials*, Ph. D. Dissertation, 1991, Tampa.

[8]  M. He and E.B.Saff, *The Zeros of Faber Polynomials for an m-Cusped Hypocycloid*, to appear in J. Approx. Theory.

[9]  A.I. Markushvich, *Theory of Analytic Function*, Vol. 3, 1967, Prentice-Hall, Englewood Cliffs, N. J.

[10]  T. Kovari and C. Pommerenke, *On Faber Polynomials and Faber Expansions*, Math, Z., 99 (1976), 193-206.

[11]  G. Szegö, *Orthogonal Polynomials*, 4th ed (1975) Amer. Math. Soc.

Department of Mathematics
Nova University
Ft. Lauderdale, FL 33314
email: hem@polaris.nova.edu

Lectures in Applied Mathematics
Volume **29**, 1993

# SYMMGRP.MAX and other Symbolic Programs for Lie Symmetry Analysis of Partial Differential Equations

WILLY HEREMAN

ABSTRACT. The MACSYMA program SYMMGRP.MAX is presented. It greatly aids in finding the Lie symmetry group of large and complicated systems of differential equations. A survey of recent Lie symmetry packages is given. Examples demonstrate the use of such programs. Directions for further research and development are indicated.

## 1. Introduction

The method of infinitesimal transformations, originated by Sophus Lie (1842-1899), brought many ad-hoc techniques for solving special classes of differential equations under a common conceptual umbrella. Symmetry analysis has many applications, including the construction of group-invariant explicit solutions of nonlinear ODEs and PDEs, the classification of such equations, asymtotic analysis of their solutions, testing of numerical computations, etc.

Lie's original ideas greatly influenced the study of physically important systems of differential equations in classical mechanics, fluid dynamics, elasticity and many other applied areas. Later on, the concepts of Lie groups and 'symmetry' evolved into one of the most explosive developments of mathematics and physics throughout the twentieth century.

The application of Lie symmetry analysis to concrete physical systems involves tedious computations. Even the calculation of the continuous symmetry group of a modest system of differential equations is prone to fail if done with pencil and paper. Symbolic packages are extremely well suited for such computations. Indeed, symbolic packages can often find the determining (or defining) equations of the Lie symmetry group, solve them in closed form, and go on to calculate

---

1991 *Mathematics Subject Classification*. Primary 68Q40, 35-04; Secondary 22E60, 34-04.
This paper is in final form and no version of it will be submitted for publication elsewhere.

the infinitesimal generators that span the Lie algebra of symmetries. The group-invariant solutions obtained via Lie's approach then provide insight into the physical models themselves, or into the accuracy of numerical and approximate treatments.

There exists a large amount of literature on Lie symmetries, see e.g. [20]. Here we just mention some books [5, 6, 14, 27, 32, 33, 50]. We particularly recommend the special issues of *Acta Applicandae Mathematicae* on 'Symmetries of Partial Differential Equations' [58], now available in book form [59].

In this paper, we present the program SYMMGRP.MAX, written in MAC-SYMA syntax, which aids in the investigation of Lie symmetries for large and complicated systems of differential equations. Following a brief review (in Sections 2 and 3) of the newest symbolic software for Lie symmetry computations, in Section 4 we discuss the purpose, the method and the algorithms of such programs. In Section 5, two examples illustrate how SYMMGRP.MAX or other symmetry packages, can be used and what type of results can be obtained. In Section 6 we indicate directions for future research and development of symbolic software for Lie symmetries. We draw a brief conclusion in Section 7.


## 2. Symbolic software for Lie symmetries

We review the newest symbolic programs. A more in-depth survey will be published shortly [20].

The programs PDELIE and SYM_DE have been added to the MACSYMA out-of-core library. PDELIE by Vafeades [55, 56, 57] attempts to produce the determining equations and the generators of the Lie group automatically. The package PDELIE tries to find the invariants of the symmetry group and dimensionally reduce the given differential equation. In cases where the reduced equation is an ODE, it tries to integrate explicitly, thus arriving at special similarity solutions of the original equation. Special subroutines within PDELIE perform computations with elements in the commutator table of structure constants of the Lie algebra. With PDELIE one can also compute some of the densities of the Noether conservation laws of systems of variational type.

The program SYM_DE by Steinberg [49] computes infinitesimal symmetry operators and the explicit form of the infinitesimal transformations for simple systems. In cases where the program cannot automatically finish the computation, the user can intervene and, for instance, ask for infinitesimal symmetries of polynomial form. The program solves some (or all) of the determining equations automatically and if needed the user can (interactively) add extra information. Currently, Steinberg is working at the extension of SYM_DE which will include the calculation of generalized symmetries.

The program SYMMGRP.MAX, written by Champagne *et al.* [11], is a modification of an earlier package [12] that has been extensively used over the last eight years at the University of Montréal and elsewhere. It has been tested on hundreds of systems of equations and has thus been solidly debugged. SYM-MGRP.MAX can be used to compute classical [11] and nonclassical (or condi-tional) Lie point symmetries [13]. The program is obtainable from the authors and through the Computer Physics Communications Program Library in Belfast. The flexibility within SYMMGRP.MAX and the possibility of using it interac-tively, allow the user to find the symmetry group of in principal arbitrarily large and complicated systems of equations on relatively small computers.

To make SYMMGRP.MAX work for *large* systems of differential equations, the designers followed the path that would be taken in manual calculations. That is, obtain in as simple a manner as possible the simplest determining equations, solve them and feed the information back to the computer. Partial information can be extracted very rapidly. For instance, one can derive a subset of the deter-mining equations, e.g. those that occur as coefficients in the highest derivatives of the independent variables. These are usually single term equations, which express that the coefficients of the vectorfield are independent of some variables or depend linearly on some of the other variables. The feedback mechanism also allows to completely solve the determining equations step by step on the computer, hence avoiding human error in the algebraic simplifications.

In [10], Carminati *et al.* present their program LIESYMM, now added to the standard library of Maple, for creating the determining equations via the Harrison-Estabrook procedure [18]. Within LIESYMM various interactive tools are available for integrating the determining equations, and for working with Cartan's differential forms.

Hickman [22] offers a collection of Maple routines that can aid in the compu-tation of Lie point symmetries, non-local symmetries, and Wahlquist-Estabrook type prolongations.

The program LIE by Head [19] comes bundled with a limited version of mu-MATH. Therefore, LIE is self contained and runs on IBM compatible PCs. LIE calculates and solves the determining equations automatically for single equa-tions and systems of differential equations. The program also computes the Lie vectors and their commutators. Interventions by the user are possible but are rarely needed. The source code of the program is available, including the heuristic routines that attempt to solve the determining equations. Due to the limitations of muMATH, the program is bounded by the 256 KB of memory for program and workspace. For a program of limited size, LIE is remarkable in its achievements. The current version 3.4 of LIE is freely available by FTP from various public domain software archives such as SIMTEL and its associates.

The newly developed REDUCE program DIMSYM by Sherring and Prince [47] was inspired by Head's program [19], but is much larger and grew inde-pendent of it during development. DIMSYM is capable of finding various types of symmetries (including nonclassical symmetries), isolating special cases, and bringing the determining equations in normal form, and aiding in the solution of group classification problems. It attempts to determine the generators and allows to check whether or not the generators are correct. It allows the user to specify the dependence of the symmetry vector field coefficients manually, which is particularly practical in the computation of Lie-Bäcklund symmetries. An attractive feature of the package is that the solver for systems of linear homogeneous differential equations works for equations that were not necessarily obtained from symmetry analysis.

In Mathematica, Herod [21] developed the program MathSym for deriving the determining equations corresponding to Lie point symmetries, including nonclassical symmetries. Most recently, the program LIE has by been added by Baumann [3] to MathSource, the Mathematica Program Library. Baumann's program is a straightforward 'translation' of the MACSYMA program SYMM-GRP.MAX [11]. The only modifications are that LIE can handle transcendental functions and reduce and simplify the determining equations a bit further. Furthermore, after the determining equations have been solved interactively by the user, LIE can continue with the computation of the structure constants and the vector basis (commutators) of the Lie algebra.

The interactive REDUCE program by Nucci [30, 31] generates determining equations for (classical, nonclassical, Lie-Bäcklund and approximate) symmetries and provides interactive tools for the user to solve them. The manual [30] gives a clear description of the various routines, with their scope and limitations, and has several worked examples.

The PL/1-based FORMAC package CRACKSTAR developed by Wolf [60] allows investigation of Lie symmetries of systems of PDEs, besides dealing with dynamical symmetries of ODEs, and the like. The successor of CRACKSTAR is the REDUCE package CRACK by Wolf and Brand [62, 63, 64]. CRACK allows to solve overdetermined systems of differential equations with polynomial terms. It also contains code for decoupling, separating and simplifying PDEs. Integration of exact PDEs and differential factorizations are possible. With CRACK one can also construct the Lagrangian for a given second order ODE, and find first integrals via the integrating factor method.

Based on the tools available within CRACK that could aid in the investigation of Lie symmetries of ODEs, Wolf developed the REDUCE program LIEPDE [61]. That new program, for Lie-point and contact symmetries of PDEs, derives and solves a few simple determining equations, before continuing with the computation of the more complicated determining equations. This idea, which makes the program highly efficient, was used in Wolf's old FORMAC program CRACKSTAR [60], and is also implemented in SYMMGRP.MAX [11]. The difference is that within LIEPDE the steps are carried out automatically, without intervention by the user.

Sarlet and Vanden Bonne [42] offer REDUCE procedures to assist in the computation of adjoint symmetries of differential equations, where determining equations of the same nature occur.

Kersten [24, 25] further perfected the software package for the calculation of the Lie algebra of infinitesimal symmetries (and Lie-Bäcklund symmetries) of exterior differential systems. Kersten's program [24], which is based on Cartan's exterior calculus [9], is accessible in REDUCE.

The well-documented program SPDE, developed by Schwarz [45] automatically derives and often successfully solves the determining equations for Lie point symmetries with minimal intervention by the user. The package SPDE is distributed with the current version of REDUCE for various types of computers, ranging from PCs to CRAYs. Schwarz also rewrote SPDE for use with AXIOM (formerly SCRATCHPAD II), a symbolic manipulation program developed by IBM. The algorithm SYMSIZE [46], for the determination of the size of the symmetry group, is available with the computer algebra systems AXIOM and REDUCE, as part of the package SPDE.

Mikhailov developed software in muMATH to verify the integrability of systems of PDEs by testing for the existence of higher symmetries. The program computes special symmetries, canonical conservation laws and carries out conformal transformations to bring PDEs into canonical form. With their PC program Mikhailov *et al.* [29] were able to produce an exhaustive list of integrable nonlinear Schrödinger type equations. In this context, integrable means that the equations have infinitely many conserved quantities and infinitely many local symmetries.

The FORMAC program *LB* [15] and its successor *LBF* by Fushchich and Kornyak [17], both create the system of determining equations for Lie-Bäcklund symmetries and attempt to solve these equations. The program *LB* is available from the Computer Physics Communications Program Library in Belfast. The program *LBF*, with its 1362 lines of PL/1-FORMAC code, is completely automatic and consists of 37 subroutines, one of which brings the determining equations in passive (Riquier-Janet) form. The above programs were designed for low memory requirements so that they could run on PCs.

The PC-package DELiA, written in Turbo PASCAL by Bocharov and his collaborators [7, 8], is a stand-alone computer algebra system for investigating differential equations. It performs various tasks based on Lie's approach, such as the computation of point and Lie-Bäcklund symmetries, canonical conserved densities and generalized conservation laws; simplification and partial integration of overdetermined systems of differential equations, etc. In order to be able to handle large problems DELiA first generates and solves first order determining equations, and then continues to generate and solve the higher-order determining equations. The analyzer/integrator, which is available as a separate tool at the user-level, includes a general algorithm for passivization, together with a set of integration rules for linear and quasi linear systems of PDEs. Currently, a MS Windows version of DELiA, called MS Win DELiA, is under development. This new version will facilitate user-level interface and allow the user to surpass the memory barrier of 640 KB under DOS.

Fushchich and Kornyak [17, 26] wrote programs in Turbo C and AMP for the computation of Lie-Bäcklund symmetries. Their programs also classify equations with arbitrary parameters and functions with respect to such symmetries. Important to note is that their programs reduce the determining equations into passive form.

Using the algorithmic language REFAL, Topunov [**52**] developed a software package for symmetry analysis that contains subroutines to reduce determining systems in passive form.

## 3. Solving the Determining Equations and Computation of the Size of the Symmetry Group

Solving the determining equations can usually be done via separation of variables, standard techniques for linear differential equations, and specific heuristic rules as described in [**19, 43, 44, 47, 49, 50**]. The only determining equations left for manual handling, should be the 'constraint' equations or any other equations whose general solutions cannot be written explicitly in closed form. Solving them on a computer may be time consuming, since the simplest approach varies greatly from case to case. Furthermore, a computer program may accidentally not catch the most general result and therefore return an incomplete symmetry group. The author is aware of this problem which occurred in testing some of the symmetry programs.

Use of SYMSIZE circumvents some of the shortcomings mentioned above. Indeed, if a differential equation has no other than trivial symmetries or if the symmetry group is small (because all generators are algebraic and of low degree), SYMSIZE will greatly help in completely solving the symmetry problem. In contrast to the *heuristic* algorithms for the explicit computation of the symmetry generators, the size of the symmetry group can always be determined with SYMSIZE in a finite number of steps. SYMSIZE accepts a systems of PDEs as input, and allows to compute *a priori* the number of free parameters if the group is finite and the number of unspecified functions if the group is infinite. In turn, SYMSIZE allows to test *a posterior* if the solution of the determining equations is complete. The mathematical background of Schwarz' algorithm [**46**] goes back to the theory of differential equations due to Riquier [**41**] and Janet [**23**]. The heart of SYMSIZE is the procedure *InvolutionSystem*, which transforms the determining system into an involutive system by means of a critical pair/completion algorithm. The algorithm SYMSIZE [**46**] is available with the computer algebra systems AXIOM and REDUCE, as part of the package SPDE, mentioned in Section 2.

Concurrently, yet independent of Schwarz, Reid [**35, 36, 37**] realized that triangularization algorithms may allow bypassing the explicit solution of the determining equations and compute the size of the symmetry group [**37**] and the commutators immediately [**38**]. Reid developed the program SYMCAL [**36**], written originally in MACSYMA, and currently available in Maple [**40**]. The program SYMCAL computes the dimension and the structure constants of the Lie symmetry algebra of any system of PDEs. An extension of the algorithm [34] also allows to classify differential equations (with variable coefficients) according to the structure of their symmetry groups. Reid's program is also able to give a series expansion of the general solution to any given finite order. Furthermore, the approach advocated by Reid applies to the determination of symmetries of Lie-contact and Lie-Bäcklund types as well as nonlocal (potential) symmetries.

Recently, Reid extended the underlying algorithm *standard_form* [36] to account for general, total degree orderings. The procedure *standard_form* can be viewed as a generalization to linear differential equations of the Gaussian reduction method for matrices or linear systems. It reduces the system of PDEs to an equivalent simplified ordered triangular system with all integrability conditions included and all redundancies (differential and algebraic) eliminated.

Reid and Wittkopf's package [40] allows automated interfacing with major symmetry packages such as DIMSYM [47], LIESYMM [10], and SYMM-GRP.MAX [11], and also with the differential Gröbner basis package DIFF-GROB [28].

Reid and McKinnon [39] just finished the development of a recursive algorithm that builds on Reid's program [37], and finds particular solutions of linear systems of PDEs using only ODE solution techniques. Applied to symmetry problems, their algorithm will find all polynomial/rational solutions of the determining equations provided the symmetry group is finite-dimensional.

## 4. Purpose, method and algorithm

**4.1. Purpose.** The application of Lie groups methods to concrete physical systems involves tedious, but rather mechanical computations. Symbolic packages can find the defining (or determining) equations of the Lie symmetry group, attempt to solve them in closed form, and go on to calculate the infinitesimal generators that span the Lie algebra of symmetries.

**4.2. Method.** There are basically three different methods to compute Lie symmetries. The first one uses prolonged vector fields [32], the second utilizes differential forms [18, 24, 25] (wedge products) due to Cartan, the third one uses the notion of 'formal symmetry' [7, 8, 29]. We restrict our discussion to the method of prolonged vector fields [32], which is used in the algorithm or our program SYMMGRP.MAX and most other Lie symmetry packages.

For notational simplicity, let us consider the case of Lie point symmetries [5, 6, 32] and follow the notations, terminology and method for symmetry analysis used in [32]. We start with a system of $m$ differential equations,

$$(4.1) \qquad \Delta^i(x, u^{(k)}) = 0, \quad i = 1, 2, ..., m,$$

of order $k$, with $p$ independent and $q$ dependent real variables, denoted by $x = (x_1, x_2, ..., x_p) \in I\!R^p$, $u = (u^1, u^2, ..., u^q) \in I\!R^q$. We stress that $m, k, p$ and $q$

are *arbitrary* positive integers. The group transformations have the form $\tilde{x} = \Lambda_g(x, u)$, $\tilde{u} = \Omega_g(x, u)$, where the functions $\Lambda_g$ and $\Omega_g$ are to be determined. Note that the subscript $g$ refers to the group parameters. The approach is an infinitesimal one; instead of looking for the Lie group $G$, we look for its Lie algebra $\mathcal{L}$, realized by vector fields of the form

$$(4.2) \qquad \alpha = \sum_{i=1}^{p} \eta^i(x, u) \frac{\partial}{\partial x_i} + \sum_{l=1}^{q} \varphi_l(x, u) \frac{\partial}{\partial u^l}.$$

The problem is now reduced to finding the coefficients $\eta^i(x, u)$ and $\varphi_l(x, u)$. In essence, the computer constructs the $k^{th}$ prolongation $pr^{(k)}\alpha$ of the vector field $\alpha$, applies it to the system (4.1), and requests that the resulting expression vanishes on the solution set of (4.1)

$$(4.3) \qquad pr^{(k)}\alpha\Delta^i \,|_{\Delta^j=0} \qquad i, j = 1, ..., m.$$

The result of implementing (4.3) is a system of linear homogeneous PDEs for $\eta^i$ and $\varphi_l$, in which $x$ and $u$ are independent variables. These are the so-called *determining equations* for the symmetries of the system. Solving these by hand or automatically will give the explicit forms of the $\eta^i(x, u)$ and $\varphi_l(x, u)$.

**4.3. Contact and generalized symmetries.** The procedure to compute generalized symmetries or Lie-Bäcklund symmetries [**2, 32**] is essentially the same as that for point-symmetries, although the calculations are lengthier and more expensive. In a generalized vector field, which still takes the form of (4.2), the functions $\eta^i$ and $\phi_l$ may also depend on derivatives of $u$. Olver [**32**] discusses various possibilities to simplify the calculations, e.g. by putting the symmetries in evolutionary form, or by fixing the order of derivation on which the $\eta$'s and $\phi$'s may depend. Contact symmetries are generalized symmetries of order one, e.g. the coefficients in the vector field include only first derivatives of the dependent variables.

**4.4. Nonclassical or conditional symmetries.** Recently, it has been shown that the 'nonclassical method of group-invariant solutions' originally introduced in [**4**], can determine new solutions of various physically significant nonlinear PDEs, see [**13, 20**].

In contrast to true Lie symmetries, the transformations corresponding to nonclassical (or conditional) symmetries do not leave the differential equation invariant, neither do they transform all the solutions into other solutions. As a matter of fact, they merely transform a subset of solutions into other solutions. Accounting for 'nonclassical symmetries', the program should automatically add the $q$ invariant surface conditions [**4**],

$$(4.4) \qquad Q^l(x, u^{(1)}) = \sum_{i=1}^{p} \eta^i(x, u) \frac{\partial u^l}{\partial x_i} - \varphi_l(x, u) = 0, \quad l = 1, ..., q,$$

and their differential consequences, to the system (4.1). However, the inclusion of nonclassical symmetries requires solving systems of determining equations which are no longer linear.

**4.5. Algorithms.** Ideally, a fully automated software package for Lie symmetries should consist of effective, powerful algorithms and fast procedures for:
(i) the derivation of the determining equations for large or complicated systems of equations;
(ii) the reduction of determining equations into so-called standard form;
(iii) finding the size of the symmetry group;
(iv) determining obvious symmetry generators by fast analysis;
(v) if not all the generators have been found, attempt simplification and integration of the defining equations to determine the generators.

Then the program should be able to execute the following steps in the order relevant to the specific application:
(a) calculation of commutator tables, based on the results of (i), (ii) and (iii);
(b) calculation of group invariant solutions.

The program should be able to handle: calculation of nonlocal (potential) symmetries [**5, 6, 33, 59**], calculation of nonclassical reductions (conditional symmetries) and resulting solutions, calculation of generalized symmetries, calculation of equivalent conservation laws. Furthermore, it should be able to handle systems with free unknown (classification) functions [**37, 47**].

So far, our code SYMMGRP.MAX, which was designed to handle large and complicated systems of differential equations, has (i) a procedure to efficiently *derive* the determining equations for classical point symmetries, and (ii) some capabilities to *solve* them interactively. In [**11**] a complete write-up of the algorithm of SYMMGRP.MAX is given, together with instructions for the user. A future version of SYMMGRP.MAX will automatically compute the determining equations for nonclassical symmetries and also generalized (velocity dependent) symmetries.

## 5. Examples

**5.1. The Harry Dym equation.** Consider the Harry Dym equation [**1**],

$$(5.1) \qquad u_t - u^3 u_{xxx} = 0.$$

Clearly, this is one equation ($m = 1$) with two independent variables ($p = 2$) and one dependent variable ($q = 1$). The assignments of the variables are as follows:

$$x \longmapsto x[1] \quad , \quad t \longmapsto x[2] \quad , \quad u \longmapsto u[1]$$

This permits to write the equation (5.1) in a standard form accepted by our program,
```
e1 : u[1,[0,1]]-u[1]^3*u[1,[3,0]].
```

Next, one selects the variable $u_t$ for elimination, i.e.,
v1 : u[1,[0,1]].

Then, SYMMGRP.MAX automatically computes the determining equations for
the coefficients $eta[1] = \eta^x$, $eta[2] = \eta^t$ and $phi[1] = \varphi^u$ of the vector field

$$(5.2) \qquad \alpha = \eta^x \frac{\partial}{\partial x} + \eta^t \frac{\partial}{\partial t} + \varphi^u \frac{\partial}{\partial u}.$$

There are only eight simple determining equations,

$$\frac{\partial eta2}{\partial u[1]} = 0, \qquad\qquad\qquad \frac{\partial eta2}{\partial x[1]} = 0,$$

$$\frac{\partial eta1}{\partial u[1]} = 0, \qquad\qquad\qquad \frac{\partial^2 phi1}{\partial u[1]^2} = 0,$$

$$\frac{\partial^2 phi1}{\partial u[1]\partial x[1]} - \frac{\partial^2 eta1}{\partial x[1]^2} = 0,$$

$$\frac{\partial phi1}{\partial x[2]} - u[1]^3 \frac{\partial^3 phi1}{\partial x[1]^3} = 0,$$

$$3u[1]^3 \frac{\partial^3 phi1}{\partial u[1]\partial x[1]^2} + \frac{\partial eta1}{\partial x[2]} - u[1]^3 \frac{\partial^3 eta1}{\partial x[1]^3} = 0,$$

$$u[1] \frac{\partial eta2}{\partial x[2]} - 3u[1] \frac{\partial eta1}{\partial x[1]} + 3\, phi1 = 0.$$

With a feedback mechanism, these determining equations can be solved explicitly. The general solution, rewritten in the original variables, is

$$(5.3) \qquad \eta^x = k_1 + k_3\, x + k_5\, x^2,$$
$$(5.4) \qquad \eta^t = k_2 - 3k_4\, t,$$
$$(5.5) \qquad \varphi^u = (k_3 + k_4 + 2k_5\, x)\, u.$$

The five infinitesimal generators then are

$$(5.6) \qquad\qquad G_1 = \partial_x,$$
$$(5.7) \qquad\qquad G_2 = \partial_t,$$
$$(5.8) \qquad\qquad G_3 = x\partial_x + u\partial_u,$$
$$(5.9) \qquad\qquad G_4 = -3t\partial_t + u\partial_u,$$
$$(5.10) \qquad\qquad G_5 = x^2\partial_x + 2xu\partial_u.$$

Clearly, (5.1) is invariant under translations ($G_1$ and $G_2$) and scaling ($G_3$ and
$G_4$). The flow corresponding to each of the infinitesimal generators can be ob-

tained via simple integration. For example, let us compute the flow corresponding to $G_5$. This requires integration of the first order system

$$\frac{d\tilde{x}}{d\epsilon} = \tilde{x}^2, \qquad \tilde{x}(0) = x,$$

$$\frac{d\tilde{t}}{d\epsilon} = 0, \qquad \tilde{t}(0) = t,$$

$$\frac{d\tilde{u}}{d\epsilon} = 2\tilde{x}\tilde{u}, \qquad \tilde{u}(0) = u,$$

where $\epsilon$ is the parameter of the transformation group. One readily obtains

$$(5.11) \qquad \tilde{x}(\epsilon) = \frac{x}{(1 - \epsilon x)},$$

$$(5.12) \qquad \tilde{t}(\epsilon) = t,$$

$$(5.13) \qquad \tilde{u}(\epsilon) = \frac{u}{(1 - \epsilon x)^2}.$$

We therefore conclude that for any solution $u = f(x, t)$ of equation (5.1), the transformed solution

$$(5.14) \qquad \tilde{u}(\tilde{x}, \tilde{t}) = (1 + \epsilon\tilde{x})^2 f\left(\frac{\tilde{x}}{1 + \epsilon\tilde{x}}, \tilde{t}\right)$$

will solve $\tilde{u}_{\tilde{t}} - \tilde{u}^3 \tilde{u}_{\tilde{x}\tilde{x}\tilde{x}} = 0$.

**5.2. The Magneto-Hydro-Dynamics equations.** As an example of a large system of differential equations, we take the equations for Magneto-Hydro-Dynamics (MHD) [16]. If we neglect dissipative effects, and thus restrict the analysis to the *ideal* case, we have

$$(5.15) \qquad \frac{\partial\rho}{\partial t} + (\vec{v} \cdot \nabla)\rho + \rho\nabla \cdot \vec{v} = 0,$$

$$(5.16) \qquad \rho\left(\frac{\partial\vec{v}}{\partial t} + (\vec{v} \cdot \nabla)\vec{v}\right) + \nabla(p + \frac{1}{2}\vec{H}^2) - (\vec{H} \cdot \nabla)\vec{H} = \vec{0},$$

$$(5.17) \qquad \frac{\partial\vec{H}}{\partial t} + (\vec{v} \cdot \nabla)\vec{H} + \vec{H}\nabla \cdot \vec{v} - (\vec{H} \cdot \nabla)\vec{v} = \vec{0},$$

$$(5.18) \qquad \nabla \cdot \vec{H} = 0,$$

$$(5.19) \qquad \frac{\partial}{\partial t}\left(\frac{p}{\rho^\kappa}\right) + (\vec{v} \cdot \nabla)\left(\frac{p}{\rho^\kappa}\right) = 0,$$

with pressure $p$, mass density $\rho$, coefficient of viscosity $\kappa$, fluid velocity $\vec{v}$ and magnetic field $\vec{H}$. Using the first equation, we eliminate $\rho$ from the last equation, and replace it by

$$(5.20) \qquad \frac{\partial p}{\partial t} + \kappa p(\nabla \cdot \vec{v}) + (\vec{v} \cdot \nabla)p = 0.$$

If we split the vector equations in scalar equations for the vector components, we have a system of $m = 9$ equations, with $p = 4$ independent variables and $q = 8$ dependent variables. For convenience, we denote the components of the vector $\vec{v}$ by $v_x, v_y$ and $v_z$, not to confused with partial derivatives of $v$. The assignment of the variables is as follows:

$$
\begin{aligned}
x &\longmapsto x[1], & v_x &\longmapsto u[3], \\
y &\longmapsto x[2], & v_y &\longmapsto u[4], \\
z &\longmapsto x[3], & v_z &\longmapsto u[5], \\
t &\longmapsto x[4], & H_x &\longmapsto u[6], \\
\rho &\longmapsto u[1], & H_y &\longmapsto u[7], \\
p &\longmapsto u[2], & H_z &\longmapsto u[8].
\end{aligned}
$$

The variables to be eliminated are selected as follows: for the first 7 variables and the ninth variable we pick the partial derivatives with respect to $t$ of $\rho, v_x, v_y, v_z, H_x, H_y, H_z$ and $p$. From the eighth equation we select $\partial H_x / \partial x$ for elimination.

We will only consider the case where $\kappa \neq 0$. Running this case on a Digital VAX 4500 with 64 MB of RAM, in about 50 minutes of CPU time the program SYMMGRP.MAX creates the 222 determining equations for the coefficients of the vectorfield

$$
\begin{aligned}
\alpha = {} & \eta^x \frac{\partial}{\partial x} + \eta^y \frac{\partial}{\partial y} + \eta^z \frac{\partial}{\partial z} + \eta^t \frac{\partial}{\partial t} + \varphi^\rho \frac{\partial}{\partial \rho} + \varphi^p \frac{\partial}{\partial p} \\
& + \varphi^{v_x} \frac{\partial}{\partial v_x} + \varphi^{v_y} \frac{\partial}{\partial v_y} + \varphi^{v_z} \frac{\partial}{\partial v_z} + \varphi^{H_x} \frac{\partial}{\partial H_x} + \varphi^{H_y} \frac{\partial}{\partial H_y} + \varphi^{H_z} \frac{\partial}{\partial H_z}.
\end{aligned}
$$

Using MACSYMA interactively, we then integrated the determining system and obtained the solution, expressed in the original variables,

$$
\begin{aligned}
(5.21) \qquad \eta^x &= k_2 + k_5\, t - k_8\, y - k_9\, z + k_{11}\, x, \\
(5.22) \qquad \eta^y &= k_3 + k_6\, t + k_8\, x - k_{10}\, z + k_{11}\, y, \\
(5.23) \qquad \eta^z &= k_4 + k_7\, t + k_9\, x + k_{10}\, y + k_{11}\, z, \\
(5.24) \qquad \eta^t &= k_1 + k_{12}\, t, \\
(5.25) \qquad \varphi^\rho &= -2\,(k_{11} - k_{12} - k_{13})\,\rho, \\
(5.26) \qquad \varphi^p &= 2\,k_{13}\,p, \\
(5.27) \qquad \varphi^{v_x} &= k_5 - k_8\, v_y - k_9\, v_z + (k_{11} - k_{12})v_x, \\
(5.28) \qquad \varphi^{v_y} &= k_6 + k_8\, v_x - k_{10}\, v_z + (k_{11} - k_{12})v_y, \\
(5.29) \qquad \varphi^{v_z} &= k_7 + k_9\, v_x + k_{10}\, v_y + (k_{11} - k_{12})v_z, \\
(5.30) \qquad \varphi^{H_x} &= k_{13}\, H_x - k_8 H_y - k_9 H_z, \\
(5.31) \qquad \varphi^{H_y} &= k_{13}\, H_y + k_8 H_x - k_{10} H_z, \\
(5.32) \qquad \varphi^{H_z} &= k_{13}\, H_z + k_9 H_x + k_{10} H_y.
\end{aligned}
$$

It is clear that there is a thirteen dimensional Lie algebra spanned by the generators:

$$(5.33) \quad G_1 \;=\; \partial_t \,,$$

$$(5.34) \quad G_2 \;=\; \partial_x \,,$$

$$(5.35) \quad G_3 \;=\; \partial_y \,,$$

$$(5.36) \quad G_4 \;=\; \partial_z \,,$$

$$(5.37) \quad G_5 \;=\; t\partial_x + \partial_{v_x} \,,$$

$$(5.38) \quad G_6 \;=\; t\partial_y + \partial_{v_y} \,,$$

$$(5.39) \quad G_7 \;=\; t\partial_z + \partial_{v_z} \,,$$

$$(5.40) \quad G_8 \;=\; x\partial_y - y\partial x + v_x\partial_{v_y} - v_y\partial_{v_x} + H_x\partial_{H_y} - H_y\partial_{H_x} \,,$$

$$(5.41) \quad G_9 \;=\; y\partial_z - z\partial y + v_y\partial_{v_z} - v_z\partial_{v_y} + H_y\partial_{H_z} - H_z\partial_{H_y} \,,$$

$$(5.42) \quad G_{10} \;=\; z\partial_x - x\partial z + v_z\partial_{v_x} - v_x\partial_{v_z} + H_z\partial_{H_x} - H_x\partial_{H_z} \,,$$

$$(5.43) \quad G_{11} \;=\; x\partial_x + y\partial_y + z\partial_z - 2\rho\partial_\rho + v_x\partial_{v_x} + v_y\partial_{v_y} + v_z\partial_{v_z} \,,$$

$$(5.44) \quad G_{12} \;=\; t\partial_t + 2\rho\partial_\rho - (v_x\partial_{v_x} + v_y\partial_{v_y} + v_z\partial_{v_z}) \,,$$

$$(5.45) \quad G_{13} \;=\; 2\rho\partial_\rho + 2p\partial_p + H_x\partial_{H_x} + H_y\partial_{H_y} + H_z\partial_{H_z} \,.$$

Thus, the MHD equations (5.15) are invariant under translations $G_2$ through $G_4$, Galilean boosts $G_5$ through $G_7$, rotations $G_8$ through $G_{10}$, and dilations $G_{11}$ through $G_{13}$. In contrast to the results obtained for the 1+1 and the 2+1 dimensional versions of the MHD problem, the dimension of the Lie algebra for (5.15) in 3+1 dimensions ($x, y, z$ and time $t$) is independent of the value of the coefficient of viscosity $\kappa$. Our result (5.33) also confirms a recent investigation [16] of the MHD system.

## 6. Further Research and Development

The most challenging part of Lie symmetry analysis by computer, involves the design of an 'integrator' for the overdetermining systems of linear homogeneous partial differential equations. This topic is also of importance in the study of so-called adjoint symmetries of differential equations [42] and in many other areas where determining equations of the same nature occur.

In the context of Lie symmetry analysis, one can aim at the design of faster and more powerful algorithms that work for large systems of determining equations, typically a few hundred, and that automatically reduce systems to where they can be handled interactively with the computer, or by hand.

Since the early developments [19, 43, 44, 49] of semi-heuristic methods to solve determining equations, substantial progress has been made in understanding the mathematics of this problem and several algorithms are now available. These algorithms attempt to close the gap between solution techniques for ODEs and PDEs (consult [48] for an impressive review and large bibliography).

Despite the innovative efforts of Reid, Schwarz, Wolf and Brand, and others, there is no general algorithm available to integrate an arbitrary (overdetermined) system of determining equations, which consists of linear homogeneous partial differential equations for the coefficients in the vector field. After searching the relevant literature it is clear that many mathematical questions remain open. However, one thing is certain: to design a reliable and powerful integration algorithm for a system of determining equations, the system needs to be brought into a canonical form. Some authors call it the normal, involutive or passive form, but the definitions vary according to their origin. The original theory of involutive systems goes back to Cartan [9], Janet [23], Riquier [41], Thomas [51] and Tresse [53, 54].

The algorithms by Topunov [52], Reid [34, 35, 40] and Schwarz [46] are clones of the Riquier-Janet-Thomas method. Roughly speaking, the methods that reduce the systems of linear homogeneous PDEs into an equivalent, but much simpler standard form, can be viewed as a generalization of triangulation by Gauss-Jordan elimination but, applied here to systems of linear PDEs. First, the original system is appended by all its' differential consequences. Second, highest derivatives are eliminated, and, if they occur, integrability conditions are added to the system. The procedure is then repeated until the new system is in involution.

In the full computer implementation of such a 'triangulation' algorithm for nonlinear systems, one takes advantage of a 'differential' generalization of Buchberger's algorithm for Gröbner bases. Buchberger's algorithm, which is included as standard package with modern symbolic manipulation programs, provides a technique for canonically simplifying polynomially nonlinear systems of algebraic equations. The 'differential' generalization of that algorithm allows to reduce systems of nonlinear (and consequently also linear) PDEs into standard form. The Maple program DIFFGBASIS by Mansfield and Fackerell [28] offers a differential Gröbner basis algorithm applicable to nonlinear PDEs, not necessary resulting from symmetry analysis. Needless to say, DIFFGROB can be applied to reduce linear and nonlinear determining equations, the latter arising in the case of nonclassical Lie symmetries [13].

## 7. Conclusion

Apart from the theoretical study of the underlying mathematics, we hope to further develop and implement effective algorithms to generate, simplify and perhaps solve the determining equations for classical, nonclassical, and generalized Lie symmetries and nonlocal symmetries.

The availability of sophisticated symbolic programs should accelerate the study of symmetries of physically important systems of differential equations in classical mechanics, fluid dynamics, elasticity and other applied areas.

## References

1. M. J. Ablowitz and P. A. Clarkson, *Solitons, Nonlinear Evolution Equations and Inverse Scattering*, Cambridge University Press, Cambridge, UK, 1991.
2. R. L. Anderson and N. H. Ibragimov, *Lie-Bäcklund Transformations in Applications*, Stud. in Appl. Math. **1**, SIAM, Philadelphia, PA, 1979.
3. G. Baumann, *Lie symmetries of differential equations*, Preprint, Abteilung für Mathematische Physik, Universität Ulm, Ulm, Germany, 1992.
4. G. W. Bluman and J. D. Cole, *The general similarity solution of the heat equation*, J. Math. Mech. **18** (1969), 1025–1042.
5. _____, *Similarity Methods for Differential Equations*, Appl. Math. Scs. **13**, Springer Verlag, New York, 1974.
6. G. W. Bluman and S. Kumei, *Symmetries and Differential Equations*, Springer Verlag, New York, 1989.
7. A. V. Bocharov, *DEliA: A system of exact analysis of differential equations using S. Lie approach*, Report, Joint Venture OWIMEX Program Systems Institute USSR, Academy of Sciences, Pereslavl-Zalessky, USSR, 1989.
8. _____, *DEliA: project presentation*, SIGSAM Bulletin **24** (1990), 37–38.
9. E. Cartan, *Les Systèmes Différentiels Extérieur et leurs Applications Géometrique*, Hermann, Paris, 1946.
10. J. Carminati, J. S. Devitt, and G. J. Fee, *Isogroups of differential equations using algebraic computing*, J. Sym. Comp. **14** (1992), 103–120.
11. B. Champagne, W. Hereman, and P. Winternitz, *The computer calculation of Lie point symmetries of large systems of differential equations*, Comp. Phys. Comm. **66** (1991), 319–340.
12. B. Champagne and P. Winternitz, *A MACSYMA program for calculating the symmetry group of a system of differential equations*, Report CRM-1278, Centre de Recherches Mathématiques, Montréal, Canada, 1985.
13. P. A. Clarkson and E. L. Mansfield, *Symmetry reductions and exact solutions of a class of nonlinear heat equations*, Preprint, Dept. of Math., Univ. Exeter, Exeter, UK, 1993.
14. N. Euler and W.-H. Steeb, *Continuous Symmetries, Lie Algebras and Differential Equations*, Bibliographisches Institut Wissenschaftsverlag, Mannheim, 1993.
15. R. N. Fedorova and V. V. Kornyak, *Determination of Lie-Bäcklund symmetries of differential equations using FORMAC*, Comp. Phys. Comm. **39** (1986), 93–103.
16. J. C. Fuchs, *Symmetry groups and similarity solutions of MHD equations*, J. Math. Phys. **32** (1991), 1703–1708.
17. W. I. Fushchich and V. V. Kornyak, *Computer algebra application for determining Lie and Lie-Bäcklund symmetries of differential equations*, J. Symb. Comp. **7** (1989), 611–619.
18. B. K. Harrison and F. B. Estabrook, *Geometric approach to invariance groups and solution of partial differential systems*, J. Math. Phys. **12** (1971), 653–666.
19. A. K. Head, *LIE: A PC program for Lie analysis of differential equations*, Preprint, CSIRO Division of Material Science and Technology, Clayton, Australia, 1992, Comp. Phys. Comm. (submitted).
20. W. Hereman, *Review of symbolic software for the computation of Lie symmetries of differential equations*, Euromath Bulletin **2**, no. 1 (1993) (to appear).
21. S. Herod, *MathSym: A Mathematica program for computing Lie symmetries*, Preprint, Program in Appl. Math., Univ. Colorado, Boulder, Colorado, 1992.
22. M. Hickman, *The use of Maple in the search for symmetries*, Preprint, Dept. of Math., Univ. Canterbury, Christchurch, New Zealand, 1993.
23. _____, *Leçons sur les Systèmes d'Equations aux Dérivées*, Gauthier-Villars, Paris, 1929; *Equations aux Dérivées Partielles*, CNRS, Paris, 1956.
24. P. H. M. Kersten, *Infinitesimal symmetries: A computational approach.* Ph.D. Thesis, Twente Univ. Tech., Enschede, The Netherlands, 1985; CWI Tract 34, Center for Mathematics and Computer Science, Amsterdam, 1987.
25. _____, *Software to compute infinitesimal symmetries of exterior differential systems, with applications*, Acta Appl. Math. **16** (1989), 207–229.

26. V. V. Kornyak and W. I. Fushchich, *A program for symmetry analysis of differential equations*, Proc. ISSAC '91 (Bonn, Germany, 1991), S. M. Watt, ed., ACM Press, New York, 1991, pp. 315–316.

27. D. Levi and P. Winternitz, Eds., *Symmetries and Nonlinear Phenomena*, World Scientific, Singapore, 1988.

28. E. L. Mansfield and E. D. Fackerell, *Differential Gröbner bases*, Preprint 92-108, School of Mathematics, Physics, Computer Science and Electronics, Macquarie Univ., Sydney, Australia, 1992, J. Symb. Comp. (submitted).

29. A. V. Mikhailov, A. B. Shabat, and V. V. Sokolov, *The symmetry approach to classification of integrable equations*, What is Integrability? (V. I. Zakharov, ed.), Springer Verlag, New York, 1990, pp. 115–184.

30. M. C. Nucci, *Interactive REDUCE programs for calculating, classical, non-classical, and Lie-Bäcklund symmetries of differential equations*, Preprint GT Math. 062090-051, School of Math., Georgia Institue of Technology, AT, Georgia, 1990.

31. _____, *Interactive REDUCE programs for calculating, classical, non-classical, and approximate symmetries of differential equations*, Proc. 13th IMACS World Congress (Dublin, Ireland, 1991), R. Vichnevetsky and J. J. H. Miller, eds., Criterion Press, Dublin, Ireland, 1991, pp. 349-350.

32. P. J. Olver, *Applications of Lie Groups to Differential Equations*, Graduate Texts in Math. **107**, Springer Verlag, New York, 1986, 1993 - 2nd edition.

33. L. V. Ovsiannikov, *Group Analysis of Differential Equations*, Academic Press, New York, 1982.

34. G. J. Reid, *A triangularization algorithm which determines the Lie symmetry algebra of any system of PDEs*, J. Phys. A: Math. Gen. **23** (1990), L853–L859.

35. _____, *Algorithmic determination of Lie symmetry algebras of differential equations*, Lie Theory, Differential Equations and Representation Theory, Proc. Ann. Sem. Can. Math. Soc. (Montréal, 1989), V. Hussin, ed., Les Publications de Centre de Recherches Mathématiques, Montréal, Québec, Canada, 1990, pp. 363–372.

36. _____, *Algorithms for reducing a system of PDEs to standard form, determining the dimension of its solution space and calculating its Taylor series solution*, Europ. J. Appl. Math. **2** (1991), 293–318.

37. _____, *Finding abstract Lie symmetry algebras of differential equations without integrating determining equations*, Europ. J. Appl. Math. **2** (1991), 319–340.

38. G. J. Reid, I. G. Lisle, A. Boulton, and A. D. Wittkopf, *Algorithmic determination of commutation relations for Lie symmetry algebras of PDEs*, Proc. ISSAC '92 (Berkeley, CA, 1992), P. S. Wang, ed., ACM Press, New York, 1992, pp. 63–68.

39. G. J. Reid and D. K. McKinnon, *Solving systems of linear PDEs in their coefficient field by recursively decoupling and solving ODEs*, Preprint, Dept. of Math., Univ. British Columbia, Vancouver, BC, Canada, 1993, J. Symb. Comp. (submitted).

40. G. J. Reid and A. Wittkopf, *Long guide to the differential algebra package*, Report, Dept. of Math., Univ. British Columbia, Vancouver, BC, Canada, 1993.

41. C. Riquier, *Les Systèmes d'Equations aux Dérivées Partielles*, Gauthier-Villars, Paris, 1910.

42. W. Sarlet and J. Vanden Bonne, *REDUCE-procedures for the study of adjoint symmetries of second-order differential equations*, J. Sym. Comp. **13** (1992), 683–693.

43. F. Schwarz, *A REDUCE package for determining Lie symmetries of ordinary and partial differential equations*, Comp. Phys. Comm. **27** (1982), 179–186.

44. _____, *Automatically determining symmetries of partial differential equations*, Computing **34** (1985), 91–106; Addendum: Computing **36** (1986), 279–280.

45. _____, *Symmetries of differential equations from Sophus Lie to computer algebra*, SIAM Review **30** (1988), 450–481.

46. _____, *An algorithm for determining the size of symmetry groups*, Computing **49** (1992), 95–115.

47. J. Sherring and G. Prince, *DIMSYM - symmetry determination and linear differential equations package*, Preprint, Dept. of Math., LaTrobe Univ., Bundoora, Australia, 1992.

48. M. F. Singer, *Formal solutions of differential equations*, J. Symb. Comp. **10** (1991), 59–94.
49. S. Steinberg, *Symmetries of differential equations*, MACSYMA Newsletter **7** (1990), 3–7.
50. H. Stephani, *Differential Equations: Their Solution using Symmetries.* Cambridge University Press, Cambridge, UK, 1989.
51. J. Thomas, *Differential Systems*, Coll. Publ. **21**, AMS, New York, 1937.
52. V. L. Topunov, *Reducing systems of linear differential equations to a passive form*, Acta Appl. Math. **16** (1989), 191–206.
53. A. M. Tresse, *Sur les invariants différentiels des groupes de transformations*, Acta Math. **18** (1894), 1–88.
54. _____, *Détermination des Invariants Ponctuel de l'Equation Différentielle Ordinaire du Second Ordre $y'' = \omega(x, y, y')$*, S. Hirzel, Leipzig, 1896.
55. P. Vafeades, *PDELIE: A partial differential equation solver*, MACSYMA Newsletter **9**, no. 1 (1992), 1–13.
56. _____, *PDELIE: A partial differential equation solver II*, MACSYMA Newsletter **9**, no. 2-4 (1992), 5–20.
57. _____, MACSYMA Newsletter **10**, no. 1 (to appear).
58. A. M. Vinogradov, Ed., *Symmetries of Partial Differential Equations*, Part I, Acta Appl. Math. **15**, nos. 1 & 2 (1989); Part II, *ibid.* **16**, no. 1 (1989); Part III, *ibid.* **16**, no. 2 (1989).
59. _____, *Symmetries of Partial Differential Equations. Conservation laws - Applications - Algorithms*, Kluwer Academic Publications, Dordrecht, The Netherlands, 1989.
60. T. Wolf, *A package for the analytic investigation and exact solutions of differential equations*, Proc. EUROCAL '87 (Leipzig, GDR, 1987), J. H. Davenport, ed., Lect. Notes in Comp. Sci. **378**, Springer Verlag, Berlin, 1989, pp. 479-491.
61. _____, *An efficiency improved program LIEPDE for determining Lie-symmetries of PDEs*, Proc. of Modern Group Analysis: Advanced Analytical and Computational Methods in Mathematical Physics (Acireale, Catania, Italy, 1992) N.H. Ibragimov, M. Torrisi, and A. Valenti, eds., Kluwer Academic Publishers, Dordrecht, The Netherlands, 1993 (to appear).
62. T. Wolf and A. Brand, *Heuristics for overdetermined systems of PDEs*, Proc. Meeting on Artificial Intelligence in Mathematics (Glasgow, UK, 1991), J. H. Johnson, S. McKee, and A. Vella, eds., Clarendon Press, Oxford, UK, 1993.
63. _____, *The computer algebra package CRACK for investigating PDEs*, Proc. ERCIM School on Partial Differential Equations and Group Theory (Bonn, 1992), J. F. Pommaret, ed., Gesellschaft für Mathematik und Datenverarbeitung, Sankt Augustin, Germany, 1992; Also: Manual for CRACK added to the REDUCE Network Library.
64. _____, *Investigating symmetries and other analytical properties of ODEs with the computer algebra package CRACK*, Exploiting Symmetry in Applied and Numerical Analysis, Proc. AMS-SIAM Summer Seminar (Fort Collins, CO, 1992), E. Allgower, K. Georg, and R. Miranda, eds., Lect. in Appl. Math., AMS, Providence, RI, 1993.

DEPARTMENT OF MATHEMATICAL AND COMPUTER SCIENCES, COLORADO SCHOOL OF MINES, GOLDEN, CO 80401-1887
*E-mail address*: whereman@flint.mines.colorado.edu

Lectures in Applied Mathematics
Volume **29**, 1993

# A Manifold Solver with Bifurcation and Symmetry

BIN HONG

Abstract: This paper presents an introduction to a manifold solver which can trace not only non–singular points but also singular points, with or without some subsymmetry of the system.

## 1. Introduction

Considerable attention has been directed toward multi–parameter dependent nonlinear equations

$$(1.1) \qquad\qquad F(z, \ \lambda) = 0,$$

where $z \in Z$ is a state variable and $\lambda \in \Lambda$ a parameter variable. In general, the set of solutions $(z, \ \lambda)$ of (1.1) represents a differentiable manifold $M$ in the product $Z \times \Lambda$. In most applications interest centers not so much on computing a few solutions $(z, \ \lambda)$ for specific values of the parameters, but rather on analysing the form and special features of the manifold.

The basic procedure for the computational analysis of the solution manifold $M$ are the continuation methods. When $M$ has dimension $p = 1$, we can obtain a trace of some segment of $M$ by a direct application of a continuation method. But when $M$ has dimension $p > 1$, these methods require an a priori restriction to some path on $M$ and then produce a sequence of points along that path (see [12]). Obviously, it is not easy to develop a good picture of a multi–dimensional manifold solely from information along the paths used by the one–dimensional continuation processes. This led to the development of methods for the computation of simplicial approximations (triangulations) of $p$–dimensional subsets of $M$ (see [1], [2], [13]).

Besides these methods for computing certain sets of points on $M$, another important class of numerical procedures concerns the determination of foldpoints of the manifold $M$ with respect to the parameter space $\Lambda$. Techniques for the computation of specific foldpoints have been proposed by several authors (see, e.g., the proceedings [10], [11], and the references found there). There is also growing interest in methods for following paths consisting of foldpoints, or more generally for computing simplicial approximations of submanifolds of foldpoints (see [5]).

Often in physical applications the system (1.1) is covariant with respect to a transformation group $\Gamma$; that is,

$$(1.2) \qquad F(T_Z(\gamma)z, \ T_\Lambda(\gamma)\lambda)) = S(\gamma)F(z, \ \lambda), \text{ for all } \gamma \in \Gamma,$$

1991 Mathematics Subject Classification, Primary 65J15, 58F14.

This paper is in final form and no version of it will be submitted for publication elsewhere.

where $T_Z$, $T_\Lambda$ and $S$ are group representations of $\Gamma$. When such a symmetry is present, it can aid considerably in the computation of the bifurcation structure, especially at multiple bifurcation points (see e.g. [6], [8]). A major advantage of applying group theoretic methods is the possibility of considering the problem in a reduced form reflecting some subsymmetry under which certain multiple bifurcation points reduce to the simple case and hence can be computed by algorithms available for such simple bifurcation points. A method is developed for the local trace of the submanifold of foldpoints of the same symmetry (see [9]).

A manifold solver called *BISYM* has been written which implements the above related processes on the solution manifold $M$ with or without symmetry. The outline of this paper is as follows. Section 2 summarizes the needed background materials. Then Section 3, details about the data structures, modules, and operating commands of *BISYM* are presented. Finally Section 4, we illustrate the behavior of this solver with examples.

## 2. Background

Throughout this paper, let $F : X \mapsto Y, X = R^n, Y = R^m, p = n - m \geq 1$, be a mapping of class $C^\infty$ on some open, connected subset $E$ of $X$. We consider the equations

$$(2.1) \qquad\qquad F(x) = 0.$$

Then it is well known that the regular solution set

$$(2.2) \qquad M = \{x \in E, F(x) = 0, \text{rank } DF(x) = m\},$$

is a $p$–dimensional $C^\infty$–manifold in $X$ without boundary. In order to avoid trivialities, $M$ is assumed to be non–empty. For simplicity, the tangent space $T_{x_0} M$ at the point $x_0 \in M$ of this manifold will be identified with the kernel of the derivative $DF(x_0)$ of $F$; that is, we set

$$(2.3) \qquad T_{x_0} M = \ker DF(x_0) = \{u : u \in X, DF(x_0)u = 0\},$$

whence

$$(2.4) \qquad N_{x_0} M = (T_{x_0} M)^\perp = (\ker DF(x_0))^\perp = \text{rge } DF(x_0)^\top$$

is the normal space at the point $x_0$. A given $p$–dimensional subspace $T$ of $X$ induces a local coordinate system of $M$ at $x_0 \in M$ if

$$(2.5) \qquad\qquad T \cap N_{x_0} M = \{0\}.$$

More specifically, if (2.5) holds then there exist open neighborhoods $V_1$ of $0 \in T$ and $V_2$ of $x_0 \in X$, respectively, as well as a unique $C^\infty$–function $w : V_1 \mapsto T^\perp, w(0) = 0$, such that

$$(2.6) \qquad M \cap V_2 = \{x \in X; x = x_0 + t + w(t), t \in V_1\},$$

(see e.g. [12]). $x_0$ is a non–singular point if (2.5) holds for $T = \Lambda$, else it is a singular point or foldpoint.

Continuation methods are always applied to an equation of the form

$$(2.7) \qquad\qquad F_0(x) = 0, \ F_0 : R^n \mapsto R^{n-1}, n \geq 2$$

involving an operator $F_0$ which is of class $C^\infty$, on some open set in $X$. As we observed in Section 1, when the solution manifold $M$ of (2.1) has dimension

$p > 1$, a priori restriction to some path on $M$ is needed in order to apply continuation methods. In other words, the equations (2.1) must be augmented by $p - 1$ suitable equations which specify the desired path on $M$. For example, this augmentation may restrict the problem to a variation of one component of the $p$–dimensional parameter vector $\lambda$ or, more generally, to a variation of all parameters as prescribed functions of a single variable.

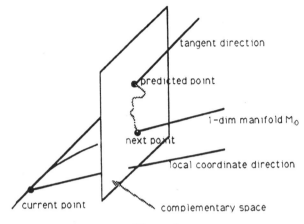

Figure 1

We denote the regular solution manifold of (2.7) as $M_0 = \{x \in X; F_0(x) = 0;\ \text{rank } DF_0(x) = n - 1\}$. All continuation processes begin from a given point $x^0 \in M_0$ and produce a sequence of points $x^k, k = 0, 1, 2, \ldots$, which approximate points of $M_0$. For any $k \geq 0$, the step from the current point $x^k$ to the next point $x^{k+1}$ corresponds to an implementation of the local coordinate representation (2.6). More specifically, if $T = \text{span} \{a\}, a \in X, a \neq 0$, is a local coordinate space of $M_0$ at $x^k$, then, for any fixed $\bar{x}^{k+1} \in X$, the Jacobian of the augmented equations

$$(2.8) \qquad \begin{pmatrix} F_0(x) \\ a^\top(\bar{x}^{k+1} - x) \end{pmatrix} = 0$$

is non-singular for all $x$ in some neighborhood of $x^k$ where $a^\top(\bar{x}^{k+1} - x) = 0$ forms a complementary space of the local coordinate space. Thus, if $y$ approximates a point of $M_0$ in that neighborhood, then it may be expected that (2.8) has a unique solution $x^{k+1} \in M_0$ which can be computed by means of a locally convergent iterative process applied to (2.8) and started, say, at $\bar{x}^{k+1}$ (see Figure 1). We shall use here the continuation package $PITCON$ (see [14]). In $PITCON$, a predicted point $\bar{x}^{k+1} = x^k + hu(x^k)$ is chosen in the direction of the tangent vector $u(x^k)$ of $M_0$ at $x^k$ and the local coordinate space $T$ is specified by a suitable natural basis vector $a = e^i$ of $X$.

The methods for obtaining simplicial approximations (triangulations) of open subsets of the solution manifold $M$ uses almost the same tools as these continuation methods. The triangulation process given in [13] begins with the choice of a simplicial decomposition $\Sigma$ of $R^p$, such as, for instance, the well–known Kuhn–triangulation, or, for $p = 2$, a triangulation of $R^2$ by means of equilateral triangles. The aim is to transfer the knots of some part of $\Sigma$, together with their connectivity information, from $R^p$ onto $M$. As in any continuation methods, a starting point $x$ on the manifold is assumed to be known. Then the triangulation process consists of two steps: First a suitable "patch" of the reference triangulation $\Sigma$ is mapped onto the affine tangent space $x + T_x M$ using an appropriate basis of $T_x M$. Thereafter, a locally convergent iterative process is applied to "project" the resulting knots of the mapped simplices from $x + T_x M$ onto the manifold $M$ (see Figure 2). These two steps are then repeated with one of the computed points on $M$ in place of the original point $x$. In order to ensure that the images of the simplices of $\Sigma$ on $M$ form a simplicial approximation of a portion of the manifold, bases on the tangent spaces $T_x M$ have to be constructed which change smoothly from point to point. In other words, we require a moving frame on some open subset $M_0$ of the manifold $M$. An algorithm which does produce moving frames of class $C^\infty$ has been presented in [13]. A package, *MATRIG*, has been written which has shown itself to be very efficient in computing triangulations even around bifurcation points. This has opened up new possibilities for determining the form and special features of the solution manifold.

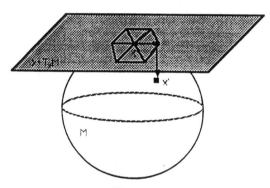

Figure 2

The computation of foldpoints generally involves some augmented equation which is then solved by a Newton–like method. For example, a class of augmented equations for computing certain foldpoints has the form (see [7])

$$(2.9) \qquad G(x, U) = \begin{pmatrix} F(x) \\ DF(x)U \\ U^\top U - I \\ c^\top U \end{pmatrix} = 0,$$

where, $x \in X$ and the $n \times p$ matrix $U$ together form the unknown, $I$ is the identity on $R^p$, and $c \in X$ is a given vector. If $G(x_0, U_0) = 0$ and the Jacobian $DG(x_0, U_0)$ is nonsingular; that is, if the solution $(x_0, U_0)$ of (2.9) is isolated, then we evidently have $x_0 \in M$ and the column vectors of $U_0$ form an orthonormal basis of $T_{x_0} M$, and therefore $c$ must be a vector in $N_{x_0} M$. In practice, it is convenient to uncouple the augmented system $G(x, U) = 0$ so that the unknowns $x$ and $U$ are obtained separately. For each $x \in M$, let $\{u_1(x), \ldots, u_p(x)\}$ be an orthonormal basis for the tangent space $T_x M$. Then with a suitable vector $c \in \Lambda$, we have

$$(2.10) \qquad G(x) = \begin{pmatrix} F(x) \\ c^\top u_1(x) \\ \vdots \\ c^\top u_p(x) \end{pmatrix} = 0.$$

Equations (2.10) represent a generalized version of an augmentation of Griewank and Reddien [7]. It is a so–called "minimal augmentation" since it does not involve any additional variables such as $U$ in (2.9). The original version of Griewank and Reddien method requires the "unfolding" vector $c$ to be a natural basis vector. A new type of minimal augmentation which does not involve an explicit a priori selection of the vector $c$ is given in [5].

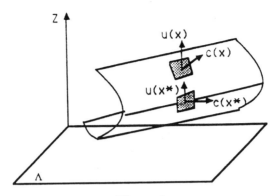

Figure 3

In [5] a method is presented for the computation of simple foldpoints of $M$ (i.e., $\dim \ker D_z F(x) = 1$). Let $x^* \in M$ be a simple non–degenerate foldpoint of $M$, $L$ orthonormal matrix whose columns span $Z$, $c^* \notin \mathrm{rge} DF(x^*)L$, then there exists an open neighborhood $U$ of $x^*$ such that: (1) for all $x \in U$, we can find two vectors $u(x), c(x)$

$$u(x) \in Z \cap \{T_x M + \mathrm{span}(DF(x)^+ c^*)\}$$

(2.11)
$$c(x) \in N_x M \cap \{\Lambda + \mathrm{span}(\gamma(x)c^*)\}$$

where $DF(x)^+$ is the pseudoinverse of $DF(x)$ and $\gamma(x) = u(x)^\top c(x)$ (see Figure 3); (2) the solutions of the augmented function

(2.12)
$$G(x) = \begin{pmatrix} F(x) \\ \gamma(x) \end{pmatrix} = 0$$

are exactly the simple foldpoints of $M$, and these solutions form a $(p-1)$–dimensional submanifold of $M$.

Let $GL(X)$ be the group of all nonsingular linear transformations of $X$ onto itself. A representation of a group $\Gamma$ on $X$ is a homomorphism $T : \gamma \mapsto T(\gamma)$ of $\Gamma$ into $GL(X)$, such that

(2.13)
$$T(\gamma_1 \gamma_2) = T(\gamma_1) T(\gamma_2), \forall \gamma_1, \gamma_2 \in \Gamma.$$

In this paper, we consider only finite groups $\Gamma$. The mapping $F$ of (2.1) is said to be equivariant under the group $\Gamma$ if the following relation holds

(2.14)
$$F(T(\gamma)x) = S(\gamma)F(x), \quad \text{for all } \gamma \in \Gamma$$

where $T$ is the group representation of $\Gamma$ on $X$, defined as the direct product of the group representation $T_Z$ of $\Gamma$ on $Z$ and the group representation $T_\Lambda$ of $\Gamma$ on $\Lambda$ such that

(2.15)
$$T(\gamma)x = (T_Z(\gamma)z, T_\Lambda(\gamma)\lambda), \forall x \in X, \forall \gamma \in \Gamma,$$

and $S$ is the group representation of $\Gamma$ on $Y$. The group representations $T_Z$ and $S$ are equivalent with an isomorphism $A : Y \mapsto Z$ of $Y$ onto $Z$ such that

(2.16)
$$S(\sigma) = A^{-1} T_Z(\sigma) A, \forall \sigma \in \Sigma.$$

For any point $x \in X$ it is well known that the set

(2.17) $$\Gamma_x = \{\gamma \in \Gamma : T(\gamma)x = x\}$$

is a subgroup of $\Gamma$. Any subgroup $\Sigma$ of $\Gamma$ such that $\Sigma = \Gamma_x$ for some $x \in X$ is called an isotropy subgroup of $\Gamma$. For any isotropy subgroup $\Sigma$ of $\Gamma$, the set

(2.18) $$X_\Sigma = \{x \in X : T(\sigma)x = x, \forall \sigma \in \Sigma\}.$$

is a subspace of $X$ called the fixed point subspace of $\Sigma$ in $X$. Analogously the fixed-point subspace of $\Sigma$ in $Y$ is defined. Let $F_\Sigma$ denote the restriction of $F$ to $X_\Sigma$, then it is computationally advantageous to consider the $\Sigma$–reduced problem

(2.19) $$F_\Sigma(x) = 0,$$

since the dimension of $X_\Sigma$ is less than the dimension of $X$. The dimension of $X_\Sigma$ and the orthogonal projection $P_\Sigma$ from $X$ onto $X_\Sigma$ are given by the following formulas:

(2.20)
$$\dim(X_\Sigma) = \frac{1}{|\Sigma|} \sum_{\sigma \in \Sigma} \mathrm{tr}\,(T(\sigma)),$$
$$P_\Sigma = \frac{1}{|\Sigma|} \sum_{\sigma \in \Sigma} T(\sigma).$$

A basis of $X_\Sigma$ can be obtained by finding an orthogonal basis $\{u_i\}$ of the nullspace of the complementary operator $I - P_\Sigma$. Such vectors $u_i, i = 1, 2, ..., n_H$, are sometimes called symmetry–coordinates or symmetry modes. There exist some efficient numerical techniques for obtaining orthonormal basis $\{u_i\}$ of $X_\Sigma$. In [3] it was shown that this can be done numerically by inverse iteration combined with matrix deflation since the basis vectors are the eigenvectors of the projection matrix $P_\Sigma$ corresponding to a repeated unity eigenvalue. Let $n_\Sigma = \dim(X_\Sigma)$, and $U_\Sigma$ be the matrix with the basis vectors $u_i$ as columns; that is,

(2.21) $$U_\Sigma = (\, u_1, u_2, ..., u_{n_H}\,),$$

and for any $a \in X_\Sigma$, set

(2.22) $$\underline{a} = U_\Sigma^\top a \in R^{n_\Sigma}.$$

Then, in the orthonormal basis $\{u_i\}$, the reduced problem has the form

(2.23) $$\underline{F}_\Sigma(\underline{x}) = (A^{-1}U_\Sigma)^\top F(U_\Sigma\, \underline{x}) = 0.$$

In recent years, a number of papers considered symmetry in connection with the computation of bifurcations through the reduced problem (see e.g. [6], [8]). But most of them are restricted to one–parameter problems, and even those concerned with multi-parameter problems single out a specific parameter for the computation and hold all other parameters at a fixed value. On the other hand, [5] only addressed the computation of the simple foldpoints on the solution manifold $M$. The computation of the multiple foldpoints (i.e., $\dim \ker D_z F(x) > 1$), in general, is difficult and costly. In [9], the reduced manifold with respect to some subsymmetry is introduced and under the reduced manifold certain multiple bifurcation points is reduced to the simple case and hence can be computed by algorithms in [5]. [9] not only extend the simple foldpoints computation on the solution manifold to the multiple foldpoints computation, but also extend the multiple bifurcation computation of one parameter to multiple parameters.

Let $\dim(X_\Sigma) = n_\Sigma \leq n$, $\dim(Y_\Sigma) = m_\Sigma \leq m$, the reduced manifold in [9] is defined as the regular solution set of $F_\Sigma$; that is,

$$(2.24) \qquad M_\Sigma = \{x : x \in X_\Sigma, F_\Sigma(x) = 0, \text{rge } DF_\Sigma(x) = Y_\Sigma\}.$$

Clearly its dimension is $p_\Sigma = n_\Sigma - m_\Sigma \leq p$. It was shown that if $x \in M \cap X_\Sigma$ then $x \in M_\Sigma$. Similar as the reduced function $F_\Sigma$ of $F$ in (2.1), we can introduce the reduced augmented function $G_\Sigma$ of $G$ in (2.12). Let $x_0 \in X_\Gamma \cap M$ be a multiple non-degenerate foldpoint of $M$, if there exists an isotropy subgroup $\Sigma$ of $\Gamma$ (including $\Sigma = \Gamma$) such that

$$(2.25) \qquad q_\Sigma = \dim(\ker D_z F(x_0) \cap X_\Sigma) = 1,$$

then locally near $x_0$, there exists a $(p_\Sigma - 1)$–dimensional submanifold of $M_\Sigma$ consisting of simple non-degenerate foldpoints of $M_\Sigma$ which are also multiple non-degenerate foldpoints of $M$. The only difference between the two algorithms in [5] and [9] is that we need to determine which isotropy subgroup $\Sigma$ of $\Gamma$ satisfies the condition (2.25), and the numerical implementation of (2.25) is discussed in [9].

## 3. Structure of the solver

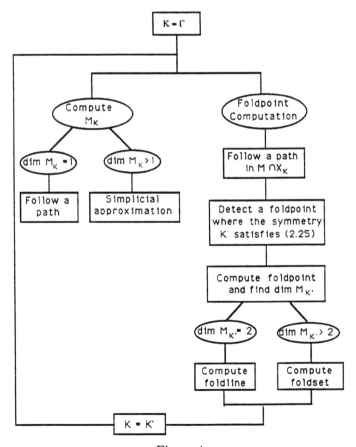

Figure 4

The results of Section 2 suggest the outline of a solver of numerical procedures shown in Figure 4 for exploiting special features of the manifold $M$ with the symmetry (including no symmetry ,i.e., $K = I$). As usual with processes of this type, we start at a point $x$ on the manifold. Without losing generality, suppose that $x \in M \cap X_\Gamma$. After setting the current group $K$ to $\Gamma$, we have two options, namely, (1) computation of an approximation of the reduced manifold $M_\Gamma$, (2) computation of foldpoints. Under the first option if $\dim M_K = 1$ then we follow a path on $M_K$ by means of the continuation method while for $\dim M_K > 1$ we compute a simplicial approximation of $M_K$ by the moving frame algorithm. Under the second option we follow a path in $M \cap X_\Sigma$ and then use the computed points on this path to detect a foldpoint by some suitable technique, for instance, by monitoring the sign or the singular values of $\det D_Z F(x)$). Then it has to be determined a symmetry $K'$ of $K$ in line with the condition (2.25) which then allows the use of the algorithm in [9] for computing this foldpoint accurately and for determining $\dim M_{K'}$. If $\dim M_{K'} = 2$ we follow a foldline on $M_{K'}$; if $\dim M_{K'} > 2$ we compute a simplicial approximation of a foldset in $M_{K'}$. Finally we set the current group $K$ to $K'$ and repeat the procedure above.

A manifold solver *BISYM* has been developed which implements the above procedures. The acronym stands for "*BI*furcation with *SYM*metry". The data structures used in *BISYM* are as follows :

(1) A tree structure of all isotropy subgroups $\Sigma$ of $\Gamma$ based on the lattice of these subgroups. Each node of this tree is associated with a record consisting of three fields : the orthogonal projection $P_\Sigma|_Z$ of $Z$ onto $Z_\Sigma$, the orthogonal projection $P_\Sigma|_\Lambda$ of $\Lambda$ onto $\Lambda_\Sigma$, and a basis of $X_\Sigma$.

(2) A code for evaluating the original function $F$ of (2.1) for any given point $x \in M$.

(3) A code for evaluating the Jacobian $DF$ of the original function $F$ of (2.1) for any given point $x \in M$.

The three records associated with the tree structure of the group are implemented as the following three user supplied subroutines:

(i) All isotropy subgroups are stored as a table of characters in the *SUBROUTINE CHOICE* which then assigns any one of these character values to the current group $K$.

(ii) The orthogonal projections $P_\Sigma$ for all isotropy subgroups $\Sigma$ of $\Gamma$ are stored in the *SUBROUTINE TDATA* which then produces for any current group $K$ the corresponding orthogonal projections $P_K|_Z$, $P_K|_\Lambda$.

(iii) The bases of $X_\Sigma$ for all the isotropy subgroups $\Sigma$ of $\Gamma$ are stored in *SUBROUTINE SDATA* which then produces for any current group $K = \Sigma$ the corresponding basis of $X_K$.

As noted earlier the evaluation of the original function $F$ of (2.1) and of the Jacobian $DF$ of $F$ at any point $x$ is performed by two further user supplied subroutines here named *SUBROUTINE FX0* and *SUBROUTINE DF0*.

There are three computational modules in *BISYM* :

(1) Bifurcation computation. This is the algorithm in [9] and is implemented in the form of the *SUBROUTINE BIFUR* which has three stages:

(i) Call *SUBROUTINE DF0* and then perform the singular value decomposition of $D_Z F(x)$. From the singular values determine the first singularity index $q$.

(ii) Select an isotropy subgroup $\Sigma$ of $\Gamma$ by calling the *SUBROUTINE CHOICE*, then call the *SUBROUTINE TDATA* to provide the corresponding orthogonal projection $P_\Sigma$. Now check to see if $\Sigma$ satisfies the condition (2.25); if so, call the *SUBROUTINE DIMCT* to determine the dimensions of $X_\Sigma$ and $\Lambda_\Sigma$, call the *SUBROUTINE SDATA* to provide the corresponding basis of $X_\Sigma$ and according to the basis of $X_\Sigma$, transform the natural coordinates of the current point $x$ to those in the basis of $X_\Sigma$.

(iii) Call the algorithm of [5] for the computation of the reduced simple foldpoint.

(2) Continuation. This is the package *PITCON* of [12]. In our implementation *PITCON* calls the *SUBROUTINE FX* and *SUBROUTINE DF* to trace the following paths :

$JOB = 0$, trace a path on the regular solution manifold $M$;

$JOB = 1$, trace a foldline on $M$;

$JOB = 2$, trace a path on the reduced manifold $M_K$ with respect to the current group $K$;

$JOB = 3$, trace a foldline on $M_K$.

(3) Simplicial approximation. This is the package $MATRIG$ of [13]. In our implementation $MATRIG$ calls the $SUBROUTINE\ FX$ and $SUBROUTINE\ DF$ to compute the following simplicial approximations :

$JOB = 0$, compute a simplicial approximation of the regular solution manifold $M$;

$JOB = 1$, compute a simplicial approximation of a foldset on $M$;

$JOB = 2$, compute a simplicial approximation of the reduced manifold $M_K$ with respect to the current group $K$;

$JOB = 3$, compute a simplicial approximation of a foldset on $M_K$.

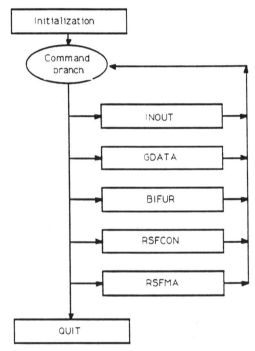

Figure 5

The entire solver $BISYM$ works interactively. A flow–chart of the permissible commands is given in Figure 5. More specifically the individual commands in Figure 5 accomplish the following tasks.

(1) *INOUT*. This command is used for input or output of the coordinates of the current point $x$.

(2) *GDATA*. This command permits a change of the current group. When it is invoked, it select an isotropy subgroup $\Sigma$ of $\Gamma$ by calling the $SUBROUTINE$ $CHOICE$. Then in turn the $SUBROUTINE\ TDATA$, $SUBROUTINE\ SDATA$, and $SUBROUTINE\ DIMCT$ are called, to determine the corresponding orthogonal projection $P_\Sigma$, the basis of $X_\Sigma$, and the dimensions of $X_\Sigma$ and $\Lambda_\Sigma$. Finally, the $SUBROUTINE\ XTOXK$ is called to transform the natural coordinates of the current point $x$ to those in the basis of $X_\Sigma$. With this we are in the setting of the reduced problem.

(3) *BIFUR*. This command calls the $SUBROUTINE\ BIFUR$.

(4) *RSFCON*. This command invokes the package $PITCON$ for various path tracing tasks. If the current group $\Sigma$ equals the identity, then this command

prompts for two options : (i) select $JOB = 0$ to trace a path on the regular solution manifold $M$; (ii) select $JOB = 2$ to trace a foldline on $M$. If the current group $\Sigma$ does not equal the identity, then this command prompts for the other two options : (i) select $JOB = 1$ to trace a path on the reduced manifold $M_\Sigma$; (ii) select $JOB = 3$ to trace a foldline on $M_\Sigma$. After the index $JOB$ is available, the *SUBROUTINE PITCON* is called.

(5) *RSFMA*. This command invokes the package *MATRIG* for various simplicial approximation tasks. If the current group $\Sigma$ equals the identity, then this command prompts for two options : (i) select $JOB = 0$ to compute a simplicial approximation of the regular solution manifold $M$; (ii) select $JOB = 2$ to compute a simplicial approximation of the foldset on $M$. If the current group $\Sigma$ does not equal the identity, then this command prompts for the other two options : (i) select $JOB = 1$ to compute a simplicial approximation of the reduced manifold $M_\Sigma$; (ii) select $JOB = 3$ to compute a simplicial approximation of the foldset on $M_\Sigma$. After the index $JOB$ has been provided, the *SUBROUTINE MATRIG* is called.

## 4. Example

In order to indicate the performance of this solver, we consider the equilibrium problem of the buckling of a simply supported flat rectangular plate [4]

$$(4.1) \qquad F(u, \lambda) = \begin{pmatrix} f_1(u, \lambda) \\ f_2(u, \lambda) \end{pmatrix} = \begin{pmatrix} -au_1^3 - bu_1u_2^2 + \lambda_1 u_1 + \lambda_2 \\ -bu_1^2 u_2 - cu_2^3 + \lambda_1 u_2 + \lambda_3 \end{pmatrix} = 0$$

where $(z, \lambda) = (u_1, u_2, \lambda_1, \lambda_2, \lambda_3)$ and $a = 3.945001 \times 10^{-4}, b = 5.007428 \times 10^{-4}, c = 1.623543 \times 10^{-4}$. The physical meanings of all parameters in this example are: $\lambda_1$ corresponds to the normal loading while $\lambda_2$ and $\lambda_3$ correspond to the transversal loads. It is evident that $F$ is equivariant under the group $\Gamma = Z_2 \times Z_2$ such that

$$(4.2) \qquad \begin{aligned} Z_2^1 : &\begin{cases} f_1(-u_1, u_2, \lambda_1, -\lambda_2, \lambda_3) = -f_1(u_1, u_2, \lambda_1, \lambda_2, \lambda_3) \\ f_2(-u_1, u_2, \lambda_1, -\lambda_2, \lambda_3) = f_2(u_1, u_2, \lambda_1, \lambda_2, \lambda_3) \end{cases} \\ Z_2^2 : &\begin{cases} f_1(u_1, -u_2, \lambda_1, \lambda_2, -\lambda_3) = f_1(u_1, u_2, \lambda_1, \lambda_2, \lambda_3) \\ f_2(u_1, -u_2, \lambda_1, \lambda_2, -\lambda_3) = -f_2(u_1, u_2, \lambda_1, \lambda_2, \lambda_3). \end{cases} \end{aligned}$$

The subgroups of $\Gamma$ are:

$$(4.3) \qquad \begin{aligned} &\Gamma = \{I, Z_2^1, Z_2^2, Z_2^1 \times Z_2^2\}, \Sigma_1 = \{I, Z_2^1\}, \\ &\Sigma_2 = \{I, Z_2^2\}, \Sigma_0 = \{I\} \end{aligned}$$

and the corresponding fixed point spaces are:

$$(4.4) \qquad \begin{aligned} &X_\Gamma = \{\vec{x} : u_1 = u_2 = \lambda_2 = \lambda_3 = 0\}, X_{\Sigma_1} = \{\vec{x} : u_1 = \lambda_2 = 0\}, \\ &X_{\Sigma_2} = \{\vec{x} : u_2 = \lambda_3 = 0\}, X_{\Sigma_0} = X. \end{aligned}$$

Using *BIFUR* in the solver *BISYM*, we found a simple foldpoint $\vec{x}_1 = \{0, -2.49354, 0.302841 \times 10^{-2}, 0, 0.503267 \times 10^{-2}\}$ and $\vec{x}_1 \in X_{\Sigma_1}$. Applying *RSFMA* with $JOB = 2$, we obtained a foldset at $\vec{x}_1$ (see Figure 6). We could also obtain the reduced manifold $M_{\Sigma_1}$ at $\vec{x}_1$ (see Figure 8) by *RSFMA* with $JOB = 1$; or the foldline on $M_{\Sigma_1}$ at $\vec{x}_1$ (see Figure 9) by *RSFCON* with $JOB = 3$. Notice that the dimension of the original solution manifold is three which makes us very hard to visualize the manifold $M$ while with the two-dimensional reduced manifold $M_{\Sigma_1}$ we are able to see the structure of solutions near $\vec{x}_1$. Comparing Figure 6 and Figure 10, we can see the bifurcation structure under $\Sigma_1$ more clearly than the bifurcation structure in general and it is cheaper to compute the symmetric foldline than the original foldset.

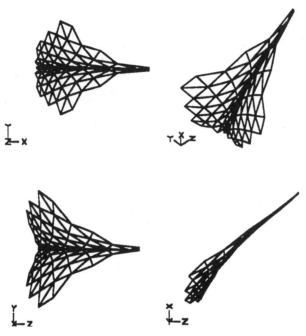

Figure 6: Foldset at $\vec{x}_1$ $(x = \lambda_1, y = \lambda_2, z = \lambda_3)$

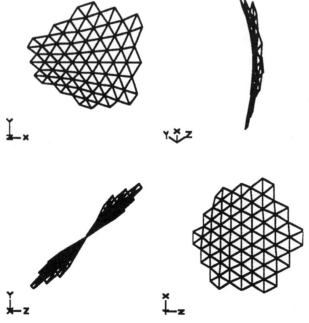

Figure 7: Reduced manifold $M_{\Sigma_1}$ at $\vec{x}_1$ $(x = u_2, y = \lambda_1, z = \lambda_3)$

Similarly, we found another simple foldpoint $\vec{x}_2 = \{-0.212519, 0, 0.534522 \times 10^{-4}, 0.757303 \times 10^{-5}, 0\}$ by $\underline{BIFUR}$ and $\vec{x}_2 \in X_{\Sigma_2}$. We could also obtain a foldset at $\vec{x}_2$ (see Figure 8), the reduced manifold $M_{\Sigma_2}$ at $\vec{x}_2$ (see Figure 9), and the foldline on $M_{\Sigma_2}$ at $\vec{x}_2$ (see Figure 11).

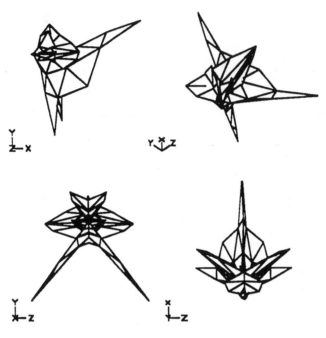

Figure 8: Foldset at $\vec{x}_2$ $(x = \lambda_1, y = \lambda_2, z = \lambda_3)$

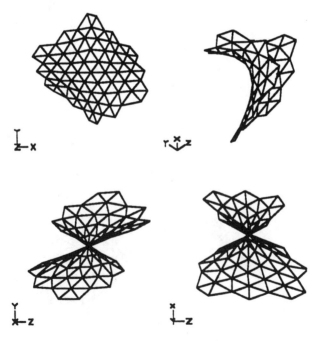

Figure 9: Reduced manifold $M_{\Sigma_2}$ at $\vec{x}_2$ $(x = u_1, y = \lambda_1, z = \lambda_2)$

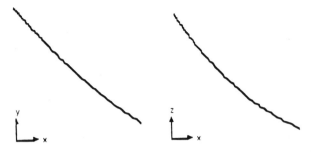

Figure 10: Foldline on $M_{\Sigma_1}$ at $\vec{x}_1$ $(x = u_2, y = \lambda_1, z = \lambda_3)$

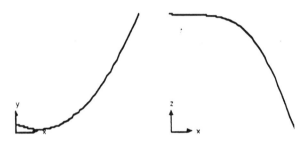

Figure 11: Foldline on $M_{\Sigma_2}$ at $\vec{x}_2$ $(x = u_1, y = \lambda_1, z = \lambda_2)$

## References

[1]. E. L. Allgower and S. Gnutzmann, An algorithm for piecewise linear approximation of implicitly defined two-dimensional surfaces, *SIAM J. Num. Anal.*, 24, 452-469 (1984).

[2]. E. L. Allgower and P. H. Schmidt, An algorithm for piecewise linear approximation of an implicitly defined manifold, *SIAM J. Num. Anal.*, 22, 322-346 (1985).

[3]. P. Chang and T. J. Healey, Computation of symmetry modes and exact reduction in nonlinear structural analysis, *Comput. & Structures*, 28, 135-142 (1988).

[4]. S. N. Chow, J. K. Hale and J. Mallet-Paret, Applications of generic bifurcation II, *Archive Rat. Mech. Anal.*, 62, 209-235 (1976).

[5]. R. X. Dai and W. C. Rheinboldt, On the computation of manifolds of foldpoints for parameter-dependent problems, *SIAM J. Num. Anal.*, 27, 437-441 (1990).

[6]. M. Dellnitz and B. Werner, In *Continuation Techniques and Bifurcation Problems*, Computational methods for bifurcation problems with symmetries–with special attention to steady state and Hopf bifurcation points, H. D. Mittelmann and D. Roose, eds. (Birkhäuser-Verlag, 1990), pp. 97-123.

[7]. A. Griewank and G. W. Reddien, Characterization and computation of generalized turning points., *SIAM J. Num. Anal.*, 21, 176-185 (1984).

[8]. T. J. Healey, A group-theoretic approach to computational bifurcation problems with symmetry, *Computer Methods in Applied Mechanics and Engineering*, 67, 257-295 (1988).

[9]. B. Hong, On the computation of multiple bifurcations with multiple parameters and symmetry, to appear in *SIAM J. Num. Anal.*, (1993).

[10]. T. Kuepper, H. D. Mittelman and H. Weber, *Numerical Methods for Bifurcation Problems* (Birkhäuser-Verlag, 1984).

[11]. T. Kuepper, R. Seydel and H. Troger, *Bifurcations: Analysis, Algorithms, Applications*, (Birkhäuser-Verlag, Basel, 1987).

[12]. W. C. Rheinboldt, *Numerical Analysis of Parametrized Nonlinear Equations* (John Wiley and Sons, Inc., 1986).
[13]. W. C. Rheinboldt, On the computation of multi-dimensional solution manifolds of parametrized equations, *Numer. Math.*, (1989).
[14]. W. C. Rheinboldt and J. V. Burkhardt, Algorithm 596: A program for a locally parametrized continuation process, *ACM Trans. on Math. Software*, 9, 236-246 (1983).

Institute for Computational Mathematics and Applications
University of Pittsburgh, Pittsburgh, Pa 15260
*Current address*: Thomson Consumer Electronics
24200 U.S. Route 23 South, Circleville, Ohio 43113

Lectures in Applied Mathematics
Volume **29**, 1993

# How to Use Symmetry to Find Models for Multidimensional Conservation Laws

BARBARA LEE KEYFITZ AND MILTON LOPES-FILHO

### Abstract

To find model equations for conservations laws in several space variables which might play the same role that Burgers' equation has done in a single space dimension, we assume that the equations respect some spatial symmetry. The theory of prolongations provides an operational calculus with which we generate some classes of models. We give some examples. Among our conclusions is an explanation of why there is no satisfactory definition of genuine nonlinearity for a scalar equation in several space variables. This approach may also be used to analyse the characteristic variety and to identify strictly hyperbolic and genuinely nonlinear systems.

## 1  Introduction

We study the structure of systems of $m$ conservation laws in two space variables:

$$(1) \qquad \frac{\partial r}{\partial t} + \frac{\partial f(w)}{\partial x} + \frac{\partial g(w)}{\partial y} = 0.$$

The notation is $w = (z, r)$; $z = (x, y, t)$ is the *independent variable* and $r(z) = (r_1, r_2, \ldots, r_m)$ is the *state variable*.

A solution $r(z)$ is a mapping from $\mathbf{R}^3$ to $\mathbf{R}^m$, the *state space*. Typically, some components of $r$ represent *fields* (velocity, electric, magnetic) in the $xy$ plane, and others represent *invariant* quantities (densities, energy, potentials). In many standard models for gas dynamics, magnetohydrodynamics (MHD) and multiphase flow, $x$ and $y$ have the interpretation of variables in a physical space

---

AMS Subject Classification Primary 35L65, Secondary 35A30, 58G35.

Research supported by the Texas Advanced Research Program under Grant 00365-2124ARP, Department of Energy grant DE-FG05-91ER25102, and NSF grant DMS-91-03560. This work was performed while the second author was visiting the University of Houston, supported by the Texas Advanced Research Program.

This paper is in final form and no version of it will be submitted for publication elsewhere.

which is considered to be invariant under Euclidean motions – translations, rotations and reflections in the plane. Translation invariance has the consequence that the fluxes $f$ and $g$ are independent of $x$ and $y$.

Rotations and reflections are identified with the orthogonal group, $\mathcal{O}(2)$. The Euler equations for gas dynamics, the shallow water equations, and the equations of MHD have these symmetries built into them. There are several reasons to study the class of equations with these symmetries. In this paper we construct simple models for two-dimensional conservation laws which respect these geometric symmetries. One result we obtain is that genuine nonlinearity in multidimensional problems is associated with the presence of equivariant components (fields) in the state variable; this can be used to understand why there is no satisfactory definition of genuine nonlinearity for a scalar multidimensional equation.

To fix ideas, regard an element $T_2$ of $\mathcal{O}(2)$ as effecting a coordinate change in $x$ and $y$. If the same *physical* point

$$(2) \qquad \vec{p} = x\vec{\imath} + y\vec{\jmath} = x'\vec{\imath}\,' + y'\vec{\jmath}\,'$$

is described by $z = (x, y, t)$ or $z' = (x', y', t)$ in two coordinate systems, then *fields* such as velocity,

$$(3) \qquad \vec{q} = u\vec{\imath} + v\vec{\jmath} = u'\vec{\imath}\,' + v'\vec{\jmath}\,',$$

are similarly described. Coordinate changes, such as (2) or (3), are determined by specifying an invertible mapping in the plane

$$(4) \qquad\qquad T_2: \qquad \vec{\imath}, \vec{\jmath} \mapsto \vec{\imath}\,', \vec{\jmath}\,'.$$

We write $z' = T_1 z$, where $T_1$ is the extension of $T_2$ which acts trivially in $t$.

The group $\mathcal{O}(2)$ is generated by the one-parameter group of rotations, $\{R_\theta\}$, and the flip, $F$. We will identify $R_\theta$ and $F$ with the matrices

$$(5) \qquad R_\theta = \begin{pmatrix} \cos\theta & -\sin\theta \\ \sin\theta & \cos\theta \end{pmatrix}, \qquad F = \begin{pmatrix} -1 & 0 \\ 0 & 1 \end{pmatrix}.$$

The "prime" coordinate vectors are rotated clockwise by $\theta$ from the original coordinates when $R_\theta$ acts by left multiplication on $(x, y)$. The notation for the flip is obvious. Throughout, we regard independent and dependent variables as column vectors. Hence

$$q' \equiv \begin{pmatrix} u' \\ v' \end{pmatrix} = R_\theta q = \begin{pmatrix} \cos\theta & -\sin\theta \\ \sin\theta & \cos\theta \end{pmatrix} \begin{pmatrix} u \\ v \end{pmatrix},$$

as well. Under this change of coordinates, the components of $r$ in (1) which represent the flow field $(u, v)$ are mapped to a new flow field $(u', v')$ while a component representing the density, for example, is mapped to itself. Variables of the first kind come in pairs, which we call *field pairs*; those of the second kind are of *potential* type. The mapping (4) induces mappings

$$q(z) \mapsto T_2 q(T_1^{-1} z), \qquad\qquad \rho(z) \mapsto \rho(T_1^{-1} z)$$

on field pairs and potentials respectively. A flow described by the functions $r = (u(z), v(z), \rho(z))$ in the old coordinate system is described by

$$r' \equiv (u'(T_1^{-1}z), v'(T_1^{-1}z), \rho(T_1^{-1}z))$$

in the new. The original planar map $T_2$ induces a map $T_1$ on the independent variables (by acting trivially in $t$) and a map $T_3$ in state space (by acting as indicated on fields and potentials).

## 2  Induced Action in the Equation

To understand how the differential equation transforms, we use the calculus of prolongations of $T_2$ on a space of 1-jets. Both Hermann's book, [1], and Olver's, [2], provide an introduction to the theory of prolongations and its application to studying symmetries of partial differential equations. The standard use of this theory is to discover the (generally nonobvious) group of symmetries acting on a given equation. This may include actions which mix domain and range, act nonlinearly, and so on. An important purpose of this effort is the construction of solutions, especially similarity solutions, of the equation. By contrast, we are looking at a simple, linear group action, and our interest is in the structure of the equation itself. Only in the last section do we touch on the effect of symmetry in conservation laws on the solutions, through the characteristic variety.

In applying the theory of prolongations, we need only a local notation. The *base space* $\Omega \subset \mathbf{R}^{3+m}$ is defined by $w = (z, r)$; and the solutions $r(z)$ of equation (1) are three-dimensional surfaces (3-surfaces) in $\Omega$. A partial differential equation gives an algebraic relation on a jet space (see [2, page 99]). If we write equation (1) in quasilinear form,

$$(6) \qquad \mathcal{P}(w; \partial)r \equiv \frac{\partial r}{\partial t} + A(w)\frac{\partial r}{\partial x} + B(w)\frac{\partial r}{\partial y} + c(w) = 0,$$

with $A \equiv \partial f/\partial r$, $B \equiv \partial g/\partial r$, and $c = f_x + g_y$, then the equation consists of $m$ linear relations among the $3m$ quantities

$$\partial_x r_1, \partial_y r_1, \partial_t r_1, \ldots, \partial_t r_m$$

and is a constraint on the tangent planes to the solution surfaces. The jet of a 3-surface in $\Omega$ can be represented uniquely (the *standard parameterization*) in the form

$$(7) \qquad r = \sigma z$$

where $r$ and $z$ are column vectors and $\sigma$ is the $m \times 3$ Jacobian matrix $\partial r/\partial z$. The $3m$ coordinates so introduced represent an open set of 3-surfaces. We identify $\sigma$ with a vector, denoted $\underset{\sim}{\sigma}$, in $\mathbf{R}^{3m}$, by

$$(8) \qquad \underset{\sim}{\sigma} = (\sigma_{11}, \sigma_{21}, \ldots, \sigma_{m1}, \sigma_{12}, \sigma_{22}, \ldots, \sigma_{m3})^\top = \begin{pmatrix} \sigma^{(1)} \\ \sigma^{(2)} \\ \sigma^{(3)} \end{pmatrix}$$

where $\sigma^{(i)}$ is the $i$-th column of $\sigma$. Thus $(w, \underset{\sim}{\sigma})$ is a point in $U \subset \mathbf{R}^{3+m+3m}$, the *total jet space*. The solutions of (6) are identified (locally, and where differentiable) as tangent to 3-planes in $\mathbf{R}^{3+m}$ – that is, as equivalent under contact equivalence to 1-jets of 3-surfaces in $\mathbf{R}^{3+m}$. (See [1].) The jet formulation of equation (6) is

$$(9) \qquad F \equiv \sigma^{(3)} + A(w)\sigma^{(1)} + B(w)\sigma^{(2)} + c(w) = 0,$$

and its solutions form a $(3m + 3)$-dimensional surface in $U$.

The action induced on jets by a diffeomorphism in the base is described by the calculus of prolongations; in our application, because the group acts linearly on domain and range separately, it can be found using the chain rule, as follows.

Suppose that $T_2$ is any $2 \times 2$ invertible matrix representing a diffeomorphism in the physical space $(x, y)$. We have seen that $T_2$ induces a diffeomorphism on $\Omega$, which we will also represent by a matrix, denoted $T$;

$$T = \begin{pmatrix} T_1 & 0 \\ 0 & T_3 \end{pmatrix} \equiv Diag(T_1, T_3),$$

where $Diag$ denotes a block diagonal matrix. The induced map $T_3$ in state space is itself block diagonal, of the form (if the $r_i$ are suitably ordered)

$$T_3 = Diag(T_2, T_2, \ldots, T_2, 1, 1, \ldots, 1).$$

We have $Tw = w' = (z', r') = (T_1 z, T_3 r)$. Since $\sigma \in \mathbf{M}_{m3}$ represents the Jacobian matrix

$$(10) \qquad \frac{\partial r}{\partial z} = \left( \frac{\partial r_i}{\partial z_j} \right) = (\sigma_{ij}),$$

we see that $\sigma'$ represents

$$\frac{\partial r'}{\partial z'} = \frac{\partial (T_3 r)}{\partial (T_1 z)} = \frac{\partial r'}{\partial r} \frac{\partial r}{\partial z} \frac{\partial z}{\partial z'} = T_3 \frac{\partial r}{\partial z} T_1^{-1}.$$

Hence,

$$(11) \qquad \sigma' = T_3 \sigma T_1^{-1}.$$

The original action $T_2$ on $(x, y)$ extends to $U$ by a prolongation given in local coordinates by

$$(12) \qquad \mathcal{T} : (w, \underset{\sim}{\sigma}) \mapsto (w', \underset{\sim}{\sigma}') = (Tw, \underset{\sim}{T}\underset{\sim}{\sigma}),$$

where $\underset{\sim}{T}$ is computed from (11) and the following Lemma.

**Lemma 1** *If $\sigma \in \mathbf{M}_{mn}$ is mapped to $\sigma' = X\sigma Y$, and if $\underset{\sim}{\sigma}$ is given by*

$$\underset{\sim}{\sigma} = (\sigma_{11}, \sigma_{21}, \ldots, \sigma_{m1}, \sigma_{12}, \ldots, \sigma_{mn})^\top$$

*then $\underset{\sim}{\sigma}' = \underset{\sim}{T}\underset{\sim}{\sigma}$ where $\underset{\sim}{T}$, the $mn \times mn$ matrix of the mapping in $\mathbf{R}^{mn}$, is*

$$(13) \qquad \underset{\sim}{T} = \begin{pmatrix} y_{11}X & y_{21}X & \cdots & y_{n1}X \\ y_{12}X & y_{22}X & & \\ \vdots & \vdots & & \\ y_{1n}X & y_{2n}X & & y_{nn}X \end{pmatrix}.$$

**Proof:** Write

$$(\sigma')_{ij} = \sum_{p=1}^{m}\sum_{l=1}^{n} x_{ip}\sigma_{pl}y_{lj}.$$

Now, $\underset{\sim}{\sigma}_k = \sigma_{ij}$ where $k = (j-1)m+i$, so $i = k - m\left[\frac{k-1}{m}\right]$; $j = 1 + \left[\frac{k-1}{m}\right]$. (Here $[\cdot]$ is the greatest integer function.) Writing

$$\underset{\sim}{\sigma}'_k = \sum_{s=1}^{mn} \underset{\sim}{t}_{ks}\underset{\sim}{\sigma}_s$$

shows that $\underset{\sim}{t}_{ks} = x_{ip}y_{lj}$ where $i$ and $j$ are functions of $k$ as above and $s$ corresponds to $(p, l)$ in the same way: $s = (l-1)m + p$; so $p = s - m\left[\frac{s-1}{m}\right]$, and $l = 1 + \left[\frac{s-1}{m}\right]$. This defines $\underset{\sim}{T} = (\underset{\sim}{t}_{ks})$. We note that $\underset{\sim}{T}$ has a block structure, since the $y_{lj}$ contribution is constant as $s$ and $k$ increase by $m-1$, while the $x_{ip}$ run through all entries of $X$. Notice also that the $y_{lj}$ are transposed in the block form. This results in the stated formula. ∎

What restrictions are placed on the coefficient matrices $A$ and $B$ in (6) or on the flux functions in the original equation (1) by requiring that solutions transform invariantly or equivariantly under the Euclidean group? Because the equation implies that a point $(w, \underset{\sim}{\sigma})$ satisfies (9) it is necessary that its image under $\mathcal{T}$ also satisfy (9). This is a formal statement of the fact that, when the partial derivatives (10) are substituted for $\sigma$, the resulting expression satisfies (6). More must be true, since the terms in (10) are not independent but satisfy integrability conditions, and one needs to know that these hold also for the image under $\mathcal{T}$. However, this follows categorically from the definitions, [1]. (See also [2, Theorem 2.71, page 165].) We conclude

**Proposition 1** *If a system of $m$ quasilinear equations (6) behaves equivariantly under the action $\mathcal{T}$ induced by a diffeomorphism $T_2$ then equation (9) must be satisfied identically for the images $(w', \underset{\sim}{\sigma}') = \mathcal{T}(w, \underset{\sim}{\sigma})$ of all points that satisfy (9).*

When the fluxes are independent of $z$ or $c(w)$ is absent from (6) for some other reason, the criterion of Proposition 1 takes a particularly simple form: we may write the algebraic relation (9) in $U$ as

(14) $$F \equiv M(w)\underset{\sim}{\sigma} \equiv [A(w)|B(w)|\,I\,]\underset{\sim}{\sigma} = 0$$

where $M(w)$ is an $m \times 3m$ matrix. Under the induced action, $F \mapsto F' = M(Tw)\underset{\sim}{\sigma}' = M(Tw)\underset{\sim}{T}\underset{\sim}{\sigma}$, and we have

**Corollary 1** *The homogeneous first-order equation*

(15) $$\mathcal{P}(w; \partial)r \equiv \frac{\partial r}{\partial t} + A(w)\frac{\partial r}{\partial x} + B(w)\frac{\partial r}{\partial y} = 0.$$

*respects the group action if and only if $F = 0$ and $F' = 0$ have the same solution space. That is,*

$$Rank\begin{bmatrix} M(w) \\ M(Tw)\underset{\sim}{T} \end{bmatrix} = Rank[M] = m.$$

Using (13) for $\underset{\sim}{T}$, with $X = T_3$ and $Y = T_1^{-1} = Diag(T_2^{-1}, 1)$, gives

$$\underset{\sim}{T} = \begin{pmatrix} t_{11}^* T_3 & t_{12}^* T_3 & 0 \\ t_{21}^* T_3 & t_{22}^* T_3 & 0 \\ 0 & 0 & T_3 \end{pmatrix}$$

where $(t_{ij}^*)$ are the elements of $(T_2^{-1})^\top$. Row-reduction of the matrix $M(Tw)\underset{\sim}{T}$ gives $2m^2$ conditions on $A$ and $B$:

(16)
$$\begin{aligned} A(w) &= T_3^{-1}(t_{11}^* A(Tw)T_3 + t_{21}^* B(Tw)T_3) \\ B(w) &= T_3^{-1}(t_{12}^* A(Tw)T_3 + t_{22}^* B(Tw)T_3) \end{aligned}$$

which must hold for every element of the symmetry group. The conditions are *nonlocal*, because each equation involves terms evaluated at both $w$ and $Tw$. For diffeomorphisms and discrete actions, such as the flip, these global conditions must be verified.

# 3   One-parameter Groups

When $T_2 = T_2(\theta)$ is a one-parameter group of diffeomorphisms, for example the group of rotations in equation (5), the computation above reduces to a local one. The infinitesimal generator of the prolongation $T(\theta)$ (equation (12)),

$$\frac{d}{d\theta} T(\theta) \mid_{\theta=0} \equiv \Xi,$$

defines a vectorfield on $U$. The following proposition is a special case of Theorem 2.31 of [2] (see also the formulas on pages 108-111 there):

**Proposition 2** *Verifying the symmetry conditions on the differential equation (6) is equivalent to verifying that $\Xi$ is tangent to the surface $F = 0$ determined by (9). That is*
(17)
$$\Xi \cdot \nabla F = 0$$

*(where tangent vectors and gradients are computed on $U \subset \mathbf{R}^{4m+3}$).*

This condition also results in $m$ linear equations for $\underset{\sim}{\sigma}$, whose coefficients involve first-order differential operators on the coefficients of $A$ and $B$. For a homogeneous equation, (15), it replaces (16) in computing the effect of rotations on $\mathcal{P}$. Once again, there will result $2m^2$ equations, now first-order differential equations (in $w$), and all quantities are evaluated at $w$. We compute $\Xi$ (the dot indicates differentiation with respect to $\theta$) from $\Xi = (\dot{T}(0)w, \underset{\sim}{\dot{T}}(0)\underset{\sim}{\sigma})$; also $\dot{T}(0)w = (\dot{T}_1(0)z, \dot{T}_3(0)r)$, and $\dot{T}_1(0) = Diag(J_2, 0)$, where

$$\dot{T}_2(0) = \begin{pmatrix} 0 & -1 \\ 1 & 0 \end{pmatrix} \equiv J_2.$$

By the same calculation,

(18)
$$\dot{T}_3(0) = Diag(J_2, J_2, \ldots, J_2, 0, 0, \ldots, 0) \equiv J.$$

We find $\dot{\underset{\sim}{T}}(0)$ from Lemma 1:

$$\dot{\underset{\sim}{T}}(0) = \begin{pmatrix} J & -I & 0 \\ I & J & 0 \\ 0 & 0 & J \end{pmatrix};$$

and

$$\Xi \cdot \nabla F = (\mathcal{K}M + \mathcal{L}M + M\dot{\underset{\sim}{T}}(0))\underset{\sim}{\sigma},$$

where $\mathcal{K} = z^T \dot{T}_1(0)\partial_z = -y\partial_x + x\partial_y$ and $\mathcal{L} = r^T J \partial_r = -r_2 \partial_{r_1} + r_1 \partial_{r_2} + \dots$ with the sum taken over all field pairs. This yields a second Corollary to Proposition 17.

**Corollary 2** *If $T_2 = T_2(\theta)$ is the one-parameter family of rotations in (5), then equation (15) respects the group action if and only if*

$$\text{Rank} \begin{bmatrix} M \\ (\mathcal{K} + \mathcal{L})M + M\dot{\underset{\sim}{T}}(0) \end{bmatrix} = \text{Rank}[M] = m,$$

*where $\mathcal{K}$, $\mathcal{L}$, and $\dot{\underset{\sim}{T}}(0)$ are defined above.*

Thus symmetry under $\mathcal{SO}(2)$ results in the $2m^2$ equations

(19)
$$\begin{aligned} (\mathcal{K} + \mathcal{L})A + AJ - JA + B &= 0 \\ (\mathcal{K} + \mathcal{L})B + BJ - JB - A &= 0. \end{aligned}$$

## 4  Examples

Equation (19) is a system of linear ordinary differential equations for the coefficients of $A$ and $B$, since all derivatives are in a single direction given by $\mathcal{K} + \mathcal{L}$. The equations decouple in blocks, with the largest block having dimension eight, so a general solution can be written down. The form of the solution makes it straightforward to impose the nonlocal constraints due to the flip, as expressed in equation (16). As a third step, if (15) comes from a conservation law, we impose compatibility conditions on the columns of $A$ and $B$ so that $A = df$, $B = dg$. Construction of $f$ and $g$ becomes a matter of linear algebra. We summarize briefly a method for organizing the calculations, and then report the results for some important sample cases.

Assume that $A$ and $B$ are independent of $z$ so that the $\mathcal{K}$ component of differentiation disappears from equation (19). Suppose the first $2p$ components of $r$ to be arranged in $p$ field pairs, the remaining $m - 2p$ entries consisting of potentials. Write (19) as a matrix equation for $[A|B]$:

(20) $$\mathcal{L}[A|B] = [A|B] \begin{pmatrix} -J & 0 \\ 0 & -J \end{pmatrix} + J[A|B] + [A|B] \begin{pmatrix} 0 & I \\ -I & 0 \end{pmatrix}$$

(where each letter represents an $m \times m$ matrix and $J$ is defined in (18)), and then rewrite this as a system of equations for a column vector, which we denote by $\underset{\sim}{\alpha}$, formed from the columns of $[A|B]$ as in Lemma 1:

$$\underset{\sim}{\alpha} = (A^{(1)}/A^{(2)}/ \ \vdots \ /A^{(m)}/B^{(1)}/ \ \vdots \ /B^{(m)})$$

where $A^{(j)}$ denotes the $j$th column of $A$ and the slanted lines indicate vertical stacking of the vectors. Use Lemma 1 to rewrite (20) as a system $\mathcal{L}\underset{\sim}{\alpha} = J_1\underset{\sim}{\alpha}$ where $J_1$ is a $2m^2 \times 2m^2$ matrix, and then rearrange the columns of $\underset{\sim}{\alpha}$ to put corresponding field pairs together:

$$\underset{\sim}{\bar{\alpha}} = (A^{(1)}/A^{(2)}/B^{(1)}/B^{(2)}/A^{(3)} \,\vdots\, /B^{(2p-1)}/B^{(2p)}/$$
$$A^{(2p+1)}/B^{(2p+1)}/ \,\vdots\, /A^{(m)}/B^{(m)});$$

then $\mathcal{L}\underset{\sim}{\bar{\alpha}} = \bar{J}_1\underset{\sim}{\bar{\alpha}}$, where $\bar{J}_1 = Diag(J_4, J_4, \dots, J_4, J_3, \dots, J_3)$ is now block diagonal, and $J_4$ and $J_3$ are $2m \times 2m$ and $4m \times 4m$ matrices:

$$J_3 = \begin{pmatrix} J & -I_m \\ I_m & J \end{pmatrix}; \quad J_4 = \begin{pmatrix} J_3 & -I_{2m} \\ I_{2m} & J_3 \end{pmatrix}.$$

Thus the four vectors $A^{(1)}$, $A^{(2)}$, $B^{(1)}$, $B^{(2)}$ are coupled in the system, as are the other field pairs, while pairs of vectors $A^{(j)}$, $B^{(j)}$ for $j > 2p$ are also coupled. Call these subsystems type I and type II respectively.

The subsystems decouple further if we order the rows by field pairs, so that, for example, the first $8 \times 8$ block corresponds to

$$\xi \equiv (a_1^{(1)}, a_2^{(1)}, a_1^{(2)}, a_2^{(2)}, b_1^{(1)}, b_2^{(1)}, b_1^{(2)}, b_2^{(2)}),$$

where $a_k^{(j)} = A_k^{(j)} = a_{kj}$ is the $k$th component of the $j$th column of $A$. Then we get $p$ $8 \times 8$ blocks

$$I_a = \begin{pmatrix} J_2 & -I_2 & -I_2 & 0 \\ I_2 & J_2 & 0 & -I_2 \\ I_2 & 0 & J_2 & -I_2 \\ 0 & I_2 & I_2 & J_2 \end{pmatrix}$$

followed by $m - 2p$ $4 \times 4$ blocks

$$I_b = \begin{pmatrix} 0 & -1 & -1 & 0 \\ 1 & 0 & 0 & -1 \\ 1 & 0 & 0 & -1 \\ 0 & 1 & 1 & 0 \end{pmatrix}$$

corresponding to $\eta \equiv (a_j^{(1)}, a_j^{(2)}, b_j^{(1)}, b_j^{(2)})$ for $j > 2p$.

Similarly, columns corresponding to potential quantities (type II blocks) decouple into $p$ $4 \times 4$ systems with matrix $II_a$ for $(a_1^{(j)}, a_2^{(j)}, b_1^{(j)}, b_2^{(j)})$, where $j > 2p$, and a remaining $m - 2p$ $2 \times 2$ systems with matrix $II_b$ for $\zeta \equiv (a_k^{(j)}, b_k^{(j)})$ with $j$, $k > 2p$. It turns out that $II_a \equiv I_b$ and $II_b \equiv J_2$.

Thus we need to solve systems $\mathcal{L}\xi = M\xi$ where $M$ is one of the matrices $I_a$, $I_b = II_a$ or $II_b$.

The eigenvalues of $I_a$ are $\pm i$ (with multiplicity three) and $\pm 3i$; those of $I_b$ are 0 (multiplicity two) and $\pm 2i$; those of $II_b$ are $\pm i$. Let $\mathcal{L} = \partial_\phi$, then for any $u$ and $v$ with $\tan\phi = v/u$, real fundamental matrices equivalent to $\exp(M\phi)$ can be found in terms of $u$ and $v$; for example, for $II_b$

$$\Omega_3 = \begin{pmatrix} -v & u \\ u & v \end{pmatrix},$$

and similarly for $\Omega_1 = \exp(I_a\phi)$ and $\Omega_2 = \exp(I_b\phi) = \exp(II_a\phi)$.

The general solution of (19) is found by assembling component parts of the form $\xi = \Omega_i c$ where $c$ is a vector of coefficients which depend on the invariant functions of $w$ ($\gamma(w)$ for which $\mathcal{L}\gamma = 0$). For the remainder of the calculation, assume a single field pair – that is, $p = 1$. Write $(u, v)$ for $(r_1, r_2)$ and abbreviate the remaining $r_j$'s by $\rho$. The invariant functions may be taken to be $s \equiv (u^2 + v^2)/2$ and $\rho$.

As one would expect, only half the columns of the $\Omega_i$ respect the reflectional symmetry. For example, in the case of a single invariant quantity, $A$ is

$$\left(\begin{array}{ccc} u(c_5 + c_7(u^2 - 3v^2)) & v(-c_1 + c_7(3u^2 - v^2)) & c_4 + c_8(u^2 - v^2) \\ v(-c_3 + c_7(3u^2 - v^2)) & u(c_1 + c_3 + c_5 - c_7(u^2 - 3v^2)) & 2uvc_8 \\ c_2 + c_6(u^2 - v^2) & 2c_6 uv & c_9 u \end{array}\right)$$

and $B$ is completely determined by $A$.

Now, conservation form is equivalent to compatibility relations

$$\frac{\partial A^{(j)}}{\partial r_k} = \frac{\partial A^{(k)}}{\partial r_j}$$

for $1 \le j, k \le 3$, $j \ne k$. The six equations serve to express all coefficients of $A$ by means of three arbitrary functions of $\rho$ and $s$, which we shall call $b$, $q$, and $d$. We obtain the following expressions for the flux matrices $f$ and $g$:

$$(21) \qquad f = \left(\begin{array}{c} b + u^2 q \\ uvq \\ ud \end{array}\right); \qquad g = \left(\begin{array}{c} uvq \\ b + v^2 q \\ vd \end{array}\right).$$

In terms of the $s$ (denoted $\dot{}$) and $\rho$ (denoted $'$) derivatives of $b$, $q$ and $d$:

$$A = \left(\begin{array}{ccc} u(\dot{b} + 2q + u^2\dot{q}) & v(\dot{b} + u^2\dot{q}) & b' + u^2 q' \\ v(q + u^2\dot{q}) & u(q + v^2\dot{q}) & uvq' \\ d + u^2\dot{d} & uv\dot{d} & ud' \end{array}\right)$$

and

$$B = \left(\begin{array}{ccc} v(q + u^2\dot{q}) & u(q + v^2\dot{q}) & uvq' \\ u(\dot{b} + v^2\dot{q}) & v(\dot{b} + 2q + v^2\dot{q}) & b' + v^2 q' \\ uv\dot{d} & d + v^2\dot{d} & vd' \end{array}\right).$$

Formulas for the fluxes and Jacobians for other values of $p$ and $m$ can be obtained using the same recipe.

## 4.1 The Invariant Scalar Equation: $m = 1$

There is only one possibility for a scalar equation in two space dimensions: the state $\rho$ must be invariant under the group action. The only symmetric fluxes are $f \equiv 0$, $g \equiv 0$. This fact helps explain why the notion of genuine nonlinearity for a scalar equation is so problematic: any nontrivial flux functions must express some sense in which the system fails to be isotropic.

## 4.2    Equivariant System under $\mathcal{O}(2)$ Action: $m = 2$

When there are two equations, for a field pair, with no invariant quantity, the calculation of (21) reduces to

$$(22) \qquad f = \begin{pmatrix} b + u^2 q \\ uvq \end{pmatrix}; \qquad g = \begin{pmatrix} uvq \\ b + v^2 q \end{pmatrix}$$

with $b = b(s)$ and $q = q(s)$. The Jacobian matrices are

$$A = \begin{pmatrix} u(\dot{b} + 2q + u^2\dot{q}) & v(\dot{b} + u^2\dot{q}) \\ v(q + u^2\dot{q}) & u(q + v^2\dot{q}) \end{pmatrix}$$

and

$$B = \begin{pmatrix} v(q + u^2\dot{q}) & u(q + v^2\dot{q}) \\ u(\dot{b} + v^2\dot{q}) & v(\dot{b} + 2q + v^2\dot{q}) \end{pmatrix}.$$

An examination of $A$ and $B$ reveals that both are identically zero at $u = v = 0$. This proves

**Proposition 3** *A system of two equations which is equivariant under the action of $\mathcal{O}(2)$ can never be strictly hyperbolic.*

## 4.3    Invariant and Equivariant Components: $m = 3$

Thus we see that the smallest symmetric system which can be both strictly hyperbolic and genuinely nonlinear consists of three equations with fluxes given by (21). Examples abound. The equations of ideal, compressible, isentropic gas flow take this form, with $u$ and $v$ the momenta in the $x$ and $y$ directions and $\rho$ the density: here $b = 1/\rho$, $d \equiv 1$ and $q = P(\rho)$ is the constitutive law (pressure-density relation) for the gas. This system is well-known to be strictly hyperbolic (for $\rho > 0$) and is also genuinely nonlinear if $P'' \neq 0$. An even simpler example is the nonlinear wave equation

$$w_{tt} = (c^2(|\nabla w|)w_x)_x + (c^2(|\nabla w|)w_y)_y$$

which becomes a symmetric system of three equations if we define $u = w_x$, $v = w_y$ and $\rho = w_t$. In this case, $q \equiv 0$, $b \equiv 1$, and $d = c^2$, a function of the invariant quantity $s$. This system, like the gas dynamics equations, is strictly hyperbolic and genuinely nonlinear if $c^2$ is convex.

# 5    Symmetry Properties of Characteristic Manifolds

To discuss hyperbolicity, strict hyperbolicity and genuine nonlinearity, we study the characteristic variety of the operator $\mathcal{P}$ of equation (6), or, rather, of its principal part, equation (15). Assume, for convenience, that $c \equiv 0$ in (6). Let $\zeta = (\xi, \eta, \tau)$ be a covector (a row vector, in our notation). Since $\mathcal{P}$ is not linear, we linearize the equation at a value $r_0$. If a small-amplitude plane wave

$$r(z) = (1 + \epsilon\varphi(\zeta z))r_0$$

is a nonconstant solution of (6), then $\mathcal{P}(z, (1+\epsilon\varphi(\zeta z))r_0; \zeta)r_0 = 0$, where $\mathcal{P}(w; \zeta)$ is the *symbol* of $\mathcal{P}(w; \partial)$

$$\mathcal{P}(w; \zeta) \equiv I\tau + A(w)\xi + B(w)\eta.$$

In the limit as $\epsilon \to 0$, nontrivial solutions exist if and only if $\det \mathcal{P}(z, r_0; \zeta) = 0$, and the covectors $\zeta$ that satisfy this equation determine the characteristic variety.

**Definition 1** *The triple $(z, \zeta; r_0)$ belongs to the characteristic variety of $\mathcal{P}$, Char $\mathcal{P}$, if and only if $\mathcal{P}(z, r_0; \zeta)r_0 = 0$.*

For linear operators, this equation is well defined on the cotangent bundle (see [3] for this and other elementary facts about characteristic varieties). In particular, under the action $T_1$ on $z$, the induced action on $\zeta$ is

$$\zeta \mapsto \zeta' = \zeta T_1^{-1}.$$

The study of characteristic varieties is standard in the theory of linear partial differential equations, and transformation formulas under group actions can be worked out. For quasilinear systems, the transformation formula is not as familiar, but can be derived using the calculus developed in the previous sections.

Our main result is the following.

**Proposition 4** *If $\mathcal{P}$ respects the symmetry of a mapping $T_2$, then*

$$(z, \zeta; r_0) \in \text{Char } \mathcal{P} \iff (T_1 z, \zeta T_1^{-1}; T_3 r_0) \in \text{Char } \mathcal{P}.$$

*In particular, if $\mathcal{P}$ is independent of $z$, then Char $\mathcal{P}$ is independent of $z$ and*

$$(\zeta; r_0) \in \text{Char } \mathcal{P} \iff (\zeta T_1^{-1}; T_3 r_0) \in \text{Char } \mathcal{P}.$$

**Proof:** If $(z, \zeta; r_0) \in \text{Char } \mathcal{P}$ then $\mathcal{P}(z, r_0; \zeta)r_0 = 0$ and this condition can be expressed (equation (14)) as

$$F \equiv M(w_0)\underset{\sim}{\sigma}_0 = 0$$

with $w_0 = (z, r_0)$ and $\underset{\sim}{\sigma}_0$ the vector corresponding to $\sigma_0 = r_0\zeta \in \mathbf{M}_{m3}$. If $\mathcal{P}$ respects the symmetry of $T_2$, then for all $(w_0, \underset{\sim}{\sigma}_0)$

(23) $$M(w_0)\underset{\sim}{\sigma}_0 = 0 \Rightarrow M(T_1 z, T_3 r_0)\underset{\sim}{T}\underset{\sim}{\sigma}_0 = 0$$

by Proposition 1. By construction, $\underset{\sim}{T}\underset{\sim}{\sigma}_0$ corresponds to $\sigma_0 = (T_3 r_0)(\zeta T_1^{-1})$, and so (23) implies

$$\mathcal{P}(T_1 z, T_3 r_0; \zeta T_1^{-1})T_3 r_0 = 0.$$

The converse implication follows from applying $T_2^{-1}$.  ∎

Even if $\mathcal{P}$ is symmetric under the group $\mathcal{O}(2)$, it does not follow that Char $\mathcal{P}$ (the characteristic cone) will have a geometric circular symmetry, since the image points $\zeta'$ correspond to the characteristic variety of a generally different state $r_0'$. However, the characteristic variety of a state that is invariant under $T_3$ will have some symmetry.

**Corollary 3** *If $\mathcal{P}$ is independent of $z$, and if $T_3 r_0 = r_0$ then $\zeta$ is a characteristic covector for the linearization of $\mathcal{P}$ at $r_0$ if and only if $\zeta T_1^{-1}$ is a characteristic covector for $\mathcal{P}$ at $r_0$.*

We conclude with some comments on the hyperbolicity and genuine nonlinearity of system (1). Recall that (1) is said to be *hyperbolic* at a point $w$ with respect to time if all the roots $\tau = \tau_i(\xi, \eta; w)$ of $\det \mathcal{P}(w; (\xi, \eta, \tau)) = 0$ are real for all values of $\xi$ and $\eta$, *strictly hyperbolic* if they are real and distinct for $(\xi, \eta) \neq 0$. *Genuine nonlinearity* at a point $w$ of a characteristic speed $\tau_i$ in the codirection $(\xi, \eta)$ means $\nabla \tau_i(\xi, \eta, w) \cdot r_0 \neq 0$, where $r_0$ is the right eigenvector of $\mathcal{P}$ corresponding to $\tau_i$. For When these conditions are checked directly from (22), we obtain

**Proposition 5** *A system of two equations consisting of a field pair is hyperbolic in time if $\dot{b}q \geq 0$. The origin is an isolated umbilic point if $\dot{b}q > 0$ and $\dot{b} + q + 2s\dot{q} \neq 0$; away from $(u, v) = (0, 0)$, the characteristic variety is a nondegenerate cone if $\dot{b}q > 0$ and $\dot{b} + q + 2s\dot{q} \neq 0$. The characteristic cone never has circular symmetry. Furthermore, for all $w$ and for each characteristic speed $\tau_1$ and $\tau_2$, there is at least one codirection $(\xi, \eta)$ where genuine nonlinearity fails.*

Many properties of the characteristic variety, including hyperbolicity itself, cannot be deduced from the imposed symmetries alone, but depend on properties of the invariant functions $b$ and $q$. For the case $m = 3$ when both field and potential quantities are present, the determination of hyperbolicity is a condition on the functions $q$, $b$ and $d$. However, we have a description of one case:

**Proposition 6** *Suppose that in (21) the coefficients $q$, $b$ and $d$ are independent of $\rho$. Then one characteristic speed is identically zero, and the other two are determined as in the case $m = 2$ and the results of Proposition 5 apply.*

This amplifies the comment on equivariant and invariant components made in the introduction: although the presence of equivariant components is necessary for a system to be nontrivial, dependence on the invariant component determines symmetry of the characteristic cone and genuine nonlinearity of the wave speeds.

# References

[1] R. Hermann, *Geometry, Physics and Systems*, Dekker, New York, 1973.

[2] P. J. Olver, *Applications of Lie Groups to Differential Equations*, Springer, New York, 1986.

[3] J. Rauch, *Partial Differential Equations*, Springer, New York, 1991.

BARBARA LEE KEYFITZ
MATHEMATICS DEPARTMENT, UNIVERSITY OF HOUSTON, HOUSTON, TEXAS
77204-3476.                    *E-mail address:* keyfitz@math.uh.edu

MILTON LOPES-FILHO
DEPARTAMENTO DE MATEMATICA, IMECC - UNICAMP, CAMPINAS, SAO
PAULO 13081-970, BRASIL          *E-mail address:* mlopes@ime.unicamp.br

Lectures in Applied Mathematics
Volume **29**, 1993

# SEMILINEAR ELLIPTIC EQUATIONS IN CYLINDRICAL DOMAINS
# - REVERSIBILITY AND ITS BREAKING -

KLAUS KIRCHGÄSSNER & KATHARINA LANKERS

ABSTRACT. Reversible elliptic equations in infinite cylindrical domains are investigated. Reversibility means that the equations are autonomous in the axial variable and exhibit a reflectional symmetry. A reduction method to a system of ODE is described, which can be used in the neighbourhood of a trivial solution to determine all bounded solutions of moderate norm. Via normal-form theory the geometry of solutions is classified for two significant cases for a specific reversible elliptic problem. The effect of reversibility-breaking on a homoclinic solution is investigated.

## 1. INTRODUCTION

The study of nonlinear elliptic equations in unbounded cylindrical domains has received much attention in the past, so much, that it is beyond the scope of this contribution to give complete reference to the existing literature. Our own interest in this field has arisen by an attempt to give a general method for constructing steady travelling waves in fluid layers, even with a free boundary [6]. In general these problems have a Galilean invariance and therefore, even if the wave speed is non-zero, the underlying equations remain reversible in the bounded space variable $x$, i.e. the orientation of the $x$-axis is free as well as an arbitrary translation. Of course this requires that there is no inhomogenity in the cylinder, i.e. explicit $x$-dependence of coefficients, due to obstacles in the layer e.g. Reversibility may be broken by such an obstacle and it seems to be of interest to analyse the situation in the reversible as well as in the broken-reversible case.

Quite a number of these phenomena have been understood for solutions of moderate amplitudes [4, 10], most of them in the realm of fluid waves, and may not be known so widely in the mathematical community. A particularly simple problem, the determination of possible crest-forms of waves in density stratified two-dimensional layers, has recently been investigated in [8]. We take it up here and analyse it on a purely mathematical basis, not touching the physical background. The main ingredient of the analysis is a reduction procedure, valid if there is a

1991 *Mathematics Subject Classification*. Primary 35J60, 35B20, 35B32; Secondary 76B15.
Research supported by the Deutsche Forschungsgemeinschaft (DFG), Ki 131/10-1.
This paper is in final form and no version of it will be submitted for publication elsewhere.

trivial, $x$-independent solution. All solutions of 'moderate' amplitude satisfy a finite-dimensional ODE-system, which inherits the symmetries - in particular the reversibility - from the original one. Normal-form-theory can be applied then to determine the qualitative behaviour of the solutions [1, 3]. The reduced system in normal form plus the reversibility classifies all elliptic systems with an equivalent solution behaviour. In fact, we could present many problems seemingly completely different by their physical background and their mathematical appearance, which are qualitatively the same.

While we expose this procedure in the first part of this contribution, we show for a particular parameter range, what happens when reversibility is broken. In a special fluid dynamical situation, namely, when there is a small obstacle in the channel, this problem has received much attention in the applied literature and one systematic resolution by MIELKE [10]. He gives a complete picture of the set of solutions and also a method which is independent of the special case treated. In particular he can answer the question, how far upstream such a small obstacle can be felt. Here, we shall apply this method to our situation and reproduce MIELKE's results in the case, where the unperturbed equation has a homoclinic solution.

The unperturbed semilinear boundary value problem we are going to study is given as follows

$$\triangle \psi + a(\lambda, y)\psi + b(\lambda, y)\psi^2 + r(\lambda, y, \psi) = 0$$

(1.1)
$$\psi(x, 0) = \psi(x, 1) = 0 \quad ,$$

where $(x, y) \in R \times (0, 1)$, $\triangle = \partial_{xx}^2 + \partial_{yy}^2$, $a$ and $b$ being smooth coefficients depending on $(\lambda, y) \in R \times [0, 1]$. It is assumed that $a(\lambda, .)$ is a strictly increasing function of $\lambda$ such that $a(\lambda, .) \geq \alpha\lambda + \beta$, for some $\alpha > 0$. $r$ is a term, smooth in its variables, bounded uniformly in $y \in [0, 1]$ and locally uniform in $\lambda$. For small values of $\psi$ we have $r = \mathcal{O}(\psi^3)$.

We could take the nonlinearity depending on $\nabla\psi$ as long as the equation is invariant under $x \mapsto x + c$, $x \mapsto -x$ (reversibility). In fact quasilinear nonlinearities are permitted in the sense that they should define a smooth mapping from $H^2 \cap H_0^1(0, 1)$ into $H^0(0, 1)$, cf. [11, 12]. In the last section we shall perturb (1.1) by some smooth function $\epsilon\, p(x)$ and thus break the reversibility. Again the choice of the perturbation has been made in order to show the principal facts on a special problem which is as simple as possible.

## 2. LINEARISATION, SYMMETRY AND REDUCTION

Equation (1.1) has the surprising property that all solutions of small amplitude can be found by solving a system of ordinary differential equations. This reduction is achieved via a method reminiscent of the center manifold procedure for dynamical systems. The key observation is the fact, that, by considering a spectral decomposition of $\psi$ in the cross-sectional direction, the modes with smallest real part modulus (center-modes) dominate the other, 'hyperbolic' modes. In effect the latter are pointwise functions of the central modes. The validity of this concept for elliptic problems has been shown in a series of papers [2, 5, 9, 11, 12], to which we refer the reader for details. We show here the main steps of the procedure and, in particular, discuss the implications for the problem under consideration.

The key step is the formulation of (1.1) as a 'dynamical' system. Although the initial value problem is not solvable in general, some of the reduction methods of this theory are still valid. Consider

$$\underline{\Psi} := \begin{pmatrix} \psi \\ \phi \end{pmatrix} \quad : \quad R \ni x \mapsto Y$$

as a mapping from $R$ into $Y$, where $Y$ is some function space to be specified later, consisting of functions living on the interval $(0, 1)$. We can rewrite (1.1) as

$$(2.1) \qquad \partial_x \underline{\Psi} = A(\lambda)\underline{\Psi} + \underline{f}(\lambda, \underline{\Psi}),$$

where

$$A(\lambda) := \begin{pmatrix} 0 & 1 \\ -\partial_{yy}^2 - a(\lambda, .) & 0 \end{pmatrix}$$

$$\underline{f}(\lambda, \underline{\Psi}) = \underline{f}(\lambda, \psi) = \begin{pmatrix} 0 \\ -b(\lambda, .)\psi^2 - r(\lambda, ., \psi) \end{pmatrix}.$$

Observe that the terms on the right side contain only derivatives in $y$-direction. We set, while using the usual Sobolev-spaces, $H^j = W^{2,j}(0, 1)$, $j \geq 0$ and $H_0^j = \{\psi \in H^j \mid \psi_{|y=0,1} = 0\}$, $j \geq 1$,

$$Y := H^0 \times H^0 , \quad D(A) := H^2 \cap H_0^1 \times H_0^1.$$

Then a solution of (1.1) is defined as a vector valued function $\underline{\Psi} \in C^1(R, Y) \cap C^0(R, D(A))$, i.e. as a curve parametrized by $x \in R$ in $D(A)$. In a standard way it can be seen, that this concept coincides with the classical understanding of a solution.

The original problem has a symmetry, which we call reversibility, which stems from the fact that (1.1) is autonomous in $x$ and invariant under $x \to -x$. This will be expressed by the fact, that the right side of (2.1) anticommutes with

$$R = \begin{pmatrix} 1 & 0 \\ 0 & -1 \end{pmatrix}, \quad \text{i.e.} \quad (A(\lambda) + \underline{f}) \circ R = -R \circ (A(\lambda) + \underline{f}).$$

$A(\lambda)$ maps $D(A)$ into $Y$, and $\underline{f}(\lambda, .)$ maps $H_0^1 \times H^0$ smoothly into $Y$. Whenever $\lambda \in K \subset R$, $K$ compact, and $\|\bar{\psi}\|_{H^1} \leq \rho < 1$, we have

$$(2.2) \qquad \|\underline{f}(\lambda, \underline{\Psi})\|_Y = \|\underline{f}(\lambda, \psi)\|_{H^0} \leq c(K, \rho)\|\psi\|_{H^1}^2 \leq c(K, \rho)\|\underline{\Psi}\|_{D(A)}^2.$$

Consider the operator $A(\lambda)\underline{\Psi}$, with $\underline{\Psi} \in D(A)$. Since it has a compact resolvent, its spectrum consists only of eigenvalues of finite multiplicities. Moreover, these eigenvalues are contained in $R \cup iR$, and accumulate only at infinity. We have $A(\lambda)\underline{\Psi} = \sigma\underline{\Psi}$, $\sigma \in C$, if

$$(2.3) \qquad \psi'' + a(\lambda, .)\psi + \sigma^2\psi = 0 , \quad \psi_{|y=0,1} = 0 , \quad ' := \partial_y$$

Let $-\rho_m(\lambda)$ denote the eigenvalues of the Sturm-Liouville problem, which one obtains from (2.3) for $\sigma = 0$, then $\rho_m(\lambda) \to \infty$ as $m \to \infty$. It follows

$$(2.4) \qquad \sigma_m(\lambda) = \pm\rho_m(\lambda)^{\frac{1}{2}} , \quad m \in N$$

We notice that the spectrum $\Sigma$ of $A(\lambda)$ is invariant under $\sigma \to -\sigma$, which is a consequence of the reversibility, and that $\Sigma_c := \Sigma \cap iR$ is finite dimensional. We call $\Sigma_c$ the 'central part' of the spectrum. If $\Sigma_c$ is empty, then the 0-solution is isolated. The solution set can change qualitatively only if $\Sigma_c$ changes. Parameter-values $\lambda$ are said to be *critical* when this happens. If $a$ increases with $\lambda$ then, there is a strictly monotone sequence $\lambda_j$ of critical values with ever increasing complexity of the solution set. The simplest situations occur for $\lambda_0 < \lambda_1$ as shown in Figure 1.

(a)                                        (b)

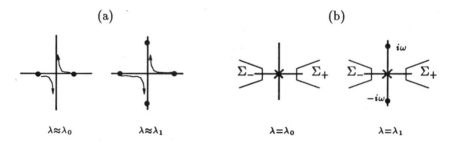

$\lambda \approx \lambda_0$        $\lambda \approx \lambda_1$        $\lambda = \lambda_0$        $\lambda = \lambda_1$

FIGURE 1. (a) Behaviour of critical eigenvalues when $\lambda$ increases through $\lambda_0$ resp. $\lambda_1$; (b) Spectral situation of $A(\lambda)$ for $\lambda = \lambda_0$ resp. $\lambda_1$ ($\times$ indicates double eigenvalues)

We sketch the reduction process. Define $\tilde{A}_j := A(\lambda_j)$, $\mu := \lambda - \lambda_j$ for any $\lambda_j$ and

$$\underline{\tilde{f}}(\mu, \underline{\Psi}) = \underline{\tilde{f}}(\mu, \psi) = (A(\lambda_j + \mu) - \tilde{A}_j)\underline{\Psi} + \underline{f}(\lambda_j + \mu, \underline{\Psi})$$

Remark that $\underline{\tilde{f}}$ depends on the first component of $\underline{\Psi}$ only. It is of order $\mathcal{O}(\|\psi\|_{H^1}^2 + \mu\|\psi\|_{H^1})$. Decompose $Y$ into $\tilde{A}_j$-invariant subspaces $Y = Y_c \oplus Y_h$ corresponding to the central resp. hyperbolic part of the spectrum, and let $P_c$ be the corresponding projection into $Y_c$, and set $\tilde{A}_j = A_c \oplus A_h$. In the following we assume $\lambda_j$ to be fixed and omit the index $j$. The dimension of $Y_c$ is 2 for $\lambda = \lambda_0$ and 4 for $\lambda = \lambda_1$. Now we can write (2.1)

$$\partial_x \underline{\Psi}_c = A_c \underline{\Psi}_c + \underline{f}_c(\mu, \underline{\Psi}_c + \underline{\Psi}_h)$$

(2.5)

$$\partial_x \underline{\Psi}_h = A_h \underline{\Psi}_h + \underline{f}_h(\mu, \underline{\Psi}_c + \underline{\Psi}_h)$$

where $\underline{\Psi}_c = P_c \underline{\Psi}$, $\underline{\Psi}_h = (1 - P_c)\underline{\Psi}$, $\underline{f}_c = P_c \underline{\tilde{f}}$ and $\underline{f}_h = (1 - P_c)\underline{\tilde{f}}$.

A solution of (2.5) exists and is uniquely determined by the additional conditions

$$\underline{\Psi}_c(0) = \underline{\xi} \in Y_c, \quad \|\underline{\Psi}_h\|_{H^1} < \epsilon,$$

when $\underline{\xi}$ and $\epsilon > 0$ are sufficiently small. If we denote this solution by $(\underline{\Psi}_c, \underline{\Psi}_h)(\mu, x, \underline{\xi})$, the mapping

(2.6)                    $$Y_c \ni \underline{\xi} \longrightarrow h(\mu, \underline{\xi}) := \underline{\Psi}_h(\mu, 0, \underline{\xi})$$

is well-defined, and its graph is a manifold of dimension $\dim Y_c + 1$. It inherits the symmetries from (2.1), in particular the reversibility. Moreover $h(\mu, \underline{\xi}) = \mathcal{O}(\mu|\underline{\xi}| + |\underline{\xi}|^2)$ holds for small $|\mu|$ and $|\underline{\xi}|$.

The existence and uniqueness result mentioned above is the key to the reduction method. It is by no means trivial and requires, at least for the quasilinear case, deeper analytical tools, cf. MIELKE [12]. Here we confine ourselves to describe the reduction result, its premises and its implications, and refer the interested reader to the original literature.

**Proposition 2.1.** *Let $Y$, $\tilde{A}_j$, $\underline{f}$ be as above, in particular* $\dim Y_c < \infty$. *Assume that*

$$\tilde{\underline{f}} : R \times D(\tilde{A}_j^\alpha) \to Y \quad , \quad \alpha \in [0,1]$$

$$\tilde{\underline{f}}(\mu, \underline{\Psi}) = \mathcal{O}(|\mu|\|\underline{\Psi}\| + \|\underline{\Psi}\|^2)$$

*near* $(\mu, \underline{\Psi}) = (0, \underline{0})$, *where the norm is taken in* $D(\tilde{A}_j^\alpha)$. *Let* $\tilde{\underline{f}}$ *be a* $C_{b,u}^k$ *mapping for some* $k \geq 1$, *where 'b,u' indicates, that the derivatives should be bounded and uniformly continuous. It suffices to fulfill this condition in a suitable neighbourhood of zero.*

*Assume further the validity of the resolvent-estimate: There exist positive constants* $\gamma$, $\beta$, $a_1$, $a_2$ *such that*

$$\|(A_h - zI)^{-1}\|_{Y_h \to Y_h} \leq a_1 \exp(a_2|z|)$$

$$\|(A_h - zI)^{-1}\|_{V \to Y_h} \leq \frac{\gamma}{1 + |z|}$$

*for all* $z \in C$ *with* $|\mathrm{Re}z| < \beta$; $V$ *is a closed subspace of* $Y_h$ *with* $\mathrm{range}(\tilde{\underline{f}}) \subset Y_c \times V$.

*Then there exist neighbourhoods* $U_\mu \in R$, $U_\xi \in Y_c$ *of* $\mu = 0$ *resp.* $\underline{\xi} = 0$ *and a function* $h : R \times Y_c \to Y_h$ *defined on* $U_\mu \times U_\xi$, *with the following properties:*

(1) *For every* $\mu \in U_\mu$ *the graph* $M_\mu := \{(\underline{\xi}, h(\mu, \underline{\xi})) \in Y | \underline{\xi} \in U_\xi\}$ *is invariant under the flow of (2.5), i.e. if there is a solution existing for all* $x \in R$, *then it must lie on* $M_\mu$.

(2) $h(\mu, \underline{\xi}) = \mathcal{O}(|\mu \underline{\xi}| + |\underline{\xi}|^2)$

(3) $h$ *inherits the* $Y$-*isometric symmetries of the system (2.1).*

In our case, *3.* implies that

$$h(\mu, R_c \underline{\xi}) = R_h h(\mu, \underline{\xi})$$

when $R = R_c \oplus R_h$ is the spectral decomposition of $R$. Moreover, *1.* says that all bounded solutions of (2.1) with sufficiently small norm satisfy $\underline{\Psi}_h(x) = h(\mu, \underline{\Psi}_c(x))$. Therefore, (2.1) can be reduced to the following ODE-system

$$(2.7) \qquad \partial_x \underline{\Psi}_c = A_c \underline{\Psi}_c + \underline{f}_c(\mu, \underline{\Psi}_c + h(\mu, \underline{\Psi}_c))$$

when we restrict our interest to solutions of small amplitudes. (2.7) is called the *reduced system.*

Estimates for the amplitudes could be easily given. They would be proportional to $\beta$. We could also construct $h$ in any algebraic order of $\underline{\xi}$ and thus obtain quantitative results by solving (2.7). In the next section another way to discuss (2.7) is proposed, which exploits only the existence part of the reduction (2.7) and then uses the theory of normal forms and persistence.

A few remarks about the validity of Proposition 2.1 for our special problem are appropriate. Since $\tilde{\underline{f}}(\mu, \underline{\Psi})$ depends on $\psi$ only, and since $\psi \in H_0^1$ implies $\psi \in C^0[0,1]$, we see that the smoothness of $\tilde{\underline{f}}$ holds. By localizing $\tilde{\underline{f}}$ via a cut-off-function we obtain the uniform boundedness and the order estimate using (2.2).

To verify the resolvent estimate we take

$$V = \{0\} \times H^0 \cap Y_h,$$

denote by $\|.\|$ the $L^2(0,1)$ norm and observe, that a constant $\beta > 0$ exists such that the spectrum of $A_1$ is in the region $|\mathrm{Re}z| > \beta$. Calculating the resolvent in $z \in C$, $|\mathrm{Re}z| \le \beta$ we have to solve

$$
\begin{aligned}
\phi - z\psi &= f \in H^0 \\
-\psi'' - a\psi - z\phi &= g \in H^0 \\
\psi_{|y=0,1} &= 0 .
\end{aligned}
$$

(2.8)

The condition $(f,g) \in V$ means $f = 0$. We have $\|\psi'\| \ge \sqrt{2}\|\psi\|$. Therefore, replacing $\phi$ by $z\psi + f$ and multiplying the $\psi$-equation by $\bar{\psi}$, one obtains

$$(2 - \|a\|_\infty - \beta^2 + (\mathrm{Im}z)^2)\|\psi\|^2 \le (|z|\|f\| + \|g\|)\|\psi\|$$

Here $\|a\|_\infty = \max |a|$. For $f = 0$ the first resolvent estimate follows for $|\mathrm{Im}z| \ge \eta > 0$ and $\eta$ sufficiently large, since $\|\psi\| \sim \mathcal{O}(|z|^{-2})$, $\|\phi\| \sim z\|\psi\|$. For $f \ne 0$ the second inequality follows with $a_2 = 0$ and $a_1$ an appropriate positive constant.

## 3. Solutions via Normal Form

For the determination of all solutions of equation (1.1) which have sufficiently small amplitude we have obtained a completely equivalent formulation in the reduced system (2.7). For $\lambda = \lambda_0$ this system is of dimension 2, for $\lambda = \lambda_1$ of dimension 4. It is known that the qualitative structure of the set of solutions near zero of a system like (2.7) is essentially determined by its linearisation at $\underline{\Psi}_c = \underline{0}$ and its symmetries; here 'essentially' means 'up to flat terms'. This is the message of normal form theory. The special problem is therefore part of a general class of equivalent problems, and its individuality is expressed through a few coefficients only. In fact, the two cases $\lambda = \lambda_0, \lambda_1$, which we are going to discuss, arise also when one discusses the appearance of nonlinear capillary-gravity waves at values of the Froude number near 1 and when the Bond number is less than one third. In this connection the cases $\lambda_0$ resp. $\lambda_1$ have been discussed in depth by IOOSS and the first author in [4]. Since all solutions found by normal-form approximation have to be shown to be persistent for the full equation (2.7), and since this task has been treated in [4], we can refer the interested reader to this paper.

Here we show for a few significant examples how the method works and what the implications are.

### Case I.

Let us take first the case $\lambda = \lambda_0$, i.e. $\Sigma_c = \{0\}$. Then $\dim Y_c = 2$. $A_c$ has a generalized eigenspace to the eigenvalue 0 spanned by $\underline{\Phi}_1 = (\psi_0, 0)$ and $\underline{\Phi}_2 = (0, \psi_0)$, where $\psi_0 \ne 0$ solves (2.3) for $\sigma = 0$ and $\lambda = \lambda_0$. Assume $\|\psi_0\| = 1$. Write $Y_c$ in the basis $\underline{\Phi}_1$, $\underline{\Phi}_2$ and identify it with $R^2$. Then $A_c$ and $R_c$ have the form

$$A_c = \begin{pmatrix} 0 & 1 \\ 0 & 0 \end{pmatrix}, \quad R_c = \begin{pmatrix} 1 & 0 \\ 0 & -1 \end{pmatrix}.$$

The projection $P_c$ is given by

$$P_c\underline{\Psi} = \psi_0 \begin{pmatrix} (\psi, \psi_0) \\ (\phi, \psi_0) \end{pmatrix}, \quad (\psi, \phi) := \int_0^1 \psi\phi .$$

Furthermore we have

$$\underline{f}_c(\mu, \underline{\Psi}) = P_c \tilde{\underline{f}}(\mu, \underline{\Psi}) = -(g(\mu, \psi), \psi_0)\,\underline{\Phi}_2\,,$$

with

$$g(\mu, \psi) = \mu\partial_\lambda a(\lambda_0, .)\psi + b(\lambda_0 + \mu, .)\psi^2 + \mathcal{O}(\mu^2\psi + \psi^3).$$

Let us set $\underline{\Psi}_c = P_c\underline{\Psi} = \alpha_1\underline{\Phi}_1 + \alpha_2\underline{\Phi}_2$ and identify $Y_c$ with $R^2$. Then (2.7) reads

(3.1)
$$\frac{d}{dx}\alpha_1 = \alpha_2$$
$$\frac{d}{dx}\alpha_2 = \mu a_{11}\alpha_1 + a_{02}\alpha_1^2 + r(\mu, \alpha_1, \alpha_2^2),$$

where

$$a_{11} = -(\partial_\lambda a(\lambda_0, .), \psi_0^2), \quad a_{02} = -(b(\lambda_0, .)\psi_0^3)$$

and $r = \mathcal{O}(\mu^2\alpha_1 + \alpha_1^3 + \alpha_1\alpha_2^2 + \alpha_2^4)$. The fact, that $r$ depends on $\alpha_2^2$ only is a consequence of the reversibility of (3.1) with respect to $R_c$. We write (3.1) in short form as

(3.2)
$$\frac{d}{dx}\underline{\alpha} = A_c\underline{\alpha} + \underline{g}^{(0)}(\mu, \underline{\alpha})$$

The theory of normal forms implies - cf. [1, 3] - that, given any $k \geq 2$, $k \in N$, there exists a polynomial transformation in $\underline{\beta}$ of the form $\underline{\alpha} = \underline{\beta} + \underline{P}(\mu, \underline{\beta})$ which determines a diffeomorphism in some neighbourhood $U_k$ of 0, such that the following holds:

(1) Equation (3.2) is transformed into

(3.3)
$$\frac{d}{dx}\underline{\beta} = A_c\underline{\beta} + \underline{N}(\mu, \underline{\beta}) + \underline{\Theta}(\mu, \underline{\beta})$$

where $\underline{\Theta}$ is of order $k + 1$ in $\mu$ and $\underline{\beta}$ and

$$\underline{N} \circ R_c = -R_c \circ \underline{N}, \quad \underline{N} = \mathcal{O}(\mu|\beta| + |\beta|^2)$$

(2)

(3.4)
$$D_\beta\underline{N}(\mu, \underline{\beta})A_c^*\underline{\beta} = A_c^*\underline{N}(\mu, \underline{\beta})$$

for all $\mu$ and $\beta \in R^2$.

The remainder term $\underline{\Theta}$ is of the form $(0, \Theta)$, and $\underline{\Theta} \circ R_c = -R_c \circ \underline{\Theta}$.

In the case under discussion, (3.4) reads

$$\beta_1\partial_{\beta_2}N_1 = 0 \quad , \quad \beta_1\partial_{\beta_2}N_2 = N_1$$

where $\underline{N} = (N_1, N_2)$ and $\underline{\beta} = (\beta_1, \beta_2)$. It follows $N_1 = N_1(\mu, \beta_1)$ and in view of reversibility $N_1(\mu, \beta_1) = -N_1(\mu, \beta_1) = 0$. Therefore $N_2 = N_2(\mu, \beta_1)$, where $N_2$ is a polynomial in $\beta_1$ of order $\mathcal{O}(\beta_1^2 + \mu\beta_1)$. Comparing this with (3.3) we see that (3.3) is in normal form in the lowest order terms.

We discuss (3.1) in lowest order by setting $r = 0$. This can be justified - cf. [4]. (3.1) has the integral

(3.5)
$$\alpha_2^2 - \mu a_{11}\alpha_1^2 - \frac{2}{3}a_{02}\alpha_1^3 = \text{const.}$$

The phase portrait of the bounded solutions - homoclinic and periodic orbits - is shown in Figure 2 for $a_{11} < 0$, $a_{02} > 0$ as a function of $\mu$. The other cases can be discussed similarly. All these solutions persist under $R_c$-reversible perturbations - cf. [4] - and thus lead, via Proposition 2.1, to solutions of the full equation (1.1).

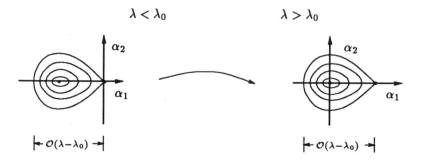

FIGURE 2. Case I: transition of phase plane picture as $\lambda$ traverses through $\lambda_0$

## Case II.
Here $\lambda = \lambda_1$, $\Sigma_c = \{0, \pm i\omega\}$ and 0 has multiplicity 2. Hence, dim $Y_c = 4$. Denote by $\psi_0$ the $L^2$-normalized solution of (2.3) for $\sigma = 0$ and by $\psi_\omega$ that for $\sigma = i\omega$ (here $\lambda = \lambda_1$). Then we can write (2.7) as $\underline{\Psi}_c = \alpha_1\underline{\Phi}_1 + \cdots + \alpha_4\underline{\Phi}_4$, where $\underline{\Phi}_1$, $\underline{\Phi}_2$ are as in Case I, $\underline{\Phi}_3 = (\psi_\omega, 0)$, $\underline{\Phi}_4 = (0, \omega\psi_\omega)$

$$A_c = \begin{pmatrix} 0 & 1 & 0 & 0 \\ 0 & 0 & 0 & 0 \\ 0 & 0 & 0 & -\omega \\ 0 & 0 & \omega & 0 \end{pmatrix} , \quad R_c = \begin{pmatrix} 1 & 0 & 0 & 0 \\ 0 & -1 & 0 & 0 \\ 0 & 0 & 1 & 0 \\ 0 & 0 & 0 & -1 \end{pmatrix}$$

The conjugate eigenvectors are $\hat{\underline{\Phi}}_j = \underline{\Phi}_j$, $j = 1, 2, 3$, $\hat{\underline{\Phi}}_4 = (1/\omega^2)\underline{\Phi}_4$, and thus we have

$$P_c\underline{\Psi} = \sum_{j=1}^{4}(\underline{\Psi}, \hat{\underline{\Phi}}_j)\underline{\Phi}_j = \psi_0 \begin{pmatrix} (\psi, \psi_0) \\ (\phi, \psi_0) \end{pmatrix} + \psi_\omega \begin{pmatrix} (\psi, \psi_\omega) \\ (\phi, \psi_\omega) \end{pmatrix} ,$$

hence

$$P_c\underline{\tilde{f}}(\mu, \underline{\Psi}) = -\, (g(\mu, \psi), \psi_0)\, \underline{\Phi}_2 - \frac{1}{\omega}\, (g(\mu, \psi), \psi_\omega)\, \underline{\Phi}_4 ,$$

$g$ as in Case I.

We determine (2.7) up to order $\mathcal{O}(\mu|\underline{\alpha}| + |\underline{\alpha}|^2)$ inclusively (i.e. $h = 0$ in (2.7)), and refer to [4] for its justification. Then we obtain

(3.6)             $$\frac{d}{dx}\underline{\alpha} = A_c\underline{\alpha} + \underline{g}^{(1)}(\mu, \underline{\alpha}) \quad , \quad \underline{g}^{(1)} = (0, p_2, 0, p_4)^T$$

where

$$p_i = \mu a_i\alpha_1 + \mu b_i\alpha_3 + c_i\alpha_1^2 + d_i\alpha_3^2 + e_i\alpha_1\alpha_3 , \; i = 2, 4$$

and

$$
\begin{aligned}
a_2 &= -(\partial_\lambda a(\lambda_1, .), \psi_0^2) \\
c_2 &= -(b(\lambda_1, .), \psi_0^3) \\
d_2 &= -(b(\lambda_1, .), \psi_0 \psi_\omega^2) \\
b_4 &= -\frac{1}{\omega}(\partial_\lambda a(\lambda_1, .), \psi_\omega^2) \\
e_4 &= -\frac{2}{\omega}(b(\lambda_1, .), \psi_0 \psi_\omega^2) = \frac{2}{\omega} d_2 .
\end{aligned}
$$

(As will be seen later, the other coefficients of $p_i$ don't play any role for the normal form of second order.)

The normal form can be constructed similarly as in Case I via the characterization (3.3), (3.4). We obtain

$$
\begin{aligned}
N_1 &\equiv 0 \\
N_2 &= F(\mu, \beta_1, K) \\
N_3 &= \beta_4 G(\mu, \beta_1, K) \\
N_4 &= -\beta_3 G(\mu, \beta_1, K)
\end{aligned}
$$

where $K = \beta_3^2 + \beta_4^2$ and $F$, $G$ are real polynomials in $\underline{\beta}$ of order $\mathcal{O}(\mu|\underline{\beta}| + |\underline{\beta}|^2)$ as $\mu \to 0$, $|\underline{\beta}| \to 0$. Therefore, the reduced system in normal form reads in lowest order

$$
\begin{aligned}
\frac{d}{dx}\beta_1 &= \beta_2 \\
\frac{d}{dx}\beta_2 &= b_{11}\mu\beta_1 + b_{02}\beta_1^2 + dK \\
\frac{d}{dx}\beta_3 &= \beta_4(-\omega + e\mu + f\beta_1) \\
\frac{d}{dx}\beta_4 &= -\beta_3(-\omega + e\mu + f\beta_1)
\end{aligned}
$$

(3.7)

In order to identify the coefficients in (3.7) by those in (3.6), we write

$$
\underline{\alpha} = \underline{\beta} + \mu \underline{P}_{11}(\underline{\beta}) + \underline{P}_{02}(\underline{\beta})
$$

where $\underline{P}_{11}$, $\underline{P}_{02}$ are homogeneous of degree 1 and 2 respectively in $\underline{\beta}$. Similarly we decompose $\underline{N}$ and $\underline{g}^{(1)}$ in (3.6). We obtain

$$
D_{\underline{\beta}} \underline{P}_{jk}(\underline{\beta}) A_c \underline{\beta} - A_c \underline{P}_{jk}(\underline{\beta}) = \underline{g}^{(1)}_{jk}(\underline{\beta}) - \underline{N}_{jk}(\underline{\beta}) ,
$$

$(j, k) = (1, 1)$ or $(0, 2)$, and conclude, by comparing coefficients

$$
b_{11} = a_2 \quad , \quad b_{02} = c_2 \quad , \quad 2d = d_2 \quad , \quad 2e = -b_4 \quad , \quad 2f = -e_4 .
$$

Equations (3.7) have the integrals

$$
\begin{aligned}
\beta_3^2 + \beta_4^2 &= K \\
\beta_2^2 - \hat{F}(\mu, \beta_1, K) &= E
\end{aligned}
$$

(3.8)

where $\partial_{\beta_1}\hat{F} = 2N_2$.

Bounded solutions of (3.7) are periodic, quasiperiodic or homoclinic orbits. The homoclinic solutions connecting a closed orbit with itself (for $K > 0$) have 'oscillatory tails' or 'ripples'. True homoclinic solutions occur for $K = 0$. Persistence of all these solutions can be shown except for the homoclinic solutions with $K = 0$, cf. [4].

For the phase portrait in the case $b_{02}d > 0$ we refer to [4], where the solutions are discussed in detail. If $b_{02}d < 0$ bounded nontrivial solutions exist for $K > 0$ independent of $\mu$, even for $\mu = 0$. The phase portrait in this case is shown in Figure 3 for $b_{11} < 0$, $b_{02} > 0$.

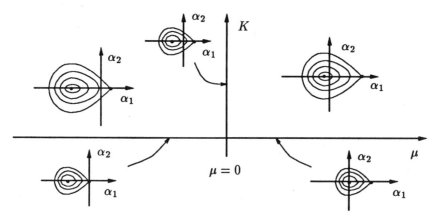

FIGURE 3. Case II: Change of phase portrait (projection onto $(\alpha_1, \alpha_2)$-plane) near $\lambda_1$ as a function of $K$ and $\mu$ if $b_{02}d < 0$

Note that each point in Figure 3 actually corresponds to a periodic motion of amplitude $\sqrt{K}$ in the $(\alpha_3, \alpha_4)$-plane.

We give an explicit description of the homoclinic solutions with ripples: Since

$$\begin{pmatrix} \psi \\ \partial_x \psi \end{pmatrix} = \underline{\Psi}_c + h(\mu, \underline{\Psi}_c) = \sum_j \beta_j \underline{\Phi}_j + \mathcal{O}(\mu|\beta| + |\beta|^2),$$

we obtain, inserting the eigenvectors $\underline{\Phi}_j$ of Case II

$$\psi = \beta_1 \psi_0 + \beta_3 \psi_\omega + \mathcal{O}(\mu|\beta| + |\beta|^2).$$

We calculate the functions $\beta_j$ for the solutions under discussion: Let

$$\tilde{\beta}_1(x) := \epsilon(\beta_1(\delta x) - \beta_1^*),$$

where

$$\beta_1^* = \frac{1}{2b_{02}}(-b_{11}\mu + \sqrt{(b_{11}\mu)^2 - 4b_{02}dK})$$

is the $\beta_1$-coordinate of the saddle point for $b_{11}\mu > 0$, and

$$\delta = \frac{1}{\nu} \quad , \quad \epsilon = -\frac{2}{3}\frac{b_{02}}{\nu^2}$$

with

$$\nu = \left((b_{11}\mu)^2 - 4b_{02}dK\right)^{\frac{1}{4}}.$$

From (3.7) we obtain for $\tilde{\beta}_1$

$$\tilde{\beta}_1''(x) = \tilde{\beta}_1(x) - \frac{3}{2}\tilde{\beta}_1^2(x)$$

and hence

$$\tilde{\beta}_1(x) = \frac{1}{\cosh^2(x/2)} = \operatorname{sech}^2(x/2)$$

Thus we have

$$\begin{aligned}
\beta_1(x) &= \frac{1}{\epsilon}\operatorname{sech}^2(\nu x/2) + \beta_1^*, \\
\beta_2(x) &= \beta_1'(x).
\end{aligned}$$

From (3.7) and (3.8) we obtain for large x

$$\beta_3(x) = \sqrt{K}\cos(\omega - e\mu - f\beta_1^*)x,$$

$$\beta_4(x) = \sqrt{K}\sin(\omega - e\mu - f\beta_1^*)x.$$

Observe that the relation between the amplitudes of the homoclinic and the oscillatory part can be arbitrary. In the case $b_{02}d > 0$ we have the condition

$$K < \frac{1}{4}\frac{(b_{11}\mu)^2}{b_{02}d}$$

but this means no restriction for the amplitude-relation.

## 4. Breaking of Reversibility

In this section the effects are studied which arise when the reversibility of (1.1) is broken. We break this symmetry in the most simple way by adding a term $\epsilon\,p(x)$ to the right side of (1.1), $p \in C_0^\infty(R)$, $\epsilon \in R$. From the variety of solutions which we have found in the previous sections, the homoclinic solution is selected, which exists for $\lambda < \lambda_0$ (see Figure 2). Since this solution corresponds to a homoclinic orbit in the infinite dimensional phase space, the effect of periodic $p$ would suit the time-spirit most. The question whether transverse homoclinic points, and thus spatial chaos exists, leads to problems of exponentially-small transversality conditions, which can be satisfied for large periods. From the point of view of applications the situation, where $p$ has compact support is of higher interest, and we are going to analyse this case here.

We denote by $(1.1)_\epsilon$ the equation (1.1) when $\epsilon\,p(x)$ is added and by $(2.1)_\epsilon$ the corresponding system ($\underline{f}$ is replaced by $\underline{f} + \epsilon\,(0, p(x))^T$). The reduction procedure can be extended to this case as was shown by MIELKE in [9]. The function $h(\mu, \underline{\xi})$ in Proposition 2.1 now has the form $h^0(\mu, \underline{\xi}) + h^1(\mu, \epsilon, x, \underline{\xi})$, where $h^0$ behaves like $h$, in particular, it inherits the reversible structure from (1.1) when $\epsilon = 0$. For small $|\epsilon|$, $h^1 = \mathcal{O}(|\epsilon|)$, uniformly in $x \in R$, and locally uniform in the parameters $\mu = \lambda - \lambda_0$, $\underline{\xi}$.

All solutions of $(2.1)_\epsilon$ of sufficiently small norm in $Y \cap D(A)$ can be found via the reduced system (2.7), in which $h$ has to be replaced by $h^0 + h^1$. The explicit form of (2.7) can be easily derived as in Case I of Section 3. We obtain ($' := \partial_x$)

$$(4.1) \qquad \alpha_1'' - \mu a_{11}\alpha_1 - a_{02}\alpha_1^2 = \epsilon[\psi_0]p(x) + r_0 + \epsilon\, r_1\,,$$

where $r_0(\mu, \alpha_1, \alpha'^2_1)$ is as in (3.1),

$$r_1(\mu, \epsilon, x, \alpha_1, \alpha_1') = \mathcal{O}(\alpha_1 + \alpha_1' + \mu + \epsilon) \quad,\quad [\psi_0] = \int_0^1 \psi_0\,.$$

Observe that $\mu$, $a_{11}$ are both negative. The scaling

$$\alpha_1(x) = \frac{\mu a_{11}}{a_{02}} A(\xi)\,,\ p(x) = P(\xi)\,,\ \xi = (\mu a_{11})^{\frac{1}{2}}x\,,\ \beta = \frac{[\psi_0]a_{02}}{a_{11}^2}$$

then yields

$$(4.2) \qquad A'' - A - A^2 = \frac{\epsilon}{\mu^2}\beta p(x) + \mathcal{O}(\mu + \frac{\epsilon}{\mu} + \frac{\epsilon^2}{\mu^2})\,.$$

To obtain the right limit-equation define the new parameter

$$\eta := \frac{\epsilon|a_{11}|^{\frac{1}{2}}}{|\mu|^{\frac{3}{2}}}\beta <p> \quad,\quad <p> := \int_R pdx$$

and assume $<p> \neq 0$. Then the right side of (4.2) reads

$$\frac{\eta}{(a_{11}\mu)^{\frac{1}{2}} <p>}p = \eta\frac{P}{<P>}\,.$$

Now $P$ concentrates its support near $\xi = 0$, when $\mu \to 0$, $\operatorname{supp}(P) \subset (-c|\mu|^{\frac{1}{2}}, c|\mu|^{\frac{1}{2}})$ for some $c > 0$. Therefore $P/<P>$ approximates the Dirac functional $\delta_0$, and we are left with the limiting equation

$$(4.3) \qquad A'' - A - A^2 = \eta\delta_0\,.$$

The remainder terms have been suppressed since one can show that all solutions of (4.3) we are going to construct can be continued in order $\mathcal{O}(|\mu|^{1/2})$ to solutions of (4.2) and thus, via (2.7) and the function $h^0 + h^1$ to a solution of $(1.1)_\epsilon$, cf. [7] for an adhoc argument and [10] for a general proof that all small solutions of $(1.1)_\epsilon$ can be found by an analysis of (4.3).

Let us finally discuss (4.3). For $\xi \neq 0$ the solutions follow the orbits of $A'' - A - A^2 = 0$. The half-orbits for $\xi < 0$ resp. $\xi > 0$ are connected continuously over $\xi = 0$, whereas the first derivative of $A$ has the jump

$$A'(+0) - A'(-0) = \eta\,.$$

Geometrically, the solutions of (4.3) can therefore be found in the way indicated in Figure 4.

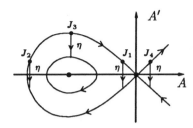

FIGURE 4. Solutions of equation (4.3) with jumps $\eta < 0$ at $J_1, \ldots, J_4$

Of course, there are values of $\eta$ for which no bounded solutions exist. If we restrict our considerations to solutions which decay to 0 as $\xi \to -\infty$, we obtain 3 types of homoclinic solutions. They have their jumps at $J_1$, $J_2$, $J_4$, whereas the solution with jump at $J_3$ has a periodic 'wake'. (Observe that the point where $\xi = 0$ holds, i.e. where the jump occurs, can be defined arbitrarily). There is a continuum of the latter type. For $\eta > 2\sqrt{5}/\sqrt{3}$ no solution exists.

The solutions described above are shown in Figure 5 for some $\eta < 0$. The first two can also be realized by changing $\eta > 0$ to $-\eta$. It is seen, that no.1 corresponds to a bifurcation from infinity. The distance of its minimum from $\xi = 0$ increases with decreasing $\eta$.

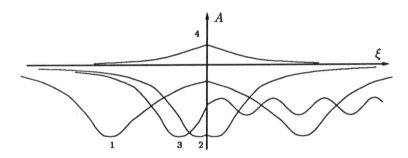

FIGURE 5. The solutions shown in Figure 4 represented in the $(\xi, A)$-plane.

## REFERENCES

1. C. ELPHICK, E. TIRAPEGUI, M.E. BRACHET, P. COULLET, G. IOOSS (1987): A simple global characterization for normal forms of singular vector fields. *Physica D* **29**, 95-127.

2. G. FISCHER (1984): Zentrumsmannigfaltigkeiten bei elliptischen Differentialgleichungen. *Math. Nachr.* **115**, 134-157.

3. G. IOOSS, M. ADELMEYER (1991). *Topics in Bifurcation Theory and Applications*. Advanced Series in Nonlinear Dynamics, Vol.3; World Scientific.

4. G. IOOSS, K. KIRCHGÄSSNER (1992): Water waves for small surface tension - an approach via normal form. *Proceedings of the Royal Society of Edinburgh* **122**A, 267-299.

5. K. KIRCHGÄSSNER (1982): Wave solutions of reversible systems and applications. *J. Differential Equations* **45**, 113-127.

6. K. KIRCHGÄSSNER (1988): Nonlinearly resonant surface waves and homoclinic bifurcation. *Adv. Appl. Mech.* **26**, 135-181.

7. K. KIRCHGÄSSNER (1989): Nonlinear surface waves under external forcing. *Equadiff 7, Proc. 7th Czechoslovak Conf. on Diff. Eq. and Appl.* (ed. J. Kurzweil), Prague, Teubner Texte zur Mathematik.

8. K. KIRCHGAESSNER, K. LANKERS (1992): Structure of permanent waves in density-stratified media. *Meccanica*, to appear.

9. A. MIELKE (1986): A reduction principle for nonautonomous systems in infinite-dimensional spaces. *J. Diff. Equations* **65**, 68-88.

10. A. MIELKE (1986): Steady flows of inviscid fluids under localized perturbations. *J. Diff. Equations* **65**, 89-116.

11. A. MIELKE (1987): Über maximale $L^p$-Regularität für Differentialgleichungen in Banach- und Hilbert-Räumen. *Math. Ann.* **277**, 121-133.

12. A. MIELKE (1988): Reduction of quasilinear elliptic equations in cylindrical domains with applications. *Math. Methods in the Applied Sciences* **10**, 51-66.

MATHEMATISCHES INSTITUT A DER UNIVERSITÄT STUTTGART, PFAFFENWALDRING 57, 7000 STUTTGART 80, GERMANY

*E-mail address*: kirchg@mathematik.uni-stuttgart.de

Lectures in Applied Mathematics
Volume **29**, 1993

# Symmetry in Rotating Plane Couette–Poiseuille Flow

GEORGE H. KNIGHTLY AND D. SATHER

ABSTRACT. The problem of rotating plane Couette-Poiseuille flow, arising from the narrow gap limit of the flow of a viscous, incompressible fluid between concentric, rotating, sliding cylinders with pressure gradients applied, is investigated. A formulation as an equation of the form

$$(*) \qquad \frac{dv}{dt} + L(\lambda, \gamma, \delta, \kappa)v + B(v) = 0$$

is obtained, where $\lambda$ is a Reynolds number, $\gamma$, $\delta$ and $\kappa$ are structure parameters, $L$ is a linear operator and $B$ is a quadratic operator. Under certain restrictions on $(\gamma, \delta, \kappa)$ it is shown that Hopf bifurcation occurs, leading to disturbance flows in the form of periodic waves that are supercritical and asymptotically stable.

## 1. Introduction

The investigation of fluid flow between rotating cylinders often leads to problems involving symmetry and equivariance in an essential way. Here we consider rotating plane Couette-Poiseuille flow (RPCPF), the narrow gap limit of the case in which a viscous incompressible fluid occupies the annular region between rigid, concentric, rotating, sliding cylinders with axial and circumferential pressure gradients applied. In particular, we investigate Hopf bifurcation for RPCPF in its setting as an abstract problem

$$(*) \qquad \frac{dv}{dt} + L(\lambda, \gamma, \delta, \kappa)v + B(v) = 0, \quad v(t) \in \mathcal{H},$$

where $\mathcal{H}$ is a Hilbert space, $L$ is a linear operator, the nonlinear operator $B$ is quadratic and $\lambda, \gamma, \delta$ and $\kappa$ are real parameters. Certain aspects of steady

1991 *Mathematics Subject Classification.* Primary 76E30, 35Q10.
The first author was supported in part by ONR Grant N00014–90–J–1031.
The second author was supported in part by ONR Grant N00014–91–J–4037.
This paper is in final form and no version of it will be submitted for publication elsewhere.

bifurcation for RPCPF are treated in [8;9] and the special case of rotating plane
Couette flow is considered in [10] .

There are two principle ways in which this paper departs from standard treatments of Hopf bifurcation for Navier-Stokes problems (e.g., see [3;4;6;13] ); (i)
the dependence of $L$ on the structural parameters, $\gamma, \delta, \kappa$, and (ii) the calculation
of expansion coefficients necessary to demonstrate the supercritical nature of the
bifurcation and the stability of the bifurcating solution. The parameter issue is
handled by requiring $\delta$ and $\kappa$ to be dominated by $\gamma$ and perturbing the linear
problem in $\gamma$ near $\gamma = 0$. (Since $\gamma$ is proportional to the Couette forces and $\delta$
and $\kappa$ are proportional to the Poiseuille forces, this is in agreement with physical
problems in which the Poiseuille forces may be assumed small relative to the
Couette forces.) The matter of calculating the crucial expansion coefficients is
also facilitated by the use of $\gamma$ as a perturbation parameter and further relies on
our ability to compute certain "cubic" integrals involving the nonlinear term $B$.
Making use of the above ideas, for each *fixed* $\gamma$ near $\gamma = 0$ we determine a unique
branch of periodic waves bifurcating supercritically from $v = 0$ at some critical
value $\lambda_c(\gamma, \delta, \kappa)$. Together with the work in [10] the developments described here
represent the first Hopf bifurcation and stability results for spiral flows (see the
remarks on p. 161 of [7]).

The operator formulation (*) is developed and properties of the various operators in (*) are summarized in §2. In §3 we outline the approach for Hopf
bifurcation and establish the existence of supercritical periodic orbits of (*). Using the equivariance of the governing partial differential equations we find that
the bifurcating orbits are actually periodic wave solutions.

## 2. Rotating Plane Couette-Poiseuille Flow

In this section we present a system of partial differential equations and boundary conditions governing RPCPF, recast the problem in the form (*) and give
some properties of the operators in (*).

RPCPF is obtained (e.g., see [1;9] and ch.VI of [7]) as a narrow gap limit
from the problem of the flow of a viscous fluid between concentric rotating cylinders when the inner cylinder also translates in the axial direction and an axial
pressure gradient $P_x$ and circumferential pressure gradient $P_\theta$ are applied. In
the limiting case the flow occurs between parallel plates. In suitably scaled rectangular coordinates (x,y,z) moving with the upper plate, the walls lie at $z = \pm\frac{1}{2}$
and a basic laminar flow with velocity $U = (U_1, U_2, 0)^\top$ is known, where

$$U_1 = U_P(\frac{1}{4} - z^2) + U_C(\frac{1}{2} - z), \quad U_2 = V_P(\frac{1}{4} - z^2) + V_C(\frac{1}{2} - z),$$

$$U_P = \hat{U}_P/\hat{U}, \quad U_C = \hat{U}_C/\hat{U}, \quad V_P = \hat{V}_P/\hat{U}, \quad V_C = \hat{V}_C/\hat{U},$$

$$\hat{U} = [\hat{U}_P^2 + \hat{U}_C^2 + \hat{V}_P^2 + \hat{V}_C^2]^{\frac{1}{2}},$$

a subscript $P$ denotes pressure-driven (Poiseuille) terms, a subscript $C$ denotes
boundary-driven (Couette) terms and $^\top$ denotes transpose. Here $\hat{U}_C$ is the

constant axial speed of the inner cylinder, $\hat{V}_C = r_2(\Omega_1 - \Omega_2)$ is a corresponding circumferential speed, $\hat{U}_P = P_x(r_2^2 - r_1^2)^2/2\nu r_2^2$, $\hat{V}_P = P_\theta(r_2 - r_1)^2/2\nu r_2$, the inner cylinder has radius $r_1$ and angular velocity $\Omega_1$, the outer cylinder has radius $r_2$ and angular velocity $\Omega_2$ and $\nu$ denotes kinematic viscosity.

If the equations (7) in [9] (see also [1;7]) for RCPCF are adjusted so that $-\frac{1}{2} < z < \frac{1}{2}$, then the disturbance velocity vector $v$ and pressure $q$ satisfy

(2.1)
$$\begin{cases} \dfrac{\partial v}{\partial t} - \Delta v + \tilde{\lambda}\tilde{L}v + v \cdot \nabla v + \nabla q = 0, \quad -\dfrac{1}{2} < z < \dfrac{1}{2}, \ y, t \in \mathbb{R} \\[2mm] \nabla \cdot v = 0 \\[2mm] v = 0 \quad \text{on} \quad z = \pm\dfrac{1}{2}. \end{cases}$$

Here $\nabla$ is the $(x, y, z)$-gradient, $\Delta = \nabla^2$,

(2.2)
$$\tilde{L} = S \begin{pmatrix} 0 & 0 & 0 \\ 0 & 0 & 1 \\ 0 & -1 & 0 \end{pmatrix} + U_1(z)\frac{\partial}{\partial x} + U_2(z)\frac{\partial}{\partial y} + \begin{pmatrix} 0 & 0 & U_1' \\ 0 & 0 & U_2' \\ 0 & 0 & 0 \end{pmatrix},$$

$S = 2(r_2 - r_1)\Omega_2/\hat{U}$, and $\tilde{\lambda} = \hat{U}(r_2 - r_1)/\nu$ is the Reynolds number.

The *basic spiral flow angle*, $\chi_0 = \chi_0(z)$, the angle the basic flow velocity $U$ makes with the $x$−axis is given by

$$\tan\chi_0 = U_2(z)/U_1(z).$$

In this paper we shall restrict our attention to the case in which the vectors $(V_P, V_C)$ and $(U_P, U_C)$ are parallel. Then $\chi_0$ is independent of $z$ and $\tan\chi_0 = V_C/U_C = V_P/U_P$.

In experimental work (e.g., see [11;12] , also ch.VI of [7]) secondary states also possessing a distinguished spiral direction seem to be observed. Thus, we seek disturbances $(v, q)$ having a constant *disturbance spiral flow angle* $\chi$ such that $(v, q)$ are independent of the coordinate in the direction $\chi$. If one rotates the $(x, y)$-coordinates so that the $x$-axis coincides with the $\chi$-direction, then $v = v(y, z, t)$ and $q = q(y, z, t)$ are independent of $x$. Proceeding as in [9, p.83] we obtain from (2.1) a system of the same form but with all quantities independent of $x$ and with the linear operator $\tilde{L}$ now given by

$$\tilde{L} = \sqrt{F}[\tilde{L}_1 + \gamma\tilde{L}_2 + \delta\tilde{L}_3 + \gamma\kappa\tilde{L}_4].$$

Here

(2.3a)
$$\tilde{L}_1 = -\begin{pmatrix} 0 & 0 & 1 \\ 0 & 0 & 0 \\ 1 & 0 & 0 \end{pmatrix}, \quad \tilde{L}_2 = \left(z - \frac{1}{2}\right)\frac{\partial}{\partial y} + \begin{pmatrix} 0 & 0 & 0 \\ 0 & 0 & 1 \\ 0 & 0 & 0 \end{pmatrix},$$

(2.3b)
$$\tilde{L}_3 = -2z\begin{pmatrix} 0 & 0 & 1 \\ 0 & 0 & 0 \\ 0 & 0 & 0 \end{pmatrix}, \quad \tilde{L}_4 = \left(z^2 - \frac{1}{4}\right)\frac{\partial}{\partial y} + 2z\begin{pmatrix} 0 & 0 & 0 \\ 0 & 0 & 1 \\ 0 & 0 & 0 \end{pmatrix},$$

(2.3c)   $\gamma = U_C \sin(\chi - \chi_0)/\cos\chi_0\sqrt{F}$,   $\delta = S\sin\chi U_P \cos(\chi - \chi_0)/\cos\chi_0\sqrt{F}$,

(2.3d)        $\kappa = U_P/U_C$,    $F = S\sin\chi[\dfrac{U_C\cos(\chi - \chi_0)}{\cos\chi_0} - S\sin\chi]$.

We assume $S$, $U_C$, $\chi$ and $\chi_0$ are such that $F > 0$. Note that $\gamma$ is a measure of $\chi - \chi_0$ and the relative magnitude of the Couette forces, $\delta$ and $\kappa$ determine the relative magnitude of the Poiseuille forces and $F$ is the Rayleigh discriminant.

To recast problem (2.1) in the form $(*)$ we seek states $v, q$ that are $2\pi/a$-periodic in $y$, with wave number $a$ to be determined. Thus, we consider $v, q$ defined in $G = \{(y, z) : y\epsilon\mathbb{R}/\frac{2\pi}{a}\mathbb{Z}, -\frac{1}{2} < z < \frac{1}{2}\}$ and introduce the following spaces:

$C_{0,\sigma}^\infty(G) = \{v \in C^\infty(G) : \nabla \cdot v = 0,\ v \equiv 0 \text{ near } z = \pm\frac{1}{2}\}$,
$\mathcal{H} = $ closure of $C_{0,\sigma}^\infty(G)$ in the norm $\|\cdot\|$ associated with the inner product

$$(u, v) = \int_{-\frac{1}{2}}^{\frac{1}{2}} \int_0^{2\pi/a} u \cdot \bar{v} \, dy \, dz,$$

$H_m(G)$ is the Sobolev space of vector fields $u$ on $G$ for which $\|u\|_{H_m(G)}$ is finite, where

$$\|u\|_{H_m(G)} = \sum_{\substack{0 \le j+k \le m \\ j \ge 0, k \ge 0}} \left[\left\|\frac{\partial^{j+k}}{\partial y^j \partial z^k} u\right\|^2\right]^{\frac{1}{2}},$$

$\overset{\circ}{\mathcal{H}}_{1,\sigma}(G) = $ closure of $C_{0,\sigma}^\infty(G)$ in $H_1(G)$,
$\mathcal{D} = \overset{\circ}{\mathcal{H}}_{1,\sigma}(G) \cap H_2(G)$, $\|\cdot\|_{\mathcal{D}} = \|\cdot\|_{H_2(G)}$,
$\mathcal{K} = \mathcal{H} \cap H_1(G)$, $\|\cdot\|_{\mathcal{K}} = \|\cdot\|_{H_1(G)}$.

The spaces $\mathcal{H}, \mathcal{D}$, and $\mathcal{K}$ are essentially those utilized in [3;4]. In particular the embeddings $\mathcal{D} \hookrightarrow \mathcal{K} \hookrightarrow \mathcal{H}$ are compact.

Letting $\Pi$ denote the orthogonal projection onto $\mathcal{H}$ in $\mathbb{L}^2(G) \equiv H_0(G)$ and using $\Pi$ to write (2.1) as an equation on $\mathcal{H}$, we find that $v$ must satisfy $(*)$ with

$$L(\lambda, \gamma, \delta, \kappa) = L_0 + (\lambda - \lambda_0)L_1 + \lambda\gamma L_2 + \lambda\delta L_3 + \lambda\gamma\kappa L_4,$$
$$B(v) = \Phi(v, v), \quad \Phi(u, v) = \Pi u \cdot \nabla v,$$

where $\lambda = \tilde{\lambda}\sqrt{F}$ and

(2.4)   $\begin{cases} L_0 = -\Pi\triangle + \lambda_0 L_1,\ L_1 = \Pi\tilde{L}_1,\ L_2 = \Pi\tilde{L}_2 = M_1 + M_2, \\[2mm] M_1 = -\dfrac{1}{2}\Pi\dfrac{\partial}{\partial y},\ M_2 = \Pi\left(z\dfrac{\partial}{\partial y} + \begin{pmatrix} 0 & 0 & 0 \\ 0 & 0 & 1 \\ 0 & 0 & 0 \end{pmatrix}\right) \\[4mm] L_3 = \Pi\tilde{L}_3,\ L_4 = \Pi\tilde{L}_4. \end{cases}$

Here the domains of $L_0$, $M_1$, $M_2$, $L_3$, and $L_4$ are $\mathcal{D}$, the domain of $L_1$ is $\mathcal{H}$ and the domain of $\Phi$ is $\mathcal{D} \times \mathcal{D}$.

We specify $\lambda_0$ as the smallest positive number for which $L_0$ has zero as an eigenvalue. The problem of determining $\lambda$ such that $L_0$ has eigenvalue zero is the classical linear Bénard problem and is well studied (e.g., see [1, ch. 2]). It requires the solution of the linear eigenvalue problem

(2.5)
$$\begin{cases} \lambda \tilde{L}_1 v = -\nabla q + \Delta v, \\ \nabla \cdot v = 0, \\ v = 0 \text{ on } z = \pm \frac{1}{2}, \quad v \text{ periodic in } y \text{ with period } 2\pi/a. \end{cases}$$

The eigenfunctions are obtained from the relations

(2.6)
$$\begin{cases} v = e^{iay}(\phi_1(z), \phi_2(z), \phi_3(z))^\top, \\ q = e^{iay}a^{-2}D^2\phi_3'(z), \\ \phi_2 = ia^{-1}\phi_3', \end{cases}$$

where $D^2 = \dfrac{d^2}{dz^2} - a^2$, a prime denotes $\dfrac{d}{dz}$, and $\phi_1, \phi_3$ satisfy

(2.7)
$$\begin{cases} D^4\phi_3 - \mu a^2 \phi_1 = 0, \\ D^2\phi_1 + \mu\phi_3 = 0, \\ \phi_1 = \phi_3' = \phi_3 = 0 \quad \text{at} \quad z = \pm\frac{1}{2}. \end{cases}$$

One can show for $a > 0$ (e.g., see [5]) that the eigenvalue problem (2.7) has a countable number of positive simple eigenvalues, $0 < \mu_1(a) < \mu_2(a) < \cdots$, depending continuously on $a$. Moreover, $\mu_1(a) \to \infty$ as either $a \to 0^+$ or $a \to \infty$. Calculations (see [1, fig. 2.2] strongly suggest $\mu_1(a)$ takes its minimum at a unique $a > 0$; we assume this is the case, choose $a$ to be the minimizing value and set $\lambda_0 = \mu_1(a)$. Then $\sigma = 0$ is an eigenvalue of $L_0$ of multiplicity two with nullspace $\mathcal{N}(L_0)$ spanned by $\psi_0 = e^{iay}\varphi(z)$ and $\overline{\psi}_0$, where

(2.8)
$$\begin{cases} \varphi = (\varphi_1, \varphi_2, \varphi_3)^\top, \varphi_2(z) = (i/a)\varphi_3'(z) \quad \text{is odd in } z, \\ \varphi_1(z), \varphi_3(z) \quad \text{are real and even in } z. \end{cases}$$

In the following lemma we collect some properties of the operators in (2.4). Here $\mathcal{N}^\perp$ denotes the orthogonal complement of $\mathcal{N}$ in $\mathcal{H}$, $K$ denotes the inverse of the restriction of $L_0$ to $\mathcal{N}^\perp$, and $\mathcal{L}(X, Y)$ denotes the Banach space of bounded linear operators from a Banach space $X$ to a Banach space $Y$.

LEMMA 2.1. *(i)* $L_0 \in \mathcal{L}(\mathcal{D}, \mathcal{H})$ *is nonnegative, selfadjoint and has nullspace* $\mathcal{N} = \mathcal{N}(L_0)$ *spanned by* $\psi_0, \overline{\psi}_0$. *The norm* $\|u\|_\mathcal{D}$ *on* $\mathcal{D}$ *is equivalent to* $[\|u\|^2 + \|L_0 u\|^2]^{1/2}$.
*(ii)* $L_1 \in \mathcal{L}(\mathcal{D}, \mathcal{K})$ *is real, selfadjoint and satisfies* $(L_1\psi_0, \overline{\psi}_0) = (L_1\overline{\psi}_0, \psi_0) = 0$ *and*

$$d_1 \equiv (L_1\psi_0, \psi_0) = -\lambda_0^{-1}\|\nabla\psi_0\|^2 < 0.$$

*(iii)* $M_1, M_2, L_3$ *and* $L_4$ *are real operators belonging to* $\mathcal{L}(\mathcal{D}, \mathcal{K})$. *Moreover,* $M_1 : \mathcal{N} \to \mathcal{N}$, $M_1 : \mathcal{D} \cap \mathcal{N}^\perp \to \mathcal{N}^\perp$, $M_2 : \mathcal{N} \to \mathcal{N}^\perp$, $L_3 : \mathcal{N} \to \mathcal{N}^\perp$,

$M_1\psi_0 = -i(a/2)\psi_0$ *and the adjoint operator,* $M_1^*$, *satisfies* $M_1^* = -M_1$.
*(iv) The bilinear operator,* $\Phi$, *is real, satisfying* $\Phi : \mathcal{D} \times \mathcal{D} \to \mathcal{K}$ *and*

$$\|\Phi(u,v)\|_\mathcal{K} \le C\|u\|_\mathcal{D}\|v\|_\mathcal{D}, \quad u,v \in \mathcal{D}, \ some \ C > 0.$$

*Furthermore,*

$$(2.9) \qquad (\Phi(u,v),w) = -(\Phi(u,\bar{w}),\bar{v}), \quad u,v,w \in \mathcal{D}$$

$$(2.10) \qquad\qquad\qquad \Phi : \mathcal{N} \times \mathcal{N} \to \mathcal{N}^\perp.$$

*(v)* $K \in \mathcal{L}(\mathcal{N}^\perp, \mathcal{D} \cap \mathcal{N}^\perp)$.

## 3. Bifurcating Periodic Waves

In this section we describe our approach to the solution of problem $(*)$, and formulate our theorem on bifurcating waves for RPCPF. In essence the method is the one for Hopf bifurcation developed in [10] (see also [3; 4; 6; 13]). One establishes, for suitable values of the parameters $\gamma, \delta, \kappa$, that there is a critical value $\lambda = \lambda_c(\gamma, \delta, \kappa)$ for which $L(\lambda_c(\gamma, \delta, \kappa), \gamma, \delta, \kappa)$ has spectrum in the left half plane except for a distinguished pair, $\zeta = \zeta(\lambda, \gamma, \delta, \kappa)$ and $\bar{\zeta}$, of complex conjugate eigenvalues on the imaginary axis. As $\lambda$ increases through $\lambda_c$ the real part of $\zeta$ passes through zero.

For $\lambda$ near $\lambda_c(\gamma, \delta, \kappa)$ one looks for a solution of $(*)$ in the form $v = u + w$ with $u = Ew$ and $v = Pw$, where $E = E(\lambda, \gamma, \delta, \kappa)$ is the projection of $\mathcal{H}$ onto the subspace spanned by the eigenfunctions $\{\psi(\lambda, \gamma, \delta, \kappa), \bar{\psi}(\lambda, \gamma, \delta, \kappa)\}$ of $L(\lambda, \gamma, \delta, \kappa)$ corresponding to $(\zeta, \bar{\zeta})$ and $P = P(\lambda, \gamma, \delta, \kappa)$ is the projection corresponding to the rest of the spectrum of $L(\lambda, \gamma, \delta, \kappa)$. One obtains $w = w(u, \lambda, \gamma, \delta, \kappa)$, e.g., by a contraction argument for the projection of $(*)$ on $P\mathcal{H}$. This leads to a problem of the form

$$(3.1) \qquad \frac{du}{dt} + L(\lambda, \gamma, \delta, \kappa)u = f(u), \quad u(t) \in E\mathcal{H}.$$

One now splits $u = \psi + U$, with $\psi$ in the nullspace $\mathcal{M}$ of the operator on the left in (3.1), and solves (3.1) for $U$, again by contraction, for small $\psi$ provided $f(u)$ has zero projection onto $\mathcal{M}$. The latter condition, the bifurcation equation, is then solved for $\psi$ and the period $T$. The proof that the bifurcating orbits thus obtained are supercritical and stable in $\mathcal{H}$ is ultimately accomplished by demonstrating the positivity of the real part of a sum of six inner products, such as $(\Phi(\psi_0, K\Phi(\psi_0, \bar{\psi}_0)), \psi_0)$, obtained from cubic terms in $u$ arising in the solution of the bifurcation equation (see (4.4) in [10]). As a consequence of the equivariance of the underlying system (2.1) in both $t$ and $y$ one finds (e.g., see §IV.3.3 of [6]) that the bifurcating states are periodic waves of the form $v = v(ay + \omega t, z)$, $2\pi$-periodic in $ay + \omega t$ with $\omega = 2\pi/T$.

A key step in the above procedure is the derivation of the required properties of $L(\lambda, \gamma, \delta, \kappa)$ based on our knowledge of $L(\lambda, 0, 0, 0)$. One can accomplish this, e.g., by setting $\delta = m_1\gamma$, $\kappa = m_2\gamma$ for fixed $m_j$, $j = 1, 2$, and using $\gamma$ as an expansion parameter. Let $L(\lambda, \gamma) \equiv L(\lambda, \gamma, m_1\gamma, m_2\gamma)$ and observe that the terms involving $L_4$ are then of order $\gamma^2$ and, by Lemma 2.1, $M_2 + m_1 L_3$ maps $\mathcal{N}$ into $\mathcal{N}^\perp$. One can now use the perturbation method developed in Lemma 3.1 of [10] to investigate properties of $L(\lambda, \gamma)$. I.e., one seeks solutions of

$$(3.2) \qquad L(\lambda, \gamma)v = \zeta v$$

in the form

$$(3.3a) \qquad v = \gamma(u + \gamma w), \quad u \in \mathcal{N}, \quad w \in \mathcal{N}^\perp,$$

$$(3.3b) \qquad \zeta = \gamma\mu = \gamma(\xi + i\eta),$$

$$(3.3c) \qquad \lambda = \lambda_0 + \gamma\tau.$$

Substituting (3.3) into (3.2) and employing splitting methods one reduces this linear problem to one of solving a linear homogeneous $2 \times 2$ system. The solvability condition for this system at $\gamma = 0$ is $0 \neq (L_1\psi_0, \psi_0)$, which is satisfied because of part (ii) of Lemma 2.1.

Proceeding in the way outlined above, we obtain the following result.

THEOREM 3.1. *Given $m_0 > 0$ there is a $\gamma_0 > 0$ such that if $0 < |\gamma| < \gamma_0$ and $|m_j| < m_0$, $j = 1, 2$, then for $\delta = m_1\gamma$, $\kappa = m_2\gamma$, problem $(*)$ for RPCPF has a branch $v(\lambda, \gamma, \delta, \kappa)$ of nontrivial periodic orbits bifurcating supercritically from $v = 0$ at $\lambda = \lambda_c(\gamma, \delta, \kappa)$. The orbits have period $T = T(\lambda, \gamma, \delta, \kappa)$ and are asymptotically stable in $\mathcal{H}$. Each periodic orbit determines a wave of the form $v = v(ay + \omega t, z)$, $2\pi$-periodic in $ay + \omega t$ with frequency $\omega = 2\pi/T$ in time.*

**Remark 3.2.** Up to terms of order $|\lambda - \lambda_0| + \gamma^2$ the frequency $\omega$ is $\tilde{\lambda}_0 \frac{a}{2}\sin(\chi - \chi_0)$, where $\tilde{\lambda}_0 = \lambda_0/\sqrt{F}$. For dimensionless time, $t\tilde{\lambda}_0$, the frequency is, to lowest order, $\frac{a}{2}\sin(\chi - \chi_0)$, in agreement with the findings of Hung, Joseph and Munson in their formal stability analysis [2] of spiral flows.

## REFERENCES

1. S. Chandrasekhar, *Hydrodynamic and Hydromagnetic Stability*, Clarendon Press, Oxford, 1961.

2. W. L. Hung, D. D. Joseph, and B. R. Munson, *Global stability of spiral flow, Part II*, J. Fluid Mech. **51** (1972), 593–612.

3. G. Iooss, *Existence et stabilité de la solution périodique secondaire intervenant dans les problèmes d'évolution du type Navier–Stokes*, Arch. Rational Mech. Anal. **47** (1972), 301–329.

4. G. Iooss, *Bifurcation and transition to turbulence in hydrodynamics*, Bifurcation Theory and Applications, (L. Salvadori, ed.), Lecture Notes in Mathematics **1057** Springer–Verlag, New York (1984), 152–201.

5. V. I. Iudovich, *On the origin of convection*, J. Appl. Math. Mech. **30** (1969), 1193–1199.

6. V. I. Iudovich,, *The onset of auto-oscillations in a fluid*, J. Appl. Math. Mech. **35** (1971), 587–603.

7. D. D. Joseph, *Stability of Fluid Motions I*, Springer Tracts in Natural Philosophy **28** Springer–Verlag, New York (1976).

8. G. H. Knightly and D. Sather, *Structure parameters in rotating Couette-Poiseuille channel flow*, Rocky Mountain J. Math. **18** (1988), 339–355.

9. G. H. Knightly and D. Sather, *Bifurcation and stability problems in rotating plane Couette–Poiseuille flow*, Contemporary Math. **108** (1990), 79–91.

10. G. H. Knightly and D. Sather, *Periodic waves in rotating plane Couette flow*, ZAMP (to appear).

11. H. Ludwieg, *Experimentelle Nachprüfung der Stabilitätstheorien für reibungsfreie Strömungen mit schrauben-linienförmigen Stromlinien*, Z. Flugwiss **12** (1964), 304–309.

12. H. M. Nagib, *On instabilities and secondary motions in swirling flows through annuli*, Ph.D. dissertation, Illinois Institute of Technology, 1972.

13. D. H. Sattinger, *Topics in stability and bifurcation theory*, Lecture Notes in Math **309** Springer–Verlag, New York, 1973.

DEPARTMENT OF MATHEMATICS, UNIVERSITY OF MASSACHUSETTS, AMHERST, MA 01003

*E-mail address*: knightly@math.umass.edu

DEPARTMENT OF MATHEMATICS, UNIVERSITY OF COLORADO, BOULDER, CO 80309–0395

Lectures in Applied Mathematics
Volume **29**, 1993

# On uniformly rotating fluid drops
# trapped between two parallel plates

H.-P. KRUSE     J.E. MARSDEN     J. SCHEURLE

ABSTRACT. This contribution is about the dynamics of a liquid bridge between two fixed parallel plates. We consider a mathematical model and present some results from the doctorial thesis [10] of the first author. He showed that there is a Poisson bracket and a corresponding Hamiltonian, so that the model equations are in Hamiltonian form. The result generalizes previous results of Lewis et al. [12] on the dynamics of free boundary problems for "free" liquid drops to the case of a drop between two parallel plates, including, especially the effect of capillarity and the angle of contact between the plates and the free fluid surface. Also, we prove the existence of special solutions which represent uniformly rotating fluid bridges, and we present specific stability conditions for these solutions. These results extend work of Concus and Finn [2] and Vogel [18],[19] on static capillarity problems (see also Finn [5]). Using the Hamiltonian structure of the model equations and symmetries of the solutions, the stability conditions can be derived in a systematic way. The ideas that are desribed will be useful for other situations involving capillarity and free boundary problems as well.

## 1   Introduction

We consider the motion of an ideal, i.e., incompressible and inviscid fluid of finite volume between two parallel flat plates. The plates are assumed to be at rest. We only take into account surface tension along the free surface of the fluid and adhesion forces along the surfaces of contact between the fluid and the plates. The influence of other forces such as gravity is neglected (cf. Concus and Finn [3]). Also, the complete separation of the fluid from one of the plates

1991 Mathematics Subject Classification.Primary 76B45, 76E05, 58F05, 58F10.
Partially supported by a Humboldt award at the University of Hamburg and by DOE Contract DE-FGO3-88ER25064.
This paper is in final form and no version of it will be submitted for publication elsewhere.

will be excluded, i.e., we assume that the fluid always forms a bridge between the two plates.

In the theory of capillary phenomena, the boundary conditions that people use for the angle of contact of the liquid with the wall is still a point of controversy if the liquid is not at rest. In the latter case, Gauss [6] gave a justification of the particular boundary conditions that people use, based on the principle of virtual work. In the present situation, we argue that there is a distinguished natural choice for these boundary conditions dictated by the Hamiltonian structure. They can be derived as part of the variational principle underlying the Hamiltonian equations. In particular, this gives some justification for the assumption of a constant contact angle even in the dynamic case.

A standard way to find Hamiltonian structures for mechanical problems is to pass from the material to the spatial representation and derive the reduced Poisson structure (see e.g. Marsden et al. [15], [16]). With regard to this Poisson structure, the equations of motion read as follows

$$(1.1) \qquad\qquad \dot{F} = \{\, F, H \,\} \quad \text{for all } F \in \mathcal{D}.$$

Here $\{\cdot, \cdot\}$ denotes the Poisson bracket on the reduced phase space $\mathcal{N}$, $\mathcal{D}$ is the class of smooth real-valued functions on $\mathcal{N}$, $\dot{F}$ denotes the derivative of $F$ along solution curves, and $H \in \mathcal{D}$ is the corresponding Hamiltonian which describes the total energy. For fluid flow problems, this method dates back to Arnold [1] who considered pure rigid wall boundary conditions (cf. also Marsden and Weinstein [17]). Lewis et al. [12] have applied this method to a free boundary value problem in fluid dynamics for the first time. But this problem does not involve capillarity. The chief difficulty with free boundary value problems is the treatment of the boundary conditions. The situation is subtle, because the free boundary must be included as a dynamic variable. Thus, the Poisson bracket picks up boundary terms and this makes the interaction with the remaining terms subtle in terms of questions like the Jacobi identity. Because of this subtleties, one cannot just quote general reduction theory here. Rather one can only use this as a guide. For a dynamic problem involving capillarity, free surfaces and the contact angle as in the present case, things are even more subtle, because the curves of contact between the free fluid surface and the plates must be included amongst the dynamic variables in addition. For a derivation, along these lines, of the Hamiltonian structure which we are going to describe here, we refer to [10].

Uniformly rotating fluid drops are relative equilibrium solutions, in the shape of surfaces of revolution around an axis perpendicular to the plates. Our representation of these drop shapes in terms of Delauny curves determines them precisely for sufficiently small angular velocities. To analyze their stability, an energy method appropriate for rotating systems is used. If one restricts the stability analysis to rotationally symmetric initial perturbations of the given drop, then the definiteness of the relevant quadratic form can be determined using Sturmian theory which leads to specific stability criteria. In such a stability analysis, one has to be careful about the choice of a potential function, as there are several candidate potential energy functions that give the relative equilibria in terms of a variational principle. These make a difference, since the stability conditions need not be optimal if one does not make the best choice. The choice is also important for locating bifurcation points correctly (cf. Lewis [11], [13]). We indeed obtain the optimal stability conditions. Our choice of the potential function is suggested by the so-called reduced energy-momentum method which is a general method to test relative equilibria of Hamiltonian systems with rotational symmetries for their (nonlinear) stability (see e.g. Marsden [14, §5]). However, we also have to take into account the volume constraint since the fluid is assumed to be incompressible. The fact that the Hamiltonian system is infinite-dimensional in the present situation makes a rigorous stability analysis subtle in terms of questions like general existence and uniqueness of solutions (cf. Kröner [9]) or consistency of various topologies which are involved. We do not address such issues here. Rather we use a formal notion of stability (cf. Holm et al. [7]).

For static drops, some results along these lines were already known before (see Concus and Finn [2], and Vogel [18], [19]). For numerical results with rotation and even with gravity included, see Hornung and Mittelmann [8].

The following part of this paper is organized as follows. In section 2, we first state and discuss the basic equations of motion in conventional form. Then we describe their Hamiltonian structure, i.e., the corresponding Hamiltonian and Poisson bracket. In section 3, we outline the construction of uniformly rotating drop solutions by means of Delauny curves and formulate the stability criteria. As a simple example, we consider cylindrical drops.

## 2   The equations of motion

We assume that at any instant of time $t$, the free surface of the fluid drop is given as the graph of a real-valued function

$$(2.1) \qquad r = \Sigma(\varphi, z; t) \qquad (r > 0,\ 0 \le \varphi < 2\pi,\ 0 \le z \le h),$$

where $r, \varphi, z$ are cylindrical coordinates in the Euclidean 3-space with origin at one plate and the $z$-axis perpendicular to the plates; $h$ denotes the distance of the two plates. Note that this assumption excludes the complete separation of the fluid drop from one of the plates. Then we have the following equations for spatial representations

$$v = v(r, \varphi, z; t), \quad p = p(r, \varphi, z; t), \quad \Sigma = \Sigma(\varphi, z; t)$$

of the velocity field, the pressure field and the free surface of the fluid:

$$(2.2) \qquad
\begin{array}{lll}
\frac{\partial v}{\partial t} + (v \cdot \nabla) v = -\nabla p & \text{in} & D_\Sigma \\[4pt]
\nabla \cdot v = 0 & \text{in} & D_\Sigma \\[4pt]
\frac{\partial \Sigma}{\partial t} = \frac{v \cdot n}{e_r \cdot n} & \text{on} & r = \Sigma \\[4pt]
p = \tau \kappa & \text{on} & r = \Sigma \\[4pt]
v \cdot n = 0 & \text{on} & \Sigma_1 \cup \Sigma_2 \\[4pt]
\cos \gamma_j = \frac{\sigma_j}{\tau} & \text{on} & c_j \quad (j = 1, 2)
\end{array}$$

Here, $\nabla$ is the Nabla operator; at any instant of time, $D_\Sigma$ denotes the region between the plates which is occupied by the fluid; for $j = 1$ and $j = 2$, $\Sigma_j$ is the region in the $j$-th plate $P_j$ which is wetted by the fluid, and $c_j$ denotes the boundary curve of $\Sigma_j$, i.e., the curve of intersection of the free surface $r = \Sigma$ with the plate $P_j$. The outer unit normal with respect to $D_\Sigma$ is always denoted by $n$. The first two equations are just Euler's equations for ideal fluids, i.e., the balance equation for linear momentum and the continuity equation which models incompressibility of the fluid. The third equation constitutes a kinematic condition for the evolution of the free surface. Here and subsequently, $e_r$ denotes the unit vector in radial direction. Along the free surface, the pressure is supposed to be balanced by surface tension. In the fourth equation, $\kappa$ denotes the mean curvature of the free surface, and $\tau > 0$ is the material constant of surface tension. As usual, along the rigid walls, we assume slip boundary conditions given by the fifth equation. Finally, $\gamma_j$ denotes the angle of contact of the fluid with the plate $P_j$, i.e., the angle between the outer unit normal $n_j$ of $\Sigma_j$ inside $P_j$ and the outer unit conormal of the free surface $r = \Sigma$ along the curve $c_j$. It is assumed to be constant and given by the sixth equation in (2.2) where $\sigma_j$ denotes the adhesion coefficient with respect to the plate $P_j$, which is another material constant. For simplicity, we have set the fluid density equal to one here.

As indicated in section 1, these equations are Hamiltonian in the sense of mechanics on Poisson manifolds. The *Hamiltonian function* is given by

$$(2.3) \quad H(\Sigma, v) \;=\; \frac{1}{2}\int_{D_\Sigma} \|v\|^2\, dV \;+\; \tau \int_\Sigma dA \;-\; \sum_{j=1,2} \sigma_j \int_{\Sigma_j} dA\,,$$

where the volume integral describes the kinetic energy of the fluid drop, $\|\cdot\|$ is the Euclidean norm, and the surface integrals describe the potential energies. The dynamic variables are the free boundary $\Sigma$ and the spatial velocity field $v$, a divergence-free smooth vector field in the region $D_\Sigma$ bounded by $\Sigma$ and the plates $P_j$. Also, $v$ is supposed to satisfy the slip boundary condition along $\Sigma_1$ and $\Sigma_2$. The surface $\Sigma$ is represented by a sufficiently smooth function $\Sigma(\varphi, z)$ as in (2.1). We assume that the volume of $D_\Sigma$ is prescribed. The (reduced) phase space $\mathcal{N}$ can be identified with all such pairs $(\Sigma, v)$. Variations of $\Sigma$ and $v$ are denoted by $\delta\Sigma$ and $\delta v$, respectively. The Poisson bracket will be defined for functions $F, G : \mathcal{N} \to \mathbb{R}$, which possess functional derivatives defined as follows.

We say that such a function $F$ has a *functional derivative with respect to $\Sigma$ at* $(\Sigma, v) \in \mathcal{N}$, if there exist maps $\frac{\delta F}{\delta \Sigma}(\Sigma, v) : \Sigma \to \mathbb{R}$ and $\frac{\delta_j F}{\delta \Sigma}(\Sigma, v) : c_j \to \mathbb{R}$ $(j = 1, 2)$, such that

$$\frac{d}{d\varepsilon}\bigg|_{\varepsilon=0} F(\Sigma_\varepsilon, v) = \int_\Sigma \frac{\delta F}{\delta \Sigma}(\Sigma, v)\, \delta\Sigma\, dA + \sum_{j=1,2} \int_{c_j} \frac{\delta_j F}{\delta \Sigma}(\Sigma, v)\, \delta\Sigma\, ds$$

holds for any curve $\varepsilon \mapsto \Sigma_\varepsilon$ of admissible surfaces with $\Sigma_0 = \Sigma$ and $\frac{d}{d\varepsilon}\big|_{\varepsilon=0}\Sigma_\varepsilon = \delta\Sigma$. Here $c_j$ denotes the curve of intersection of $\Sigma$ with the plate $P_j$ as above; for path integrals, the element of integration is denoted by $ds$. Similarly, we say that a function $F$ has a *functional derivative with respect to v* at $(\Sigma, v) \in \mathcal{N}$, if there exists a divergence-free vector field $\frac{\delta F}{\delta v}(\Sigma, v)$ in $D_\Sigma$, the normal component of which vanishes along $\Sigma_1$ and $\Sigma_2$, such that

$$\frac{d}{d\varepsilon}\Big|_{\varepsilon=0} F(\Sigma, v_\varepsilon) = \int_{D_\Sigma} \frac{\delta F}{\delta v}(\Sigma, v)\delta v \, dV$$

holds for any curve $\varepsilon \mapsto v_\varepsilon$ of admissible vector fields with $v_0 = v$ and $\frac{d}{d\varepsilon}\big|_{\varepsilon=0} v_\varepsilon = \delta_v$.

Let $\mathcal{D}$ be the set of all functions $F : \mathcal{N} \to \mathbb{R}$, which have functional derivatives as defined above at any point $(\Sigma, v) \in \mathcal{N}$. We have $H \in \mathcal{D}$. In fact,

(2.4)
$$\begin{aligned}
\frac{\delta H}{\delta \Sigma}(\Sigma, v) &= (\frac{1}{2}\|v\|^2 + \tau\kappa)\, e_r \cdot n \\
\frac{\delta_j H}{\delta \Sigma}(\Sigma, v) &= (\tau \cos\gamma_j - \sigma_j)\, e_r \cdot n_j \quad (j = 1, 2) \\
\frac{\delta H}{\delta v}(\Sigma, v) &= v \quad.
\end{aligned}$$

We now define a *Poisson bracket on $\mathcal{N}$* as follows. For functions $F, G \in \mathcal{D}$, we set

$$\begin{aligned}
\{F, G\} = &\int_{D_\Sigma} (\nabla \times v) \cdot (\frac{\delta F}{\delta v} \times \frac{\delta G}{\delta v}) \, dV \\
&+ \int_\Sigma \left[\frac{\delta F}{\delta \Sigma}(\frac{\delta G}{\delta v} \cdot n) - \frac{\delta G}{\delta \Sigma}(\frac{\delta F}{\delta v} \cdot n)\right] \frac{1}{e_r \cdot n} \, dA \\
&+ \sum_{j=1,2} \int_{c_j} \left[\frac{\delta_j F}{\delta \Sigma}(\frac{\delta G}{\delta v} \cdot n) - \frac{\delta_j G}{\delta \Sigma}(\frac{\delta F}{\delta v} \cdot n)\right] \frac{1}{e_r \cdot n} \, ds.
\end{aligned}$$

With this Poisson bracket, using the divergence theorem, it is not hard to show that for any solution $(v, p, \Sigma)$ of the basic equation (2.2), the relation (1.1) is satisfied along the curve $t \mapsto (\Sigma, v)$ in $\mathcal{N}$. Conversely, given any such curve for which the relation (1.1) is satisfied, one can construct a pressure field $p$ in $D_\Sigma$, such that $(v, p, \Sigma)$ is a solution of (2.2) (see Kruse [10]). The pressure field satisfies the following boundary value problem, where $\Delta$ is the Laplace operator.

(2.5)
$$\begin{aligned}
\Delta p &= -\nabla \cdot ((v \cdot \nabla) v) && \text{in } D_\Sigma \\
p &= \tau\kappa && \text{on } \Sigma \\
\nabla p \cdot n &= -((v \cdot \nabla) v) \cdot n && \text{on } \Sigma_j \ (j = 1, 2)
\end{aligned}$$

In that sense, the equations in (2.2) are equivalent to the Hamiltonian equation (1.1).

## 3   Uniformly rotating liquid drops

Now we look for special solutions of (2.2), for which the fluid drops rigidly rotate around the $z$-axis with constant angular velocity $\omega$, in the shape of surfaces of revolution

$$\Sigma = f(z) \quad (0 \le z \le h).$$

Obviously, such solutions are invariant under the group of rotations around the $z$-axis which is a symmetry group of the problem. With respect to the material representation, they are relative equilibria. For that kind of solutions, the basic equations reduce to the following boundary value problem for a nonlinear second-order ODE:

(3.1)
$$\tau\kappa \;-\; \frac{1}{2}\omega^2 f^2 \;=\; c \quad (c \in \mathbb{R})$$
$$f'(0) \;=\; \frac{-\sigma_1}{\sqrt{\tau^2 - \sigma_1^2}} \;=:\; \rho_1$$
$$f'(h) \;=\; \frac{\sigma_2}{\sqrt{\tau^2 - \sigma_2^2}} \;=:\; \rho_2$$

Here, the mean curvature $\kappa$ of the free surface is given by

$$\kappa = \kappa_f = \frac{1}{f(1+(f')^2)^{1/2}} - \frac{f''}{(1+(f')^2)^{3/2}}$$

in terms of $f$; $f'$ and $f''$ are derivatives with respect to $z$, and $c$ is an arbitrary real constant which is related to the pressure field as follows:

(3.2)
$$p = c + \frac{1}{2}\omega^2 r^2$$

Note that, in terms of $f$, the volume of the drop is given by

(3.3)
$$vol(f) = \pi \int_0^h f^2 \, dz.$$

**Theorem 3.1** *Let the constants $h, \tau, \rho_1, \rho_2$ as well as a number $K > 0$ be given. Then there exists a constant $\omega_0$ such that for all $\omega \in [-\omega_0, \omega_0]$, (3.1) has a solution $(f, c)$ with $vol(f) < K$.*

Proof: For $\omega = 0$, equation (3.1) says that the surface of revolution generated by $f$ must have constant mean curvature. There is a simple construction of such functions $f$ due to Delauny [4]. One rolls up an ellipse along the $z$-axis inside a plane without sliding. Then each focus describes a curve which generates a surface of constant mean curvature through rotation about the $z$-axis. Thus, using the parameters of the ellipse to fit the boundary conditions as well as the volume constraint, we can construct a solution $f$ of (3.1) for $\omega = 0$. Furthermore, for small $|\omega| \neq 0$, we can use Delauny curves as a first approximation for the solution. Thus, the theorem follows by a perturbation argument.

To fit the boundary conditions, we use the shooting method. To this end, we consider the following initial value problem:

$$
\begin{array}{rcl}
\tau \kappa_\phi - \dfrac{1}{2} \omega^2 \phi^2 &=& \dfrac{c}{r} \\[2mm]
\phi(0) &=& r\, f(0) \\[2mm]
\phi'(0) &=& \rho_1
\end{array}
$$

(3.4)

Here $r$ is a parameter, $r \in (0,1)$, and $c = c(f) = \tau \kappa_f$, where $f = f(z), z \in \mathbb{R}$, represents a Delauny curve given by the parametrization

$$
f^2 = a^2 \left( 1 - 2e \cos \frac{s}{a} + e^2 \right).
$$

Here

$$
s = s(z) = \int_0^z (1 + f'(z)^2)^{\frac{1}{2}} dz + s_0, \qquad s_0 \in \mathbb{R},
$$

denotes the arc length parameter of the Delauny curve, $a$ is the length of the major semiaxis of the underlying ellipse and $e$ the numerical excentricity of this ellipse. Note that $0 \leq e < 1$ and $0 < f < 2a$. Also, $f$ is a periodic function of $z$ with period equal to the circumference of the underlying ellipse. Taking the derivative with respect to $z$ in the formula for $f^2$, we immediately find

$$
f' = e\, \frac{\sin \left( \frac{s}{a} \right)}{1 - e \cos \left( \frac{s}{a} \right)}.
$$

This function is $2\pi$-periodic and odd in $x = \frac{s}{a}$ and attains arbitrarily large positive and negative values near $x = 0$, as $e$ approaches 1. Therefore, $e$ can be chosen such that

$$
\max_{\mathbb{R}} |f'| > \max(|\rho_1|, |\rho_2|)
$$

holds. Then we have $f'(0) = \rho_1$ provided that the constant $s_0$ is chosen appropriately, and we can find values $z_0, z_1 > h$ such that $f'(z_0) > \rho_2$ and $f'(z_1) < \rho_2$. Without loss of generality, we assume that $z_0 > z_1$ holds. Finally we assume that $a$ is chosen such that $4\pi h a^2 < K$ is satisfied.

Then a straightforward computation shows that for $\omega = 0$ and $r \in (0,1)$,

$$
\phi(z) = r\, f(\frac{z}{r}), \qquad z \in \mathbb{R},
$$

is a solution of (3.4) such that in addition, the following inequalities hold:

$$
\begin{array}{rcl}
\phi'(r\, z_0) &>& \rho_2 \\[1mm]
\phi'(r\, z_1) &<& \rho_2 \\[1mm]
vol(\phi) &<& K
\end{array}
$$

(3.5)

Note that scaling $f$ by $r$ as above is equivalent to replacing $a$ by $ra$. Finally we use the well known theorem of continuous dependence on parameters for the solution of an initial value problem such as (3.4) to argue that, in particular for $\frac{h}{z_0} \leq r \leq \frac{h}{z_1}$ and $0 \leq z \leq 2h$, there is a solution $\phi = \phi(z)$ of (3.4) and (3.5) even for sufficiently small $|\omega| \neq 0$, say for $|\omega| \leq \omega_0$. This solution depends continuously on $r$, and by (3.5), $\phi'(h) = \phi'(rz_0) > \rho_2$ for $r = \frac{h}{z_0}$ and $\phi'(h) = \phi'(rz_1) < \rho_2$ for $r = \frac{h}{z_1}$. Hence, by the intermediate value theorem, for any $\omega \in [-\omega_0, \omega_0]$ there exists a value of $r \in [\frac{h}{z_0}, \frac{h}{z_1}]$ such that the corresponding function $f = \phi(z)$ solves the boundary value problem (3.1) for a certain value of $c \in \mathbb{R}$. This proves the theorem.

To analyze the stability of these special solutions, we use the *augmented potential function* $V_\omega : \mathcal{N} \to \mathbb{R}$ given by

$$V_\omega = V_p(\Sigma) - \frac{1}{2} I(\Sigma) \omega^2,$$

where $V_p$ denotes the total potential energy of a drop (cf. 2.3), and $I(\Sigma) = \int_{D_\Sigma} r^2 \, dV$ is its moment of inertia about the $z$-axis. For axially symmetric drop shapes given by a function $f(z)$, it follows that

$$(3.6) \quad V_\omega = V_\omega(f) \;=\; 2\pi\tau \int_0^h f\sqrt{1 + (f')^2} \, dz \;-\; \sigma_1 \pi f(h)^2 \;-\; \sigma_2 \pi f(0)^2$$

$$-\; \frac{\pi}{4} \omega^2 \int_0^h f^4 \, dz \;.$$

Moreover, if $f$ is a solution of (3.1), then $f$ is a critical point of the functional $\tilde{V}_\omega = V_\omega - c\,vol$ on the function space $C^1[0, h]$ (see Kruse [10]), where the values of $\omega$ and $c$ are the same as in (3.1). This modification of $V_\omega$ is consistent with the volume constraint $vol(f) = const.$.

The potential function $V_\omega$ is appropriate for our situation. In fact, let $f$ be a solution of (3.1) and denote the corresponding surface of revolution by $\Sigma_f$ and the velocity field corresponding to the angular velocity $\omega$ by $v_\omega$. Define an *augmented Hamiltonian (energy-momentum functional)* on $\mathcal{N}$ by

$$H_\omega = H - \omega(J - J(\Sigma_f, v_\omega)) \,,$$

where $J = J(\Sigma, v)$ is the momentum map that assigns to each drop state $(\Sigma, v) \in \mathcal{N}$ the corresponding angular momentum about the $z$-axis. This is a conserved quantity for our system due to the rotational symmetry. Hence, $H_\omega$ is also a conserved quantity. Furthermore, $(\Sigma_f, v_\omega)$ is a critical point of $H_\omega$. Also, if $f$ is a strict minimum of $V_\omega$ subject to the volume constraint, then $(\Sigma_f, v_\omega)$ is a strict minimum of $H_\omega$ restricted to the subset of pairs $(\Sigma, v) \in \mathcal{N}$, such that $\Sigma$ is axially symmetric. Hence, $H_\omega$ is a kind of Liapunov function for $(\Sigma_f, v_\omega)$. This is a consequence of the fact that $H_\omega$ can be rewritten as

$$(3.7) \qquad\qquad H_\omega = K_\omega + V_\omega + \omega\, J(\Sigma_f, v_\omega)$$

with the *augmented energy functional* $K_\omega$ given by

$$K_\omega = \frac{1}{2} \int_{D_\Sigma} \|v - v_\omega\|^2 \, dV \;;$$

cf. Marsden [14, §5]. By this reference, one expects to obtain sharper (optimal) stability conditions if instead of $V_\omega$ one works with the so-called *amended potential function* $V_\mu : \mathcal{N} \to \mathbb{R}$ given by

$$V_\mu = V_p(\Sigma) + \frac{1}{2} I(\Sigma)^{-1} \mu^2$$

with $\mu = J(\Sigma_f, v_\omega)$. However, as far as the stability analysis below is concerned, these two potential functions lead to the same results. Because $V_\mu$ is singular at $\Sigma = \Sigma_f$ with $f \equiv 0$, we have chosen to work with $V_\omega$ here.

These ideas motivate the following stability criterium. A solution $(\Sigma_f, v_\omega)$ is said to be *formally stable with respect to rotationally symmetric initial perturbations*, if the second variation of $V_\omega$ at $f \in C^1[0, h]$ is positive definite subject to the volume constraint, which amounts to saying (cf. Vogel [18]) that the quadratic form

(3.8)
$$\beta(\phi) = \int\limits_0^h (P(\phi')^2 - Q(\phi)^2) \, dz$$

with

$$P = \frac{\tau h}{(1 + (f')^2)^{3/2}} \quad , \quad Q = \frac{\tau}{f(1 + (f')^2)^{1/2}} + \omega^2 f^2$$

is positive definite on the function space

$$C^1_\perp [0, h] = \left\{ \phi \in C^1 [0, h] \; \middle| \; \int\limits_0^h f \, \phi \, dz = 0 \right\}.$$

For $\omega = 0$, this stability criterium agrees with that in Vogel [18]. According to Vogel [18, Theorem 3.1], $\beta(\phi)$ is positive definite on $C^1_\perp [0, h]$, if the associated Sturm-Liouville eigenvalue problem

(3.9)
$$L[\phi] = \lambda \phi \quad , \quad \phi'(0) = \phi'(h) = 0$$

with

$$L[\phi] = -(P \, \phi')' + Q \, \phi$$

has eigenvalues $\lambda_1 < 0 < \lambda_2 < \dots$ , and the solution $\psi$ of the boundary-value problem

(3.10)
$$L[\psi] = f \quad , \quad \psi'(0) = \psi'(h) = 0$$

has the property $\int_0^h \psi \, dz < 0$. Thus, one can use Sturmian theory to verify the definiteness of the relevant quadratic form.

We conclude with a simple example. Suppose that $\sigma_1 = \sigma_2 = 0$ holds. Then $f(z) \equiv d$ is a solution of (3.1) for any $\omega$, any $d > 0$ and a certain constant $c$. These solutions represent rigidly rotating cylindrical drops. In this case, the eigenvalues of the associated Sturm-Liouville eigenvalue problem (3.9) can be computed explicitly, since the operator $L[\phi]$ has constant coefficients. The smallest eigenvalue $\lambda_1 = -(\tau/d + \omega^2 d^2)$ is always negative. The other eigenvalues are given by $\lambda_{n+1} = \tau d n^2 \pi^2/h^2 - \tau/d - \omega^2 d^2$, $n \in \mathbb{N}$. Hence, the condition $\lambda_2 > 0$ leads to the explicit stability criterium

$$(3.11) \qquad \frac{h^2}{\pi^2 d^2} + \frac{\omega^2 h^2 d}{\pi^2 \tau} < 1 \ .$$

The solution $\psi = -\frac{d}{\tau/d + \omega^2 d^2}$ of (3.10) clearly has the required property. .

## REFERENCES

1.   V.I. Arnold, Sur la géometrie differentielle des groupes de Lie de dimension infinie et ses applications à l'hydrodynamique des fluids parfaits, Ann. Inst. Fourier, Grenoble **16** (1966), 319–361.

2.   P. Concus and R. Finn, The shape of a pendent liquid drop, Philos. Trans. Roy. Soc. London Ser. A, **292** (1979), 307–340.

3.   P. Concus and R. Finn, Capillary surfaces in microgravity, in *Low-Gravity Fluid Mechanics and Transport Phenomena*, J.N. Koster and R.L. Sani edts., Progress in Astronautics and Aeronautics **130** AIAA, Washington, DC, (1990), 183–206.

4.   C. Delauny, J. Math. Pures Appl. **6** (1841), 309– .

5.   R. Finn, *Equilibrium Capillary Surfaces*, Springer-Verlag, New York, 1986.

6.   C.F. Gauss, *Werke (Collected Works)*, Vols. **1–12**, Göttingen, 1863–1929.

7.   D.D. Holm, J.E. Marsden, T. Ratiu and A. Weinstein, Nonlinear stability of fluid and plasma equilibria, Physics Reports **123** (1985), 1–116.

8.   U. Hornung and H.D. Mittelmann, Bifurcation of axially symmetric capillary surfaces, J. Colloid and Interface Sci. **146 (1)** (1991), 219–225.

9.   D. Kröner, The flow of a fluid with a free boundary and dynamic contact angle, Z. angew. Math. Mech. **67** (1987), T 304–T 306.

10.   H.-P. Kruse, *Flüssigkeitstropfen zwischen parallelen Platten: Hamiltonsche Struktur, Existenz von Lösungen und Stabilität*, Doctorial thesis, Universität Hamburg, 1992.

11.   D.R. Lewis, Nonlinear stability of a rotating planar liquid drop, Arch. Rat. Mech. Anal. **106** (1989), 287–333.

12.   D.R. Lewis, J.E. Marsden, R. Montgomery and T.S. Ratiu, The Hamiltonian structure for dynamic free boundary problems, Physica D **18** (1986), 391–404.

13.   D.R.Lewis, J.E. Marsden and T.S. Ratiu, Stability and bifurcation of a rotating liquid drop, J. Math. Phys. **28** (1987), 2508–2515.

14.   J.E. Marsden, *Lectures on Mechanics*, London Math. Soc. Lect. Note Ser. **174**, Cambridge University Press, Cambridge, 1992.

15.   J.E. Marsden, T.S. Ratiu and A. Weinstein, Semi-direct products and reduction in mechanics, Trans. Am. Math. Soc. **281** (1984), 147–177.

16.   J.E. Marsden, T.S. Ratiu and A. Weinstein, Reduction and Hamiltonian structures on duals of semidirect product Lie algebras, Cont. Math. AMS **28** (1984), 55–100.

17.   J.E. Marsden and A. Weinstein, Coadjoint orbits, vortices and Clebsch variables for incompressible fluids, Physica D **7** (1983), 305–323.

18.   T.I. Vogel, Stability of a liquid drop trapped between two parallel planes, SIAM J. Appl. Math. **47** (1987), 516–525.

19.   T.I. Vogel, Stability of a liquid drop trapped between two planes II: general contact angles, SIAM J. Appl. Math. **49** (1989), 1009–1028.

INSTITUT FÜR ANGEWANDTE MATHEMATIK, UNIVERSITÄT HAMBURG, BUNDESSTRASSE 55, D-W-2000 HAMBURG 13, GERMANY
*E-mail address:* AM90020@DHHUNI4.BITNET

DEPARTMENT OF MATHEMATICS, UNIV. OF CALIFORNIA, BERKELEY, CALIFORNIA 94720,USA
*E-mail address:* marsden@math.berkeley.edu

Lectures in Applied Mathematics
Volume **29**, 1993

# The symmetry group of the integro-partial differential equations of Poisson-Vlasov

POL V.A.J. LAMBERT

ABSTRACT. The Poisson-Vlasov equations are relevant to collisionless plasmas and to stellar dynamics. The main result of this paper is the determination of the connected symmetry group of the three-dimensional integro-partial differential system of Poisson-Vlasov and at the same time of this slightly modified system (formulas (14) and (13) respectively). We developed and used therefore a minor extension of the Lie-method, which was necessary in order to cope with the presence of integrals.

## 1. Introduction

Referring to the abstract we can specify that the Poisson-Vlasov equations describe for ex. the evolution of a three-dimensional collisionless ionized plasma moving under the influence of self-induced electrostatic forces.
The Poisson-Vlasov equations for the considered plasma moving in $I\!R^3$ are :

$$
(1) \qquad
\begin{cases}
\dfrac{\partial f}{\partial t} + \displaystyle\sum_{i=1}^{3} \left( v^i \dfrac{\partial f}{\partial x^i} - \dfrac{1}{m} \dfrac{\partial \varphi}{\partial x^i} \dfrac{\partial f}{\partial v^i} \right) = 0, \ \text{(Vlasov)} \\[2mm]
\Delta \varphi = -\rho_f, \ \text{(Poisson)}
\end{cases}
$$

where $t$ in the time, $x = (x^1, x^2, x^3)$ the space-variable, $v = (v^1, v^2, v^3)$ the velocity-variable, $f(v, t, x)$ the plasma density in position-velocity space $I\!R^6$ at time $t$, $e$ the ion charge, $m$ the ion mass and $\rho_f$ the charge determined by $f$, i.e.

$$
(2) \qquad \rho_f(t, x) := e \int\int\int_{I\!R^3} f(v, t, x) dv.
$$

The function $\varphi$ must further satisfy the restriction that the electric field grad $\varphi$ depends only on $t$ and $x$. For an application to stellar dynamics, where

1991 Mathematics Subject Classification. Primary 45K05, Secondary 53C21.

This paper is in final form and no version of it will be submitted for publication elsewhere.

the attractive forces replace the repulsine forces, one has to change the sign of $f$ but the symmetry group remains the same. We are considering here the motion of a cloud of charged ions of a single species for simplicity. The generalization to several species is easy. For $e = 0$ we get Liouiville's equation for free particles in $\mathbb{R}^3$. In what follows we choose natural units in which $e = m = 1$. Moreover, we shall identity the velocity $v$ with the momentum $p = mv$, so we consider the position-momentum phase-space $\mathbb{R}^6$ with coordinates $(p, x)$, $p = (p^1, p^2, p^3)$, $x = (x^1, x^2, x^3)$ and the system (1) becomes then :

$$(3) \quad \begin{cases} \left( \dfrac{\partial f}{\partial t} + \sum_{i=1}^{3} \left( p^i \dfrac{\partial f}{\partial x^i} - \dfrac{\partial \varphi}{\partial x^i} \dfrac{\partial f}{\partial p^i} \right) \right)(p, t, x) = 0 \\ \displaystyle\sum_{i=1}^{3} \dfrac{\partial^2 \varphi}{\partial (x^i)^2}(t, x) + \int\int\int_{\mathbb{R}^3} f(p, t, x)\, dp = 0 \end{cases}$$

We determined the connected symmetry group of the system (3) by using an appropriate version of the Lie-method. The main difficulty is that the presence of an integral in (3) makes that most of the classical, finite-dimensional versions of the Lie-method, as exposed in ref. [3], cannot be applied straightforwardly here. We had first applied an infinite-dimensional extension of the Lie-method, which is a generalization to the three-dimensional case of a work of W.B. Baranow on a one-dimensional system of Maxwell-Vlasov type, ref. [1]. But this method is mathematically unsatisfactory and its use did not yield the full symmetry group of (3). W. B. Baranow dit not obtain it either in the easier one dimensional case of ref. [1], where he did not give any detail of his calculations. But later a work of Mrs. Dana Roberts on the one-dimensional Vlasov-Maxwell equations, ref. [4], inspired us partially to an appropriate extension of the Lie-method to our three-dimensional integro-partial differential system (3), which led us to the determination of the full connected symmetry group of that system. We give a summary of the method and of the results in the following paragraph.

## 2. The connected symmetry group of the three-dimensional integro-partial differential system of Poisson-Vlasov

1. Assumptions on the density function

$$f : \mathbb{R}^7 \ni (p, t, x) \rightarrow f(p, t, x) \in \mathbb{R}$$

Let $\mathbb{N}$ be the set of natural numbers $0,1,2,\ldots$, and, for any $n$-tuple $q = (q_1, \ldots, q_n) \in \mathbb{N}^n$, define the length $|q|$ of $q$ by $|q| := \sum_{i=1}^{n} q_i$. For any local coordinates $y = (y^1, \ldots, y^n) \in \mathbb{R}^n$ define the differential operator $(\partial/\partial y)^q$ of order $|q|$ by $(\partial/\partial y)^q := (\partial/\partial y^1)^{q_1} \ldots (\partial/\partial y^n)^{q_n}$. For $p = (p^1, p^2, p^3) \in \mathbb{R}^3$

we use the euclidian norm $|p| = [\sum_{i=1}^{3}(p^i)^2]^{1/2}$ and we put $y = (y^1, \ldots y^7) = (p^1, p^2, p^3, t, x^1, x^2, x^3)$. Now we assume $f(p, t, x)$ to be $C^\infty$ on $\mathbb{R}^7$ and moreover such that for any $\alpha \in \mathbb{N}$, any $q$ and any $(t, x) \in \mathbb{R}^4$ there is a neighborhood $U(t, x)$ of $(t, x)$ in $\mathbb{R}^4$ with the property : $\lim\limits_{|p|\to\infty} |p|^\alpha |((\partial/\partial y)^q f)(y)| = 0$ uniformly on $U(t, x)$. These assumptions seems to be very reasonable since the plasma density decreases very rapidly to 0 when the velocity becomes large. We could even have assumed that the function $p \to f(p, t, x)$ has compact support independently of $(t, x)$. These assumptions guarantee the local uniform convergence and hence the infinite differentiability (by applications of Leibnitz's rule) of all the integrals which will be considered.

2. In order to find the symmetry group of the system (3) we shall see that it is not necessary to assume that the electric field components $\dfrac{\partial\varphi}{\partial x^i}(t, x)$, $1 \le i \le 3$, derive from a potential $\varphi$. So we consider the new dependent variables $E^i(t, x)$, $1 \le i \le 3$, without assuming first that $E = (E^1, E^2, E^3) = grad\ \varphi$. This reduces the calculations under the application of the Lie-method. So we replace temporarily the system (3) by the following modified Poisson-Vlasov system :

$$(4) \quad \begin{cases} \Delta_1(p, t, x) \equiv \left(\dfrac{\partial f}{\partial t} + \sum_{i=1}^{3}\left(p^i\dfrac{\partial f}{\partial x^i} - E^i\dfrac{\partial f}{\partial p^i}\right)\right)(p, t, x) = 0 \\ \Delta_2(t, x) \equiv (div\ E)(t, x) + \int\int\int_{\mathbb{R}^3} f(p, t, x)dp = 0 \end{cases}$$

where $div\ E := \sum_{i=1}^{3}\dfrac{\partial E^i}{\partial x^i}$, just as the electric field $E$, depends only on $t$ and $x$.
The independent variables are $y = (y^1, \ldots, y^7) = (p^1, p^2, p^3, t, x^1, x^2, x^3)$.
The dependent variables are $u = (u^1, \ldots, u^4) = (f, E^1, E^2, E^3)$ to which we add the auxiliary dependent variable $I(f, t, x) = \int\int\int_{\mathbb{R}^3} f(p, t, x)dp$, which is not a new dependent variable since it is determined by $f$, but which facilitates the computations. It is easy to check that the system (4) is of maximal rank 2 in the partial derivatives. If we assume that $I(f, t, x)$ is analytic the system (4) in also locally solvable and in that case the Lie-method furnishes its full connected symmetry group (ref. [3], Theorem 2.71).
Following more or less the notations of ref. [3] we consider the value spaces :
$Y = \mathbb{R}^7$ for the independent variables $y = (p, t, x)$.
$U = \mathbb{R}^4$ for the dependent variables $u = (u^1, \ldots, u^4) = (f, E^1, E^2, E^3)$.
$W = \mathbb{R}$ for the auxiliary dependent variable $I$. $M = Y \times U \times W$.
A general vector field on $M$ has the form

$$(5) \quad \begin{cases} \theta = \xi + \eta + \chi, \text{ with} \\ \xi = \sum_{i=1}^{7}\xi^{y^i}\dfrac{\partial}{\partial y^i}, \eta = \eta^f\dfrac{\partial}{\partial f} + \sum_{i=1}^{3}\eta^{E^i}\dfrac{\partial}{\partial E^i}, \chi = \eta^I\dfrac{\partial}{\partial I}, \end{cases}$$

where all the components depend on $(y, u) \in Y \times U$ but not on the auxiliary variable $I$. The flow generated by this vector field is a one-parameter $C^\infty$-group action on $M$ of the type :

(6)    $g : I\!\!R \ni \epsilon \to ((y, u, I) \to (\tilde{y}, \tilde{u}, \tilde{I}) = (g^1(y, u, \epsilon), \ldots, g^n(y, u, \epsilon)))$

where $n = 12$, $g(0) = Id_{I\!\!R^n}$ and $g$ does not depend on $I$. The way of determining the vector fields $\theta$, which generate group actions for which the system $\Delta$ is invariant, by using a suitable jet space of $M$, is well known, (see for ex. ref. [3] §2.3) except for the determination of the coefficient $\eta^I$, which is new. We shall now explain the headlines of this last determination. In order to study the action of $g$ on functions we have to identify them with their graphs. Considering then the second order Taylor approximation of the action of $g$ on vector functions on $M$ we have (see (6) for the notations) :

(7)          $(\tilde{y}, \tilde{f}) = (y, f) + \epsilon(\xi + \eta)(y, f)(y, u) + O(\epsilon^2).$

where $O(\epsilon^2)$ is a smooth vector function of $(y, u)$ and $\epsilon$. This yields :

$$\tilde{f}(\tilde{y}) = f(y) + \epsilon(\xi + \eta)(f)(y, u) + O(\epsilon^2)$$

(8)          $$= f(y) + \epsilon[\xi(f)(y, u) + \eta^f(y, u)] + O(\epsilon^2)$$

We determine now $\eta^I = \dfrac{dI}{d\epsilon}|_{\epsilon=0} = \lim\limits_{\epsilon \to 0, \epsilon \neq 0} \dfrac{1}{\epsilon}(\tilde{I} - I)$, where $I = \int \int \int_{I\!\!R^3} f(y)dp$
and $\tilde{I} = \int \int \int_{I\!\!R^3} \tilde{f}(\tilde{y})d\tilde{p}$. We determine $\tilde{I} - I$ by considering $I\!\!R^3$ as an orientable 3-dimensional Riemannian $C^\infty$-manifold $M$ inbedded in $I\!\!R^n$, $3 < n$ $(n = 7)$, with Riemannian metric induced by the standard in-product in $I\!\!R^n$. For any local chart $(W, \varphi)$ on $M$ call $U = \varphi[W]$, $\psi = \varphi^{-1}$, so that $W$ is parametrized by

(9)          $\psi : I\!\!R^3 \supseteq U \ni \tau \to \psi(\tau) = (y^1(\tau), \ldots, y^n(\tau)) \in I\!\!R^n.$

Then the integral $I = I(f, t, u)$ is generalized (provided analogous convergence assumptions) to :

(10)                $I_{f,W} = \int \int \int_{\varphi[W]} f(y(\tau))V_{n,3}(\tau)d\tau,$

where $V_{n,3}(\tau)$ is the volume element (first fundamental form) on this chart (see for ex. ref. [2], §V.7). By the application of any group action $g_\epsilon$ the manifold $M$ is transformed into another $p$-dimensional manifold $M_\epsilon$ in $I\!\!R^n$ and in particular the local chart $(W, \psi)$ for $M$ is transformed into the local chart $(g_\epsilon[W], \tilde{\psi})$, $\tilde{\psi} = g_\epsilon \circ \psi$, for $M_\epsilon$ with the same parameter definition domain $U = \varphi[W]$ in $I\!\!R^3$. Therefore the integral $I_{f,W}$ of (10) is transformed into the integral

(11)      $\tilde{I} := I(\epsilon) = g_\epsilon \circ I_{f,W} = \int \int \int_{\varphi[W]} \tilde{f}(\tilde{y}(\tau))\tilde{V}_{n,3}(\tau)d\tau.$

Using (8) and the same kind of Taylor approximation for $\tilde{V}_{n,3}$ and $\tilde{I}$ successively, we find after quite long and technically complicated calculations :

$$(12) \quad \eta^I = \frac{dI}{d\epsilon}\Big|_{\epsilon=0} = \int\int\int_{I\!\!R^3}(\xi(f) + \eta^f + f\sum_{i=1}^{3}\frac{\partial\xi^i}{\partial p^i})dp^1dp^2dp^3$$

3. Using the known Lie-method together with (5) and (12) (see for ex. ref [3], §2.3) we determined then the first polongation $pr^{(1)}\theta$ of the vector field $\theta$, to the first order jet space of $M$, which is relevant to (4). We required then that $(pr^{(1)}\theta)(\Delta_1) = 0 = pr^{(1)}\theta)(\Delta_2)$ in the points where $\Delta_1 = 0 = \Delta_2$ by substituting for $\dfrac{\partial f}{\partial t}$ its value from $\Delta_1 = 0$ and for $\dfrac{\partial E^3}{\partial x^3}$ its value from $\Delta_2 = 0$. In both equations $(pr^{(1)}\theta)(\Delta_1) = 0$ and $(pr^{(1)}\theta)(\Delta_2) = 0$ we used then the fact that the remaining partial derivatives are independent free variables, also under the integral signs, and we used also once the fact that our assumptions on the function $f$ allow us to differentiate under the integral-sign. Having performed those long and patient calculations, which can not be reproduced here because of the limited scope of this paper, we were able to determine exactly the Lie-algebra of vector fields, which generates the connected symmetry group of the system (4). This group is generated by the one-parameter groups $(i = 1, 2, 3)$ :

$$(13) \quad \begin{cases} G_1 : \epsilon \to t + \epsilon \quad \text{(translation in time)} \\ G_2^i : \epsilon^i \to x^i + \epsilon^i, \quad \text{(translations in space)} \\ G_{\alpha_i}^i : \delta^i \to (x^i + \delta^i(\int\alpha_i)(t), p^i + \delta^i\alpha_i(t), E^i(t,x) - \delta^i(\alpha_i)'(t)), \end{cases}$$

where $\alpha_i$ is any real $C^\infty$-function of the time $t$, $\int\alpha_i$ any primitive of $\alpha_i$ and $(\alpha_i)'$ the first derivative of the function $\alpha_i$.

One can check very easily that these groups leave the system (4) invariant. Finally we want now to deduce the symmetry groups of the system (3) where it is assumed that $E = grad\ \varphi$ and where $E$ depends only on $(t, x)$ but $\varphi$ may depend also on $p$. By the transition from the system (4) to the system (3) any group action on $E^i$, resp. $\dfrac{\partial E^i}{\partial x^i}$, has to be interpreted as a group action on $\dfrac{\partial\varphi}{\partial x^i}$, resp. $\dfrac{\partial^2\varphi}{(\partial x^i)^2}$, while for any real $C^\infty$-function $\beta(p,t)$ the vector field $\beta\dfrac{\partial}{\partial\varphi}$ generates a symmetry group action on (3). Hence for any real $C^\infty$-functions $\beta(p,t)$, $\alpha_i(t)$, $i = 1, 2, 3$, the following one-parametergroups are symmetry groups of the system (3) (we use here the knowledge of (13)) :

$$(14) \quad \begin{cases} G_1, G_2^i, i = 1, 2, 3. \\ G_\beta : \delta \to \varphi(p, t, x) + \delta\beta(p, t) \\ \tilde{G}_{\alpha_i}^i : \delta^i \to (x^i + \delta^i(\int\alpha_i)(t), p^i + \delta^i\alpha_i(t), \varphi - \delta^i x^i(\alpha_i)'(t)) \end{cases}$$

Conversely if $g$ is any one-parameter symmetry group of the system (3), then replacing for every $\epsilon \in I\!\!R$ $\quad g(\epsilon)(\varphi)$ by $\partial/\partial x^i(g(\epsilon)(\varphi))$ must yield the action on $E^i$ of a one-parameter symmetry group of the system (4). Hence by (13) we must have

$$\partial/\partial x^i(g(\epsilon)\varphi)(t,x) = E^i(t,x) - (\alpha_i)'(t), i = 1,2,3$$

for some $\alpha = (\alpha_1, \alpha_2, \alpha_3) \in C^\infty(I\!\!R; I\!\!R^3)$. Hence $g$ must be a composition of finitely many groups from (14).
We conclude that the connected symmetry group of the Poisson-Vlasov system (3) is generated by the one-parameter groups of (14).

## 4. General conclusions

The connected symmetry group of the Poisson-Vlasov system (generated by (14)) and the analogous one of the modified Poisson-Vlasov system (generated by (13)) are both infinite-dimensional, i.e. their Lie-algebras of vector fields are both infinite-dimensional (which is not a priori known since the systems are not linear), and the actions of both groups are strongly projectable, i.e. their actions depend only on the independent variables. Moreover there is no action on the dependent variable $f$, which is typical for the Poisson-Vlasov systems which are not one-dimensional. For the action on $f$ in the one-dimensional case, see for ex. ref. [1].
For the classification of the symmetry subgroups and the determination of the corresponding group-invariant solutions for both systems we refer the reader to a following paper, which we intend to publish later.

## REFERENCES

[1] W.B. Baranow, Symmetry of one-dimensional, high frequency movement of collisionless plasma, (in Russian). Journal of Technical Physics (1976), Vol. 46, No-6, pp. 1271-1277.

[2] W.M. Boothby, An introduction to Differential Manifolds and Riemannian Geometry. Academic Press, Inc., Second Edition, 1986.

[3] P. Olver, Applications of Lie Groups to Differential Equations. Springer Verlag, 1986.

[4] D. Roberts, The general Lie group and similarity solutions for the one-dimensional Vlasov-Maxwell equations. J. Plasma Physics (1985), Vol. 33, part 2, pp. 219-236.

Department of Mathematics, Physics and Informatics, Limburgs Universitair Centrum, Diepenbeek, Belgium, B-3590
Current address : Department of Math. Phys. and Inform., Limburgs Universitair Centrum, Universitaire Campus, Diepenbeek, Belgium, B-3590.
E-mail address : lambert@bdiluc∅1.bitnet

Lectures in Applied Mathematics
Volume **29**, 1993

# Explicit Symplectic Splitting
# Methods Applied to PDE's

ROBERT I. MCLACHLAN

ABSTRACT. The symplectic integration of Hamiltonian partial differential equations with constant symplectic structure is discussed, with a consistent, Hamiltonian approach. The stability, accuracy, and dispersion of different explicit splitting methods are analyzed, and we give the circumstances under which the best results can be obtained. Many different treatments and examples are compared.

## 1. Introduction

A standard method of developing integrators for PDE's is to derive an *ad hoc* discretization in space and time, and then study the properties (convergence, stability) of that method. However, because symplectic integrators have shown an ability to capture the long-time dynamics of Hamiltonian ODE's, one would like to apply them to Hamiltonian PDE's as well. A natural approach is to discretize both the Hamiltonian function (an integral) and the Hamiltonian (Poisson) structure, then form the resulting ODE's. In principle this deals with all Hamiltonian PDE's at once, and provides a simple framework for incorporating symmetries of the phase space (Casimir functions), spatial symmetries (in the Hamiltonian function), and temporal symmetries (such as reversibility, a property of some symplectic integrators). Unfortunately some symmetries, such as those giving rise to integrals of more than second degree, cannot be preserved in the discrete system, and furthermore, symplectic integrators can be difficult to construct for exotic Poisson structures.

The main difference from the low-dimensional ODE case is that the equations are stiff—they contain widely different time-scales. Hence any numerical dissipation will have a severe effect; but standard conservative schemes (Crank-Nicolson, leapfrog on $u_t = u_x$) have undesirable attributes such as implicitness,

1991 *Mathematics Subject Classification.* Primary 65M20, 58F05; Secondary 35L70, 35Q20.
This paper is in final form and no version of it will be submitted elsewhere.

parasitic waves, and low order. Hamiltonian methods have no parasitic waves, extend easily to any order, and are often explicit. However, note that if one wants to compute $u(T)$ accurately for fixed $T$ then conventional methods will always do better—this only depends on the truncation error of the method used. This is not an appropriate test for symplectic integrators: one should concentrate instead on phase space structures (e.g. the shape of a traveling wave) and not on temporal errors (its speed). Whether or not accumulating phase errors (e.g. of angles on Liouville tori) corrupt the dynamics depends on the particular system studied, and on the measured property being structurally stable in the space of Hamiltonian systems.

We outline Hamiltonian systems in §2, and symplectic splitting methods in §3. These may be applied when $H = T(p) + V(q)$, ("P-Q splitting") but the best methods are possible when the nonlinear terms in $H$ may be integrated exactly, as often happens ("L-N splitting"). Proposition 1 proves sufficient conditions for the much more accurate Runge-Kutta-Nyström methods to apply in this case. Sections 4 and 5 analyze the stability and dispersion of the two different splittings with different integrators—L-N splitting turns out to be more accurate and more stable, and if the equation is linear in its highest derivatives, dispersion errors are almost eliminated. Examples appear throughout.

## 2. Hamiltonian partial differential equations

Olver [12] is a good introduction to the structure of Hamiltonian ODE's and PDE's. Here we give a brief overview. A Hamiltonian dynamical system consists of a triple $(M, \{\cdot, \cdot\}, H)$ where $M$ is a smooth manifold (the *phase space*), $H : M \to R$ is the Hamiltonian function, and $\{\cdot, \cdot\}$ is a Poisson bracket, a bilinear, skew-adjoint operator satisfying the Jacobi identity and the Leibniz rule. The bracket can be written in coordinates $x_i$ as

$$\{F, G\} = (\nabla F)^T J(x)(\nabla G)$$

where $F$, $G : M \to R$ and $J$ is called the Poisson tensor. A change of variables $x \to X = \phi(x)$ induces a bracket in the new variables by

$$\{F \circ \phi, G \circ \phi\}_X = \{F, G\}_x \circ \phi$$

or, in coordinates,

$$(D_x\phi)J(x)(D_x\phi)^T = \widetilde{J}(X).$$

If $J = \widetilde{J}$, $\phi$ is called a Poisson map; the time-map of the Hamiltonian dynamical system

$$\dot{x} = \{x, H\} = J(x)\nabla H(x)$$

is a Poisson map. A symplectic (or Poisson) integrator is one for which a time step is a Poisson map. Casimirs are functionals $C$ such that $\{C, F\} = 0 \; \forall F$, hence integrals of the motion for any $H$. When the phase space is infinite dimensional, we write the triple as $(\mathcal{M}, \{\cdot, \cdot\}, \mathcal{H})$, and the Poisson operator as $\mathcal{J}$. Typically $\mathcal{M}$ consists of sets of smooth functions on a finite dimensional space $Z$ i.e. an element in $\mathcal{M}$ is $u(x)$, $x \in Z$. The Hamiltonian $\mathcal{H} : \mathcal{M} \to R$ is a functional on this space, and the bracket can be written as

$$\{\mathcal{F}, G\}[u] = \int_Z \frac{\delta \mathcal{F}}{\delta u} \mathcal{J}(u) \frac{\delta G}{\delta u} \, dx$$

where $\delta \mathcal{F}/\delta u$ is the variational derivative. When $\mathcal{J}$ is constant over $\mathcal{M}$, the Jacobi identity is trivially satisfied, and one need only check skew-adjointness. The most common cases are the canonical $\mathcal{J} = \begin{bmatrix} 0 & 1 \\ -1 & 0 \end{bmatrix}$ and the Gardner–Zakharov–Faddeev operator $\mathcal{J} = \partial_x$, which appears for example in the Korteweg–de Vries equation.

To reduce a Hamiltonian PDE to a set of Hamiltonian ODE's which can be symplectically integrated, our approach is to discretize $\mathcal{J}$ and $\mathcal{H}$ separately and then form the resulting dynamical system. $\mathcal{H}$ is an integral which can be discretized in any (suitably accurate) way, being careful to maintain the symmetry of any derivatives in $\mathcal{H}$. If $\mathcal{J}$ is constant, it may be discretized by replacing the differential operators by any (matrix) difference operator $D$, for example, central or pseudo-spectral differences. In what follows $D$ will be any appropriate matrix difference operator. Some points to remember are that $\partial_x \partial_x = \partial_{xx}$ usually breaks down when discretized, $\mathcal{H}$ may be integrated by parts as necessary to get compact differences, and that for equations involving odd derivatives we may get 1-point-more-compact differences using staggered grids.

Consider periodic boundary conditions. Finite differences introduce excessive dispersion and will usually be inadequate, but full-spectral, pseudo-spectral and anti-aliased schemes are all possible, by treating the nonlinear terms in $\mathcal{H}$ (not $\dot{u}$) appropriately. The time-continuous dynamics are the same whether one works in Fourier or physical variables; usually the choice is made to minimize the number of Fourier transforms required per time-step.

EXAMPLE 1. $\mathcal{H} = -\int (u_x)^2 \, dx = \int u u_{xx} \, dx$ may be discretized as $H = \sum_{i,j} u_i D_{ij} u_j$ where $D$ is a matrix difference operator approximating $\partial_{xx}$. Then $(dH)_i = \sum_j (D_{ij} + D_{ji}) u_j$. This approximates $2u_{xx}$ only if $D_{ij} = D_{ji}$, i.e. the matrix $D$ must be symmetric. So Chebyshev-spectral or finite differences skewed at boundaries are not suitable.

## 3. Symplectic integration

Suppose the Hamiltonian may be split into two parts—the "P-Q" splitting:

$$H = T(p) + V(q).$$

This was first considered by Feng [4]. Later we will also consider "L-N" splitting

$$H = L(u) + N(u)$$

where $L$ has linear dynamics and $N$ is nonlinear. Let $X_T = JdT$, etc., be the associated vector fields, and $e^{kX}$ be the time-$k$ flow of the vector field $X$. If

$J = \begin{bmatrix} 0 & K \\ -K^T & 0 \end{bmatrix}$ then the following map is an explicit, first-order approximation of the true flow $e^{kX_H}$ [13]:

$$(3.1) \qquad\qquad e^{kX_T} e^{kX_V} = e^{kX}$$

which is computed as

$$p^{n+1} = p^n - k \left( K^T V'(q^n) \right), \qquad q^{n+1} = q^n + k \left( KT'(p^{n+1}) \right).$$

In (3.1),

$$(3.2) \qquad\qquad X = X_T + X_V + k[X_T, X_V] + \mathcal{O}(k^2) = Jd\widehat{H}$$

where

$$(3.3) \qquad\qquad \widehat{H} = H + H_0 + \mathcal{O}(k^2), \qquad H_0 = -k\{T, V\}$$

where the error (due to the noncommutativity of $T$ and $V$) is expanded using the Campbell-Baker-Hausdorff (BCH) formula, and (3.3) follows because Hamiltonian vector fields form a Poisson algebra. The series in (3.2) is only an asymptotic series in $k$—it does not generally converge, although it will for linear systems for small enough $k$. Despite this problem, one might call $H_0$ the "autonomous Hamiltonian truncation error" (Yoshida [17]).

The leapfrog ("LF2") method extends the method (3.1) to second-order:

$$\varphi(k) = e^{\frac{1}{2}kX_T} e^{kX_V} e^{\frac{1}{2}kX_T} = e^{kX_{\widehat{H}}},$$

$$(3.4) \qquad \widehat{H} = H + \frac{k^2}{24} \left( 2\{V, \{T, V\}\} - \{T, \{T, V\}\} \right) + \mathcal{O}(k^4)$$

It is *symmetric*, that is, $\varphi(k)\varphi(-k) = 1$. Suzuki [15] and Yoshida [16] use this property to construct schemes of arbitrary order by concatenating $(2s + 1)$ leapfrog stages and preserving the symmetry:

$$(3.5) \qquad \varphi(w_s k) \ldots \varphi(w_1 k) \varphi(w_0 k) \varphi(w_1 k) \ldots \varphi(w_s k).$$

where $w_0 = 1 - 2(w_1 + \ldots + w_s)$. Particular schemes are given in Table 1. Other compositions, analogous to (3.4), can be made which preserve specified reversibilities of the continuous system [14]. A fourth-order scheme which has been rediscovered many times ([2], [7], [15], [16]) is LF4a, which has $s = 1$ (see Table 1). However, this method takes a large backwards step of $1.70k$, leading to poor accuracy and stability. A better fourth order method LF4b [15] whose largest step is $-0.66k$, has $s = 2$, and one can show that this is close to the most accurate fourth-order method of this type. The best sixth-order method, LF6a, is Yoshida's Method A which has $s = 3$.

If $T(p)$ is quadratic (i.e. one may write $\ddot{q} = f(q)$), one can do significantly better by simply concatenating several stages of (3.1): $\varphi = \prod_{i=s}^{1} e^{a_i kX_T} e^{b_i kX_V}$. These are known as Runge-Kutta-Nyström (RKN) methods. The most accurate

**Table 1.** Symplectic Integrators

**I.** General methods

$$\varphi(w_s k)\ldots\varphi(w_1 k)\varphi(w_0 k)\varphi(w_1 k)\ldots\varphi(w_s k), \qquad w_0 = 1 - 2(w_1 + \ldots + w_s)$$

where $\varphi(k)$ is any symmetric method, such as leapfrog

(LF2:) $\qquad\qquad e^{\frac{1}{2}kX_A}e^{kX_B}e^{\frac{1}{2}kX_A}, \qquad H = A + B$

LF4a: $\quad s = 1, \quad w_1 = (2 - 2^{1/3})^{-1}$
LF4b: $\quad s = 2, \quad w_1 = w_2 = (4 - 4^{1/3})^{-1}$
LF6a: $\quad s = 3, \quad w_1 = -1.17767998417887, \; w_2 = 0.235573213359357,$
$$w_3 = 0.78451361047756$$

**II.** Runge-Kutta-Nyström methods, $\varphi = \prod_{i=s}^{1} e^{a_i kX_T}e^{b_i kX_V}$.

LF4c: $\quad s = 4,$

$a_1 = 0.5153528374311229364 \quad b_1 = 0.1344961992774310892$
$a_2 = -0.085782019412973646 \quad b_2 = -0.2248198030794208058$
$a_3 = 0.4415830236164665242 \quad b_3 = 0.7563200005156682911$
$a_4 = 0.1288461583653841854 \quad b_4 = 0.3340036032863214255$

LF6b: $\quad s = 8,$

$b_1 = 0$
$a_1 = -1.0130879789881764712 \quad b_2 = 0.00016600692650939825$
$a_2 = 1.18742957380274263478 \quad b_3 = -0.379624214274416219$
$a_3 = -0.018335852095646462 \quad b_4 = 0.68913741186280925274$
$a_4 = 0.34399425728108029845 \quad b_5 = 0.38064159097019513586$
$a_i = a_{9-i}, \; i = 5,6,7,8 \qquad b_i = b_{10-i}, \; i = 6,7,8$

4th- and 5th-order methods (in the sense of minimizing the Hamiltonian truncation error) are due to McLachlan and Atela [**10**]; the 4th-order one, LF4c, has $s = 4$ stages. Okunbor and Skeel [**11**] give sixteen 8-stage, 6th-order methods. Their method 13 (LF6b) has the smallest truncation error, about 0.02 times that of LF6a.

If both $X_L$ and the nonlinear vector field $X_N$ can be integrated exactly, then one may use the same composition methods with L-N splitting ($\varphi = e^{kX_L}e^{kX_N}$, etc.) This will usually be superior in that more (or all) or the derivatives in $H$ will be treated exactly, and for weak nonlinearities, the truncation error will be asymptotically smaller. Furthermore, the more accurate RKN methods may still sometimes be used:

PROPOSITION 1. *Let* $J = \begin{bmatrix} 0 & K \\ -K^T & 0 \end{bmatrix}$ *and* $H = L(q,p) + N(q)$ *where* $L$ *is a quadratic polynomial in* $p$ *and* $q$. *Then any canonical Runge-Kutta-Nyström method of order five or less, or any symmetric (i.e.* $\varphi(k)\varphi(-k) = 1$*) method of order six or less, maintains its order of accuracy when applied to this splitting.*

PROOF. The special requirement of RKN for the splitting $H = T(p) + V(q)$ is that certain terms in the expansion of $e^{a_i k X_T} e^{b_i k X_V} \ldots$ contain a factor $T''''$ and hence vanish identically. We compare these terms to those appearing in the expansion via the BCH formula, namely higher-order commutators of

$$\{T, V\} = -W_j T_j, \qquad T_j = \frac{\partial T}{\partial p_j}, \qquad W_j = \sum_i \frac{\partial V}{\partial q_i} K_{ij}.$$

The first vanishing term is at $\mathcal{O}(k^4)$, which corresponds to a 4th-order method. It is $\{V, \{V, \{T, V\}\}\} = T_{ijk} W_i W_j W_k$. For the L-N splitting this term is $L_{ijk} W_i W_j W_k$ which is also identically zero (although $\partial L / \partial q$ does enter in the other terms). At fifth order, the two zero terms (in P-Q) are the two commutators of this one, hence also zero in L-N. At sixth order this simplicity breaks down: the twelve terms in the BCH expansion reduce to eight for both the P-Q and the L-N splitting. These eight contain five distinct terms in the P-Q case but eleven in the L-N case (the extra terms containing $\partial L / \partial q$, etc); hence the order conditions in the two cases are different. But if the method is symmetric, the sixth-order terms are identically zero, so the RKN methods do then carry over to the L-N splitting.

EXAMPLE 2. De Frutos, Ortega and Sanz-Serna have given two treatments of the Boussinesq equation $\mathcal{J} = \begin{bmatrix} 0 & \partial_x \\ \partial_x & 0 \end{bmatrix}$, $\mathcal{H} = \int \frac{1}{2}(p^2 + (q_x)^2 + q^2) + \frac{1}{3}q^3 \, dx$. The first method [5] is unconditionally stable; this is achieved by time-averaging the stiffest $(D^4 q)$ term:

$$(3.6) \quad (q^{n+1} - 2q^n + q^{n-1})/k^2 = -\frac{1}{4} D^4 (q^{n+1} + 2q^n + q^{n-1}) + Dq^n + D((q^n)^2)$$

where superscripts denote time-levels and $D$ is the pseudo-spectral difference operator, but could just as well be the (diagonal) spectral difference operator or even $\partial_x$. Rearranging terms, this can be written as a map $\varphi$:

$$p^{n+\frac{1}{2}} = p^{n-\frac{1}{2}} + k(I + \frac{k^2}{4} D^4)^{-1}(-D^3 q^n + Dq^n + D(q^n)^2)$$

$$(3.7) \qquad \equiv p^{n-\frac{1}{2}} + k \qquad E \qquad N(q^n)$$

$$q^{n+1} = q^n + kD p^{n+\frac{1}{2}}$$

showing that stability is achieved by braking the high modes severely; in fact $k = O(h^2)$ is required for consistency. Secondly, a direct calculation of $\varphi' J \varphi'^T$ shows that $\varphi$ is a Poisson map iff $EN' = N'E$, which is not true here. It is true if $N(q)$ is linear; in this case the method is equivalent to leapfrog with the high modes braked in the Hamiltonian. (A similar method of gaining unconditional stability is used in Dai [3] for the variable-coefficient Schrödinger equation.)

Their second method (de Frutos et al. [6]) is equivalent to (3.7) with $E = I$. In our framework this is P-Q splitting with time-stepping $e^{k X_T} e^{k X_V}$, which is second-order if the unknowns are staggered in time $(q^n, p^{n+\frac{1}{2}})$. They prove convergence and nonlinear stability for this method. Consider instead the equivalent LF2; the Hamiltonian truncation errors (3.4) for the P-Q and L-N splittings are

$$\text{P-Q:} \qquad -\frac{k^2}{24} \int -pp_{xx} + p^2 + 2qp^2 + 2(-q_{xx} + q + q^2)^2 \, dx$$

$$\text{L-N:} \qquad -\frac{k^2}{24} \int \qquad\qquad 2qp^2 \qquad\qquad + 2q^4 \, dx$$

or their corresponding discretizations. For strong nonlinearities, both are $\mathcal{O}(q^4)$; for $q \sim p \sim 1$, L-N has two terms against nine, and no derivatives (which can be larger); and for weak nonlinearities $(q \sim p \ll 1)$, L-N is $\mathcal{O}(q^3)$ whereas P-Q is $\mathcal{O}(q^2)$. In addition, one may use the optimal RKN integrators; and the L-N splitting gains a factor $\frac{\pi}{2}$ in the stability criterion.

Integrals of the system are conserved if they are integrals of each part of the Hamiltonian separately. This is clearly the case for linear integrals (conserved by any consistent scheme anyway) and for bilinear integrals under both the P-Q and L-N splittings. When $J$ is constant, Casimirs are linear functions and hence conserved.

## 4. Behavior of P-Q splitting

With finite differences, P-Q splitting must be used, because computing $e^{kX_L}$ requires a Fourier transform. Even if L-N splitting is feasible, there is still the question of how the splitting acts on any derivatives remaining in $N$. For a simple analysis, we consider the linear wave equation $\dot{q} = p$, $\dot{p} = q_{xx}$ with P-Q splitting, and investigate the above methods with regard to stability and dispersion. The time-stepping is identical if one works in real or in Fourier space; choose the latter, so that the modes uncouple, and a change of scale reduces each to a linear oscillator:

$$\dot{q} = p, \qquad \dot{p} = -q.$$

Write one time step of the method as an explicit linear map

$$\begin{pmatrix} q^1 \\ p^1 \end{pmatrix} = A(k) \begin{pmatrix} q^0 \\ p^0 \end{pmatrix} = \begin{pmatrix} A_{11}(k) & A_{12}(k) \\ A_{21}(k) & A_{22}(k) \end{pmatrix} \begin{pmatrix} q^0 \\ p^0 \end{pmatrix}$$

where the polynomials $A_{ij}(k)$ can be found explicitly. One has by induction in the number of stages that $A_{11}$ and $A_{22}$ are even functions and $A_{12}$ and $A_{21}$ are odd. For symmetric methods, writing out the symmetry condition shows that $A_{11} = A_{22}$. Because the methods are symplectic, $\det A = 1$, and thus by standard stability analysis, the method is stable iff $|\text{tr} A(k)| < 2$. The exact solution for the linear oscillator is $A_0 = \begin{pmatrix} \cos k & \sin k \\ -\sin k & \cos k \end{pmatrix}$, so $A(k)$ for a method of order $p$ will agree with this up to terms of order $k^p$; thus the first wrong term in $\text{tr} A$ is of order $k^{p+2}$ for even-order methods.

PROPOSITION 2. *Consider $q_{tt} = q_{xx}$ discretized with time-step $k$ and spatial mesh size $h$ and a symplectic integrator with matrix $A(k)$ defined above. Let $k^*$ be the least positive root of $|\mathrm{tr} A(k)| = 2$. Then the stability criterion, depending on the spatial discretization, is*

$$(a)\ \textit{Pseudo-spectral differences:} \qquad \frac{k}{h} \leq \frac{1}{\pi} k^*$$

$$(b)\ \textit{Second-order finite differences:} \qquad \frac{k}{h} \leq \frac{1}{2} k^*$$

$$(c)\ \textit{Fourth-order finite differences:} \qquad \frac{k}{h} \leq \frac{\sqrt{3}}{4} k^*$$

PROOF. For the spectral discretization with Fourier modes $-\frac{M}{2}+1 \leq m \leq \frac{M}{2}$, we require stability for each oscillator ($\ddot{q}_m = -m^2 q_m$) separately; rescaling leads to $mk \leq k^*$ for $0 \leq m \leq \frac{M}{2}$. Then $M = 2\pi/h$ gives (a). (b) and (c) follow from standard von Neumann stability analysis.

EXAMPLE 3. For leapfrog, $\mathrm{tr} A/2 = 1 - k^2/2$ so $k^* = 2$. We therefore have stability in the spectral approximation for $k/h \leq \frac{2}{\pi} \sim 0.6366$. LF4a is worse: we find $2k^{*2} = 12 - 6(w + w^2) + 3\sqrt{-8 + 2w + 4w^2}$, where $w = \sqrt[3]{2}$; $k^* \sim 1.5734$ and we need $k/h \leq 0.5008$.

For the other methods the roots of the polynomials must be found numerically, and the corresponding stability criteria are given in Table 2. Notice that they are quite good—for non-symplectic methods (e.g. three-time-level leapfrog), one typically needs Courant numbers $k/h$ near 1 with finite differences, and near $1/\pi$ with spectral differences. The results apply to any linear PDE with P-Q splitting: if the time-continuous problem has eigenvalues $i\sigma_m$, then the stability criterion is $k\sigma_m < k^*$.

---

**Table 2.** Stability criteria, P-Q splitting

$k/h$ is the stability criterion for spectral differencing of the linear wave equation; $a$ is the first ($\mathcal{O}(k^p)$) term in the expansion of the phase speed error.

| Method | $k^*$ | $k/h$ | Phase error $a$ |
|--------|-------|-------|-----------------|
| LF2    | 2      | 0.6366 | $-4.2 \times 10^{-2}$ |
| LF4a   | 1.5734 | 0.5008 | $6.6 \times 10^{-2}$ |
| LF4b   | 2.7210 | 0.8661 | $9.3 \times 10^{-4}$ |
| LF4c   | 3.0389 | 0.9673 | $1.1 \times 10^{-4}$ |
| LF6a   | 2.2691 | 0.7223 | $-3.8 \times 10^{-3}$ |
| LF6b   | 3.0674 | 0.9764 | $-1.3 \times 10^{-6}$ |

---

No general time-integrator can be free of dispersion in general. Historically this has led to schemes which introduce artificial dissipation of the high modes to prevent "wiggles." Indeed Crank-Nicolson (symplectic for a linear PDE) is often frowned on for just this reason. Now we have expressly disallowed numerical dissipation. Does dispersion mean that we cannot expect good long-time behavior from symplectic integrators? Certainly it does in the case of the linear wave equation, for which any initial condition will eventually disperse into its constituent modes as all phase accuracy is lost. However, for nonlinear and particularly for near-integrable equations, we can hope that phase locking inherent in the system will prevent this. (Consider the ODE case of coupled oscillators, for example.) In addition, it turns out that some integrators (e.g. LF4c) and L-N splitting have negligible dispersion errors.

We take $k \leq k^*$ and calculate the eigenvectors of $A$; separating real and imaginary parts show that the phase space is foliated by similar invariant ellipses, of which one is

$$B \begin{pmatrix} \cos \alpha \\ \sin \alpha \end{pmatrix} = \begin{pmatrix} A_{12} & 0 \\ \cos \theta - A_{11} & \sin \theta \end{pmatrix} \begin{pmatrix} \cos \alpha \\ \sin \alpha \end{pmatrix}.$$

Applying the map to this ellipse and using $\det A = 1$ gives

$$A B \begin{pmatrix} \cos \alpha \\ \sin \alpha \end{pmatrix} = B \begin{pmatrix} \cos(\alpha - \theta) \\ \sin(\alpha - \theta) \end{pmatrix}$$

where $\cos \theta = \operatorname{tr} A / 2$. Thus the map moves a point an angle $\theta$ around the ellipse each time-step, giving a dispersion relation $\theta$. The exact map has $\theta = k$, and we are only considering a single wave, so the most natural error measure is the relative phase speed of that wave, $c = \theta / k$. Because $\cos \theta = \cos k + a k^{p+2} + o(k^{p+2})$ for even-order methods, we have $c \sim 1 - a k^p$ for small $k$ (see Table 2.) At $k = k^*$, $c = \pi / k^*$. The figure shows $c$ for different methods and also illustrates their stability limits.

## 5. Stability of L-N splitting

There are two approaches to the linear stability of splitting methods. Firstly, one can make general statements based on the generic bifurcations of symplectic vector fields and maps, giving sufficient conditions for linear stability. Secondly, the eigenvalues of the time-map can be computed explicitly for particular examples; the generic sufficient conditions turn out to be often necessary as well.

Here we are thinking of integrating $H = L + N$ with one of the composition methods in Table 1 with the resulting map $\widehat{\varphi}$ linearized about some steady state. Now $\widehat{\varphi} = e^{kX_{\widehat{H}}} = C$ is linear and hence is the time-$k$ map of some autonomous linear Hamiltonian $\widehat{H}$ which can be found directly: $\widehat{H} = u^T B u$ where $C^T B C = B$. In this case the asymptotic series

$$(5.1) \qquad \widehat{H} = H + k^p H_0 + \dots$$

(cf. (3.4)) will converge to $\widehat{H}$ for $k$ small enough. However, one cannot use the series to examine stability because near the onset of instability, typically all its terms are the same order in $k$. Examining the first term in (5.1) can determine

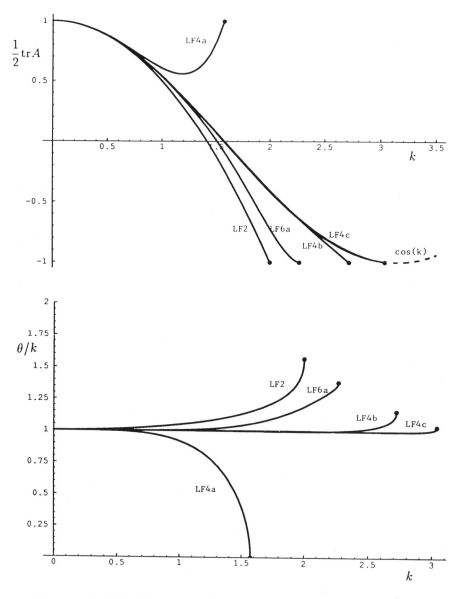

FIGURE. Stability & accuracy for explicit P-Q splitting. Top: $\frac{1}{2}\mathrm{tr}A$ for five explicit methods applied to the linear oscillator; the true solution is $\cos k$. The • shows the stability limit for each method. LF4c and LF6b are indistinguishable here. Bottom: Relative phase speed $c$ for each method.

when the conditions of the following proposition are *not* satisfied, and can help in choosing a good splitting of $H$. Then (roughly, if dispersion errors are $o(1)$ as $k \to 0$) two time-steps per period of the fastest wave are sufficient for stability:

PROPOSITION 3. *Suppose $X_H$ has pure imaginary eigenvalues $\{\pm i\sigma_m\}_{m=1}^M$, $\sigma_1 \leq \ldots \leq \sigma_M$, any multiple eigenvalues have positive signature, and any zero eigenvalues are associated with zeros of both $X_L$ and $X_N$. Let $k \to 0$ with $k\sigma_M = k^*$ held fixed, and assume that in this limit $\widehat{\varphi}$ is a small perturbation of $\varphi = e^{kX_H}$. (One may need to rescale the independent variables to get $\varphi \sim 1$ first.) Then, for $M$ sufficiently large, the method $\widehat{\varphi}$ is generically linearly stable for $k^* < \pi$.*

PROOF. We are investigating the stability of the fixed point at the origin to small symplectic perturbations. The nonunit eigenvalues of $\varphi$ are $e^{ik\sigma_m}$ which are are bounded away from $-1$ if $k^* < \pi$. Because the eigenvalues of the vector field have positive signature, so do those of its time-$k$ flow. These are just the requirements for generic stability of the origin when $\varphi$ is perturbed to $\widehat{\varphi}$ (Arnol'd [1], MacKay [8], [9]). Finally, if zero eigenvalues of $X_H$ come from zeros in $X_L$ and $X_N$, then there is a corresponding zero in $X_{\widehat{H}}$, so the $+1$ eigenvalues of $\varphi$ are fixed and do not split.

Notes:
1. If $H = T(p) + V(q)$, eigenvalues are guaranteed to have positive signature [9].
2. At $\pm k\sigma_m = \pi$, $\varphi$ has a double eigenvalue at $-1$. In the perturbed map $\widehat{\varphi}$ this generically splits into a real pair, signaling loss of stability in this mode. It may be a bubble of instability or a permanent loss.
3. The proposition applies if $H = L(q,p) + N(q)$ and $N$ has fewer derivatives of $q$ than $L$—as in the nonlinear wave equation $q_{tt} = q_{xx} + V'(q)$ and the Boussinesq equation (example 2).

EXAMPLE 4: A NONLINEAR WAVE EQUATION. Sine-Gordon $q_{tt} - q_{xx} + \sin(q) = 0$ linearized about $q = 0$ is Klein-Gordon. Consider LF2 with L-N splitting on this equation: each mode decouples into a linear map

$$A = e^{\frac{1}{2}kX_L}e^{kX_N}e^{\frac{1}{2}kX_L}$$

$$= \begin{pmatrix} \cos(mk) - \frac{k}{2m}\sin(mk) & \frac{k}{2m^2}(\cos(mk) - 1) + \sin(mk)/m \\ -\frac{k}{2}(\cos(mk) + 1) - m\sin(mk) & \cos(mk) - \frac{k}{2m}\sin(mk) \end{pmatrix}$$

(5.2) $$\cos\theta = \frac{1}{2}\text{tr}A = \cos(mk) - \frac{k}{2m}\sin(mk)$$

The stability limit for large $m$ is indeed $mk < \pi$, but beyond this there are bubbles of instability. Consider $k$ small with $mk = x$ held fixed, so $\frac{1}{2}\text{tr}A = \cos x - \frac{1}{2}k^2 \sin(x)/x$. Solving by series for $x$ near $l\pi$ shows that this is larger than 1 in absolute value for $x \in (l\pi - k^2/l\pi, l\pi)$. However, these instabilities are unlikely to be triggered as they are only $k/l\pi$ times as wide as the spacing between modes, hence one can easily choose $k$ so as to avoid them. The maximum growth rate in the bubbles is only $k^2/2\pi$.

Numerical experiments show that the formal stability limit can indeed be exceeded by a factor of three or four without the nonlinear terms triggering any instability, even for strong nonlinearities.

Expanding (5.2) for small $k$ and any $m$ gives the numerical dispersion relation:

$$\theta/k = \sqrt{1+m^2} + \frac{k^2}{24\sqrt{1+m^2}} + \mathcal{O}(k^4)$$

which is $\mathcal{O}(k^2)$ away from the true relation, uniformly in $m$. This, and the smaller truncation errors, are the great advantages of the L-N splitting.

## 6. A numerical example

For a numerical test of the above analyses consider the nonlinear wave equation

$$\dot{q} = p, \quad \dot{p} = -q_{xx} + q^3$$

on $[0, 2\pi]$ with periodic boundary conditions; the Hamiltonian discretized by pseudo-spectral differences and the (spectral) trapezoidal rule for the integral; and the time-integrator LP4c. (The advantage of this equation is that it is not integrable but it is known that $C^\infty$ initial data stays $C^\infty$ for all time.) The initial conditions considered are $p = 0$, $q = a\cos x$ for various amplitudes $a$. Standard properties of symplectic integrators were confirmed: (a) the energy error did not increase secularly with time; (b) the bilinear momentum integral $\int pq_x \, dx$ ($\sum m p_m q_{-m}$ for the ODE's) was conserved within round-off error; and (c) the above accuracy and stability analyses were confirmed (nonlinear terms could destabilize the calculation only when the solution was extremely poorly resolved spatially).

When $|q|$ is small the relative truncation error of P-Q splitting is $\mathcal{O}(1)$, against $\mathcal{O}(q^2)$ for L-N, so the latter is clearly superior then. Its maximum energy error was 2.4 times smaller at $a = 0.5$, and 3.2 times smaller at $a = 2$.

For all amplitudes $a$, P-Q splitting was stable for Courant numbers $c = k/h <\sim 0.94$ (cf. $k/h < 0.9673$ in Table 2). L-N was stable for $c <\sim 14$ when $a = 0.5$ (but only useful for $c < 5$) and for $c < 3$ when $a = 2$—L-N has no *linear* stability limit here. Clearly linear–nonlinear splitting is preferred when feasible.

*Acknowledgements.* I would like to thank Jim Curry and Harvey Segur for their helpful comments on an early version of this paper, and also Mark Ablowitz and Connie Schober for useful discussions.

## References

1. V. Arnol'd, *Mathematical Methods of Classical Mechanics*, Springer-Verlag, New York, 1988.

2. J. Candy and W. Rozmus, *A symplectic integration algorithm for separable Hamiltonian functions*, J. Comp. Phys. **92** (1991), 230–256.
3. Dai Weizhong, *An unconditionally stable three-level explicit difference scheme for the Schrödinger equation with a variable coefficient*, SIAM J. Num. Anal. **29(1)**, 174.
4. Feng Kang, *Difference schemes for Hamiltonian formalism and symplectic geometry*, J. Comp. Math. **4** (1986), 276–289.
5. J. de Frutos, T. Ortega, and J. M. Sanz-Serna, *Pseudospectral method for the "good" Boussinesq equation*, Math. Comp. **57(195)** (1991), 109–122.
6. _____, *A Hamiltonian, explicit algorithm with spectral accuracy for the "good" Boussinesq equation*, Comp. Meth. Appl. Mech. Eng. **80** (1990), 417–423.
7. J. de Frutos and J. M. Sanz-Serna, *An easily implementable fourth-order method for the time-integration of wave problems*, J. Comp. Phys. **103(1)** (1992), 160.
8. R. S. MacKay, *Some aspects of the dynamics and numerics of Hamiltonian systems*, The dynamics of numerics and the numerics of dynamics (D. S. Broomhead and A. Iserles, eds.), Oxford University Press, Oxford, 1992.
9. _____, *Stability of equilibria of Hamiltonian systems*, Nonlinear phenomena and chaos (Sarben Sarkar, ed.), Hilger, Bristol; Boston, 1986, pp. 254–270.
10. R. I. McLachlan and P. Atela, *The accuracy of symplectic integrators*, Nonlinearity **5** (1992), 541–562.
11. D. I. Okunbor and R. D. Skeel, *Canonical Runge-Kutta-Nyström methods of orders 5 and 6*, Math. Comp. (to appear).
12. P. J. Olver, *Applications of Lie groups to differential equations*, Springer-Verlag, New York, 1986.
13. R. D. Ruth, *A canonical integration technique*, IEEE Trans. Nucl. Sci. **NS-30** (1983), 2669–2671.
14. J. C. Scovel, *Symplectic numerical integration of Hamiltonian systems*, The Geometry of Hamiltonian Systems (T. Ratiu, ed.), Springer-Verlag, New York, 1991.
15. M. Suzuki, *General theory of fractal path integrals with applications to many-body theories and statistical physics*, J. Math. Phys. **32(2)** (1991), 400–407.
16. H. Yoshida, *Construction of higher order symplectic integrators*, Phys. Lett. A **150** (1990), 262–269.
17. _____, *Conserved quantities of symplectic integrators for Hamiltonian systems*, preprint.

PROGRAM IN APPLIED MATHEMATICS, UNIVERSITY OF COLORADO AT BOULDER, BOULDER, CO 80309-0526

*E-mail address*: rxm@boulder.colorado.edu

Lectures in Applied Mathematics
Volume **29**, 1993

# Symmetric Capillary Surfaces in a Cube
# Part 2. Near the Limit Angle

HANS D. MITTELMANN

*Dedicated to Robert Finn on the occasion of his 70th birthday.*

March 9, 1993

ABSTRACT. Numerical experiments are described by which stable capillary surfaces are calculated. The surfaces in question are determined by the following data: the container is a cube in space; the contact angle is near the critical angle of $45^0$; the Bond number is zero; only symmetric configurations are taken into consideration.

## 1. The Mathematical Problem

Let $\Phi$ be the unit cube in $\mathbf{R}^3$. We are considering subdomains $\Omega \subset \Phi$ having a piecewise smooth boundary $\partial\Omega = \Gamma \cup \Sigma$, where $\Gamma$ is a subset of the interior of $\Phi$ and $\Sigma$ is a subset of the boundary $\partial\Phi$ of $\Phi$. We are looking for those subdomains $\Omega$ which solve the following variational problem: The energy functional

$$E = \int_\Gamma d\Gamma - \cos\vartheta \int_\Sigma d\Sigma$$

is minimal under the restriction that the volume

$$V = \int_\Omega d\Omega$$

attains a prescribed value. It is a well known fact - going back to K. F. Gauß - that solutions of this variational problem must be such that the *capillary surface* $\Gamma$ has constant mean curvature

$$2H = p$$

---

1991 *Mathematics Subject Classification.* Primary 65N30, 76B45; Secondary 49Q10, 35R35.

This work was supported in part by the U.S. Air Force Office of Scientific Research Grant AFOSR #90-0080.

and the *contact angle* between $\Gamma$ and $\Sigma$ equals to $\vartheta$ (see, e.g., [6]). The number $p$ is the Lagrange multiplier of the variational problem. It turns out that it is equal to the difference of the pressures of two liquids (fluid or gas) occupying the domains $\Omega$ and $\Phi \setminus \Omega$, resp.

So far, for the problem studied here, namely capillary surfaces in a cube under zero gravity conditions, there are no existence nor uniqueness proofs, except for special cases. Nevertheless, the results presented in this paper are consistent with laboratory experiments that have been performed up to now.

There are several papers dealing with similar problems. The problem of determining shapes of capillary surfaces experimentally, mathematically, and numerically for zero gravity conditions has been studied in [5] and [9], see also [14]. In [4] the package EVOLVER was used to determine the shape of equilibrium capillary surfaces for exotic containers. The same package was used for the numerical studies of the present paper, see section 5. A numerical method that allows solving the Euler-Langrange equation for the variational problem described above has been presented in [7]. This method, and also the EVOLVER, allows calculating surfaces that are not simple graphs of functions over a planar domain; i.e., more complicated geometrical configurations can be treated rather than only projectable surfaces. The following is known for a semi-infinite cylindrical tube of general cross section with gravity zero (or positive): If the boundary of the free surface lies entirely on the cylindrical walls, then the surface is determined uniquely by its contact angle and volume, see [13]. Therefore, in that case the standard numerical methods determine a uniquely defined solution, if it exists, see [10]. In the more general case of a finite container such as a cube, there is no uniqueness result of that type. This lack of uniqueness makes the numerical computations especially important.

A more recent study of bifurcation phenomena for problems with axial symmetry is given in [8]. There, for drops that are entrapped between two parallel planes, numerical methods were used that depend strongly on path-following techniques, see [1]. The situation considered here was first addressed in [11]. In that paper a contact angle of $70^0$ was chosen and symmetric solutions were computed. For each surface the values of the total energy, of its surface area, and of the pressure inside the liquid were graphed as a function of volume. Symmetry was a condition imposed on the non-contiguous distributions of the liquid. One of the interesting results was that over a large range of the volume a contiguous configuration ("pumpkin") yielded the global energy minimum and thus represented the physically most stable solution. In the computations of this surface no symmetry requirements were imposed and, in fact, during the course of the iteration unsymmetric shapes were observed which after convergence disappeared resulting in a completely symmetric surface.

It is well-known, see [6], that beyond certain critical angles, here $45^0$ and $135^0$, solutions cease to be bounded. Therefore, one purpose of this work is to specifically study the behaviour of the symmetric capillary surfaces in a cube under zero gravity near these critical angles, The emphasis is on the explicit and reproducable numerical solution of these problems and on the verification of nonlinear phenomena inherent in the capillarity problem. Some of these phenomena can be observed particularly well. Here, especially the unboundedness of surfaces for contact angles below $45^0$ manifests itself very clearly due to the boundedness and the shape of the container. The emphasis of this work is less on investigating to a fuller extent the bifurcation phenomena of the capillary surfaces considered. The principal reason for this is that a numerical approach, cf. section 5, was chosen with which only minima of the total energy can be computed. The approaches utilized in [**7, 11**] do not have this restriction, and in [**11**] a complete bifurcation diagram could be obtained.

## 2. Symmetric Configurations

In this paper we restrict ourselves to domains that share symmetries with the cube. It is obvious that after prescribing the contact angle $\vartheta$ and the volume $V$ and then finding a solution $\Omega$ of the variational problem, the set $\tilde{\Omega} = \Phi \backslash \Omega$ solves the variational problem for the data $\tilde{\vartheta} = \pi - \vartheta$ and $\tilde{V} = 1 - V$. Conversely, if we prescribe $\tilde{\vartheta} = \pi - \vartheta$ and $\tilde{V} = 1 - V$ and find the set $\tilde{\Omega}$, we get a solution of the original problem as $\Omega = \Phi \backslash \tilde{\Omega}$. We call this the *complementary configuration*.

In the following we are going to describe the various topological situations and show plots of figures that were obtained numerically for the contact angles $\vartheta = 50^0$ and $\vartheta = 40^0$. While the case $\vartheta = 70^0$ was considered in [**11**] it is worthwhile to consider angles close to 45 (135) degrees which is the limit angle for unbounded solutions due to the interior angles of $90^0$ (see, e.g., [**6**]). What will happen is that for angles smaller than $45^0$ the liquid will extend infinitely along the edges of the cube. In those cases where the volume $\Omega$ is the union of several unconnected symmetric sets we will in general show only one of these sets. In order to indicate the container, we also show the bottom face of the cube. First, the case $\vartheta = 50^0$ is considered and for ease of comparison the same configurations are described and depicted as in [**11**]. Then we point out the differences in the case $\vartheta = 40^0$.

(i) **The Corners:** For this situation we assume that $\Omega$ is the union of eight symmetric sets $\Omega_i$, $i = 1, ..., 8$ which are attached to the eight corners of the cube. This case is denoted by "Ce". If $\tilde{\Omega}$ is this union, the complementary configuration $\Omega$ is denoted by "Ci". Obviously, these solutions make sense only as long as the individual sets $\Omega_i$ do not touch each other. Hence, they are taken into consideration only for a range of volumes of the form $0 < V < V_{Ce}$ and $V_{Ci} < V < 1$, resp., with some values $V_{Ce}$ and $V_{Ci}$. Figure 1 shows one eighth of $\Omega$ for the case "Ce" with $V = 0.1$.

(ii) **The Edges:** For this situation we assume that $\Omega$ is the union of four symmetric sets $\Omega_i$, $i = 1, ..., 4$ which are attached to four of the twelve

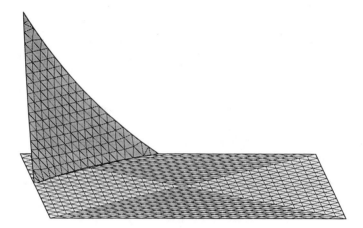

FIGURE 1. Ce: Exterior Corner, $V = 8 * 0.0125 = 0.1$

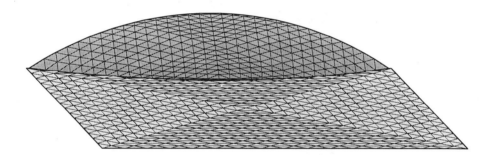

FIGURE 2. Le: Exterior Lemon, $V = 4 * 0.006 = 0.024$

edges of the cube. Each of the $\Omega_i$ is assumed to cover a part of the edge not touching the neighbouring corners. This case is denoted by "Le" (lemon). The complementary situation is denoted by "Li". The domains of existence are $0 < V < V_{Le}$ and $V_{Li} < V < 1$, resp. Figure 2 shows one fourth of $\Omega$ for the case "Le" with $V = 0.024$.

(iii) **The Faces:** For this situation we assume that $\Omega$ is the union of six symmetric sets $\Omega_i$, $i = 1, ..., 6$ which are attached to the six faces of the cube without touching one of the edges or each other. This case is denoted by "Oe" (orange). The complementary situation is denoted by "Oi". Here we have $0 < V < V_{Oe} = 0.5769$ and $0.6695 = V_{Oi} < V < 1$, resp., see section 3. Figure 3 shows one sixth of $\Omega$ for the case "Oi" with $V = 0.4$.

(iv) **Bridges between Two Corners:** For this situation we assume that $\Omega$ is the union of four symmetric sets $\Omega_i$, $i = 1, ..., 4$ which are attached

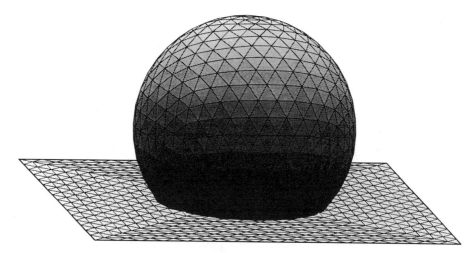

FIGURE 3. Oi: Interior Orange, $V = 1 - 6 * 0.1 = 0.4$

to four of the twelve edges of the cube. Each of the $\Omega_i$ is assumed to cover the whole edge including the neighbouring corners. This case is denoted by "Se" (sausage). The complementary situation is denoted by "Si". Here we have $V_{Se}^- < V < V_{Se}^+$ and $V_{Si}^- < V < V_{Si}^+$, resp. Figure 4 shows one fourth of $\Omega$ for the case "Se" with $V = 0.12$.

(v) **Dry Spots:** For this situation we assume that $\Omega$ is the union of two symmetric sets $\Omega_i$, $i = 1, 2$ which are attached to two of the six faces of the cube. Each of the $\Omega_i$ is assumed to cover completely the edges that belong to the face including the neighbouring corners but to leave out an uncovered fraction in the interior of the face. This case is denoted by "De". The complementary situation is denoted by "Di". Here we have $V_{De}^- < V < V_{De}^+$ and $V_{Di}^- < V < V_{Di}^+$, resp. Figures 5 and 7 show one half of $\Omega$ for the cases "De" with $V = 0.20$ and "Di" with $V = 0.76$, resp.

(vi) **The Pumpkin:** For this situation we assume that $\Omega$ is a connected set which covers all eight corners of the cube and also all twelve edges but which leaves out interior parts of all six faces and also a certain volume in the interior of the cube itself. This case is denoted by "Pe". The complementary situation is denoted by "Pi". Here we have $V_{Pe}^- < V < V_{Pe}^+$ and $V_{Pi}^- < V < V_{Pi}^+$, resp. Figures 6, 8, 9, 10, 11, and 12 show the set $\Omega$ for the cases "Pe" with $V = 0.2, 0.5, 0.6$ and "Pi" with $V = 0.35, 0.5, 0.85$, resp.

(vii) **The Cylinder:** For this situation we assume that $\Omega$ is the union of two symmetric sets $\Omega_i$, $i = 1, 2$ which are attached to two opposite ones

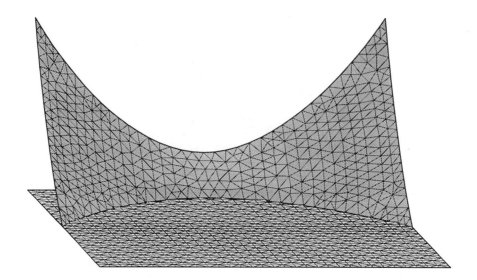

FIGURE 4. Se: Exterior Sausage, $V = 4 * 0.03 = 0.12$

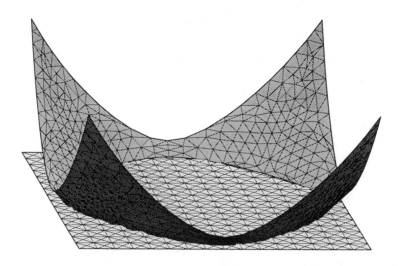

FIGURE 5. De: Exterior Dry Spot, $V = 2 * 0.05 = 0.2$

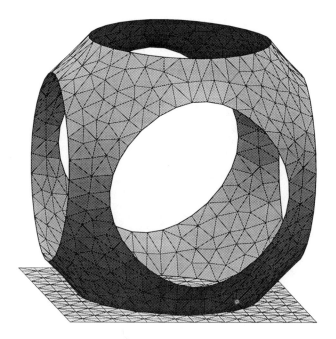

FIGURE 6. Pe: Exterior Pumpkin, $V = 0.2$

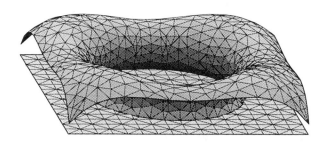

FIGURE 7. Di: Interior Dry Spot, $V = 1 - 2 * 0.12 = 0.76$

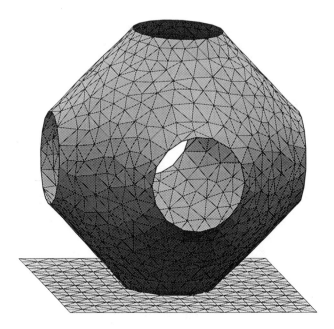

FIGURE 8. Pe: Exterior Pumpkin, $V = 0.5$

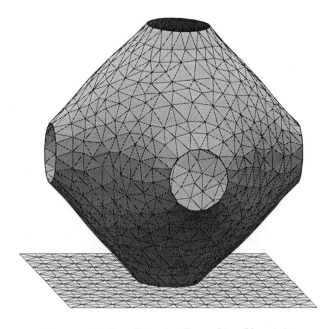

FIGURE 9. Pe: Exterior Pumpkin, $V = 0.6$

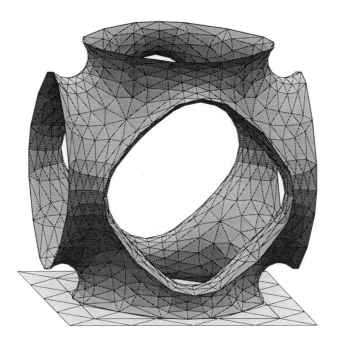

FIGURE 10. Pi: Interior Pumpkin, $V = 0.35$

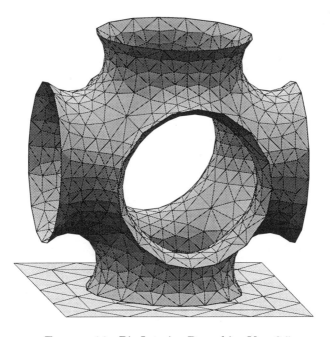

FIGURE 11. Pi: Interior Pumpkin, $V = 0.5$

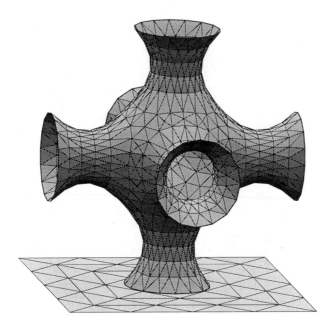

FIGURE 12. Pi: Interior Pumpkin, $V = 0.85$

of the six faces of the cube covering these faces completely including the neighbouring edges and parts of the neighbouring faces. This case is denoted by "Te" (tub). The complementary situation is denoted by "Ti". Here we have $V_{Te}^- < V < V_{Te}^+$ and $V_{Ti}^- < V < V_{Ti}^+$, resp. Figure 13 shows the set $\Omega$ for the case "Te" with $V = 0.3$.

(viii) **The Ball:** Here we assume that $\Omega$ is the exterior of a ball that has no contact to any of the six faces of the cube. This case is denoted by "Be". Here we have $V_{Be} < V < 1$.

While for a contact angle of $50^0$ all the same configurations exist as for $70^0$ there are important differences which will be apparent from the graphs given in the next section. If, however, an angle below $45^0$, say, $40^0$ is chosen then the following must be true. Any distribution that has liquid covering part of an edge will lead to a complete covering of all edges and corners of the cube, no matter how small the volume. This configuration was called pumpkin above and thus the only other configurations left for such angles are the orange and the ball. While the interior pumpkin has a similar shape and range of existence as for the case $\vartheta = 50^0$, we depict two cases of the exterior pumpkin, one for a small volume $V = .01$ in figure 14 and in figure 15 for $V = .59$ just before collapsing to an exterior ball, or, in other words, to a bubble surrounded by liquid.

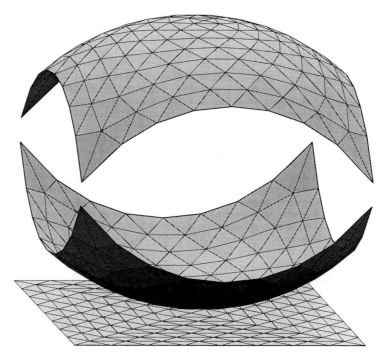

FIGURE 13. Te: Exterior Tub, $V = 2 * 0.15 = 0.3$

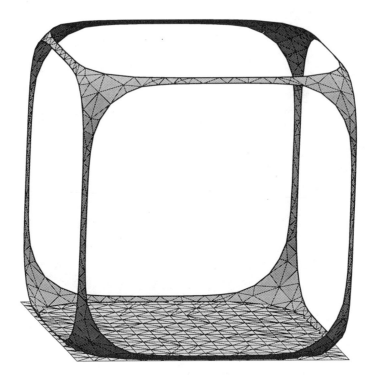

FIGURE 14. Pe4: Exterior Pumpkin, $V = 0.01$

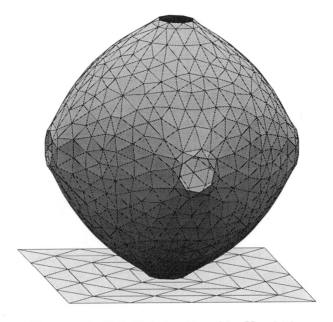

FIGURE 15. Pe5: Exterior Pumpkin, $V = 0.59$

## 3. Results in Closed Form

For simple geometric configurations one can give the volume $V$, the energy $E$, the pressure $p$, and the area $A = \int_\Gamma d\Gamma$ in closed form.

- **The Ball:** If the domain $\Omega$ is the complement with respect to the cube $\Phi$ of a ball having radius $r$, i.e., in situation "Be", we have the following relations. The volume $V$ and the area $A$ are given by

$$V = 1 - \frac{4\pi}{3} r^3, \quad A = 4\pi r^2.$$

Therefore, the energy $E$ is

$$E = A - \cos \vartheta |\partial\Phi|,$$

where $|\partial\Phi| = 6$ is the area of the surface of the container. The pressure $p$ is equal to $2H$, where $H = \frac{1}{r}$ is the mean curvature of the surface of the ball, hence

$$p = \frac{2}{r}.$$

- **The Orange:** If $\Omega$ is one slice of height $h < r$ of a ball which has radius $r$, we have the following relations. With

$$h = r(1 - \cos \vartheta), \quad s = r \sin \vartheta$$

we have

$$V = \frac{\pi}{3} h^2 (3r - h), \quad A = 2\pi r h$$

and

$$p = \frac{2}{r}.$$

Hence, in the situation "Oe" we get for the six slices

$$V_e = 6V, \quad A_e = 6A, \quad \text{and} \quad E_e = A_e - 6\pi s^2 \cos \vartheta.$$

Simple geometric considerations lead one to the restriction

$$0 < r < r_{max} = \frac{1}{2} \min\{\frac{1}{\sqrt{2} - \cos \vartheta}, \frac{1}{\sin \vartheta}\}.$$

If $\Omega$ is the complement with respect to the ball of a slice cut from the ball, we get the following relations.

$$V = \frac{4\pi}{3} r^3 - \frac{\pi}{3} h^2 (3r - h), \quad A = 2\pi r (2r - h)$$

and

$$p = -\frac{2}{r}.$$

Hence, in the situation "Oi" we get for the complement of the six sets of this type

$$V_i = 1 - 6V, \quad A_i = 6A, \quad \text{and} \quad E_i = A_i - 6(1 - \pi s^2) \cos \vartheta.$$

Here the restrictions to be satisfied are

$$0 < r < r_{max} = \frac{1}{2}\frac{1}{\sqrt{2} + \cos\vartheta}.$$

## 4. Energy, Area, and Pressure Graphs

For an analysis of the surfaces presented above graphs were produced from quantities that are part of the output of the EVOLVER program. First, the case $\vartheta = 50^0$ is considered. There are curves of the following three relations: Figure 16 shows $E$ versus $V$, figure 17 shows $p$ versus $V$, and figure 18 shows $A$ versus $V$. All these curves are obtained by calculating the solutions using the EVOLVER package. For each of the solutions, one gets not only the surface $\Gamma$ described by its vertices, edges, and faces, but also the values for $V$, $E$, $A$, and $p$. These numerical values were used to draw polygonal lines that connect the points in the $V$-$E$-plane, the $V$-$p$-plane, and the $V$-$A$-plane. Since obviously $p \to \infty$ for $V \to 0$ in the cases "Le" and "Oe", and $p \to -\infty$ for $V \to 1$ in the cases "Be", "Li", "Ci", and "Oi", the $V$-$p$-curves were cut at $p = 10$ and $p = -10$, resp. Also, the curves for "Ci" and "Oe" were cut at the extreme values of the volume and the energy curve for "Ti" was scaled by .5 to blow up the interesting portions of the graphs.

Since the case $\vartheta = 70^0$ has already been solved in [11] it is interesting to compare the present case of a near-limiting angle with that. While some surfaces look similar in both cases, for example, those of the "Pumpkin" there are also significant differences. Figure 1 of the exterior corner shows that the surface is concave and not convex as one might think in light of the fact that the contact angle is larger than $45^0$. As a result, the pressure, being equal to the mean curvature, is negative and decreases with $V$. Similarly the exterior lemon has a pressure tending to $\infty$ for $V \to 0$. How the liquid starts creeping along the edges of the cube can be observed comparing figures 4,5, 7, and 13 with the corresponding ones in [11]. In addition to curvatures having an opposite sign to that of the same figures for $\vartheta = 70^0$, the volume ranges for which these surfaces exist are greatly reduced. As was already stated in section 2, for a contact angle less than the critical angle only 3 configurations exist, the pumpkin, the orange, and the ball. In figures 19, 20, 21 the corresponding graphs are given.

**4.1. Special Properties.** If one considers an experiment in which the cube is partially filled with a liquid leaving the complement as void and if one increases the value of the volume $V$ from 0 to 1, the following sequence of stable configurations will result for $\vartheta 50^0$: 1) The corners "Ce", then from $V = 0.075$ on 2) the sausage "Se", after this from $V = 0.125$ on 3) the pumpkin "Pe", and finally from $V = 0.6$ on 4) the ball "Be". This observation is based on the assumption that one considers only configurations with the symmetries described in this paper, and that one looks only for those cases which have the absolute minimum of the energy functional $E$. This means that the other cases for which figure

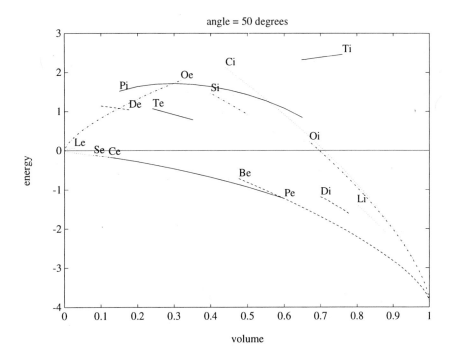

FIGURE 16. Energy versus Volume

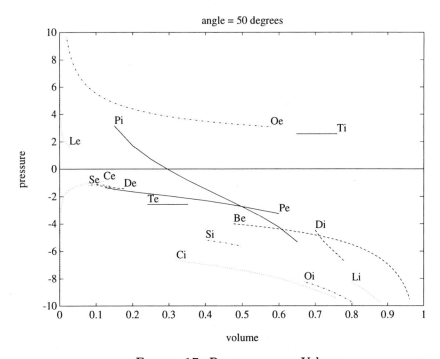

FIGURE 17. Pressure versus Volume

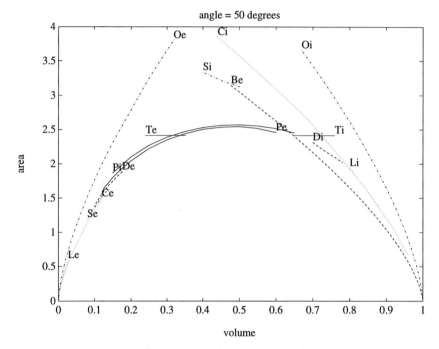

FIGURE 18. Area versus Volume

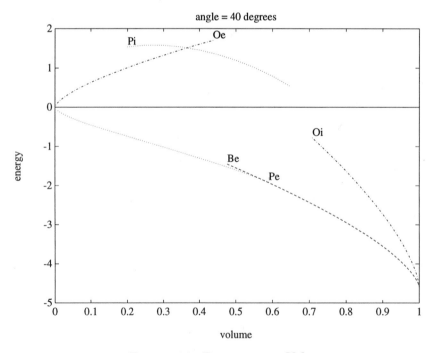

FIGURE 19. Energy versus Volume

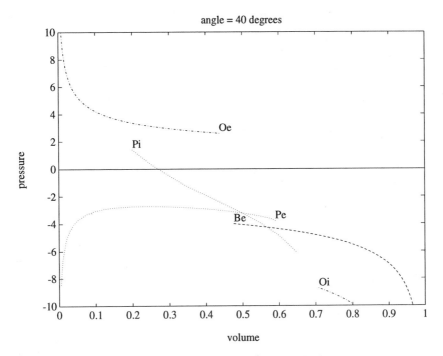

FIGURE 20. Pressure versus Volume

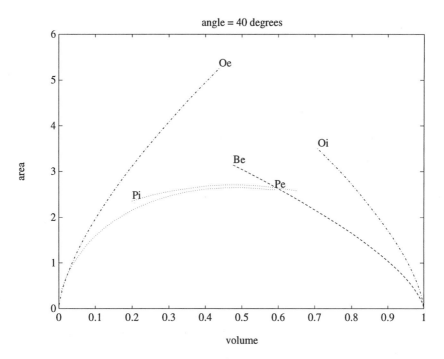

FIGURE 21. Area versus Volume

16 shows larger values of the energy $E$ represent only local minima of the variational problem. In principle, they can be obtained using careful experimental manipulations; but they are only locally stable.

For certain values of the volume $V$ there are several different local minima. As an example, let us look at $V = 0.49$. There are five different possible configurations for this volume. Here, the values of the energy increase in the following order: "Pe", "Be", "Si", "Pi", "Ci". It depends on the history of an experiment which of the configurations will actually be observed in a given situation. This is a pronounced *hysteresis* effect.

The curves $p$ versus $V$ are the derivatives $p = \frac{dE}{dV}$ of the curves $E$ versus $V$. This is due to the fact that the Lagrange multiplier of a variational problem is the derivative of the objective function with respect to the value of the constraint. For these curves one can see that for large values of the volume $V$ the pressure $p$ is always negative in this case, except for "Ti". For those cases in which the pressure happens to be zero, one gets *minimal surfaces* for $\Gamma$, since then the mean curvature $H$ is zero. This is the case here only for "Pi" at $V = .3$.

The curves $A$ versus $V$ seem to indicate that for the two cases "Pe" and "Pi" one has $A_i(V) = A_e(V)$, i.e., that the areas $A$ are (nearly) the same for the same volume $V$ prescribed in the two cases, cf. [11].

From the basic concept of the package EVOLVER it is clear that one can get only local minima of the variational problem. It is not possible to calculate other stationary points of the energy functional using this program package. One way of doing that would be to use an approach that has been studied previously in [7]. Therefore, the curves in the $V$-$E$-plane leave the branches out which have saddle-points. In this sense, we have calculated only a part of the full bifurcation diagrams.

## 5. The Numerical Method

Although the results displayed in graphical form above are of a certain aesthetic value and of physical interest, there are several other questions that should be addressed. One is certainly concerned with the continuation of the various curves shown in the previous section. While it is clear that at the endpoints of these curves the surfaces either cease to exist and give rise to a topologically different configuration or, at least, they cease to be local minima of the total energy. It would be of interest to investigate what is happening in each case but this could not be done with the approach used. For such an investigation in a similar case see [8].

Another question is for details of the numerical procedure used. As was claimed in section 1 all the above results are reproducable. A software package [3] was used which can be retrieved as given in the citation. In fact, the utilization of this package for the computations has led to several improvements of the code. While a detailed user's manual is part of the distribution it is still useful

FIGURE 22. Initial configuration for dry spot

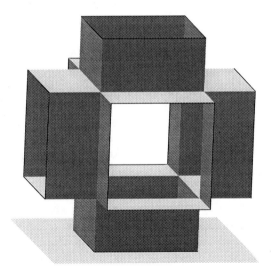

FIGURE 23. Initial configuration for pumpkin

to outline how the code works, how it needs to be called, and to report about experiences. The EVOLVER program is based on minimizing a rather general energy functional under a very general set of constraints. Ideally, an initial configuration that is topologically equivalent to the expected solution is provided by the user. It is beyond the scope of this work to describe how the energy functional has to be defined in general and reference is made again to the user's manual. Since, however, the computations needed for this work use only some of the facilities of the package an exemplary case will be given here in more detail. The complete input file for the dry spot is given below. The initial configuration for the dry spot is shown in figure 22. The user has to define the vertices with their (x,y,z)- coordinates, the edges through their endpoints, the faces each given by an oriented loop of its edges, and the liquid volume through its faces. Furthermore, the vertices and edges that are constrained to move along certain sides of the cube get these constraints as attributes.

The constraints themselves are defined geometrically, the bottom corresponds to z=0 etc. and the energy functionals originally defined on $\Gamma$ respectively $\Sigma$ (cf. section 1) have to be expressed differently. In fact, the program handles the integrals over the facets on $\Gamma$ automatically, while for a reason explained in the user's guide the integrals over $\Sigma$ are reduced via Stoke's theorem to integrals along the edges of these facets. If, for example, the left side, $y = 0$, is considered, there one has to find a vector field $w$ such that curl $w = Tj$, where $j$ is the unit vector in $y$-direction. From the definition of $T$ it is then clear that the energy associated with the edges of the facets of the body on the left side is the same that was originally associated with this facet. One vector field which satisfies the above relation is $w = (Tz, 0, 0)$. From Stoke's theorem then the integral of $Tj$ over the facet is equal to that of $w$ over its edges which is defined in the first constraint in the input file. Similarly, all other constraints can be verified. To finally illustrate which initial configuration was used for the pumpkin figure 23 is included. The input file is not given due to its length, but it may be derived analogously to the one for the dry spot given above.

The initial surface is coarsely triangulated and then one proceeds by alternately iterating and refining. The minimization is accomplished either through steepest descent or conjugate gradient steps. On the average, for surfaces that have about 1000 vertices one needs 10 steepest descent iteration steps followed by at least 50 conjugate gradient steps to obtain results with sufficient accuracy. But this was only true if the triangulation was very carefully chosen with the principal goal of obtaining near-equilateral triangles of sufficiently small side-length. Between 500 and 5000 vertices where needed to represent the surfaces. The method underlying this program package is in principle similar to the technique described in [12].

```
// dry.fe
// Evolver data for liquid of prescribed volume inside box.
// Contact angle can be varied between 45 and 90 degrees.
```

```
// Initially there is a dry spot on bottom

PARAMETER angle = 50  // interior angle between plane and surface,
degrees

GRAVITY_CONSTANT:        0

#define T  (-cos(angle*pi/180))  // virtual tension of facet on plane

constraint 1   /* the left side */
function:  y = 0
energy:  // for contact angle //
e1: T*z
e2: 0
e3: 0

constraint 2   /* the right side */
function:  y = 1
energy:  // for contact angle //
e1: -T*z
e2: 0
e3: 0

constraint 3   /* the front side */
function:  x = 1
energy:  // for contact angle //
e1: 0
e2: T*z
e3: 0

constraint 4   /* the back side */
function:  x = 0
energy:  // for contact angle //
e1: 0
e2: -T*z
e3: 0

constraint 5   /* the bottom */
function:  z = 0
energy:  // for contact angle //
e1: T*y
e2: 0
e3: 0

vertices   /* given by coordinates and attribute */
```

```
1     0.0   0.0   0.25   constraint 1,4
2     1.0   0.0   0.25   constraint 1,3
3     1.0   1.0   0.25   constraint 2,3
4     0.0   1.0   0.25   constraint 2,4
5     .25   .25   0.0    constraint 5
6     .75   .25   0.0    constraint 5
7     .75   .75   0.0    constraint 5
8     .25   .75   0.0    constraint 5
9     0.0   0.0   0.0    fixed
10    1.0   0.0   0.0    fixed
11    1.0   1.0   0.0    fixed
12    0.0   1.0   0.0    fixed


edges   /* given by endpoints and attribute */
1    1   2     constraint 1
2    2   3     constraint 3
3    3   4     constraint 2
4    4   1     constraint 4
5    1   5
6    2   6
7    3   7
8    4   8
9    5   6     constraint 5
10   6   7     constraint 5
11   7   8     constraint 5
12   8   5     constraint 5
13   9  10     fixed
14  10  11     fixed
15  11  12     fixed
16  12   9     fixed


faces   /* given by oriented edge loop */
1    1    6   -9   -5
2    2    7  -10   -6
3    8  -11   -7    3
4    5  -12   -8    4
5  -16  -15  -14  -13    fixed


bodies  /* one body, defined by its oriented faces */
1    1 2 3 4    volume 0.1  density 1
```

### REFERENCES

1. E. L. Allgower and K. Georg, *Numerical Continuation Methods*, Springer, Berlin (1990).

2. F. Almgren, *Computing soap films and crystals*, Computing Optimal Geometries (J. E. Taylor, ed.), AMS, Providence, RI (1991), 3–5.
3. K. A. Brakke, *The Surface Evolver*, Experimental Mathematics **1** (1992), 141–165, Program and manual available per anonymous ftp from geom.umn.edu (128.101.25.31).
4. M. Callahan, P. Concus, and R. Finn, *Energy minimizing capillary surfaces for exotic containers*, Computing Optimal Geometries, (J. E. Taylor, ed.), AMS, Providence, RI (1991), 13–15.
5. P. Concus and R. Finn, *Capillary surfaces in microgravity*, Low-Gravity Fluid Mechanics and Transport Phenomena, (J. N. Koster and R. L. Sani, eds.), Progress in Astronautics and Aeronautics **130**, AIAA, Washington, DC, (1990), 183–206.
6. R. Finn, *Equilibrium capillary surfaces*, Springer, New York (1986).
7. U. Hornung and H. D. Mittelmann, *A Finite Element Method for Capillary Surfaces with Volume Constraints*, Journal of Computational Physics **87** (1990), 126–136.
8. U. Hornung and H. D. Mittelmann, *Bifurcation of Axially Symmetric Capillary Surfaces*, Journal of Colloid and Interface Science **146** (1991), 219–225.
9. D. Langbein, *Problems in fluid statics and fluid dynamics under microgravity conditions*, Free Boundary Problems: Theory and Applications I (K.-H. Hoffmann and J. Sprekels, eds.), Pitman, Boston (1990), 110–137.
10. H.D. Mittelmann, *Die Approximation der Lösungen gemischter Randwertprobleme quasilinearer elliptischer Randwertprobleme*, Computing **13** (1974), 252–265.
11. H.D. Mittelmann, *Symmetric capillary surfaces in a cube*, Mathematics and Computers in Simulation (to appear).
12. S. Saha, P. Das, and N. N. Kishore, *A shape-optimization technique for the capillary surface problem*, Numer. Math. **58** (1991), 737–757.
13. T. I. Vogel, *Uniqueness for certain surfaces of prescribed mean curvature*, Pacific J. Math. **134** (1988), 197–207.
14. H. U. Walter, ed., *Fluid Science and Material Science in Space*, Springer, Berlin (1987).

DEPARTMENT OF MATHEMATICS, ARIZONA STATE UNIVERSITY, TEMPE, AZ 85287-1804
*E-mail address*: mittelmann@math.la.asu.edu

Lectures in Applied Mathematics
Volume **29**, 1993

# Factorization and Completely Integrable Systems

D.H. SATTINGER & J.S. SZMIGIELSKI

## 1. Introduction

Factorization theory and Lie algebra decompositions, (called the Adler-Kostant-Symes method) play a central role in the theory of completely integrable systems. We begin with a simple illustratation in the case of the Toda flow.

The Toda flow,

$$\dot{a}_n = a_n(b_{n+1} - b_n), \qquad \dot{b}_n = 2(a_n^2 - a_{n-1}^2),$$

has the Lax representation $\dot{A} = [A, B]$ where

$$A = \begin{pmatrix} a_1 & b_1 & 0 & \dots & & \\ b_1 & a_2 & b_2 & 0 & \dots & \\ 0 & b_2 & a_3 & b_3 & 0 & \\ & \ddots & \ddots & & b_{n-1} & \\ & & & b_{n-1} & a_n & \end{pmatrix}, \qquad B = \begin{pmatrix} 0 & b_1 & 0 & & & 0 \\ -b_1 & 0 & b_2 & 0 & & \\ 0 & -b_2 & 0 & b_3 & & \vdots \\ & & \ddots & \ddots & & b_{n-1} \\ & & & -b_{n-1} & & 0 \end{pmatrix}.$$

The invariants of the motion are $H_k = \operatorname{tr} A^k$.

LEMMA. *Any real non-singular matrix $A$ can be uniquely factored as $A = QR$ where $Q$ is orthogonal and $R$ is lower triangular with positive diagonal entries.*

*Any matrix $X$ can be uniquely decomposed as $X = S + T$ where $S$ is skew symmetric and $T$ is lower triangular.*

The proof is a simple exercise in linear algebra. We denote the projections by $S = (X)_s$, $T = (X)_l$. For example, the decomposition of $A$ in the Toda flow is $A_s = B$, so the Toda flow can be written as

$$\dot{A} = [A, A_s].$$

1991 *Mathematics Subject Classification.* 35Q55, 38F07.

This research was supported in part by N.S.F. Grant DMS-8901607 and DMS-9123844

This paper is in final form and no version of it will be submitted elsewhere for publication.

THEOREM. [6] *Let $A(t)$ be a trajectory of the Toda flow and suppose $e^{tA(0)} = Q(t)R(t)$. Then $A(t) = Q^\dagger(t)A(0)Q(t)$; i.e. the family $A(t)$ is unitarily equivalent to $A(0)$ via the family of orthogonal transformations $Q(t)$.*

PROOF. Differentiating the identity $Q(t)R(t) = e^{tA(0)}$ with respect to $t$ we get

$$\dot{Q}R + Q\dot{R} = A(0)e^{tA(0)} = A(0)QR,$$

hence

$$Q^\dagger\dot{Q} + \dot{R}R^{-1} = Q^\dagger A(0)Q. \tag{1}$$

Now $Q^\dagger\dot{Q}$ is skew symmetric and $\dot{R}R^{-1}$ is lower triangular, so (1) is the unique decomposition of $Q^\dagger A(0)Q$ as the sum of a skew symmetric matrix and a lower triangular matrix. We write this decomposition as

$$Q^\dagger\dot{Q} = (Q^\dagger A(0)Q)_s, \qquad \dot{R}R^{-1} = (Q^\dagger A(0)Q)_l$$

where $(\ )_s$ and $(\ )_l$ denote the projections onto the skew symmetric and lower triangular parts of a matrix.

Hence

$$\frac{d}{dt}Q^\dagger A(0)Q = \dot{Q}^\dagger A(0)Q + Q^\dagger A(0)\dot{Q}$$

$$= [Q^\dagger A(0)Q, Q^\dagger\dot{Q}] = [Q^\dagger A(0)Q, (Q^\dagger A(0)Q)_s].$$

Hence $A(t)$ and $Q^\dagger A(0)Q$ satisfy the same equations and the same initial values. $\square$

## 2. AKNS Systems

We now turn to an infinite dimensional case and consider a similar procedure of factorization and Lie algebra decomposition.

Consider the commuting family of operators

$$\frac{\partial}{\partial x} - zJ, \qquad \frac{\partial}{\partial t_k} - z^k\mu, \qquad k = 1, 2, \ldots, \tag{2}$$

where $J$ and $\mu$ are constant diagonal matrices, and $z$ is a complex parameter (the spectral parameter). We have

$$[\frac{\partial}{\partial x} - zJ, \frac{\partial}{\partial t_k} - z^k\mu] = 0$$

Consider a function $m(x, t, z)$ such that

$$\left(\frac{\partial}{\partial x} - u(x, t, z)\right)m = m\left(\frac{\partial}{\partial x} - zJ\right),$$

$$\left(\frac{\partial}{\partial t_k} - G_k(x, t, z)\right)m = m\left(\frac{\partial}{\partial t_k} - z^k\mu\right). \tag{3}$$

$m$ is said to *dress* the bare operators to the non trivial operators.

Now let $g$ be a matrix-valued function defined on the unit circle in the $z$ plane and suppose we can solve the Riemann-Hilbert Factorization problem:

$$e^\theta g(z) e^{-\theta} = m_-^{-1} m_+, \qquad \theta = xzJ + t_k z^k \mu, \tag{4}$$

where

$$m_- = \sum_{j=0}^\infty m_j(x,t) z^{-j}, \qquad m_0 = I, \qquad \det m_- \equiv 1$$

and $m_+$ is an entire function of $z$.

Now let

$$D_0 = \frac{\partial}{\partial x} - z \mathrm{ad}J, \qquad D_k = \frac{\partial}{\partial t_k} - z^k \mathrm{ad}\mu,$$

and note that

$$D_k e^\theta g(z) e^{-\theta} = 0.$$

The intertwining relations (3) can be written

$$u^\pm = \frac{\partial m_\pm}{\partial x} m_\pm^{-1} + m_\pm z J m_\pm^{-1}, \qquad G_k^\pm = \frac{\partial m_\pm}{\partial t_k} m_\pm^{-1} + m_\pm z^k \mu m_\pm^{-1}, \tag{5}$$

where $u^\pm$, $G_k^\pm$ are the values of $u, G_k$ in $|z| < \infty$ and $|z| > 1$.

We leave it to the reader to derive the identities on $|z| = 1$:

$$m_-(D_k e^\theta g(z) e^{-\theta}) m_+^{-1} = \begin{cases} G_k^+ - G_k^- & k > 0 \\ u^+ - u^- & k = 0 \end{cases}$$

Consequently $u$ and $G_k$ are continuous across the unit circle and are therefore entire functions of $z$. Since $m \to I$ as $z \to \infty$, $u = O(z)$ and $G_k = O(z^k)$ as $z \to \infty$, so $u$ is linear in $z$ and $G_k$ is a polynomial of degree $k$ in $z$.

Denoting the projection onto non-negative powers of $z$ by $(\cdot)_+$ we find

$$u = zJ + q = (mzJm^{-1})_+, \qquad q = [m_1, J];$$

and

$$G_k = (z^k F)_+ = z^k \mu + B_k(z)$$

where

$$F = m \mu m^{-1}$$

and $B_k$ is a polynomial of degree $k - 1$ in $z$.

The role of the Lax pair of the Toda flow is played by the connection with components

$$D_x = \frac{\partial}{\partial x} - u, \qquad D_{t_k} = \frac{\partial}{\partial t_k} - G_k.$$

The situation here is slightly different than in §1; in this case the solutions of the factorization problem ($m_\pm$) are used to dress the bare operators to obtain the non-trivial connection. The projection in this case is $(\cdot)_+$ .

The entries of $F$ are differential polynomials in $q$ ([2]), and $F$ satisfies the differential equation

$$[\frac{\partial}{\partial x} - u, F] = 0, \qquad F_0 = \lim_{z \to \infty} F(z) = \mu.$$

A hierarchy of nonlinear flows is given by the zero curvature conditions

$$[D_x, D_{t_k}] = 0, \qquad D_x = \frac{\partial}{\partial x} - u, \qquad D_{t_k} = \frac{\partial}{\partial t_k} - [z^k F]_+.$$

Expanding

$$F = \sum_{j=0}^{\infty} F_j z^{-j}, \qquad F_0 = \mu,$$

the zero curvature condition is equivalent to

$$[J, F_{j+1}] = [\frac{\partial}{\partial x} - q, F_j], \quad j < k, \qquad q_{t_k} = [J, F_{k+1}] \tag{6}$$

In general the factorization (3) is an infinite dimensional problem (Riemann-Hilbert problem). But in certain cases it simplifies to an algebraic problem.

*Example.* As an example we construct rational solutions of the modified KdV equation. These may be obtained by taking $\mu = J = \sigma_3$ and

$$D_x = \frac{d}{dx} - z\sigma_3 - u\sigma, \qquad \sigma = \begin{pmatrix} 0 & 1 \\ 1 & 0 \end{pmatrix} \qquad \sigma_3 = \begin{pmatrix} 1 & 0 \\ 0 & -1 \end{pmatrix}.$$

From the recursion relations (6) one finds that

$$[z^3 F]_+ = z^3 \sigma_3 + z^2 F_1 + z F_2 + F_3$$

where

$$F_1 = q = u\sigma, \qquad F_2 = \frac{1}{2} \begin{pmatrix} -u^2 & u_x \\ -u_x & u^2 \end{pmatrix}, \qquad F_3 = \left( \frac{u_{xx}}{4} - \frac{u^3}{2} \right) \sigma,$$

The Modified KdV flow is the flow at $t_3$, *viz.*

$$q_t = \frac{\partial F_3}{\partial x} - [q, F_3], \qquad \text{or} \qquad u_t = \tfrac{1}{4} u_{xxx} - \tfrac{3}{2} u^2 u_x.$$

However, the computations are simplified by working in a basis in which $\sigma$, rather than $J$ is diagonal. We will see this again in §3 in dealing with the general Gel'fand-Dikki flows. We therefore conjugate by

$$P = \frac{1}{\sqrt{2}} \begin{pmatrix} 1 & -1 \\ 1 & 1 \end{pmatrix}.$$

In the new basis,

$$D_x = \frac{d}{dx} + z\sigma - u\sigma_3.$$

We note the symmetry $\sigma_3 D_x(z, q)\sigma_3^{-1} = D_x(-z, q)$. We take $g$ to be a function whose determinant is identically 1 and which satisfies the symmetry $\sigma_3 g(z)\sigma_3^{-1} = g(-z)$. Such a function is given by

$$g = \left( I + \frac{1}{z} \begin{pmatrix} 0 & 0 \\ 1 & 0 \end{pmatrix} \right) \left( I + \frac{1}{z} \begin{pmatrix} 0 & 1 \\ 0 & 0 \end{pmatrix} \right)$$

$$= I + \frac{1}{z} \begin{pmatrix} 0 & 1 \\ 1 & 0 \end{pmatrix} + \frac{1}{z^2} \begin{pmatrix} 0 & 0 \\ 0 & 1 \end{pmatrix}$$

Taking $\theta = (xz + tz^3)\sigma$, the factorization problem is solved if we can find an $m_-$ of the form

$$m_- = I + \frac{1}{z}m_1 + \frac{1}{z^2}m_2$$

such that

$$m_- e^\theta g(z) \text{ is an entire function of } z.$$

We leave it to the reader to show that

$$e^\theta g(z) = \sum_{j \geq -2} \Lambda_j z^j$$

where

$$\Lambda_{-2} = \begin{pmatrix} 0 & 0 \\ 0 & 1 \end{pmatrix} \qquad \Lambda_{-1} = \begin{pmatrix} 0 & 1+x \\ 1 & 0 \end{pmatrix}$$

$$\Lambda_0 = \begin{pmatrix} 1+x & 0 \\ 0 & 1+x+x^2/2 \end{pmatrix} \qquad \Lambda_1 = \begin{pmatrix} 0 & x^2/2+x+\omega \\ x^2/2+x & 0 \end{pmatrix}$$

where $\omega = t + x^3/6$.

The factorization problem reduces to the linear system of algebraic equations

$$m_2\Lambda_{-2} = 0, \qquad m_2\Lambda_{-1} + m_1\Lambda_{-2} = 0,$$
$$m_2\Lambda_0 + m_1\Lambda_{-1} + \Lambda_{-2} = 0, \qquad m_2\Lambda_1 + m_1\Lambda_0 + \Lambda_{-1} = 0.$$

The symmetry $\sigma_3 m(x, z)\sigma_3^{-1} = m(x, -z)$ implies $\sigma_3 m_j \sigma_3^{-1} = (-1)^j m_j$, hence

$$m_j = \begin{cases} \begin{pmatrix} a_j & 0 \\ 0 & d_j \end{pmatrix}, & j \text{ even}, \\[2mm] \begin{pmatrix} 0 & b_j \\ c_j & 0 \end{pmatrix}, & j \text{ odd}. \end{cases}$$

The equations are overdetermined; nevertheless, one can solve them to obtain

$$c_1 = -\frac{1}{x+1}, \qquad b_1 = \frac{(x+1)^2}{\frac{-x^3}{3} - x - x^2 + t - 1}.$$

The potential $q$ in this basis is given by

$$q = [\sigma, m_1] = (c_1 - b_1)\sigma_3,$$

and

$$u(x, t) = \frac{-1}{x+1} + \frac{(x+1)^2}{1 + x^2 - t + x + \frac{x^3}{3}}.$$

## 3. The Gel'fand-Dikii flows

Gel'fand and Dikii [3] constructed a hierarchy of isospectral flows of the $n^{th}$ order scalar differential operator

$$L = \sum_{j=0}^{n} u_j(x) D^{n-j}; \quad D = \frac{1}{i}\frac{d}{dx}; \quad u_0 = 1, \quad u_1 = 0.$$

where $u_j = u_j(x)$, $j > 2$, are elements of the Schwartz class $\mathcal{S}(\mathbb{R})$. The flows are given by

$$\dot{L} = [L_+^{k/n}, L], \quad k = 1, 2, \dots \tag{7}$$

where $k \neq 0 \mod n$, and $L_+^{k/n}$ denotes the differential part of $L^{k/n}$ considered as a pseudodifferential operator. We shall refer to the coefficients $u_j$ of $L$ as the potential. The coefficients of $L_+^{k/n}$ are obtained by solving formal recursion relations and are universal differential polynomials in the $\{u_j\}$. The scattering theory of $L$ proceeds by converting the $n^{th}$ order scalar equation $Lv = \lambda v$ to a first order system. Let $\phi_j = D^{j-1}v$ for $j = 1, \dots, n$. Then the column vector $\phi = (\phi_1, \phi_2, \dots, \phi_n)^{\dagger}$ satisfies the first order system

$$(D - J_\lambda - q)\phi = 0, \tag{8}$$

where

$$J_\lambda = \begin{pmatrix} 0 & 1 & 0 & \dots & 0 \\ 0 & 0 & 1 & \dots & 0 \\ & & \dots & & \\ 0 & 0 & 0 & \dots & 1 \\ \lambda & 0 & 0 & \dots & 0 \end{pmatrix}, \quad q = -\begin{pmatrix} 0 & 0 & \dots & 0 & 0 \\ 0 & 0 & \dots & 0 & 0 \\ & & \dots & & \\ 0 & 0 & \dots & 0 & 0 \\ u_n & u_{n-1} & \dots & u_2 & 0 \end{pmatrix}. \tag{9}$$

The linear space of all such $q$ constitutes a special class of potentials, which Drinfeld and Sokolov call "canonical potentials". Another reduction to a first order system is obtained by factoring $L = (D + v_n)(D + v_{n-1}) \cdots (D + v_1)$; then the equation $Lv = \lambda v$ can be written as a first order system (8) with $q$ a diagonal matrix with entries $v_1, \dots, v_n$. ( $\operatorname{tr} q = 0$ since the coefficient of $D^{n-1}$ vanishes.) This reduction corresponds to the modified Gel'fand-Dikki flows, a generalization of the modifed KdV flow.

Let $\alpha_1, \dots, \alpha_n$ be the $n^{th}$ roots of unity, and let

$$J = \operatorname{diag}(\alpha_1, \dots, \alpha_n), \quad d(z) = \operatorname{diag}(1, z, \dots, z^{n-1}), \quad \Lambda_z = d(z)\Lambda(\alpha),$$

where

$$\Lambda(\alpha) = \begin{pmatrix} 1 & 1 & 1 & \dots & 1 \\ \alpha_1 & \alpha_2 & & \dots & \alpha_n \\ \alpha_1^2 & & \dots & & \alpha_n^2 \\ & & \dots & & \\ \alpha_1^{n-1} & & & \dots & \alpha_n^{n-1} \end{pmatrix}.$$

Then $\Lambda_z J_\lambda \Lambda_z^{-1} = zJ$.

When $q$ is diagonal, $\Lambda_z^{-1} q \Lambda_z = \tilde{q}$ is independent of $z$. Hence we can write the isospectral problem as

$$(D - zJ + \tilde{q})\tilde{\psi} = 0.$$

This is the isospectral problem for an AKNS system, whose flows were constructed in §2, corresponding to the modified Gel'fand-Dikii flows.

The coefficients of $L$ are differential polynomials in the entries of $q$. When $n = 2$,

$$D^2 + u = (D + v)(D - v), \qquad u = -v_x - v^2;$$

and the transformation is known as the Miura transformation.

Other reductions to first order systems are also possible, and all reductions are gauge equivalent. Drinfeld and Sokolov considered the flows generated by the isospectral operator

$$D_x = \frac{\partial}{\partial x} - J_\lambda - q$$

where $q$ is a lower triangular matrix. In each case, the entries of $q$ are differential polynomials in the $u_j$. In [4] we considered factorization problems associated with three special cases of the Drinfeld-Sokolov flows, corresponding to the modified, potential, and standard Gel'fand-Dikii flows. We describe those results briefly here.

Consider the space of analytic functions of $\lambda$ with values in $M_n(C)$. We can expand any such matrix in one of two ways:

$$a = \sum_k a_k J_\lambda^k = \sum_k g_k \lambda^k,$$

where $a_k \in \mathcal{D}$, the space of diagonal matrices in $M_n$; and $g_k \in M_n(C)$.

Define projections $P_1$ and $P_2$ (denoted respectively by $P_+$ and $P^+$ in [2]) by

$$P_1 a = \sum_{k \geq 0} g_k \lambda^k, \qquad P_2 a = \sum_{k \geq 0} a_k J_\lambda^k.$$

Consider the subgroups

$$G_1^- = \{g : \det g = 1, \quad g = \sum_{j \leq 0} g_j \lambda^j, \quad g_0 = I\}$$

$$G_1^+ = \{g : \det g = 1, \quad g = \sum_{j \geq 0} g_j \lambda^j\},$$

and

$$G_2^- = \{g : \det g = 1, \quad g = \sum_{j \leq 0} g_j J_\lambda^j, \quad g_j \in \mathcal{D}, \quad g_0 = I\}$$

$$G_2^+ = \{g : \det g = 1, \quad g = \sum_{j \geq 0} g_j J_\lambda^j, \quad g_j \in \mathcal{D}\}.$$

As in §2 we solve a factorization problem, this time

$$e^{i(xJ_\lambda + tJ_\lambda^k)} g(\lambda) e^{-i(xJ_\lambda + tJ_\lambda^k)} = m_-^{-1}(x, t, \lambda) m_+(x, t, \lambda), \quad m_- \in G_j^- \qquad (10)$$

*Remark:* Note that we again find ourselves working in a basis in which $J_\lambda$ is not diagonal. It is considerably more convenient, in treating the Drinfeld-Sokolov flows, to work in the present basis, rather than one in which $J_\lambda$ is diagonalized.

As in §2 we have $(m = m_-)$

$$(\frac{\partial m}{\partial x})m^{-1} + mJ_\lambda m^{-1} = J_\lambda + q, \qquad (\frac{\partial m}{\partial t_k})m^{-1} + mJ_\lambda^k m^{-1} = G_k \qquad (11)$$

where $q = q(x,t)$ is independent of $\lambda$ and $G_k$ is a polynomial in $\lambda$:

$$G_k = J_\lambda^k + B_{n,k}(\lambda).$$

We may obtain $q$ and $G_k$ as the projections

$$J_\lambda + q = P_j(mJ_\lambda m^{-1}) \qquad G_k = P_j(mJ_\lambda^k m^{-1}).$$

THEOREM. *[4] For $j = 1,2$ the connections constructed above by solving the factorization problem are flat: $[D_x, D_t] = 0$, and $G_k$ is a differential polynomial in $q$.*

*For $j = 2$, $q$ is a diagonal matrix of trace zero, and the corresponding flows are the $(n, k)$ modified Gel'fand-Dikii flows.*

*For $j = 1$ $q$ has the structure*

$$q = \begin{pmatrix} m_{1n} & 0 & \cdots & 0 \\ m_{2n} & 0 & \cdots & 0 \\ & \vdots & & \\ m_{nn} - m_{11} & -m_{12} & \cdots & -m_{1n} \end{pmatrix},$$

*where $m_{jk}$ are the entries of the coefficient matrix $m_1$ in the Laurent expansion of $m_-$.*

The flows for the case $j = 1$ are called the "potential Gel'fand-Dikii flows." [7]. In these two cases, the factors lie in a subgroup, and the connection potentials (i.e. $q$ and $G_k$) are obtained as Lie algebra decompositions (the projection $P_j$).

The construction of the Gel'fand-Dikii flows themselves is complicated by the fact that the submanifold of values of the wave function $m_-$ does does not have a group structure when $n \geq 4$ [4]. The explicit description of that submanifold is somewhat technical and is given in [4]; we denote it by $G_3^-$. When the factorization problem (10) is solved for $m_- \in G_3^-$, we proved that $q$ has the canonical form given in (9). Since we no longer have a loop algebra decomposition, we cannot obtain $G_k$ as a projection.

Instead we argue as follows. The wave function $\psi = m_- \exp[xJ_\lambda + tJ_\lambda^k]$, where $m_- \in G_3^-$ is a solution of the factorization problem, satisfies the simultaneous equations

$$(D - J_\lambda - q)\psi = 0, \qquad (D_t - G_k)\psi = 0,$$

where $G_k = J_\lambda^k + B_{n,k}$. From the first equation we see that $\psi$ is a Wronskian: $\psi = ||D^{j-1}v_i||$, $i,j = 1,\ldots,n$, and that the entries of the first row of $\psi$ satisfy $Lv = \lambda v$.

From the second equation we find that each of the $v_j$ satisfies

$$\frac{\partial v}{\partial t} = \sum_{j=1}^{n} G_{1j} D^{j-1} v$$

where the $G_{1j}$ denote the entries of the first row of $G_k$. Since $Lv_j = \lambda v_j$, we can replace $\lambda^s v$ by $L^s v$. The result is that each $v_j$ satisfies an equation of the form $v_t = P_k v$ where $P_k = D^k + \ldots$ is a differential operator of order $k$. Since these equations hold for a complete set of independent solutions $v_1, \ldots, v_n$ we must have

$$[L, \frac{\partial}{\partial t} - P_k] = 0, \quad \text{or} \quad \dot{L} = [P_k, L].$$

These equations imply that the commutator $[P_k, L]$ is a differential operator of order $n - 2$. It follows that the coefficients of $P_k$ are differential polynomials of the coefficients of $L$ and that in fact $P_k = L_{+}^{k/n}$ (cf. Proposition 2.3 [2]).

Thus factorizations with $m_{-} \in G_3^{-}$ yield the Gel'fand-Dikii flows.

One interesting class of flows is the rational solutions. For $n = 2$ the rational solutions are governed by the Calogero-Moser equations for the KdV equation. It would be interesting to construct rational solutions of the entire Gel'fand-Dikii hierarchy, as well as some of the other Drinfeld-Sokolov flows.

The reduction of the $n^{th}$ order scalar equation to a first order system by the standard method (Wronskians) used in scattering theory [1] and by Drinfeld and Sokolov [2]. As noted above this conversion does not lead to a Lie algebra decompostion of the loop algebra. Nevertheless, it is still possible to carry through the dressing method. Schilling [5] has introduced a different conversion, which he calls the "natural conversion", given by

$$\phi_j = (L^{(j-1)/n})_{+} v, \quad j = 1, \ldots, n,$$

where $Lv = \lambda v$. The two conversions differ by a lower triangular gauge transformation; and the natural conversion leads to a *bona fide* Lie algebra decomposition. The flows obtained, however, are not the Gel'fand- Dikii flows, but are related to them by a differential subsitution, just as the Modified Gel'fand-Dikii flows are related by a generalized Miura transformation. It would be interesting to determine the corresponding factorization and flows.

### REFERENCES

1. R. Beals, P. Deift, and C. Tomei, *Direct and Inverse Scattering on the Line*, Amer. Math. Soc., Providence, R.I., 1989.
2. V.G. Drinfel'd and V.V. Sokolov, *Lie algebras and equations of Korteweg-deVries type*, Journal of Soviet Mathematics **30** (1985), 1975-2036.
3. I.M. Gelfand, and L.A. Dikii, *Fractional powers of operators and Hamiltonian systems*, Functional Analysis and its Applications. **10** (1976), 259-273.
4. D. H. Sattinger and J. Szmigielski, *Factorization and the Dressing Method for the Gel'fand-Dikii Hierarchy*, Physica D (to appear).
5. R. Schilling, *A loop algebra decomposition for Korteweg-deVries equations*, Integrable and Superintegrable Systems (B. Kupershmidt, eds.), World Scientific Press, Singapore, 1990.

6. Symes, W., *The QR algorithm and scattering for the finite nonperiodic Toda lattice*, Physica D **4** (1982), 257-280.

7. J. Szmigielski, *Infinite dimensional homogeneous manifolds with translational symmetry and nonlinear partial differential equations*, Disseration, University of Georgia-Athens, Ga., 1987.

DEPARTMENT OF MATHEMATICS, UNIVERSITY OF MINNESOTA, MINNEAPOLIS, MINNESOTA 55455

*E-mail address*: dhs@math.umn.edu

UNIVERSITY OF SASKATCHEWAN, SASKATOON, SASKATCHEWAN, CANADA S7N 0W0

*E-mail address*: szmigiel@snoopy.usask.can

Lectures in Applied Mathematics
Volume **29**, 1993

# Hopf/Steady-state Mode Interaction for a Fluid Conveying Elastic Tube with D₃-symmetric Support

A. STEINDL

ABSTRACT. The loss of stability of the straight downhanging equilibrium of a tube conveying fluid is investigated. The tube has circular cross-section and is elastically supported by visco-elastic springs arranged at the corners of a regular triangle. For certain parameter values the loss of stability occurs by the simultaneous occurence of a Hopf and a steady state bifurcation. Center manifold reduction and normal form theory lead to a low dimensional system of equivariant bifurcation equations. By restricting to isotropy subspaces the solutions with nontrivial isotropy subgroups are classified. The invariant subspaces are used to calculate the stability properties of the solution branches.

## 1. Introduction

The dynamics of tubes conveying fluid has been studied intensively over a long period of time from quite different points of view. While the linear models were used to calculate the critical flow rate ([**2, 8, 11**]), nonlinear investigations determined the post-critical behaviour of the pipes. Breaking the mirror reflection symmetry of a planar system at a Hopf/Steady-state interaction is studied in ([**9**]). The transition from planar to spatial models with rotational symmetry showed that planar oscillations and rotating solutions are possible ([**1**]). By adding a rotational symmetric elastic support of appropriate stiffness more degenerate bifurcations could be generated, leading to a large variety of interacting branches of solutions. All solutions are quite easily classified and visualized by

1991 *Mathematics Subject Classification.* Primary: 35B32, 73H10, Secondary: 73K05, 58G28.

*Key words and phrases.* Hopf bifurcation, Steady state bifurcation, mode interaction, bifurcation, equivariance, tube.

This paper is in final form and no version of it will be submitted for publication elsewhere

This research project has been supported by the Austrian Science Foundation (Fonds zur Förderung der wissenschaftlichen Forschung) under project P07003.

their symmetry properties. For the mathematical treatment a strong similarity
to the bifurcation analysis of the Taylor-Couette problem([4]) is given.

In [11] comparisons are reported between theoretical and experimental results
for the stability boundary of elastically supported tubes. As far as available we
work with the parameter values used in that report.

In order to get rid of the artificial rotational symmetry of the support and to
gain further insight into problems with less symmetry we started our investigations for a support with discrete symmetry $\mathbf{D}_4$, which is generated by 4 identical
springs located at the endpoints of a square. Results of this investigation are
reported in [10].

Hopf bifurcations with dihedral symmetry $\mathbf{D}_n$ are discussed in [4], indicating
qualitatively quite different branching behavior for even and odd values of $n$.
The case $n = 3$ is somewhat exceptional, since the lowest order nonlinear terms
in the Hopf/Steady state mode interaction are quadratic, but the stability of
the standing wave oscillations depends on fifth order terms, while the branching
behavior of the steady state bifurcation is determined by the coefficients of the
quadratic and cubic terms. Since the second order terms lead to unstable solution
branches, we regard the $\mathbf{D}_3$-symmetry of the support as small perturbation of the
rotational symmetric model. As perturbation parameter we may choose the ratio
$\varepsilon_s = \ell/\ell_s$, where $\ell$ is the length of the tube and $\ell_s$ is the length of the springs.
If we make $\ell_s$ infinitely large, we recover a rotational symmetric problem.

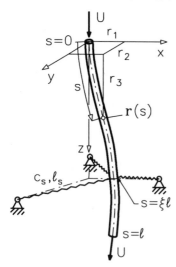

FIGURE 1.    Fluid conveying
elastic tube with $\mathbf{D}_3$-symmetric
elastic support.    One spring is
arranged along the positive $x$-
axis. The position of the centerline is denoted by $\boldsymbol{r}(s)$.

## 2. Mechanical model and equations of motion

The tube is treated as an inextensible thin linearly visco-elastic rod of uniform annular cross-section. The effects of rotatory inertia and shear deformation
are ignored, that is plane sections before deformation remain plain during deformation (Bernoulli-Euler theory). The tube is clamped at the upper end $s = 0$

and free at the lower end $s = \ell$. At the location $s = \xi\ell$ the tube is supported
by $n$ identical linearly visco-elastic springs, arranged at the corners of a regular
$n$-gon. One spring is located at the positive $x$-axis. The fluid enters the pipe at
the upper end and flows with constant speed $U$ through the pipe.

The configuration of the tube is given by the position of its centerline $r(t, s)$
and the orientation of the cross-sections, expressed by an orthogonal matrix
$\mathbf{B}(t, s) \in \mathbf{SO}(3)$. The columns of $\mathbf{B}$ can be visualized as a tripod which is
attached to the centerline of the tube and coincides with the spatial unit vectors
$e_i$ in the straight downhanging configuration. The third column of $\mathbf{B}$ coincides
with the tangent of the centerline $\partial r / \partial s$.

In nondimensional terms the equations of motion are (derivatives with respect
to arclength $s$ are denoted by $(\cdot)'$ and time derivatives by $(\dot{\cdot})$.)

$$
\begin{aligned}
r' &= \mathbf{B}e_3, \\
\mathbf{B}' &= \mathbf{B} \cdot \widehat{\mathbf{\Omega}}, \\
T' &= T \times \mathbf{\Omega} + F \times e_3, \\
F' &= F \times \mathbf{\Omega} + \mathbf{B}^T \left( \ddot{r} + 2\sqrt{\beta}\rho\dot{r}' + \rho^2 r'' + \alpha_e \dot{r} - \gamma e_3 \right), \\
r(0) &= 0, \\
\mathbf{B}(0) &= \mathbf{E}, \\
F(\xi_+) - F(\xi_-) &= -\mathbf{B}^T(\xi)f_s, \\
F(1) &= 0, \\
T(1) &= 0,
\end{aligned}
$$

(1)

with the material law relating the resultant couple vector $T$ to the bending and
twisting deformations $\kappa_i$:

$$
\begin{aligned}
T_1 &= \kappa_1 + \alpha_1 \dot{\kappa}_1, \\
T_2 &= \kappa_2 + \alpha_1 \dot{\kappa}_2, \\
T_3 &= \gamma_3(\kappa_3 + \alpha_3 \dot{\kappa}_3),
\end{aligned}
$$

(2)

and

$$
\widehat{\mathbf{\Omega}} = \begin{pmatrix} 0 & -\kappa_3 & \kappa_2 \\ \kappa_3 & 0 & -\kappa_1 \\ -\kappa_2 & \kappa_1 & 0 \end{pmatrix}.
$$

$T$ and $F$ are the resultant couple and force vectors in body coordinates, the
skewsymmetric matrix $\widehat{\mathbf{\Omega}}$ and its associated vector $\mathbf{\Omega}$ contain the bending and
twisting deformations $\kappa_i$. The parameter $c$ denotes the stiffness of the support,
$\gamma$ is proportional to the weight of the tube and fluid, $\rho$ is proportional to the
flow rate $U$ and the $\alpha_i$ are damping coefficients. $\mathbf{E}$ denotes the unit matrix, $\gamma_3$
measures the torsional rigidity and $\beta$ is the mass ratio $m_F/(m_F + m_T)$, where
$m_F$ and $m_T$ are the mass of the fluid and tube per unit length, respectively.

Equations $(1_1)$ and $(1_2)$ are geometric relations between the configuration
variables $r$ and $\mathbf{B}$ and the bending and twisting deformations $\mathbf{\Omega}$. The loading of
the tube is due to the acceleration of the fluid and the tube, external damping
and gravity. Balance of forces and momenta yields $(1_4)$ and $(1_3)$, respectively.

The system (1) is a vector valued geometric nonlinear three-dimensional version of the scalar partial differential equations in [11], taking also the gravitational effects into account.

The 2nd order expansion of the restoring force created by the three springs, which leads to the jump condition ($1_7$) for the resultant force $\boldsymbol{F}$ at $s = \xi$, is given by

$$
(3) \quad \boldsymbol{f}_s = -c \left(
\begin{array}{c}
r_1 + \alpha_s \dot{r}_1 + \varepsilon_s \left( \dfrac{3(r_1^2 - r_2^2)}{4} + \alpha_s(\dot{r}_1 r_1 - \dot{r}_2 r_2) \right) \\[2ex]
r_2 + \alpha_s \dot{r}_2 - \varepsilon_s \left( \dfrac{3(r_1 r_2)}{2} + \alpha_s(\dot{r}_1 r_2 + \dot{r}_2 r_1) \right) \\[2ex]
0
\end{array}
\right)_{s = \xi}.
$$

Due to the circular cross-section of the tube and the symmetric arrangement of the support the differential equations (1) are equivariant under the representation of $\mathbf{D}_n$:

$$
(4) \qquad
\begin{array}{ll}
\boldsymbol{r} \mapsto \mathbf{D}\boldsymbol{r}, & \mathbf{B} \mapsto \mathbf{D} \cdot \mathbf{B} \cdot \mathbf{D}^T, \\[1ex]
\boldsymbol{T} \mapsto (\det \mathbf{D}) \, \mathbf{D}\boldsymbol{T}, & \boldsymbol{F} \mapsto \mathbf{D}\boldsymbol{F},
\end{array}
$$

where $\mathbf{D} \in \mathbf{D}_n$ is a (possibly improper) rotation matrix

$$
(5) \quad \mathbf{D} = \mathbf{R}_3(\zeta) = \left(
\begin{array}{ccc}
\cos \zeta & -\sin \zeta & 0 \\
\sin \zeta & \cos \zeta & 0 \\
0 & 0 & 1
\end{array}
\right) \quad \text{or} \quad \mathbf{D} = \mathbf{diag}(1, -1, 1) \cdot \mathbf{R}_3(\zeta),
$$

with $\zeta = 2k\pi/n$.

The transformations (4) with $\mathbf{D} = \mathbf{R}_3(\zeta)$ are obtained by rotating the tube rigidly around the vertical axis and counterrotating the cross-sections about the central axis by the same amount.

## 3. Linearization

The straight downhanging configuration

$$
(6) \qquad
\begin{array}{ll}
\boldsymbol{r}_0 = s\boldsymbol{e}_3, & \mathbf{B}_0 \equiv \mathbf{E}, \\[1ex]
\boldsymbol{T}_0 \equiv \mathbf{0}, & \boldsymbol{F}_0 = (1 - s)\gamma \boldsymbol{e}_3
\end{array}
$$

is a solution of (1) for all parameter values. It is called the "trivial solution" and is invariant under $\mathbf{D}_n$.

The linearization of (1) at the trivial solution is given by the differential equations

$$
(7) \qquad
\begin{array}{rcl}
\boldsymbol{r}' &=& \mathbf{B}\boldsymbol{e}_3, \\[1ex]
\mathbf{B}' &=& \widehat{\Omega}, \\[1ex]
\boldsymbol{T}' &=& \boldsymbol{F} \times \boldsymbol{e}_3, \\[1ex]
\boldsymbol{F}' &=& \boldsymbol{F}_0 \times \Omega + \ddot{\boldsymbol{r}} + 2\sqrt{\beta}\rho\dot{\boldsymbol{r}}' + \rho^2 \boldsymbol{r}'' + \alpha_e \dot{\boldsymbol{r}} - \gamma \mathbf{B}^T \boldsymbol{e}_3,
\end{array}
$$

with homogenous boundary conditions for $r$, $B$, $T$, $F$ and the jump condition at the support

$$(8) \qquad F(\xi_+) - F(\xi_-) = -c \begin{pmatrix} r_1 + \alpha_s \dot{r}_1 \\ r_2 + \alpha_s \dot{r}_2 \\ 0 \end{pmatrix}_{s=\xi} .$$

From $(7_1)$ and $(7_2)$ we get the linearization of $\Omega$ to

$$(9) \qquad \Omega = (-r_1'', r_2'', \chi')^T,$$

where $\chi$ measures the twisting angle. Using the material law (2) and eliminating $F$ by $(7_3)$ the system decouples into two identical bending equations and one simple equation for $\chi$

$$(10) \qquad \ddot{r}_i + r_i'''' + \alpha_i \dot{r}_i'''' + 2\sqrt{\beta}\rho\dot{r}_i' + \rho^2 r_i'' + \alpha_e \dot{r}_i - (\gamma(1-s)r_i')' = 0,$$

$$(11) \qquad \gamma_3(\chi'' + \alpha_3 \dot{\chi}'') = 0,$$

for $i = 1, 2$. Also the boundary and jump conditions separate

$$(12) \qquad \begin{aligned} r_i(0) &= 0, & (r_i'' + \alpha_1 \dot{r}_i'')(1) &= 0, \\ r_i'(0) &= 0, & (r_i''' + \alpha_1 \dot{r}_i''')(1) &= 0, \\ (r_i''' + \alpha_1 \dot{r}_i''')|_{\xi_-}^{\xi_+} + c(r_i + \alpha_s \dot{r}_i)(\xi) &= 0, \\ \chi(0) &= 0, & \gamma_3(\chi' + \alpha_3 \dot{\chi}')(1) &= 0. \end{aligned}$$

Since every solution of (11) decays exponentially, the linear stability investigation may be restricted to one bending equation, corresponding to an equivalent planar problem. But every critical eigenvalue of the planar problem will appear twice in the full 3-dimensional system.

The stability limit is computed by looking for pure imaginary and zero eigenvalues using the ansatz functions

$$r_1(t, s) = \cos \omega t \, \phi_1(s) + \sin \omega t \, \phi_2(s), \qquad \text{or} \qquad r_1(t, s) = \phi_3(s),$$

respectively and solving the resulting nonlinear boundary value problems for $\phi_j$, $\omega$ and the critical flow rate $\rho_c$ by the multiple shooting method BOUNDSCO ([7]).

Fig. 2 shows a typical stability chart in $(c, \rho)$-space for $\xi = 0.5$. For stiff support the trivial state loses its stability by a zero eigenvalue, whereas for soft support generally purely imaginary eigenvalues occur. For a certain critical stiffness $c_c$ a codimension 2 bifurcation occurs with simultaneous appearance of zero and imaginary eigenvalues $\pm i\omega$.

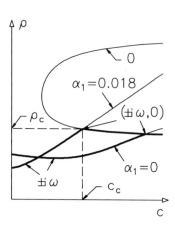

FIGURE 2.    Stability boundary in $(c, \rho)$-parameter space for $\xi = 0.5$ and two different values of internal damping $\alpha_1$. The material parameters are chosen according to pipe 1 in [11]. For stiff support ($c > c_c$) the loss of stability occurs by a zero eigenvalue, for soft support ($c < c_c$) purely imaginary eigenvalues lead to flutter instabilities. At $c = c_c$ a Steady-state/Hopf mode interaction occurs. For soft springs the figure shows the destabilizing effect of internal damping.

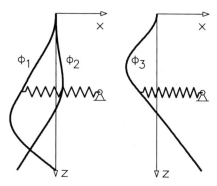

FIGURE 3.    Shape of the eigenfunctions $\phi_i(s)$ corresponding to the critical eigenvalues $\pm i\omega$ (left) and 0 (right).

## 4. Nonlinear bifurcation equations

The six-dimensional eigenspace at the parameter values $(\rho_c, c_c)$ is spanned by

$$
\begin{aligned}
(13) \quad r_1(t,s) &= x_1(t)\phi_1(s) + x_2(t)\phi_2(s) + x_3(t)\phi_3(s), \\
r_2(t,s) &= y_1(t)\phi_1(s) + y_2(t)\phi_2(s) + y_3(t)\phi_3(s), \\
\dot{r}_1(t,s) &= -\omega x_2(t)\phi_1(s) + \omega x_1(t)\phi_2(s), \\
\dot{r}_2(t,s) &= -\omega y_2(t)\phi_1(s) + \omega y_1(t)\phi_2(s).
\end{aligned}
$$

Center manifold reduction yields a system of six ordinary differential equations for the critical variables $x_j(t)$ and $y_j(t)$.

The center manifold reduction requires some information about the remaining eigenvalues of the linearized system. For spatially high oscillating modes the linear differential equations are dominated by

$$
\begin{aligned}
(14) \quad &\ddot{r} + r'''' + \alpha_1 \dot{r}'''' = 0, \\
&r(0) = 0, \qquad\qquad r''(1) + \alpha_1 \dot{r}''(1) = 0, \\
&r'(0) = 0, \qquad\qquad r'''(1) + \alpha_1 \dot{r}'''(1) = 0.
\end{aligned}
$$

Setting $r = \exp(\lambda t)\, r_0(s)$ and $\mu^4 = -\lambda^2/(1 + \alpha_1 \lambda)$, (14) has nontrivial solutions if

(15) $$\cos\mu\cosh\mu + 1 = 0.$$

The solutions $\mu_k$ of (15) converge to $(k + 1/2)\pi$ for $k \to \infty$. Solving for $\lambda$ we obtain 2 families of eigenvalues, one of which converges to $-\infty$ like $-\alpha_1\mu_k^4$ whereas the second one accumulates at $\lambda = -1/\alpha_1$ due to the appearance of material damping in the highest order derivatives. The location of the intermediate eigenvalues has to be calculated numerically, a rough approximation is obtained by discretizing the spatial derivatives by finite differences and computing the eigenvalues of a large sparse matrix. These eigenvalues provide an initial guess for a subsequent solution of a nonlinear boundary value problem for the eigenfunctions and corresponding eigenvalues by multiple shooting ([7]). For material data according to pipe 1 in [11] only two pairs of stable complex eigenvalues are found at the critical parameter values; all further eigenvalues are real and strictly negative.

According to the separation of the linear system into bending and twisting equations the approximation of the center manifold splits into two parts.

In order to obtain the cubic terms in the bifurcation equations, a trivial ansatz suffices for the stable components in the *bending* system. It remains to show that there exists a sufficiently smooth center manifold.

Since the highest order derivatives are the same as in the panel flutter problem (page 355ff in [6]), and the Coriolis force $-2\sqrt{\beta}\rho\dot{r}'$ is dissipative

$$\int_0^1 -\dot{r}' \cdot \dot{r}\, ds = -\frac{1}{2}\dot{r}^2\Big|_0^1,$$

the linear equations generate a contractive semiflow on $(\mathbf{H}^2 \times \mathbf{L}^2)^2$. The further results for the panel flutter problem cannot be applied to our system, because contrary to the panel flutter problem ([6, 5]) the fourth order spatial derivatives appear in the nonlinear equations.

The bifurcation equations on the center manifold will be governed by quadratic and cubic terms. Since the variables corresponding to *twisting* deformations appear quadratically, these are removed by the following 3 steps. By $(1_3)$ the torque $T_3$ satisfies the quadratic differential equation

(16) $$T_3' = \kappa_1 T_2 - \kappa_2 T_1 = \alpha_1(\kappa_1\dot{\kappa}_2 - \kappa_2\dot{\kappa}_1),$$

with boundary condition $T_3(1) = 0$. Using the ansatz

$$T_3 = \sum_{i<j} T_{3,ij}(s)(x_i y_j - x_j y_i)$$

and (9) and (13), we find

$$T_{3,12} = \alpha_1 \omega \int_1^s \left( \phi_2''^2 - \phi_1''^2 \right) ds,$$

$$T_{3,13} = \alpha_1 \omega \int_1^s \phi_2'' \phi_3'' ds,$$

$$T_{3,23} = -\alpha_1 \omega \int_1^s \phi_1'' \phi_3'' ds.$$

Now the material law $(2_3)$ can be regarded as ordinary differential equation for $\kappa_3$. Inserting $\kappa_3 = \sum_{i<j} \kappa_{3,ij}(s)(x_i y_j - x_j y_i)$ we obtain the equations

$$(17) \qquad \gamma_3 \begin{pmatrix} 1 & 0 & 0 \\ 0 & 1 & \alpha_3 \omega \\ 0 & -\alpha_3 \omega & 1 \end{pmatrix} \begin{pmatrix} \kappa_{3,12} \\ \kappa_{3,13} \\ \kappa_{3,23} \end{pmatrix} = \begin{pmatrix} T_{3,12} \\ T_{3,13} \\ T_{3,23} \end{pmatrix}.$$

Finally the twisting angle $\chi$ is calculated from $(1_2)$.

As stated in [4] and [3] the normal form may be chosen to commute with $\mathbf{S}^1$, where $\mathbf{S}^1$ acts as flow of the linearized equations of motion. Since the bifurcation equations also inherit the $\mathbf{D}_n$-symmetry of the original system, the bifurcation equations are equivariant wrt. the group $\mathbf{D}_n \times \mathbf{S}^1$. If we choose complex coordinates

$$\begin{aligned} z_1 &= x_1 - y_2 + i(y_1 + x_2), \\ z_2 &= x_1 + y_2 - i(y_1 - x_2), \\ z_3 &= x_3 + iy_3, \end{aligned}$$

according to [12], by (4) and (13) the symmetry transformations operate as follows:

$$(18) \qquad \begin{aligned} \zeta \in \mathbf{Z}_n : & \quad (z_1, z_2, z_3) \mapsto (e^{i\zeta} z_1, e^{-i\zeta} z_2, e^{i\zeta} z_3), \\ \kappa \in \mathbf{Z}_2 : & \quad (z_1, z_2, z_3) \mapsto (z_2, z_1, \overline{z}_3), \\ \theta \in \mathbf{S}^1 : & \quad (z_1, z_2, z_3) \mapsto (e^{i\theta} z_1, e^{i\theta} z_2, z_3). \end{aligned}$$

It's a straightforward task to find those nonlinear functions $\mathbf{g}(\mathbf{z})$, which commute with any element $\sigma \in \mathbf{D}_n \times \mathbf{S}^1$. If the terms are written in multi-index notation

$$g_j(\mathbf{z}) = \sum_{\mathbf{m}} g_{j,\mathbf{m}} \mathbf{z}^{\mathbf{m}} = \sum_{\mathbf{m}} g_{j,m_1 m_2 m_3 m_4 m_5 m_6} z_1^{m_1} \overline{z}_1^{m_2} z_2^{m_3} \overline{z}_2^{m_4} z_3^{m_5} \overline{z}_3^{m_6}$$

equivariance under $\mathbf{Z}_n$ requires

$$(19) \qquad m_1 - m_2 - m_3 + m_4 + m_5 - m_6 \equiv 1 \pmod{n} \quad \text{for } j = 1 \text{ and } 3.$$

Equivariance wrt. $\mathbf{S}^1$ yields the conditions

$$(20) \qquad m_1 - m_2 + m_3 - m_4 = 1 \qquad \text{for } j = 1,$$

$$(21) \qquad m_1 - m_2 + m_3 - m_4 = 0 \qquad \text{for } j = 3.$$

The 2nd component of $g(z)$ is calculated by exploiting the flip symmetry

$$(22) \qquad\qquad g_2(z) = g_1(\kappa \cdot z).$$

The numerical computation of the coefficients is performed by calculating all $\mathbf{D}_n$-equivariant terms up to the required order. From these terms all coefficients are eliminated, which violate the equivariance conditions wrt. $\mathbf{S}^1$.

The bifurcation equations truncated at 3rd order read

$$
(23) \qquad
\begin{aligned}
g_1 &= (\lambda + i\omega + A_1|z_1|^2 + A_2|z_2|^2 + A_3|z_3|^2)z_1 \\
&\quad + A_4 z_2 z_3^2 + \varepsilon_s A_6 z_2 \bar{z}_3, \\
g_3 &= (\mu + A_7|z_1|^2 + A_8|z_2|^2 + A_9|z_3|^2)z_3 \\
&\quad + A_{10}z_1\bar{z}_2\bar{z}_3 + \varepsilon_s A_{11}\bar{z}_1 z_2 + \varepsilon_s A_{12}\bar{z}_3^2,
\end{aligned}
$$

with $A_j = c_j + id_j$, $d_8 = -d_7$ and $d_9 \ldots d_{12} = 0$. The equation $g_2$ is obtained by applying the flip $\kappa$ to $g_1$. The parameters $\lambda$ and $\mu$ are the mathematical unfolding parameters. The length ratio $\varepsilon_s$ is considered as a small perturbation parameter. For easy comparison of the equations with the $\mathbf{D}_4 \times \mathbf{S}^1$ equivariant system in [10], where each equation contains six nonlinear terms, the coefficient $A_5$ is skipped.

## 5. Equivariant bifurcation equations

In order to find stationary and periodic solutions of (23), we introduce additional unknown phase parameters $\nu_j$ in (23) ([4]), forming

$$h_j = g_j - i\,\nu_j\,z_j, \qquad \text{for} \quad j = 1, 2$$

and look for zeroes of $h(z)$. This method is similar to the splitting of (23) into amplitude and phase equations by introducing polar coordinates, but avoids the singularity at the origin.

Under quasistatic variation of $\lambda$ and $\mu$, which depend on the parameters $\rho$ and $c$, the possible bifurcation sequence can be seen from the isotropy lattice. At the top we find the trivial solution with full symmetry, at lower levels there are less symmetric solutions. The symmetry of a particular solution type is given by its conjugacy class of isotropy subgroups: the symmetry of a statically buckled solution is the mirror reflection through the buckling plane; since there are $n$ such planes for odd $n$, we obtain an orbit of $n$ conjugate solutions. The solution $z_0 = (0, 0, b)$ buckles in the $(x, z)$-plane and has isotropy subgroup $\Sigma_0 = \mathbf{Z}_2(\kappa) \times \mathbf{S}^1$, while the isotropy subgroup of the solution $z_\zeta = (0, 0, \exp(i\zeta)b)$ is $\Sigma_\zeta = \mathbf{Z}_2(\kappa\zeta) \times \mathbf{S}^1$. Since all conjugate solutions have the same properties, it is convenient to investigate one representative of each orbit. Table 1 lists the solution types and their (conjugacy classes of) isotropy subgroups.

The isotropy lattice for odd $n$ is quite different from the rotational symmetric case, because the standing wave solutions have submaximal isotropy subgroups, for odd values of $n$ there are no purely planar oscillations about the trivial state;

TABLE 1.  Isotropy Subgroups $\Sigma$ and Fixed-point Subspaces $V_\Sigma$ of $\mathbf{D}_n \times \mathbf{S}^1$ acting on $\mathbb{C}^3$, when $n$ is odd. The standing wave 1 oscillates in a vertical plane containing a spring, while the standing wave 2 oscillates normal to that plane.

| Orbit Type | $\Sigma$ | $V_\Sigma$ | Dim. | |
|---|---|---|---|---|
| $(0,0,0)$ | $\mathbf{D}_n \times \mathbf{S}^1$ | $\{(0,0,0)\}$ | 0 | trivial state |
| $(0,0,b)$ | $\mathbf{Z}_2(\kappa) \times \mathbf{S}^1$ | $\{(0,0,x)\}$ | 1 | planar stat. state |
| $(a,0,0)$ | $\widetilde{\mathbf{Z}_n} = \{(\zeta,-\zeta)\}$ | $\{(z_1,0,0)\}$ | 2 | rotating wave |
| $(a,a,b)$ | $\mathbf{Z}_2(\kappa)$ | $\{(z_1,z_1,x)\}$ | 3 | standing wave 1 |
| $(a,-a,b)$ | $\mathbf{Z}_2(\kappa,\pi)$ | $\{(z_1,-z_1,x)\}$ | 3 | standing wave 2 |
| $(0,0,z_3)$ | $\mathbf{S}^1$ | $\{(0,0,z_3)\}$ | 2 | spatial stat. state |
| $(a,z_2,z_3)$ $z_2 \neq \pm a, 0$ | $1$ | $\mathbb{C}^3$ | 6 | tori |

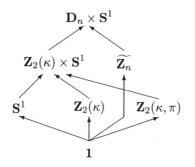

FIGURE 4. Isotropy lattice for odd $n$. Solutions with isotropy subgroup $\widetilde{\mathbf{Z}_n} = \{(\zeta,-\zeta)\}$ are discrete rotating waves. Standing wave solutions with isotropy $\mathbf{Z}_2(\kappa)$ and $\mathbf{Z}_2(\kappa,\pi)$ are in-plane and out-of-plane oscillations about a planar buckled configuration with isotropy $\mathbf{Z}_2(\kappa) \times \mathbf{S}^1$. The standing wave solutions may bifurcate from the trivial solution.

every standing wave oscillates about a buckled configuration. It may bifurcate from the trivial solution or from the planar stationary buckled configuration.

**5.1. Branching equations and stability.** The branching equations and stability properties of the various solution branches are obtained by evaluating the bifurcation equations at the invariant subspaces and exploiting the isotropy subgroups to decompose the 6-dimensional Jakobi matrices into smaller blocks, which determine the stability properties of the isotypic components.

Generically we expect to find the following solution branches in the neighborhood of the bifurcation point

$\mathbf{D}_n \times \mathbf{S}^1$: $z = 0$. Trivial solution.
Eigenvalues:

$$\lambda \pm i\omega \text{ (twice)}, \quad \mu \text{ (twice)}.$$

$\mathbf{Z}_2(\kappa) \times \mathbf{S}^1$: $z = (0,0,b)$ Stationary buckling in $x$-direction.
Branching equation: $\mu + c_9 b^2 + \varepsilon_s c_{12} b = 0$.

Eigenvalues:

$$\sigma_0 = 2c_9 b^2 + \varepsilon_s c_{12} b,$$
$$\sigma_1 = -3\varepsilon_s c_{12} b,$$
$$\sigma_{2,3} = \lambda + i\omega + A_3 b^2 \pm (A_4 b^2 + \varepsilon_s A_6 b),$$
$$\sigma_{4,5} = \overline{\sigma}_{2,3}.$$

The solution branch is a skew-symmetric pitchfork; due to the term $\varepsilon_s c_{12} b$ both branches of the solution are initially unstable. If the stationary state is stable for the rotational symmetric system ($c_9 < 0$), the subcritical branch becomes stable after a secondary fold bifurcation at $b = -\varepsilon_s c_{12}/2c_9$, $\mu = \varepsilon_s^2 c_{12}^2/4c_9$.

$\widetilde{\mathbf{Z}}_3$: $z = (a, 0, 0)$ Rotating wave solution. The tube rotates like a turbine with almost constant shape around the vertical axis. A phase shift by $1/3$ of a period yields the same state as a rigid rotation of the pipe by $\zeta = 2\pi/3$. By applying the flip $\kappa$ we obtain a rotation in the reverse direction.
Branching equations:

$$\lambda + c_1 a^2 = 0,$$
$$\omega + d_1 a^2 = \nu_1.$$

The stability is governed by the eigenvalue

$$2c_1 a^2$$

and the complex-valued matrices

$$M_1 = \begin{pmatrix} (A_2 - A_1)a^2 & \varepsilon_s A_6 a \\ \varepsilon_s A_{10} a & \mu + A_7 a^2 \end{pmatrix} \quad \text{and} \quad \overline{M}_1.$$

For $n > 3$ the off-diagonal terms in $M_1$ vanish, since $\widetilde{\mathbf{Z}}_n$ acts differently on $z_2$ and $z_3$. If $M_1$ has simple eigenvalues for $\varepsilon_s = 0$, the influence of $\varepsilon_s$ on the eigenvalues is of second order in $\varepsilon_s$. Close to $\mu = -c_7 a^2$ a secondary bifurcation to a modulated rotating wave occurs.

$\mathbf{S}^1$: $z = (0, 0, z_3)$ Stationary asymmetric buckled solution.
Its investigation requires at least fifth order terms. Since that solution exists close to a degenerate steady state bifurcation ($c_{12} = 0$), we do not expect it to occur generically.

$\mathbf{Z}_2(\kappa)$: $z = (a, a, b)$ Standing in-plane oscillation in $x$-direction.
The tube swings like a pendulum about a planar buckled state; the oscillation and buckling occur in the same plane.
Branching equations:

$$(24) \quad \begin{aligned} \lambda + (c_1 + c_2)a^2 + (c_3 + c_4)b^2 + \varepsilon_s c_6 b &= 0, \\ (\mu + (2c_7 + c_{10})a^2 + c_9 b^2 + \varepsilon_s c_{12})b + \varepsilon_s c_{11} a^2 &= 0, \\ \omega + (d_1 + d_2)a^2 + (d_3 + d_4)b^2 + \varepsilon_s d_6 b &= \nu_1 = \nu_2. \end{aligned}$$

Comparing (24) with the $\mathbf{O}(2)$-symmetric problem ($\varepsilon_s = 0$), we see that this solution contains the mixed mode solution $(a, a, b)$ and the pure standing wave $(a, a, 0)$. The mixed mode solution bifurcates from the standing wave at

$$\lambda + (c_1 + c_2)a^2 = 0,$$
$$\mu + (2c_7 + c_{10})a^2 = 0$$

by a pitchfork bifurcation. The $\mathbf{D}_3$-equivariant terms ($\varepsilon_s > 0$) break the pitchfork, leading to two solution branches close to the unperturbed solutions.

The Jakobian splits into two blocks governing the stability in the invariant subspace $V_0 = \{(\xi, \xi, \eta) | \xi \in \mathbb{C}, \eta \in \mathbb{R}\}$

$$M_0 = \begin{pmatrix} 2(c_1 + c_2)a^2 & 2(c_3 + c_4)ab + \varepsilon_s c_6 a \\ 2(2c_7 + c_{10})ab + 2\varepsilon_s c_{11}a & 2c_9 b^2 + \varepsilon_s c_{12} b - \varepsilon_s c_{11} a^2 / b \end{pmatrix}$$

and in the complementary subspace $V_1 = \{(\xi, -\xi, i\eta) | \xi \in \mathbb{C}, \eta \in \mathbb{R}\}$

$$M_1 = \begin{pmatrix} 2(c_1 - c_2)a^2 - 2c_4 b^2 \\ \quad - 2\varepsilon_s c_6 b & 2d_4 b^2 + 2\varepsilon_s d_6 b & -2d_4 ab + \varepsilon_s d_6 a \\ 2(d_1 - d_2)a^2 - 2d_4 b^2 \\ \quad - 2\varepsilon_s d_6 b & -2c_4 b^2 - 2\varepsilon_s c_6 b & 2c_4 ab - \varepsilon_s c_6 a \\ 4d_7 ab & 2c_{10}ab - 2\varepsilon_s c_{11}a & \begin{array}{c} -2c_{10}a^2 - 3\varepsilon_s c_{12}b \\ - \varepsilon_s c_{11}a^2/b \end{array} \end{pmatrix}.$$

For $\varepsilon_s = 0$ the matrix $M_1$ has a zero eigenvalue with eigenvector $\boldsymbol{v} = (0, a, b)^T$, due to the rotational symmetry of the model. In order to calculate the eigenvalue for $\varepsilon_s \neq 0$ we have to distinguish two cases.

If the solution is close to the mixed mode solution $\mathbf{Z}_2(\kappa)$ of the rotational symmetric problem, the adjoint eigenvector for the zero eigenvalue is given by

$$\boldsymbol{v}^* = \alpha \begin{pmatrix} (c_1 - c_2)a^2 - c_4 b^2 \\ (d_1 - d_2)a^2 - d_4 b^2 \\ 2d_7 ab \end{pmatrix} \times \begin{pmatrix} d_4 b^2 \\ -c_4 b^2 \\ c_{10}ab \end{pmatrix},$$

where the factor $\alpha$ is chosen such that $\boldsymbol{v}^* \cdot \boldsymbol{v} = 1$. The leading expansion of the perturbed eigenvalue is given by

$$(25) \qquad \sigma = \varepsilon_s \boldsymbol{v}^* \cdot \partial M_1/\partial \varepsilon_s \, \boldsymbol{v} = 3\varepsilon_s \boldsymbol{v}^* \cdot \begin{pmatrix} d_6 ab \\ -c_6 ab \\ -c_{11}a^2 - c_{12}b^2 \end{pmatrix}.$$

If the solution approximates the pure standing wave $(a, a, 0)$, we have to calculate the eigenvectors of the matrix $M_1$, evaluated at $\boldsymbol{z} = (a, a, 0)$

$$M_1 = \begin{pmatrix} 2(c_1 - c_2)a^2 & 0 & 0 \\ 2(d_1 - d_2)a^2 & 0 & 0 \\ 0 & 0 & \mu + (2c_7 - c_{10})a^2 \end{pmatrix}.$$

The kernel is spanned by $e_2$ and the adjoint eigenvector satisfies $v_3^* = 0$. Using (25) we find $\sigma = \mathcal{O}(\varepsilon_s^2)$. Therefore the cubic expansion of the bifurcation equations is insufficient to calculate the stability of the standing wave solutions, as it is already explained in [4].

$\mathbf{Z}_2(\kappa, \pi)$: $z = (a, -a, b)$ The tube performs an out-of-plane oscillation about a planar buckled state. The plane of oscillation is normal to the buckling plane, which contains a spring.

The branching equations are similar to (24)

$$(26) \quad \begin{array}{rcl} \lambda + (c_1 + c_2)a^2 + (c_3 - c_4)b^2 - \varepsilon_s c_6 b & = & 0, \\ (\mu + (2c_7 - c_{10})a^2 + c_9 b^2 + \varepsilon_s c_{12})b - \varepsilon_s c_{11} a^2 & = & 0, \\ \omega + (d_1 + d_2)a^2 + (d_3 - d_4)b^2 - \varepsilon_s d_6 b & = & \nu_1 = \nu_2. \end{array}$$

The solutions of equations (26) contain the perturbed branches of purely standing waves and the mixed mode solution $z = (a, -a, b)$ of the rotational symmetric model. The calculations of the stability properties are very similar to the $\mathbf{Z}_2(\kappa)$-symmetric solution.

Again the Jakobian splits into 2 blocks governing the stability in the invariant subspace $V_0 = \{(\xi, -\xi, \eta) | \xi \in \mathbb{C}, \eta \in \mathbb{R}\}$

$$M_0 = \begin{pmatrix} 2(c_1 + c_2)a^2 & 2(c_3 - c_4)ab - \varepsilon_s c_6 a \\ 2(2c_7 - c_{10})ab - 2\varepsilon_s c_{11} a & 2c_9 b^2 + \varepsilon_s c_{12} b + \varepsilon_s c_{11} a^2 / b \end{pmatrix}$$

and in the complementary subspace $V_1 = \{(\xi, \xi, i\eta) | \xi \in \mathbb{C}, \eta \in \mathbb{R}\}$

$$M_1 = \begin{pmatrix} 2(c_1 - c_2)a^2 + 2c_4 b^2 \\ + 2\varepsilon_s c_6 b & -2d_4 b^2 - 2\varepsilon_s d_6 b & 2d_4 ab - \varepsilon_s d_6 a \\ 2(d_1 - d_2)a^2 + 2d_4 b^2 \\ + 2\varepsilon_s d_6 b & 2c_4 b^2 + 2\varepsilon_s c_6 b & -2c_4 ab + \varepsilon_s c_6 a \\ 4d_7 ab & -2c_{10} ab + 2\varepsilon_s c_{11} a & \begin{array}{c} 2c_{10} a^2 - 3\varepsilon_s c_{12} b \\ + \varepsilon_s c_{11} a^2 / b \end{array} \end{pmatrix}.$$

As before the stability calculation of the perturbed standing wave solution requires higher order terms and the small eigenvalue of the perturbed mixed mode solution is given by

$$\sigma = 3\varepsilon_s v^* \cdot (-d_6 ab, c_6 ab, c_{11} a^2 - 3c_{12} b^2)^T,$$

where $v^*$ is the normalized adjoint null-eigenvector of $M_1$, evaluated at the unperturbed mixed mode solution branch.

**5.2. Numerical Results.** Due to the large number of possible branching sequences and stability conditions it seems impossible to perform a complete classification of all possible bifurcation diagrams. Instead we calculate the coefficients $A_1 \dots A_{12}$ for certain choices of physical parameters and derive the corresponding bifurcation diagram. Table (2) displays the results for tubes 1 and 2 in [11] for $\xi = 0.5$. Figure 5 shows a qualitatively drawn possible bifur-

TABLE 2.  Parameter values and normal form coefficients for tubes 1 and 2 in [11] for common parameter values $\xi = 0.5$, $\gamma_3 = 1/1.4$ and $\alpha_s = 0$.

|  | Tube 1 | Tube 2 |  | Tube 1 | Tube 2 |
|---|---|---|---|---|---|
| $\alpha_1 = \alpha_3$ | 0.015 | 0.019 | $\alpha_e$ | 0.012 | 0 |
| $\beta$ | 0.249 | 0.505 | $\gamma$ | 21.4 | 9.49 |
| $\rho_c$ | 10.35 | 11.53 | $c_c$ | 705.7 | 289.5 |
| $\partial\lambda/\partial\rho$ | 0.333 | 0.447 | $\partial\lambda/\partial c$ | $-1.32 \cdot 10^{-3}$ | $-2.91 \cdot 10^{-3}$ |
| $\partial\mu/\partial\rho$ | 0.675 | 0.0896 | $\partial\mu/\partial c$ | $5.88 \cdot 10^{-4}$ | $3.53 \cdot 10^{-3}$ |
| $\omega_c$ | 38.3 | 29.4 |  |  |  |
| $c_1$ | $-0.0548$ | $-0.0479$ | $d_1$ | $-0.0447$ | $-0.0186$ |
| $c_2$ | $-0.0244$ | $-0.0333$ | $d_2$ | 0.0918 | 0.0841 |
| $c_3$ | 0.00495 | 0.0324 | $d_3$ | 0.0517 | 0.0760 |
| $c_4$ | $-0.00222$ | 0.02907 | $d_4$ | 0.0554 | 0.0546 |
| $c_6$ | 0.0382 | 0.0943 | $d_6$ | 0.0499 | $-0.0412$ |
| $c_7$ | $-0.164$ | $-0.0324$ | $d_7$ | 0.0532 | 0.0373 |
| $c_9$ | $-0.0743$ | $-0.0730$ | $c_{10}$ | $-0.0749$ | $-0.0893$ |
| $c_{11}$ | $-0.0258$ | $-0.0346$ | $c_{12}$ | $-0.0085$ | $-0.0571$ |

FIGURE 5.  Possible bifurcation diagram for tube 1 in [11] for $c > c_c$. Only representatives of solutions with nontrivial isotropy subgroups are depicted. Bold solid segments indicate stable branches; dashed curves denote unstable solutions. Dotted curves correspond to solutions branches, whose stability properties depend on higher order terms. Since the stationary solution is slightly asymmetric, the branches $b > 0$ and $b < 0$ have to be drawn separately. The branch $b < 0$ bifurcates subcritically from the trivial state.

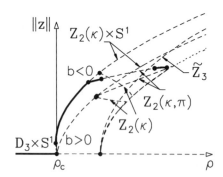

cation sequence if the flow rate $\rho$ is increased quasistatically and the stiffness of the support $c$ is fixed at a sligthly larger value than $c_c$. First the trivial solution loses its stability at a zero eigenvalue; the tube jumps to a stable stationary state. Due to a secondary Hopf bifurcation the tube starts to oscillate normal to the buckling direction. By a tertiary steady state bifurcation this solution becomes unstable, but gains stability at a higher flow rate for a small parameter range. The stability of the large amplitude planar oscillations cannot be determined from the cubic truncation of the bifurcation equations.

## Concluding remarks

Comparing the results for the rotational symmetric case and the $D_n$-symmetric problem with even $n$, we find that the case $n = 3$ behaves quite differently: the planar oscillations occupy a lower level in the isotropy lattice and there are fewer solution types. Also the branching equations and stability calculations are more complicated due to the appearance of quadratic terms and the vanishing eigenvalue of the standing wave branches.

In the rotational symmetric case heteroclinic orbits may exist, connecting the steady states to standing waves. The behaviour of these orbits under $D_3$-symmetric perturbations should be very complicated.

**Acknowledgement.** I would like to thank the referee for his valuable comments.

## REFERENCES

1. A. K. Bajaj and P. R. Sethna, *Flow induced bifurcations to three-dimensional oscillatory motions in continous tubes*, SIAM J. Appl. Math. **44** (1984), 270–286.

2. T. B. Benjamin, *Dynamics of a system of articulated pipes conveying fluid*, Proceed. Roy. Soc. London **261** (1961), 457 – 486.

3. C. Elphik, E. Tirapegui, M. E. Brachet, P. Coullet, and G. Iooss, *A simple global characterization for normal forms of singular vector fields*, Physica **29D** (1987), 95–127.

4. M. Golubitsky, I. Stewart, and D. Schaeffer, *Singularities and Groups in Bifurcation Theory*, Applied Math. Sciences, vol. 51 and 69, Springer-Verlag, New York – Heidelberg – Berlin, 1985, 1988.

5. P. Holmes and J Marsden, *Bifurcation to divergence and flutter in flow induced oscillations; an infinite dimensional analysis*, Automatica **14** (1978), 367–384.

6. J. E. Marsden and Th. J.R. Hughes, *Mathematical foundations of elasticity*, Prentice-Hall, Inc., 1983.

7. H. H. Oberle, W. Grimm, and E. Berger, *BNDSCO, Rechenprogramm zur Lösung beschränkter optimaler Steuerungsprobleme*, Benutzeranleitung TUM-M8509, TU-München, Math. Inst., 1985.

8. M. P. Paidoussis and N. T. Issid, *Dynamic stability of pipes conveying fluid*, Journal of Sound and Vibration **33** (1974), 267–294.

9. P. R. Sethna and S. W. Shaw, *On codimension-three bifurcations in the motion of articulated tubes conveying a fluid*, Physica **24D** (1987), 305–327.

10. A. Steindl, *Hopf/Steady-state Mode Interaction for a Fluid Conveying Elastic Tube with $D_4$-symmetric Support*, Bifurcation and Symmetry (Marburg) (E. Allgower, K. Böhmer, and M. Golubitsky, eds.), International Series of Numerical Mathematics, vol. 104, Birkhäuser Verlag, Basel, Boston, Berlin, 1992, pp. 305–316.

11. Y. Sugiyama, Y. Tanaka, T. Kishi, and H. Kawagoe, *Effect of a Spring Support on the Stability of Pipes Conveying Fluid*, J. Sound and Vibration **100** (1985), 257–270.

12. H. Troger and A. Steindl, *Nonlinear Stability and Bifurcation Theory: An Introduction for Engineers and Applied Scientists*, Springer-Verlag, Wien – New York, 1991.

INST. FOR MECHANICS, TECHN. UNIV. VIENNA, WIEDNER HAUPTSTRASSE 8–10, A-1040 WIEN

*E-mail address*: asteindl@mch2ws2.tuwien.ac.at

Lectures in Applied Mathematics
Volume **29**, 1993

# Dependence of bifurcation structures on the approximation of O(2) symmetry

EMILY STONE AND MICHAEL KIRBY

April 22, 1993

ABSTRACT. We present a investigation of the bifurcation structure of a set of ODEs approximating the Kuramoto-Sivashinsky equation. We apply the Karhunen-Loève transform using a Sobolev metric to optimize the approximation of the differential operator. We find that the approximation of the O(2) symmetry possessed by the original equation, by the system on the KL-basis, has important effect on the appearance of the stable limit cycle seen at $\alpha = 83.25$ in the PDE.

## 1. Minimal Systems

This study continues our investigation of the use of a weighted Sobolev metric and the K-L expansion for modeling partial differential equations (PDEs) by minimal systems of ordinary differential equations (ODEs). We are concerned with PDEs which are equivalent to a finite dimensional set of ODEs and our object is to find a system which reflects the intrinsic dimensionality of the attractor. In this study we focus on the 1-dimensional scalar Kuramoto-Sivashinsky equation which has been rigorously shown to possess a finite dimensional inertial manifold for the case of Neumann boundary conditions [5]. The K-S equation is also a convenient model equation given the substantial number of numerical and theoretical studies devoted to it [**1, 10, 11, 12, 13, 14, 15**].

The approach we consider here for the numerical simulation of the K-S equation is the Galerkin method [**9**]. It projects the phase space of the solution onto a subspace spanned by some set of basis functions which are chosen for their analytical properties and are typically the solutions to some Sturm-Louiville differential equation.

1991 *Mathematics Subject Classification.* 35A40,41A58,65M60.
*Key words and phrases.* dynamical system, bifurcation, symmetry, optimal expansion.
This paper is in final form and no version of it will be submitted elsewhere.

The main issue of this paper is how do we choose the *appropriate* basis in the Galerkin procedure such that the resulting system of ODEs has minimal size. Clearly the basis will be problem dependent and typically we will not expect the functions to have a closed analytical form. We note the exceptions are the sinusoids for periodic problems with translational symmetry and Legendre functions for rotational symmetry [16]. In general a natural candidate for a set of optimal basis functions is based on the Karhunen-Loève (K-L) expansion [8] (also referred to as proper orthogonal decomposition (POD) [17, 2, 19]).

As described in [15], we propose that the most efficient numerical simulations must use eigenfunctions which optimally approximate not only the the flow field but also its derivatives. The procedure may be viewed as a modification of the optimality criterion of the K-L expansion which optimizes convergence of the flow field alone. In Section 2 we present the framework of the K-L expansion with weighted Sobolev metric. In Section 3 we develop this for the K-S equation and discuss the symmetries possessed by the equation and their relevance in orthogonal expansions. In Section 4 we present the results of the numerical simulation of the K-S equation based on the eigenfunctions generated by the weighted Sobolev metric, as well as an AUTO [7] assisted bifurcation analysis, and discuss the effects of truncation order on capturing the behavior of the converged system in the context of symmetry considerations. Section 5 analyzes the symmetry of the system of ODEs as a function of the number of equations retained in the expansion.

We must mention here a complimentary study by Aubry, Lian and Titi [3] which presents a method for using the K-L expansion to create a basis that respects a particular symmetry of the governing PDE. We note that symmetry properties of eigenfunctions have been described in [19]. For more details see also [16]. As an example Aubry *et al.* consider the K-S equation restricted to odd Fourier modes in a chaotic parameter regime and discuss the construction of K-L modes that preserve the discrete $Z_2$ symmetry (i.e. a flip symmetry) in the equation. We will compare our results (carried out in a different parameter regime and for the O(2) equivariant form of the K-S equations) in section 5.

## 2. The eigenfunction approach

We assume we are given a PDE of the form

$$(1) \qquad\qquad\qquad u_t = D(u)$$

where $D$ is taken to be a nonlinear operator and $u, x, t$ are all scalars. We remark that we might also consider the operator to be a nonlinear function of the variable and its derivatives. A typical numerical procedure used to compute solutions to this equation will approximate each of these terms at every time step. In the case we study we will be concerned with derivatives up to $u_{xxxx}$. We assume that the solution $u(x,t)$ can be written without approximation as

$$(2) \qquad u(x,t) = u_M(x,t) = \sum_{n=1}^{M} a_n(t)\phi_n(x).$$

and assert that the flow is actually embedded in an $M$-dimensional space where $M$ is much larger than the intrinsic dimensionality of the flow.

The *optimal basis approach* expands the flow $u(x,t)$, namely

$$(3) \qquad u_N(x,t) = \sum_{n=1}^{N} a_n(t)\phi_n(x).$$

where the basis functions $\{\phi_n\}$ are chosen such that $N << M$ is as small as possible. The Karhunen-Loève expansion characterizes eigenfunctions by requiring that the mean-square error of the truncation be minimal. In its standard form it employs the normal Euclidean metric $d_E(f,g) = \| f - g \|^2$ where the norm is induced by the usual Euclidian inner product, i.e., $\| \, . \, \| = (.,.)^{1/2}$. Thus we seek the minimum of

$$(4) \qquad \epsilon_E = \min\langle d_E(u, u_N)\rangle,$$

over the expansion functions, where the brackets denote a time average.

Alternatively, it is equivalent to choose $\phi_1$ to maximize the quantity $\lambda_E^{(1)}$ where

$$(5) \qquad \lambda_E^{(1)} = \langle(\phi_1, u)^2\rangle$$

For this to make sense we add the side constraint that $(\phi_1, \phi_1) = 1$. This may be interpreted as maximizing the projection of the flow onto the first coordinate direction. The rest of the eigenfunctions are obtained by applying the procedure inductively.

It is well known that this variational problem is equivalent to solving a Fredholm type integral equation

$$(6) \qquad \int K(x,y)\phi(y) = \lambda\phi(x)$$

where the kernel $K(x,y) = \langle u(x,t)\overline{u(y,t)}\rangle$. This problem falls within the framework of Hilbert-Schmidt theory and the eigenfunctions form a complete orthogonal set with non-negative eigenvalues. We have written the kernel in its most general form to prepare for deliberations in Fourier-space.

**2.1. Metric choices.** The motivation for this study comes from the observation that the *optimal* (in the K-L sense) expansion for spatio-temporal signal

$$(7) \qquad u(x,t) \approx u_N(x,t) = \sum_{n=1}^{N} a_n(t)\phi_n(x)$$

is not optimal for the derivatives of the signal. Namely, the optimal expansion for the spatial derivatives is not given in terms of the spatial derivatives of the eigenfunctions. In other words the best basis for the derived ensemble $\{u_x(x,t)\}$ *is not* given by $\{\phi_x\}$ and the corresponding expansion

$$(8) \qquad \frac{\partial u}{\partial x}(x,t) \approx \frac{\partial u_N}{\partial x}(x,t) = \sum_n^N a_n(t)\phi'(x)$$

in general will not be that which leads to a minimal system of ODEs.

In what follows we develop the K-L expansion within the framework of a *weighted Sobolev metric* which we define to be

$$(9) \qquad d_S^{(k)}(f,g) = \sum_{j=0}^k w_j \parallel f^{(j)} - g^{(j)} \parallel^2$$

where the $w_i$ are the weights which are required to normalize the magnitudes of the derivatives and the norm is as defined before. Physically one can interpret the optimality criterion as best approximating the flow, the flux, the dissipation, and so on in the mean square sense.

We begin by modifying the variational problem (3) to

$$(10) \qquad \epsilon_S = \min d_S(u, u_N)$$

where we assume $k$ is fixed and determined by the smoothness of the solutions and that the minimization is being done over the expansion functions. In what follows we take $k = 2$. We will use the expansions

$$\frac{\partial u_N}{\partial x} = \sum_{n=1}^N b_n \phi_n, \quad \frac{\partial^2 u_N}{\partial x^2} = \sum_{n=1}^N c_n \phi_n$$

in addition to (6). Then

$$(11)$$
$$\epsilon_S = \min\langle w_0 \parallel \sum_{n=N+1}^M a_n\phi_n \parallel^2 + w_1 \parallel \sum_{n=N+1}^M b_n\phi_n \parallel^2 + w_2 \parallel \sum_{n=N+1}^M c_n\phi_n \parallel^2\rangle$$

which equals

$$(12) \qquad \epsilon = \min\langle w_0 \sum_{i=N+1}^M a_i^2 + w_1 \sum_{i=N+1}^M b_i^2 + w_2 \sum_{i=N+1}^M c_i^2\rangle.$$

Since this equation is maximized for each $N$ we take $N = 1$ to give

$$(13) \qquad \epsilon = \min\langle w_0 a_1^2 + w_1 b_1^2 + w_2 c_1^2\rangle.$$

The Fourier coefficients above are simply $a_1 = (\phi_1, u)$, $b_1 = (\phi_1, u_x)$ and $c_1 = (\phi_1, u_{xx})$. Inserting these into the previous equation we obtain

$$(14) \qquad \lambda_S^{(1)} = \langle w_0(\phi_1, u)^2 + w_1(\phi_1, u_x)^2 + w_2(\phi_1, u_{xx})^2\rangle.$$

For the maximization to make sense we add the constraint $(\phi_1, \phi_1) = 1$. As with the K-L eigenfunctions, the rest of the $\{\phi_n\}$ are obtained by induction.

In this case the variational equation leads to the eigenfunction problem

$$(15) \qquad \int K_S(x,y)\phi(y)dy = \lambda\phi(x)$$

where we have the modified kernel

$$(16) \qquad K_S = \langle w_0 u \otimes \overline{u} + w_1 u_x \otimes \overline{u}_x + w_2 u_{xx} \otimes \overline{u}_{xx} \cdots \rangle$$

and $\otimes$ denotes an outer product. The determination of the basis is now reduced to choosing the weighting constants in an optimal way. We remark that it is once the fundamental kernels $K_j = \langle u^{(j)} \otimes \overline{u}^{(j)} \rangle$ have been found, various weightings can be readily tested by forming the new kernel

$$(17) \qquad K_S^{(4)} = w_0 K_0 + w_1 K_1 + w_2 K_2 + w_3 K_3 + w_4 K_4.$$

In other words these kernels can be computed once and stored. Clearly we do not expect an arbitrary set of weights to work for a truncated expansion. It is left to determine a best set and we might consider the lower bound to be the standard K-L decomposition with $w_0 = 1$ with all the other weights set to zero. While in the current investigation we have chosen the weights manually we feel it is possible to do this algorithmically by applying another optimality criterion using the weights as parameters, possibly using some type of adjoint method [21]. This approach will be pursued elsewhere. Also, we remark that the actual computations of the eigenfunctions are carried out in transform space, see[15] for details. It is worth noting that this weighting procedure is only appropriate for truncated expansions. If we include all $M$ terms the bases are just rotated versions of each other.

## 3. The Kuramoto-Sivashinsky equation

We recall the Kuramoto Sivashinsky equation (K-S)

$$(18) \qquad u_t + 4u_{xxxx} + \alpha(u_{xx} + \frac{1}{2}(u_x)^2) = 0$$

where $u_{xx}, u_{xxxx}, u_x^2$ correspond to unstable diffusion, stablizing viscosity, and nonlinear coupling terms respectively. It was originally derived as an amplitude equation [20] and its solutions are characterized by coherent spatial structure with complex temporal dynamics. We have made no attempt to include all or even most of the numerous studies concerning this equation but have restricted ourselves to those most relevant to the current investigation.

Let us first consider a general orthogonal decomposition where we do not specify the form the expansion functions. This will allow us considerable freedom to experiment with various bases. We begin in the usual manner, i.e. by decomposing the velocity field via the expansion

$$(19) \qquad u(x,t) = \sum_{n=1}^{\infty} a_n(t)\phi^{(n)}(x)$$

where the $\phi^{(n)}(x)$ are any complete orthogonal family. Thus

(20)
$$\dot{a}_l = -4\sum_n a_n(\phi^{(l)}, \phi^{(n)}_{xxxx}) - \alpha\sum_n (\phi^{(l)}, \phi^{(n)}_{xx}) - \frac{\alpha}{2}\sum_{mn}(\phi^{(l)}, \phi^{(m)}_x\phi^{(n)}_x)a_m a_n.$$

As in [15], we decompose the orthogonal set into its Fourier coefficients, i.e., let $\phi^{(l)} = \sum_{k=-N}^{N}\alpha_k^{(l)}e^{ikx}$. Then a general N term eigenfunction expansion takes the form

$$(21) \qquad \dot{a}_l = \sum_{n=1}^{N} q_{ln}a_n + \sum_{m,n=1}^{N} p_{lmn}a_m a_n,$$

with $q_{ln} = \sum_{k=-M}^{M}(-4k^4 - k^2)\alpha_k^{(l)}\overline{\alpha}_k^{(n)}$, $p_{lmn} = -\alpha\sum_{k,k'=-M}^{M}kk'\alpha_{k+k'}^{(l)}\overline{\alpha}_k^{(m)}\overline{\alpha}_{k'}^{(n)}$ and the restriction that $\mid k + k' \mid\le M$. In our computer investigations we used this representation to allow us to read the basis functions from a generic file.

We consider the K-S equation (18) with periodic boundary conditions, $u(0,t) = u(h,t)$, $u_x(0,t) = u_x(h,t),\ldots$. The periodic length $h$ is subsequently normalized to 1 and $\alpha$ is used as a bifurcation parameter. The equation is invariant under arbitrary translations $x \to x + \beta$, and reflections $x \to -x$. So for any solution $u(x,t)$, $u(x+\beta,t)$ and $u(-x,t)$ are also solutions. These two symmetry operations taken together comprise the group O(2).

For the special case where the expansion functions are complex Fourier modes, i.e.

$$u(x,t) = \sum_{k=-\infty}^{\infty} a_k(t)e^{ikx},$$

the set of ODEs resulting from a Galerkin projection are:

$$(22) \qquad \dot{a}_l = (-4l^4 + \alpha l^2)a_l + \sum_{k=-\infty}^{\infty} a_k a_{l-k}k(l-k).$$

Due to the special form of the Fourier basis functions, to all truncation orders the ODEs thus produced will be O(2) equivariant, and

$$u(x,t) = \sum_{k=-\infty}^{\infty} a_k(t)e^{ikx} \to u(x+\beta,t) = \sum_{k=-\infty}^{\infty} a_k(t)e^{ik(x+\beta)}$$

so if $a_k$ is a fixed point then so is $a_k e^{ik\beta}$, for any rotation $\beta$. Hence any fixed point is accompanied by an entire circle of fixed points (referred to as the group orbit) possessing the same eigenvalue spectrum, up to a rotation of eigenvectors.

Similarly the reflection symmetry yields:

$$u(x,t) = \sum_{k=-\infty}^{\infty} a_k(t)e^{ikx} \rightarrow u(-x,t) = \sum_{k=-\infty}^{\infty} a_k(t)e^{ik(-x)}$$

and hence

$$a_k \rightarrow a_{-k} = a_k^*,$$

since we require that the solution $u(x,t)$ be real. These facts can be verified by direct substitution into the governing ODEs (22).

We distinguish between two types of periodic solutions:

(i) Limit cycles in the amplitudes of the complex Fourier modes (sometimes referred to as standing waves), which are actually a whole torus of solutions under the action of the group O(2).

(ii) Solutions in which the amplitudes of the modes remain constant, but the angle in the complex plane of each increases at a constant speed given by $\omega_k = \omega k$, (for some constant $\omega$) for the kth mode. This is a traveling wave and is a constant velocity drift around the group orbit. (A limit cycle solution in the amplitudes accompanied by a such a drift is called a modulated traveling wave.)

The Fourier-Galerkin projection possesses invariant subspaces. The primary one is the real subspace: $a_k \in \mathcal{R}$. Solutions restricted to this subspace are even waveforms in physical space, that is, solutions to the problem with the more restrictive Neumann boundary conditions. It is the problem for which much work concerning bounds on dimensions of inertial manifolds has been conducted. (See, for example, [5].) It is also restricted to this subspace that all AUTO [7] calculations must be carried out. We further note that in $\mathcal{R}$ bifurcations to solutions with time dependence on the group orbit, i.e. traveling waves and modulated traveling waves, can not occur. Hence these bifurcations will not be captured by an AUTO study of the Fourier-Galerkin ODEs. But AUTO will give all stationary branches, since each fixed point must possess an image on the group orbit that is real.

Any rotation of the real subspace is also invariant, and similarly any pure mode subspace will be invariant since it can be rotated to the real subspace, i.e. for subspace where $a_n = 0$ for all $n \neq m, a_m \neq 0$, there exists some $\phi$ such that $a_m e^{im\phi} \in \mathcal{R}$ and $a_n e^{in\phi} = 0$, for the rest. Finally, because of the special form of the quadratic term the subspaces given by $(a_0, a_2, a_4, a_6, ...), (a_0, a_3, a_6, a_9, ...)$, or more generally the set $(a_n)$, $n \in L\mathcal{Z}, L \in \mathcal{Z}$ are invariant, as can be easily verified. By examining the form of the linear term the sequence of bifurcations off the trivial solution is evident. These occur when $(\alpha - 4l) = 0$, with pairs of double zero eigenvalues:

$$\lambda_l = l^2(\alpha - 4l^2).$$

This multiplicity is a result of the O(2) equivariance of the ODEs, see [6]. For $\alpha \in (l^2, (l+1)^2)$, the zero solution has $2l$ real positive eigenvalues while all the

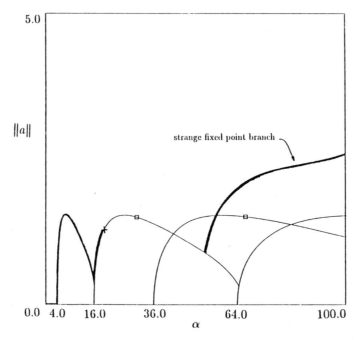

FIGURE 1. Fourier Galerkin system: AUTO bifurcation diagram.

others are negative. A center manifold reduction [1] demonstrates that these are pitchfork bifurcations that arise again from the O(2) equivariance. Of course each pitchfork leads to an entire circle of fixed points, and as $\alpha$ increases there is an infinite sequence of supercritical bifurcations to invariant circles of fixed points.

As an illustration see Figure 1, an AUTO generated bifurcation diagram of 9 mode Fourier-Galerkin projection, restricted to the real subspace. This and all following AUTO bifurcation diagrams are sets of fixed points branches computed by AUTO plotted against the bifurcation parameter $\alpha$. The branches with the real part of all eigenvalues negative are denoted by a heavier line, and Hopf bifurcations are marked with boxes. Anywhere a branch of limit cycles from a Hopf bifurcation has been calculated, it is denoted by a line of interlocked circles. Keep in mind that the norm of the amplitudes of the modes is being plotted, so that the pitchfork bifurcation from the trivial solution to 2 symmetric branches is seen as one branch in Figure 1, and 2 branches crossing on the diagram does not necessarily indicate that they are interacting in any way. Bifurcations of branches are denoted by crosses.

## 4. K-L Sobolev results

In this paper we focus our study on the limit cycle solution at $\alpha = 84.25$. The dynamics for this case are more fully described in [10, 13, 15]. The calculations

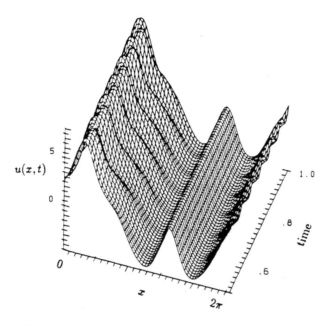

FIGURE 2. Limit cycle solution in the physical space.

of the eigenfunctions for various weights are described in [**15**]. We show the
original simulation at this parameter value in Figure 2. In this simulation of
the K-S equation we used a 10 complex mode Fourier-Galerkin procedure with
initial conditions $u(x, 0) = \cos x + \sin 2x + \sin 3x + \cos 4x$, and periodic boundary
conditions. Given the extreme sensitivity of the K-S equation to numerical accuracy, all the simulations were carried out in double precision and the solution
was advanced in time using Gear's method with the error tolerance set at $10^{-10}$.
We observed that this simulation remained unchanged when using more complex
modes but became unstable when using 9 or fewer.

The K-L expansion with 18 real modes is also unstable, a rather surprising
result as we expect the dimensionality in the neighborhood of a Hopf bifurcation
to be substantially smaller than this. As the dynamics seemed to be qualitatively accurate for short times it is reasonable to speculate that the energy
dissipation mechanism is not being accurately modeled as we reduce the number
of equations. The modes which control this are very low energy as the first 3
eigenfunctions correspond to over 99.9% of the energy. In view of the numerical
sensitivity of the K-S equation we speculate that the dissipative term $u_{xxxx}$ is
being poorly approximated and not providing the appropriate means for energy
removal from the equation. Thus, while the energy of the system is dominated
by a very few large scale structures, the actual simulation of the system in the
K-L framework appears relatively high-dimensional.

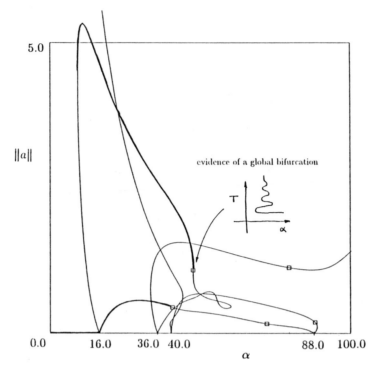

FIGURE 3. 7 Sobolev-KL mode AUTO bifurcation diagram.

To test our hypothesis we applied the K-L procedure within the context of a Sobolev metric and found that if the weights were chosen appropriately we could recapture the limit cycle with only 7 real modes. The simulation results for this case are shown in [15]. While this appeared to substantiate our hypothesis we found that the bifurcation at $\alpha = 83.25$ was not a simple Hopf which led to further complications.

In Figures 3 and 4 we show bifurcation diagrams for 7 and 14 mode truncations of the equation projected onto the Sobolev basis mentioned above. At 14 modes the bifurcation sequence through $\alpha = 83.0$ is as seen in the full simulations, a sub-critical Hopf from a stable fixed point. At 7 modes we have the existence of the limit cycle, but it appears abruptly at $\alpha = 81.0$; prior to that the solution diverges. Though we have yet to investigate the nature of this bifurcation fully, it appears to be global. At truncation orders higher than 14 the limit cycle results from a local Hopf bifurcation, but its stability depends on truncation order. After 19 modes the simulation converges, as expected, since the 10 complex Fourier mode simulation, from which the eigenmodes were extracted, also converged.

To analyze this difficulty in capturing the limit cycle we refer again to the Fourier-Galerkin projection. In particular, note that in the AUTO bifurcation diagram (Figure 1), though we show the "strange fixed point branch" (which does not possess a mirror-image below the $\alpha$ axis, as the branches bifurcating from the trivial solution do), there is no Hopf bifurcation at $\alpha = 83.25$. This is

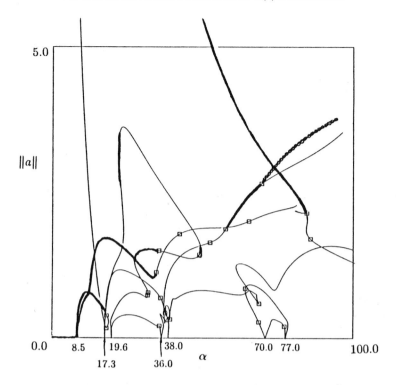

FIGURE 4. 14 Sobolev-KL mode AUTO bifurcation diagram.

a result of conducting our AUTO investigations on the real subspace, since the bifurcation occurs in the $\mathcal{C}$ half of the full space. This means that the pair of complex eigenvalues that cross the imaginary axis in the Hopf bifurcation have eigenvectors in complex directions. Loosely speaking, the Hopf bifurcation is "off" the real subspace. Hence AUTO cannot detect it. Once the limit cycle bifurcates off the real subspace it can no longer be a stationary solution, it must possess a non-zero velocity along the group orbit [18]. Therefore the bifurcation is directly to a modulated traveling wave. This explains why the limit cycle is difficult to recover in our truncations without symmetry and then only partially recovered when the truncation order is large enough that the ODEs possess "approximate" symmetry. Full O(2) symmetry is needed for solutions such as traveling waves and modulated traveling waves to exist.

## 5. Recovering the symmetry with added modes

The eigenfunctions were extracted from a simulation of 10 complex Fourier modes, and in principle the system of ODEs from a Galerkin projection onto the full set of 20 KL eigenfunctions should possess identical behavior, including O(2) equivariance. As the truncation order of the projection is increased the equivariance is observed to return gradually. Primarily, we observe that the bifurcations off the trivial solution move together in pairs as the truncation order is increased, until at 20 modes the Fourier bifurcation structure is recovered (eigenvalues with multiplicity 2 that pass through zero at $\alpha = 4.0, 16.0, 36.0, 64.0, 100.0...$). Initially the truncations are transcritical (the branches are distinct), they grow more like pitchforks (the branches have identical amplitudes and cannot be distinguished with this representation).

To study this more closely consider the form of the ODEs in the N mode truncations of the K-L Galerkin projection:

$$(23) \qquad \dot{a}_l = \sum_{n=1}^{N} q_{ln} a_n + \sum_{m,n=1}^{N} p_{lmn} a_m a_n.$$

Referring to the ODEs in the Fourier-Galerkin projection (22), we can see that in substituting $a_l e^{il\theta}$ for $a_l$, the equations would remain unchanged. This is clearly not the case with (23) above, since the linear and quadratic terms are completely general and hence the multiplication by the angle cannot be pulled out of the sums. Also, the linear term will not necessarily be diagonal, (as in the Fourier-Galerkin system) and except when $N = 20$ will not have eigenvalues of multiplicity 2. As the truncation order is increased, however, the eigenvalues of $r_{ln}$ move together in pairs, as mentioned above.

The second manifestation of this gradual recovery of O(2) symmetry is in the shift of the bifurcations from transcritical to pitchforks. Analytically we can perform a center manifold reduction of the vector field about such a bifurcation point to understand how this works. We follow the analysis applied to the Fourier-Galerkin projection in [1].

First assume a linear transformation of variables $(a \to b)$ is found to uncouple the linear term, make it diagonal. After the transformation (23) becomes:

$$(24) \qquad \dot{b}_l = \sum_{n}^{N} Q_{ln} b_n + \sum_{n,m}^{N} P_{lmn} b_n b_m.$$

Now the matrix $Q$ is diagonal and $P$ is still completely general. Consider the bifurcation from zero of the decoupled $i$th mode governed by:

$$\dot{b}_i = Q(\alpha) b_i + \sum_{n,m}^{N} P_{imn} b_n b_m.$$

We construct a center manifold tangent to the eigenspace of the zero eigenvalue, the 1-D eigenspace spanned by $b_i$. We seek some set of functions $h_j$:

$$b_j = h_j(b_i), \quad i \neq j, \quad h_j = \mathcal{O}(|b_i|^2)$$

that are invariant for the flow. The reduced system on the center manifold is then:

$$\dot{b}_i = Q(\alpha) b_i + \sum_{n,m=1}^{N} P_{imn} h_m(b_i) h_n(b_i).$$

The nonlinear term will determine the type of bifurcation that occurs, and looks at first glance to be quartic, but there are cubic terms, namely:

$$2 \sum_{n=1}^{N} P_{iin} b_i h_n(b_i)$$

and a quadratic term: $P_{iii} b_i b_i$ For lower order truncations this term is non-zero and the bifurcations are transcritical. As the truncation order increases we see the term tends to zero, until at $N = 20$ it is small enough that the bifurcations appear to be pitchforks. Because of the special form of the quadratic term in the Fourier-Galerkin ODEs, the lowest order terms are cubic and the bifurcations are pitchforks, reflecting the O(2)-equivariance.

In Figures 5 and 6 we see another manifestation of this "approximate symmetry", or, in the reverse sense, a breaking of symmetry, as the truncation order decreases from 20 to 19 equations. In Figure 5 are shown selected branches from the 19 mode AUTO bifurcation diagram, and it is clear that form of the converged Fourier bifurcation structure is emerging. But in Figure 6 we show more (but not all) of the solution branches found by AUTO. Our speculation is that AUTO is finding fixed points that are remnants of the continuous circle of fixed points found in the 20 mode system. Hence they have nearly the same magnitude, and we find bundles of similar solutions branches. It would be interesting to study this as a problem in symmetry breaking, though the break occurs in a discrete step, rather than an infinitesimal perturbation as is usually considered. One of us (E.S.) is continuing this study on a lower dimensional version of the system (in order to facilitate direct analysis) with the assistance of Sue Campbell, who has studied O(2) symmetry-breaking to a system with D4 symmetry [4], i.e.: a circle of fixed points becomes 4 distinct fixed points upon symmetry breaking.

In the work of Aubry et al [3] a basis of K-L modes (for a different parameter regime:$\alpha = 33.005$) is constructed which preserves the $Z_2$ symmetry in the K-S equation restricted to odd Fourier modes. A comparison between a Galerkin projection onto this basis and the usual K-L basis is made using AUTO bifurcation diagrams of the steady state branches. We draw the reader's attention to the branch splitting phenomenon seen in Figures 2,3,4 of that paper, which is seen in Figure 7 here, as well as the change in the bifurcation points off the trivial solution as the truncation order is increased. They achieve considerable improvement in the convergence of the diagrams by using their symmetric K-L modes, though there are still discrepancies that are to varying extents understood.

We note here that applying their technique for constructing symmetric eigenfunctions to our system with O(2) symmetry would result in constructing a basis of Fourier modes. The translation symmetry, applied to the data, will create a translation invariant kernel in the eigenvalue problem which then admits Fourier modes as solutions. The optimal basis for a system possessing O(2) symmetry, created from data that also possesses O(2) symmetry, is the Fourier basis.

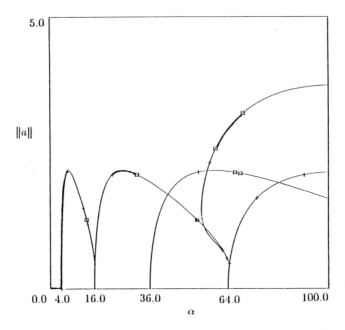

FIGURE 5. 19 Sobolev-KL mode AUTO bifurcation diagram,
selected branches.

## 6. Conclusions

We conclude by noting that we have here a case in which the addition of low
energy modes can dramatically effect dynamics. While the energy optimization
criterion is suitable and yields good results for systems with large scale coherent
structures, here adding very low energy modes will change the bifurcation structure significantly (compare Figures 2 and 3). In forming a low dimensional model
for the KS equation, approximating the O(2) symmetry appears to be more important than capturing the most energy. Our work also serves as a warning: we
observed apparently converged behavior for the 7 mode truncation, but upon
investigation it proved to arise via a different bifurcation sequence than in the
converged system.

While we don't expect to get the bifurcation structure over the entire range
of parameter space using eigenfunctions constructed at one value of $\alpha$, we had
hoped to capture the at least the nature of the bifurcations near that value of $\alpha$.
At least for this system, the KL decomposition can only be expected to reproduce
behavior within the range of parameter space for which it is used, that is, up to,
but not including any bifurcation points in that range. This is particularly true
when the solution used in the KL-decomposition does not possess the symmetry
of the governing PDE. The ergodicity assumption, (the data sampled covers the
entire solution space at that parameter value), has not been satisfied. If we

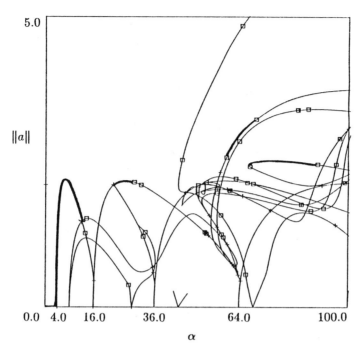

FIGURE 6. 19 Sobolev-KL mode AUTO bifurcation diagram, more branches.

had used all representations of the strange fixed point in the decomposition the resulting eigenfunctions would have been sines and cosines, which produce an O(2) equivariant basis for the KS equation at any truncation order.

We see that while the Sobolev metric procedure greatly reduces the number of modes needed for a given $\alpha$, the bifurcation structure is not preserved. In fact it is apparent that we have broken the translation symmetry of the system of equations. This will always be the case, even if we truncate only one mode. (Figures 4 and 5.) This will also be true for the standard K-L procedure. Hence it is clearly unwise to use such an approach to model the full system (over a large range of $\alpha$) based on a truncation which will necessarily always break the symmetry of the original equations. The nature of this discrete step symmetry breaking (as opposed to symmetry breaking via an infinitesimal) is currently being investigated. We hope also to find a more suitable system for examining the performance of the Sobolev norm and establishing an optimality criterion for choosing the weights.

**Acknowledgements:** We would like to thank Dieter Armbruster and Mary Silber for helpful discussions. E.S. would like to acknowledge the hospitality of Institut Non Lineaire de Nice, Université de Nice Sophia Antipolis and the Dept. of Mathematics, Colorado State University during this project.

## References

1. D. Armbruster, J. Guckenheimer and P. Holmes (1989), *Kuramoto-Sivashinsky Dynamics on the Center-Unstable Manifold*, SIAM J. Appl. Math., **49**, 676.
2. N. Aubry, P. Holmes, J.L. Lumley, E. Stone (1988), *The dynamics of coherent structures in the wall region of a turbulent boundary layer*, J. Fluid Mech. **192**, 115.
3. N. Aubry, W. Lian, E. Titi (1992), *Preserving Symmetries in the Proper Orthogonal Decomposition*, preprint.
4. S. Campbell (1992), *Heteroclinic cycles and modulated traveling waves in a system with D4 symmetry*, to appear in Physica D.
5. P. Constantin, C.Foias, B.Nicolaenko and R.Témam (1988), *Integral Manifolds and Inertial Manifolds for Dissipative Partial Differential Equations*, Applied Mathematics Sciences, **70**, Springer-Verlag, Berlin.
6. G. Dangelmayr (1986), *Steady State mode interaction in the presence of O(2)-symmetry*, Dynamics and Stability of Systems, **1**, 159-185.
7. E.J. Doedel (1981), *AUTO: A program for the bifurcation analysis of autonomous systems*, Cong. Num. **30**, 265.
8. K. Fukunaga (1972), *Introduction to Statistical Pattern Recognition*. Academic Press, New York.
9. D. Gottlieb and S.A. Orszag (1977), Numerical Analysis of Spectral Methods: Theory and Applications, SIAM-CBMS, Philadelphia.
10. J.M. Hyman, B. Nicolaenko and S. Zaleski (1986), *Order and complexity in the Kuramoto-Sivashinsky model of weakly turbulent interfaces*, Physica 23 D, 265-292.
11. M.S. Jolly, I.G. Kevrekidis and E.S. Titi, (1990) *Approximate inertial manifolds for the Kuramoto-Sivashinsky equation: Analysis and computations*, Physica 44 D, 38.
12. I.G.Kevrikidis, B.Nicolaenko, and C.Scovel (1990), *Back in the saddle again: a computer assisted study of the Kuramoto-Sivashinsky equation*, SIAM J. Appl. Math., **50**, 760.
13. M. Kirby, D. Armbruster and W. Güttinger, (1991) *An approach for the analysis of spatially localized oscillations*, Bifurcations and Chaos: Analysis, Algorithms and Applications, Intl. Ser. of Num. Math., **Vol 97** 183. Editors: R. Seydel, F.W. Schneider, T. Küpper, H. Troger, Birkhäuser Verlag Basel.
14. M. Kirby and D. Armbruster, (1992) *Reconstructing phase-space from PDE simulations*, Z angew Math Phys *43*, 999-1022.
15. M. Kirby (1992), *Minimal dynamical systems from partial differential equations using Sobolev eigenfunctions*, Physica D 57, 466-475.
16. M. Krupa (1990), *Bifurcations of Relative Equilibria*, SIAM J. Appl. Math., **21** No. 6, 1453.
17. R. Lenz (1990), *Group Theoretical Methods in Image Processing*, Lecture Notes in Computer Science (Vol. 413). Springer-Verlag, Heidelberg, Berlin, New York.
18. J.L. Lumley, (1967) *The structure of inhomogeneous turbulent flows*,In: *Atmospheric Turbulence and Radio Wave Propagation*, (A.M. Yaglom and V.I. Tatarski, eds), 166-178, Moscow: Nauka.
19. L. Sirovich (1987), *Turbulence and the dynamics of coherent structures* Parts I-III, Quarterly of Applied Mathematics, Vol. XLV, **3**, 561.
20. G.I. Sivashinsky (1977), *Nonlinear analysis of hydrodynamic stability of laminar flames-Derivation of basic equations*, Acta Astr. 4, 1177.
21. G. Taylor (1992), personal communication.

DEPT. OF MATHEMATICS, ARIZONA STATE UNIVERSITY, TEMPE, AZ 85287-1804
*E-mail address*: stone@hilbert.la.asu.edu


DEPT. OF MATHEMATICS, COLORADO STATE UNIVERSITY, FT. COLLINS, CO 80523
*E-mail address*: kirby@ritz.math.colostate.edu

Lectures in Applied Mathematics
Volume **29**, 1993

# A Generalization of the Discrete Fourier Transformation

JOHANNES TAUSCH

Feb. 1993

ABSTRACT. The discrete Fourier transformation is an efficient tool to diagonalize circulant matrices. This paper introduces a generalization of this idea for the block diagonalization of a matrix that is equivariant with respect to an arbitrary group of permutations. It turns out that the resulting blocks are identical to the reduced systems obtained by the symmetry exploiting methods of Allgower, Georg and Miranda. Thus an alternative approach to their symmetry reduction is presented.

## 1. Introduction

In many applications matrices arise that commute with a group $\Gamma$ of permutation matrices:

$$(1) \qquad A\Pi_\gamma = \Pi_\gamma A \quad \text{for} \quad \gamma \in \Gamma,$$

where $\Pi_\gamma = (e_{\gamma(1)}, \ldots, e_{\gamma(n)})$, $\Gamma$ is a group of permutations of $\mathcal{N} := \{1, 2, .., n\}$, and $A = \{A(i,j)\}_{1 \le i,j \le n}$. A matrix with the above property is called *equivariant*.

To simplify the following discussion, we assume that the group acts without fixed point on $\mathcal{N}$, i.e., if $\gamma(j) = j$ for some $\gamma \in \Gamma$ and $j \in \mathcal{N}$, then $\gamma = e$. The definition of equivariance (1) can also be stated as

$$(2) \qquad A(i,j) = A(\gamma(i), \gamma(j)) \quad \text{for} \quad \gamma \in \Gamma.$$

final version

1991 *Mathematics Subject Classification.* 65F05, 20C15.

*Key words and phrases.* Equivariant linear operator equations, representation theory, symmetry groups, linear equation solvers.

Partially supported by the National Science Foundation under grant number DMS-9104058

A subset $S$ of $\mathcal{N}$ which contains exactly one index from each orbit (under the group action of $\Gamma$) is called a *selection* of indices. It follows from (2) that the entries of the matrix are determined by the columns ( or the rows ) indexed by $S$. It is possible to exploit this structure to save computational effort when solving a linear system $Ax = b$. Recently, much work on using symmetry for solving linear systems has been done by various authors. For examples we refer the reader to the following articles: [1, 2, 3, 5, 6, 7, 8, 12]. The basic tool of these methods is the use of all irreducible representations of the group $\Gamma$. These are linear transformations $R(\gamma): \mathbb{C}^d \to \mathbb{C}^d$ , $\gamma \in \Gamma$ with the additional properties:

$$(3) \qquad R(\alpha\gamma) \;=\; R(\alpha)R(\gamma) \quad \text{for } \alpha, \gamma \in \Gamma,$$

$$(4) \qquad R(e) \;=\; \text{Id},$$

and

$$(5) \qquad R(\gamma)V \;\subset\; V \quad \text{for a subspace } V \subset \mathbb{C}^d \text{ implies}$$
$$V = \{0\} \text{ or } V = \mathbb{C}^d.$$

The number $d$ is also called the dimension of the irreducible representation $R$. For the following discussion we need some facts from group representation theory, for more details we refer the reader to [9].

(i) The matrices $R(\gamma)$ of the irreducible representation $R$ can be chosen to be unitary. In the following we abbreviate "unitary irreducible representation" as "irrep".

(ii) The number of non-equivalent irreps is finite. We assume $\{R_p\}$ to be a maximal number of non-equivalent irreps. Then the following relation holds between their dimensions and the group order:

$$(6) \qquad \sum_p d_p^2 = \#\Gamma.$$

(iii) The following orthogonality relation between two irreps $R_p, R_q$ holds:

$$(7) \qquad \sum_{\gamma \in \Gamma} R_p^{i,j}(\gamma)\overline{R_q^{k,l}}(\gamma) = \frac{\#\Gamma}{d_p}\delta_{p,q}\delta_{i,k}\delta_{j,l}.$$

(iv) The irreps of all important groups are known and listed in the literature.

## 2. Circulant Matrices

To motivate the following discussion we look at a well-known example of equivariance, namely the circulant matrices, see, e.g., [4, 10] for details. We consider the cyclic group $\Gamma = \{e, \pi, \pi^2, \dots, \pi^{n-1}\}$ of order $n$ and define the group action on the index set $\mathcal{N}_0 = \{0, 1, \dots, n-1\}$ by $\pi^k(i) = i + k$ where all indices are taken modulo $n$. It can be easily seen that an $n$-by-$n$ matrix $A$ which is equviariant with respect to this group is circulant, i.e., $A$ has the form:

(8)
$$\begin{pmatrix} a_0 & a_{n-1} & a_{n-2} & \cdots & a_1 \\ a_1 & a_0 & a_{n-1} & & \\ \vdots & & & \ddots & \\ a_{n-1} & & & & a_0 \end{pmatrix}.$$

Note that the matrix $A$ is determined by the elements in the first column $a$; the other elements are determined by the equivariance (2). The multiplication of the matrix $A$ with a vector $x$ can also be viewed as a discrete convolution with $a$, since we have

$$(Ax)_k = \sum_{\nu \in \mathcal{N}_0} a_{k-\nu} x_\nu =: (a * x)_k \text{ for } k \in \mathcal{N}_0.$$

The linear system $a * x = b$ is efficiently solved by the discrete Fourier transform: The vector $x$ is mapped on $\hat{x}$ via

$$\hat{x}_k = \frac{1}{\sqrt{n}} \sum_{\nu \in \mathcal{N}_0} \omega_\nu^k x_\nu \qquad \text{where} \qquad \omega_j^k = \exp\left(\frac{2\pi i}{n} jk\right)$$

or, in matrix notation, $\hat{x} = Fx$ where

(9)
$$F = \frac{1}{\sqrt{n}} \begin{pmatrix} 1 & 1 & \cdots & 1 \\ 1 & \omega_1^1 & \cdots & \omega_{n-1}^1 \\ \vdots & \vdots & & \vdots \\ 1 & \omega_1^{n-1} & \cdots & \omega_{n-1}^{n-1} \end{pmatrix}.$$

Since the Fourier matrix $F$ is unitary, the transformation is bijective and the inverse can be easily computed by a multiplication with the Hermitian transpose: $x = F^H \hat{x}$. In the transformed space the convolution reduces to a component-wise product, since we have:

(10)
$$\begin{aligned} \widehat{(a * x)}_k &= \sum_{\nu \in \mathcal{N}_0} \omega_\nu^k (a * x)_\nu \\ &= \frac{1}{\sqrt{n}} \sum_{\nu \in \mathcal{N}_0} \sum_{\mu \in \mathcal{N}_0} \omega_\nu^k a_{\nu-\mu} x_\mu \\ &= \frac{1}{\sqrt{n}} \sum_{\mu \in \mathcal{N}_0} \sum_{\nu \in \mathcal{N}_0} \omega_{\nu+\mu}^k a_\nu x_\mu \\ &= \frac{1}{\sqrt{n}} \sum_{\nu \in \mathcal{N}_0} \omega_\nu^k a_\nu \sum_{\mu \in \mathcal{N}_0} \omega_\mu^k x_\mu \\ &= \sqrt{n} \, \hat{a}_k \hat{x}_k. \end{aligned}$$

Thus the linear system is diagonal in the transformed space and can be solved in $\mathcal{O}(n)$ operations. The solution $x$ can be computed from $\hat{x}$ in $\mathcal{O}(n^2)$ operations, or even in only $\mathcal{O}(n \log n)$ operations, if the Fast Fourier Transformation (FFT) is applied (for large $n$).

In matrix notation the process of solving $Ax = b$ can be written as:

$$(11) \qquad Ax = b \Leftrightarrow FAx = Fb \Leftrightarrow FAF^H \hat{x} = \hat{b},$$

where the above calculations imply that the Fourier transformation of the matrix $A$ is diagonal:

$$(12) \qquad FAF^H = \sqrt{n} \, \text{diag} \, (\hat{a}_0, \dots , \hat{a}_{n-1})$$

Hence, the Fourier transformation is a fast diagonalization: The numbers $\sqrt{n} \, \hat{a}_k$ are the eigenvalues and the columns of $F$ are the eigenvectors of $A$.

The method can be extended to block circulant matrices, i.e., matrices of the type (8) whose entries $a_k$ consist of $m$-by-$m$ matrices. In this case we define the transformation matrix in the following way:

$$F_m = I_m \otimes F$$

where $I_m$ is the $m$ - dimensional identity and $\otimes$ denotes the usual tensor product. The transformation now yields a block diagonalization.

## 3. Extension to Finite Groups

We now address the problem of block diagonalizing a matrix that is equivariant with respect to an arbitrary finite group $\Gamma = \{\gamma_1, \dots , \gamma_g\}$ of order $g$. To simplify our notation we assume that the selection of indices consists of the first $m$ elements of the index set $\mathcal{N}$, i.e., $\mathcal{S} = \{1, 2, \dots , m\}$ and $\mathcal{N} = \{\gamma\mathcal{S} \mid \gamma \in \Gamma\}$. A vector $x \in \mathbb{C}^n$ splits into the components

$$(13) \qquad x = (x(\gamma_1), x(\gamma_2), \dots , x(\gamma_g))^T \qquad \text{where} \qquad x(\gamma) \in \mathbb{C}^m, \ \gamma \in \Gamma.$$

The components of $x(\gamma)$ consist of the components of $x$ indexed by $\gamma\mathcal{S}$. Since we assumed a fixed point free group action, all sets $\{\gamma\mathcal{S}\}_{\gamma \in \Gamma}$ are disjoint and have the same size. Therefore the group order divides the dimension of $A : n = mg$. The blocks of the matrix $A$ are indexed in the same way:

$$A = (A(\alpha, \gamma))_{\alpha, \gamma \in \Gamma} \qquad \text{where} \qquad A(\alpha, \gamma) \in \mathbb{C}^{m \times m}.$$

In this notation the equivariance of $A$ reads

$$A(\alpha, \gamma) = A(\gamma^{-1}\alpha, e) =: a(\gamma^{-1}\alpha).$$

As in the circulant case, the entries of $A$ are determined by one $n$-by-$m$ block column $a = (a(\gamma_1), \dots , a(\gamma_g))^T$. The matrix vector product now resembles the convolution formula as follows:

$$(Ax)(\gamma) = \sum_{\nu \in \Gamma} A(\gamma, \nu)x(\nu) = \sum_{\nu \in \Gamma} a(\nu^{-1}\gamma)x(\nu) =: (a * x)(\gamma)$$

Now we have to find an appropriate transformation $F_\Gamma$ which must satisfy two conditions:

(i) The inverse must be known explicitly.

(ii) The convolution $a*x$ must transform to a block component-wise product.

The key to finding $F_\Gamma$ lies in the observation that the rows of the Fourier matrix $F$, see (9), consist of the irreducible representations of the cyclic group. As a natural generalization we choose the matrix entries of the irreps of the group $\Gamma$: A typical row of $F_\Gamma$ has the form

$$\sqrt{\frac{d_p}{g}}\left(R_p^{i,j}(\gamma_1),\ldots,R_p^{i,j}(\gamma_g)\right) \qquad 1 \le i,j \le d_p \ .$$

For instance, the dihedral group $D_4$ has four irreps of degree one and one irrep of degree two, thus the Fourier matrix assumes the form:

$$F_{D_4} = \frac{1}{\sqrt{8}}\begin{pmatrix} R_1(\gamma_1) & \cdots & R_1(\gamma_8) \\ \vdots & & \vdots \\ R_4(\gamma_1) & \cdots & R_4(\gamma_8) \\ \sqrt{2}R_5^{1,1}(\gamma_1) & \cdots & \sqrt{2}R_5^{1,1}(\gamma_8) \\ \sqrt{2}R_5^{1,2}(\gamma_1) & \cdots & \sqrt{2}R_5^{1,2}(\gamma_8) \\ \sqrt{2}R_5^{2,1}(\gamma_1) & \cdots & \sqrt{2}R_5^{2,1}(\gamma_8) \\ \sqrt{2}R_5^{2,2}(\gamma_1) & \cdots & \sqrt{2}R_5^{2,2}(\gamma_8) \end{pmatrix} \ .$$

Let us note three points: First, we obtain a square matrix, since each representation contributes $d_p^2$ rows. These numbers add up to the group order, see (6). Second, $F_\Gamma$ is in fact unitary, because of the orthogonality relations of the irreps described in (7). Thus it is easy to switch from the vector to its transform and vice versa, since the inverse of $F_\Gamma$ is the hermitian transpose: $F_\Gamma^{-1} = F_\Gamma^H$. Therefore our first condition is satisfied. Third, there is no restriction on the way the group elements and the representations have to be arranged, since interchanging rows and columns does not affect orthogonality. Let us introduce another numbering $\{\gamma_p^{i,j}\}_{1 \le i,j \le d_p}$ of the group elements in $\Gamma$ taken with respect to an ordering of the entries of the irreps. For instance, in the above example the group element $\gamma_5$ obtains the number $\gamma_5^{1,1}$ as well.

Since we are dealing with $m$-dimensional blocks, we inflate $F_\Gamma$ by taking the tensor product with the $m$-by-$m$ identity, which will provide us with the desired transformation:

$$\hat{x} = F_{\Gamma,m}x \quad \text{and} \quad x = F_{\Gamma,m}^H\hat{x} \quad \text{where} \quad F_{\Gamma,m} = I_m \otimes F_\Gamma \ .$$

A block component of the vextor $x$ transforms to

$$\hat{x}(\gamma_p^{i,j}) = \sqrt{\frac{d_p}{g}}\sum_{\alpha \in \Gamma} R_p^{i,j}(\alpha)x(\alpha)$$

and similarly, an $m$-by-$m$ block of the matrix A transforms to

$$\hat{a}(\gamma_p^{i,j}) = \sqrt{\frac{d_p}{g}} \sum_{\alpha \in \Gamma} R_p^{i,j}(\alpha) a(\alpha) .$$

Using the representation property (3) of $R_p$

$$R_p^{i,j}(\alpha\gamma) = \sum_{k=1}^{d_p} R_p^{i,k}(\alpha) R_p^{k,j}(\gamma),$$

we study the behavior of the convolution under $F_{\Gamma,m}$:

$$
\begin{aligned}
\hat{b}(\gamma_p^{i,j}) = \widehat{(Ax)}(\gamma_p^{i,j}) \ &= \ \widehat{(a*x)}(\gamma_p^{i,j}) \\
&= \ \sqrt{\frac{d_p}{g}} \sum_{\alpha \in \Gamma} R_p^{i,j}(\alpha) \sum_{\gamma \in \Gamma} a(\gamma^{-1}\alpha) x(\gamma) \\
&= \ \sqrt{\frac{d_p}{g}} \sum_{\gamma \in \Gamma} \sum_{\alpha \in \Gamma} R_p^{i,j}(\alpha) a(\gamma^{-1}\alpha) x(\gamma) \\
&= \ \sqrt{\frac{d_p}{g}} \sum_{\gamma \in \Gamma} \sum_{\beta \in \Gamma} R_p^{i,j}(\gamma\beta) a(\beta) x(\gamma) \ \text{where } \beta = \gamma^{-1}\alpha \\
&= \ \sqrt{\frac{d_p}{g}} \sum_{\gamma \in \Gamma} \sum_{\beta \in \Gamma} \sum_{k=1}^{d_p} R_p^{i,k}(\gamma) R_p^{k,j}(\beta) a(\beta) x(\gamma) \\
&= \ \sqrt{\frac{d_p}{g}} \sum_{k=1}^{d_p} \sum_{\beta \in \Gamma} R_p^{k,j}(\beta) a(\beta) \sum_{\gamma \in \Gamma} R_p^{i,k}(\gamma) x(\gamma) \\
&= \ \sum_{k=1}^{d_p} \sqrt{\frac{g}{d_p}} \hat{a}(\gamma_p^{k,j}) \, \hat{x}(\gamma_p^{i,k})
\end{aligned}
$$

Hence, the linear system $Ax = b$ transforms to a block diagonal system:

$$
\begin{aligned}
(14) \qquad\qquad\qquad \hat{A}\hat{x} \ &= \ \hat{b} \\
\text{where } \ \hat{A} \ &= \ F_{\Gamma,m} A F_{\Gamma,m}^H \\
\text{and } \ \hat{A}(\gamma_p^{i,j}, \gamma_q^{k,l}) \ &= \ \delta_{p,q} \delta_{i,k} \sqrt{\frac{g}{d_p}} \hat{a}(\gamma_p^{l,j}).
\end{aligned}
$$

Varying the indices in the above formula provides the information about the structure of the transformed system: An irrep of dimension $d_p$ contributes $d_p$ linear systems of size $d_p m$. Note that these $d_p$ matrices are identical.

For instance, the symmetry of the dihedral group $D_4$ reduces the linear system to the following form:

| $\hat{x}(\gamma_1)$ | $\hat{x}(\gamma_2)$ | $\hat{x}(\gamma_3)$ | $\hat{x}(\gamma_4)$ | $\hat{x}(\gamma_5^{11})$ | $\hat{x}(\gamma_5^{12})$ | $\hat{x}(\gamma_5^{21})$ | $\hat{x}(\gamma_5^{22})$ | |
|---|---|---|---|---|---|---|---|---|
| $\tilde{a}(\gamma_1)$ | | | | | | | | $\hat{b}(\gamma_1)$ |
| | $\tilde{a}(\gamma_2)$ | | | | | | | $\hat{b}(\gamma_2)$ |
| | | $\tilde{a}(\gamma_3)$ | | | | | | $\hat{b}(\gamma_3)$ |
| | | | $\tilde{a}(\gamma_4)$ | | | | | $\hat{b}(\gamma_4)$ |
| | | | | $\tilde{a}(\gamma_5^{11})$ | $\tilde{a}(\gamma_5^{21})$ | | | $\hat{b}(\gamma_5^{11})$ |
| | | | | $\tilde{a}(\gamma_5^{12})$ | $\tilde{a}(\gamma_5^{22})$ | | | $\hat{b}(\gamma_5^{12})$ |
| | | | | | | $\tilde{a}(\gamma_5^{11})$ | $\tilde{a}(\gamma_5^{21})$ | $\hat{b}(\gamma_5^{21})$ |
| | | | | | | $\tilde{a}(\gamma_5^{12})$ | $\tilde{a}(\gamma_5^{22})$ | $\hat{b}(\gamma_5^{22})$ |

where the blocks of the reduced matrix $\hat{A}$ are given by:

$$(15) \qquad \tilde{a}(\gamma_p^{i,j}) := \sum_{\alpha \in \Gamma} R_p^{i,j}(\alpha)\, a(\alpha)\,,$$

since the constant factors cancel.

It is important to remark that these blocks are identical to those obtained by the symmetry reduction approaches in [2, 6, 7].

The above discussion leads us to the following description of the Generalized Fourier Transformation (GFT) for solving a linear system $Ax = b$ with an equivariant matrix:

(i) Find a selection of indices $\mathcal{S}$ and reorder the indices such that the vector $x$ can be written in the form (13).

(ii) Find the irreps of the group.

(iii) Calculate the Fourier Transformation of $b$: $\hat{b} = F_{\Gamma,m} b$

(iv) Calculate the blocks $\tilde{a}(\gamma_p^{i,j})$ of the matrix $\hat{A}$.

(v) Solve the subproblems of the block diagonalized system (14).

(vi) Calculate the solution $x$ from $\hat{x}$ using the inverse Fourier Transformation: $x = F_{\Gamma,m}^H \hat{x}$.

The matrix $F_{\Gamma,m}$ contains $ng$ non-zero entries. This implies the following computational expense for the overhead of the GFT method: Initially we have to transform $b$ into $\hat{b}$ ($gn$ flops) and the block vector $a$ into $\hat{a}$ ($n^2$ flops). Then the subproblems have to be solved (not considered as overhead). Finally the solution vector $\hat{x}$ has to be transformed back into $x$ ($gn$ flops).

If $g \ll n$, then by far the largest computational expense is involved in the solution of the subproblems. Let us assume that a direct method like the QR or LR decomposition is applied. The work to factorize a system of size $N$ is essentially $cN^3$ with a constant $c$ independent of the size. Since we have one matrix of size $d_p m$ for each irrep $R_p$, we obtain:

$$\sum_p c d_p^3 m^3 = cn^3 \frac{1}{g^3} \sum_p d_p^3$$

Hence, the expense $cn^3$ for solving the full linear system is reduced by the factor

$$\frac{1}{g^3} \sum_p d_p^3.$$

This factor is very small: The dihedral group of our previous example yields a reduction of $\frac{1+1+1+1+8}{8^3} \approx 0.023$. The symmetry group of the cube yields a factor $128/48^3 \approx 0.00116$, see [6]. Numercal experiments confirm these estimates, see [2, 11].

## REFERENCES

1. E. L. Allgower, K. Böhmer, K. Georg, and R. Miranda, *Exploiting symmetry in boundary element methods*, SIAM J. Numer. Anal. **29** (1992), 534–552.
2. E. L. Allgower, K. Georg, and J. Walker, *Exploiting symmetry in 3D boundary element methods*, Contributions in Numerical Mathematics (Singapore) (R. P. Agarwal, ed.), World Scientific Series in Applicable Analysis, vol. 2, World Scientific Publ. Comp., Singapore, 1992, to appear.
3. A. Bossavit, *Symmetry, groups, and boundary value problems. A progressive introduction to noncommutative harmonic analysis of partial differential equations in domains with geometric symmetry*, Computer Methods in Applied Mechanics and Engineering **56** (1986), 165–215.
4. P. J. Davis, *Circulant matrices*, John Wiley and Sons, New York, 1979.
5. A. Fässler and E. Stiefel, *Group theoretical methods and their applications*, Birkhäuser, Boston, 1992.
6. K. Georg and R. Miranda, *Symmetry aspects in numerical linear algebra with applications to boundary element methods*, these proceedings.
7. _____, *Exploiting symmetry in solving linear equations*, Bifurcation and Symmetry (Basel, Switzerland) (E. L. Allgower, K. Böhmer, and M. Golubitsky, eds.), ISNM, vol. 104, Birkhäuser Verlag, 1992, pp. 157–168.
8. T. J. Healey, *Numerical bifurcation with symmetry: Diagnosis and computation of singular points*, Bifurcation Theory and its Numerical Analysis (Xi'an, P. R. China) (L. Kaitai, J. Marsden, M. Golubitsky, and G. Iooss, eds.), Xi'an Jiaotong University, Xi'an Jiaotong University Press, 1989, pp. 218–227.
9. J.-P. Serre, *Linear representations of finite groups*, Graduate Texts in Mathematics, vol. 42, Springer Verlag, Berlin, Heidelberg, New York, 1977.
10. C. Van Loan, *Computational frameworks for the fast fourier transform*, Frontiers in Applied Mathematics, vol. 10, SIAM, 1992.
11. J. Walker, *Numerical experience with exploiting symmetry groups for bem*, Exploiting Symmetry in Applied and Numerical Analysis (Providence, RI) (E. L. Allgower, K. Georg, and R. Miranda, eds.), Lectures in Applied Mathematics, vol. 28, American Mathematical Society, 1993.
12. B. Werner, *Eigenvalue problems with the symmetry of a group and bifurcations*, Continuation and Bifurcations: Numerical Techniques and Applications (Dordrecht, The Netherlands) (D. Rose, B. de Dier, and A. Spence, eds.), NATO ASI Series C, vol. 313, Kluwer Academic Publishers, 1990, pp. 71–88.

DEPARTMENT OF MATHEMATICS, COLORADO STATE UNIVERSITY, FT. COLLINS, COLORADO 80523

*E-mail address*: Tausch@Math.ColoState.edu

Lectures in Applied Mathematics
Volume **29**, 1993

# Symmetry Methods in Symmetry-broken Systems

E. VAN GROESEN

March, 1993

ABSTRACT. A Poisson system (generalized Hamiltonian system) with a symmetry related to a (momentum) integral has relative equilibria that can be found variationally and define a two-dimensional manifold (MRE) of group orbits with different values of the momentum.

We will show that the parameters describing this MRE can be efficiently used as a convenient set of collective coordinates to approximate the distortion of the relative equilibria when the equation is disturbed in such a way that the groupinvariance is lost.

The two explicit examples to be considered are surface waves over an uneven bottom and the motion of Bloch walls in ferromagnetic crystals. The loss of translation symmetry leads to dynamical equations for the parameters of the MRE that are Hamiltonian for a one-degree of freedom system.

## 1. Introduction

In this paper we deal with continuous dynamical systems which in the "unperturbed" situation have translation symmetry. In the "perturbed" situation the symmetry is broken by spatial inhomogeneity. The two specific examples to which the general ideas are illustrated are surface waves over a flat and a sloping bottom, and the motion of domain-walls in homogeneous and inhomogeneous crystals.

The unperturbed systems allow special solutions as a consequence of a symmetry, which are travelling waves (solitons and kinks) in the specific examples,

---

1991 *Mathematics Subject Classification*. Primary 58F05, 35Q53, 58F30; Secondary 35L05, 35B20, 58F39.

This paper is in final form and no version of it will be submitted for publication elsewhere.

which are no longer exact solutions in the perturbed case. Nevertheless we show how the family of these special solutions can be used as a set of base functions with the aid of which the deformation due to the inhomogeneity of an initially travelling wave can be studied.

The special solutions are recognized to be the relative equilibria of the systems, both of which have a Poisson (Hamiltonian) structure. As such, the relative equilibria can be characterized variationally, and the parameters entering the constrained extremum formulation of the energy can be used efficiently as a kind of collective coordinates in the perturbation analysis. The dynamical equations for the parameters turn out to have a Hamiltonian structure as a consequence of inhereted energy conservation.

The method described here is quite generally applicable. The specific systems to be described were investigated together with S.R. Pudjaprasetya (see [8], [15]) and E. Fledderus ([14]). A somewhat related approach has been used with B. van de Fliert to study the static states of swirling flows in expanding pipes, [7]. See Van Groesen e.a. [6] and Derks [3] for the use of manifolds of relative equilibria to describe large deviations in systems with dissipation.

The first system deals with unidirectional surface waves on a layer of fluid above an even or uneven bottom. For even bottom the governing equation is the well known *Korteweg-de Vries equation* for the surface elevation $\eta$, given by

$$(1.1) \qquad \partial_t \eta = -c_0 \, \partial_x \left( \eta + \epsilon \left( c_0^4 \, \eta_{xx} + \frac{3}{c_0^2} \eta^2 \right) \right).$$

Here $c_0$ is related to the constant depth $h_0$ by $c_0 = \sqrt{h_0}$ (where $g$, the acceleration of gravity, has been normalized to unity), and the contribution due to dispersion and nonlinearity is measured by the small quantity $\epsilon$.

For uneven bottom, with $h(x)$ the depth of the fluid describing the bottom topography, let $c(x) = \sqrt{h(x)}$. Assuming slow bottom variations it was shown in [8] that the resulting equation for unidirectional waves is the following modification of the KdV-equation:

$$(1.2) \quad \partial_t \eta = -\tfrac{1}{2}(c(x)\partial_x + \partial_x c(x)) \left( \eta + \epsilon \left( \partial_x(c^4(x) \, \eta_x) + \frac{3}{c^2(x)} \eta^2 \right) \right).$$

**Remark** Various other equations of KdV-type have been proposed in the literature, and several approximating techniques for the deformation of waves have been investigated; see e.g. [5], [9], [10], [11], [12], [13]. Equation (1.2) is closely related to the equation derived by Newell [13], see [8]. The formulation given here is choosen because of its explicit character as a Hamiltonian system, as we shall see shortly.

The second system describes the motion of Bloch walls in a ferromagnetic crystal, i.e. the places where the magnetization vector changes from a position up to a position down. A one dimensional model, with $\theta$ measuring the angel of the spin vector with the $x$-axis, the governing equation is the familiar Sine-Gordon equation:

$$(1.3) \qquad\qquad \theta_{tt} = \theta_{xx} - K\sin 2\theta.$$

In a homogeneous crystal, the positive magnetic anisotropy coefficient $K$ is a constant, while in inhomogeneous crystals it is space dependent. To allow for at least one Bloch wall, boundary conditions are taken to be

$$\begin{cases} \theta(-\infty,\cdot) &=& 0 \text{ (spin-up)} \\ \theta(+\infty,\cdot) &=& \pi \text{ (spin-down)} \end{cases}.$$

## 2. Underlying Poisson structure

Consider an (infinite) dimensional Poisson manifold with structure map $\Gamma$. The Poisson bracket for functionals $F$ and $G$ is given by

$$\{F,G\}(u) := \langle \delta F(u), \Gamma \delta G(u) \rangle$$

where $\delta F$ denotes the variational derivative of the functional $F$ (with respect to a suitable innerproduct, which will be the $L_2$-innerproduct in the applications). For given Hamiltonian $H$ the evolution equation is

$$(2.1) \qquad\qquad \partial_t u = \Gamma \delta H(u),$$

the flow of which will be denoted by $\Phi_t^H$.

The systems considered in this paper are all of this form.

The KdV-equation for waves above an even bottom can be written as

$$(2.2) \qquad\qquad \partial_t \eta = \Gamma_0 \, \delta_\eta H_0(\eta)$$

where $\Gamma_0 = -c_0 \partial_x$ is the structure map and the Hamiltonian $H_0(\eta)$ is explicitly given by

$$(2.3) \qquad H_0(\eta) = \int \tfrac{1}{2}\eta^2 + \epsilon\left(-\tfrac{1}{2}c_0^4 \, \eta_x^2 + \frac{1}{c_0^2}\eta^3\right).$$

For uneven bottom, the governing equation (1.2) has the Poisson structure

$$(2.4) \qquad\qquad \partial_t \eta = \Gamma(x) \, \delta_\eta H(\eta)$$

where now the structure map $\Gamma(x) = -\tfrac{1}{2}(c(x)\partial_x + \partial_x c(x))$ depends explictly on $x$, just like the Hamiltonian:

$$(2.5) \qquad H(\eta) = \int \tfrac{1}{2}\eta^2 + \epsilon\left(-\tfrac{1}{2}c^4(x) \, \eta_x^2 + \frac{1}{c^2(x)} \, \eta^3\right).$$

We will restrict in the following to functions on the real line, vanishing at infinity; all integrals are understood to be over the whole real line.

The Sine-Gordon equation can be brought to a canonical Hamiltonian form by introducing the canonical pair of variables $\theta$ and $p = \theta_t$. Then the equation is written like (2.1) with $\Gamma = J = \begin{pmatrix} 0 & 1 \\ -1 & 0 \end{pmatrix}$ the standard symplectic matrix and with the total energy (sum of potential and kinetic energy) as Hamiltonian:

$$(2.6) \qquad H(\theta, p) = \int \left( \tfrac{1}{2}p^2 + \tfrac{1}{2}\theta_x^2 + K \sin^2 \theta \right).$$

With functions satisfying the required boundary conditions, the integrals are over the whole real line.

## 3. Exploiting Symmetry: MRE

Having recognized the underlying Poisson structure of the equations, we can exploit in a standard way the presence of a symmetry.

Let $I$ be an integral of the Poisson system, i.e. $\{H, I\} = 0$. Then the corresponding symmetry, when non-trivial, is given by the Hamiltonian $I$-flow $\Phi_\phi^I$. (This is a variant of Noether's theorem). The Hamiltonian is invariant for this flow, and the equation equivariant.

**Proposition:**
*With $I$ an integral independent of the Hamiltonian $H$, consider for given value $\gamma$ the constrained optimization problem*

$$(3.1) \qquad Extr\{H(u) \mid I(u) = \gamma\}.$$

*If $U = U(\gamma)$ is an extremizer, then so is each point on the orbit under the $I$-flow: $U(\gamma, \phi) = \Phi_\phi^I U(\gamma)$. Each element satisfies for some multiplier $\lambda = \lambda(\gamma)$ the equation*

$$(3.2) \qquad \delta H(U) = \lambda \delta I(U).$$

*In fact, $\lambda$ is the derivative of the value function $\mathcal{H}(\gamma)$:*

$$(3.3) \qquad \mathcal{H}(\gamma) = H(U(\gamma)), \quad and \quad \lambda(\gamma) = \frac{d\mathcal{H}}{d\gamma}.$$

*$U$ is called a relative equilibrium, and the translation along the $I$-orbit with speed determined by the multiplier is a dynamical solution of the Poisson system:*

$$\Phi_t^H U(\gamma, \phi_0) = \Phi_{\lambda t}^I U(\gamma, \phi_0) = U(\gamma, \phi_0 + \lambda t).$$

This proposition is a standard result for Poisson systems; see e.g. Abraham & Marsden, [1].

Note that the proposition makes it possible to consider the *Manifold of Relative Equilibria* (MRE), the two-parameter manifold of functions $U(\gamma, \phi)$ for all values of $\gamma$ for which extremizers exist:

$$(3.4) \qquad\qquad \mathcal{M} = \{U(\gamma, \phi) \mid \gamma, \phi\}.$$

For the unperturbed systems under consideration, the additional integral is a momentum integral for which the corresponding symmetry is spatial translation. For the KdV equation above an even bottom it is given by

$$I(\eta) = \tfrac{1}{2} \int \frac{\eta^2}{c_0} \text{ with } \Phi_\phi^I \eta(x) = \eta(x + \phi),$$

and for Sine-Gordon with constant $K$ by

$$I(\theta, p) = \int p\theta_x.$$

**Remark** In fact, both systems (1.1) and (1.3) with $K$ constant, are completely integrable and admit infinitely many integrals. We will not exploit the complete integrability of the equations under consideration; the results hold true for more general equations which have at least one additional integral.

For KdV the resulting MRE consists of solitary wave profiles, and the dynamical solutions are the solitary waves. Expressed in the parameters introduced above, they are given by

$$(3.5) \qquad\qquad S(\gamma, \phi) = a \operatorname{sech}^2 (b(x - \phi))$$

with

$$(3.6) \qquad a \quad = \quad a(\gamma, \phi) = \tfrac{1}{2} c^{-4/3} (3\gamma)^{2/3}$$
$$(3.7) \qquad b \quad = \quad b(\gamma, \phi) = \tfrac{1}{2} c^{-11/3} (3\gamma)^{1/3}.$$

For Sine-Gordon the MRE consists of the familiar kink-profiles

$$(3.8) \qquad\qquad \{\hat{\theta} \mid \sin \hat{\theta} = \frac{1}{\cosh Bx}\},$$

with

$$(3.9) \qquad\qquad B = \sqrt{\frac{2K}{1 - \lambda^2}} \quad \text{and} \quad \lambda = \frac{\gamma}{2B}.$$

## 4. Broken symmetry: quasi-static approximation

For a layer of fluid over an uneven bottom, and for spatially inhomogeneous crystals, the translation symmetry is no longer present. The momentum functionals considered above are no longer constants of the motion, and the travelling waves are no longer exact solutions of the inhomogeneous systems.

We will investigate the distortion of such a travelling wave and show that the distortion can well be described in a quasi-static way as a succession of suitable elements from MRE's. It will become clear that the parameters describing these intermediate states are most suited to use as collective coordinates since their dynamic evolution is easily obtained.

For the two examples above it is intuitively obvious that when the physical quantities, $c$ resp. $K$, depend on $x$ but vary only slowly with $x$, they may be well approximated by a constant $c(\varphi)$ resp. $K(\varphi)$ when the wave is positioned at $\varphi$. This idea leads one to consider the problem with "frozen" coefficients. That is, instead of the Hamiltonian and momentum with spatial dependent densities, we consider the functionals $\bar{H}(u, \varphi)$, $\bar{I}(u, \varphi)$ in which the variable quantities are replaced by their constant value at the place $\varphi$. Then the constrained optimization problems (with $\varphi$ as parameter)

$$(4.1) \qquad\qquad Extr\{\bar{H}(u, \varphi) \mid \bar{I}(u, \varphi) = \gamma\}$$

define for each $\varphi$ a MRE. From this manifold we now select the element positioned at $\varphi$, and denote it again by $U(\gamma, \varphi)$. In this way we have obtained a two-parameter manifold

$$(4.2) \qquad\qquad \mathcal{N} = \{U(\gamma, \varphi) \mid \gamma, \varphi\}$$

as a section of a one-parameter family of two-dimensional MRE's.

An approximation for the distortion of an initial relative equilibrium can now be sought as a quasi-static succession of these elements:

$$(4.3) \qquad\qquad t \rightarrow U(\gamma(t), \varphi(t))$$

for an appropriate choice of the dynamics

$$(4.4) \qquad\qquad t \rightarrow (\gamma(t), \varphi(t)).$$

For the two systems under consideration, in this approximation the solution consists of a quasi-static succession of solitons, resp. kinks, the profile at a certain place only depending on the depth, resp. value of the anisotropy constant, at that place and the current value of the momentum.

To find the appropriate dynamics for the parameters, we exploit the following observation. Since the energy (in contrast to the momentum) is conserved for any solution of the inhomogeneous systems, we require energy conservation for the approximation in the frozen energy-expression. Explicitly, define the value of the constrained optimization problem $\mathcal{H}(\gamma, \varphi)$ as

$$(4.5) \qquad \mathcal{H}(\gamma, \varphi) = \overline{H}(\varphi, U(\gamma, \varphi)).$$

With $H$ the Hamiltonian of the inhomogeneous system, exact energy conservation for the approximation would require $H(U(\gamma, \varphi))$ to be constant. This condition is replaced by the more convenient requirement

$$(4.6) \qquad \frac{d}{dt} \mathcal{H}(\varphi, \gamma) = 0.$$

This equation already determines the dynamics for $\gamma$ and $\varphi$ up to some time scaling: from $\partial_\varphi \mathcal{H} \partial_t \varphi + \partial_\gamma \mathcal{H} \partial_t \gamma = 0$ it follows that

$$\frac{\partial_t \varphi}{\partial_t \gamma} = -\frac{\partial_\gamma \mathcal{H}}{\partial_\varphi \mathcal{H}}.$$

The time scaling is determined uniquely from

$$(4.7) \qquad \partial_t \gamma = \frac{d}{dt} \bar{I}(S(\gamma(t), \varphi(t))).$$

With this it is found that the dynamical equations for $\gamma$ and $\varphi$ are a Hamiltonian system with Hamiltonian $\mathcal{H}$, i.e.

$$(4.8) \qquad \partial_t \begin{pmatrix} \gamma \\ \varphi \end{pmatrix} = \begin{pmatrix} 0 & -1 \\ 1 & 0 \end{pmatrix} \begin{pmatrix} \mathcal{H}_\gamma \\ \mathcal{H}_\varphi \end{pmatrix}.$$

Note in particular the equation for the position: with $\lambda(\gamma) = \frac{d\mathcal{H}}{d\gamma}$ it follows, as is to be expected, that the velocity of the posity equals the velocity of the soliton, resp. kink, at that place:

$$(4.9) \qquad \partial_t \varphi = \frac{\partial \mathcal{H}}{\partial \gamma} \equiv \lambda(\gamma).$$

For the wave problem the value function whose level sets determine the complete dynamics, is given by

$$(4.10) \qquad \mathcal{H} = c\gamma + \tfrac{3}{5}\epsilon c^{-7/3}(3\gamma)^{5/3}$$

where $c = c(\varphi)$.

The evolution during run-up on a shoal shows the known phenomena of decreasing velocity and wavelength and increasing amplitude, see [8] and [15].

**Remark** For the inhomogeneous equation (1.2) there is a generalization for the mass functional of (1.1). This functional is a Casimir functional for (1.2), given by

$$\int \frac{\eta(x,t)}{\sqrt{c(x)}}.$$

The approximation described above suffers from the fact that this quantity is not conserved. This can be modified by introducing a "shelf" which defines a tail behind the soliton. See Newell [13] and [15].

For the Sine-Gordon equation, the correponding expression reads

$$(4.11) \qquad \mathcal{H}^2(\gamma, \varphi) = \gamma^2 + 8K(\varphi).$$

From this result it follows that the Bloch walls behave quite analogous to a particle in a potential, with potential given by $K$. In particular, for convex functions $K$, the wall experiences a periodic motion for each energy; for periodic functions $K$ this is only true for restricted values of the energy. See Fig. 1 (from [14]).

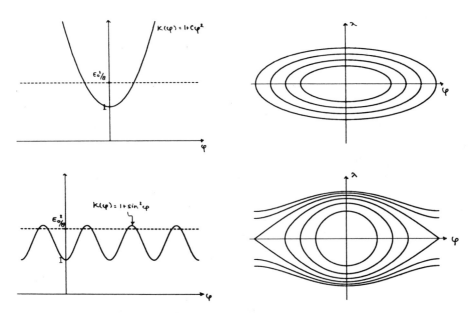

FIGURE 1. For convex (upper) and periodic (lower) space in-homogenieties, the characteristic motion of a Bloch wall follows from the level line of the value function $\mathcal{H}$: closed level lines correspond to time-periodic motions.

From (4.11) it is possible to find an explicit expression for the period. In terms of the maximal devation $\varphi_{max}$ related to the prescribed value of the energy $E_0$ as

$$E_0^2 = 8K(\varphi_{max}),$$

the period $T$ is given by

$$(4.12) \qquad T = 4 \int_0^{\varphi_{max}} \frac{d\varphi}{\sqrt{1 - \dfrac{K(\varphi)}{K(\varphi_{max})}}}$$

For the simple function $k(x)$ given by

$$(4.13) \qquad\qquad k(x) = 1 + cx^2$$

the result can be written explicitly:

$$(4.14) \qquad\qquad T = \frac{\pi}{\sqrt{2}} \frac{E_0}{\sqrt{c}}.$$

## 5. Justification and numerical verification

The validity of the quasi-static approximation described in the previous section will now be considered in some detail.

It must be remarked that a complete analytical verification is difficult to obtain. This observation is based on very detailed analysis performed by Derks [3] for problems that are somewhat comparable. No easy, generally applicable method seems to be available, and the analysis depends on the detailed behaviour of the specific system under consideration. Some general comments can be made, however.

Since the base functions used in the approximation are perturbed by the spatial inhomogeneity, it is intuitively clear, and easy to verify, that an *a priori* necessary condition is that the base functions, i.e. the solitary wave and kink solution, are stable as solutions of the unperturbed, spatially homogeneous, systems. A natural norm for stability presents itself from the extremal characterization of these solutions. Variational arguments, based on the analysis of the second variation, lead to energy type norms (Sobolev norms) which are independent for a spatial shift of the functions. Stated differently, the solitary waves are constrained, orbitally stable solutions of the unperturbed system, which is related to the fact that for the reduced dynamics obtained by deviding out the spatial translations, these solutions are (constrained) stable in the usual sense (see Derks [3] for more details).

Furthermore, the residual which measures how well the approximation satisfies the complete equation for the perturbed system, can be shown to be of the order of the *rate* of the variations in the spatial inhomogeneity. Since this rate is supposed to be small, local estimates can be found in a rather direct way. However, the total variations and changes are large, as we have seen in the examples. A comparison between exact solution and approximation, for long times in which the solution travels over distances for which the variations are large, needs completely different methods for estimation (as in Derks [3]). We will not dwell upon these analyical methods here.

Instead, we will show some numerical results that indicate the validity of the approximation, and that can stimulate to continue research for suitable methods to prove such results analytically.

The numerical results to be presented concern the motion of Bloch walls; results for the waves on a shoal will be reported elsewhere.

The spatial inhomogeneity is described by the function (4.13) with $c = 0.2$. The Sine-Gordon equation (1.3) is discretised using a method described by Strauss & Vázquez [16]: finite difference for the spatial and time derivatives, and the nonlinearity is written as a difference quotient of the potential evaluated for the solution at the next and the previous time step. This method turns out to be quite stable and robust if the spatial step and time step are related according to the characteristic velocity, i.e are taken to be equal. The (discretized) energy is exactly conserved in all cases; the momentum is quite well conserved in the homogeneous case. The numerical results to be presented below were obtained by Blommers & Booij [2].

As initial data an exact kink solution is taken and its evolution is followed. The position of the calculated solution is measured by the center of gravity of the energy density. (This functional is in fact a variable that is canonically conjugate to the momentum functional and can therefore be well compared with the variable $\varphi$ used above; see [3], [4]). Fig. 2 shows the plot of position versus value of the momentum, for the calculated solution and for the quasi-static approximation determined by (4.8). In this display the difference of the analytic approximation and the numerical calculation is invisible. In Fig. 3 the period as function of the initial energy is displayed for some numerical solutions (crosses); the analytically computed period (4.14) is given by the solid line. These results clearly show that the quasi-static approximation performs quite well. Taken into account that during one period the spatial variations are quite large (the value of $k$ running from 1 up to approximately 6 in one of the examples), it can be said that the qualitative behaviour is captured precisely, and that the quantitative results are better than may have been expected. Note, however, that the error in the period, however small, will lead for increasing time to large errors in the position (phase mismatch). This is related to the "orbital" stability of the kinks as mentioned above.

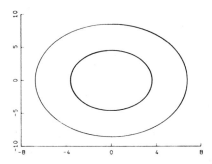

FIGURE 2. The periodic motion of a Bloch wall in an inhomogeneous material described by (4.13) with $c = 0.2$ for two different values of initial energy. The plots show the position of the kink (horizontally) versus its value of the momentum (vertically). The plots of the orbit for the numerically calculated solution and the quasi-static approximation coincide and the difference is invisible.

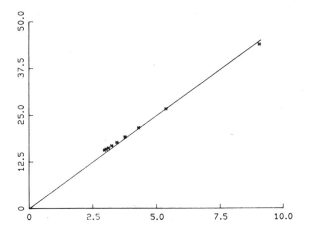

FIGURE 3. Plot of the period of the motion versus initial energy. Crosses denote results from numerical calculations, the straight line is the analytic result (4.14) for the quasi-static approximation.

## References

1. R. Abraham and J.E. Marsden, *Foundations of Mechanics*, 2nd edition, Benjamin/Cummings Publishing Company, 1978.
2. B. Blommers and W. Booij, *Periodic soliton solutions for an inhomogeneous Sine-Gordon equation*, internal report Univ. Twente, 1992.
3. G. Derks, *Coherent structures in the dynamics of perturbed Hamiltonian systems*, Ph.D.-thesis, December 1992, Univ. of Twente, Applied Mathematics.
4. G. Derks and E. van Groesen, *Energy propagation in dissipative systems, part II, Centro-velocity for nonlinear wave equations*, Wave Motion **15** (1992), 159–172
5. R.H.J. Grimshaw and N. Smyth, *Resonant flow of a stratified fluid over topography*. J. Fluid Mech. **169** (1986), 429–464.
6. E. van Groesen, F.P.H. van Beckum and T. Valkering, *Decay of travelling waves in dissipative Poisson systems*, ZAMP **41** (1990), 501–523.
7. E. van Groesen and B. van de Fliert, *Quasi-homogeneous critical swirling flows in expanding pipes*, submitted 1992.
8. E. van Groesen and S.R. Pudjaprasetya, *Uni-directional waves over slowly varying bottom; Part I: Derivation of a KdV-type of equation*, submitted 1992.
9. R.S. Johnson, *On the development of a solitary wave moving over an uneven bottom*, Proc. Cambridge. Philos. Soc. **73** (1973), 183–203.
10. O.S. Madsen and C.C. Mei, *The transformation of a solitary wave over an uneven bottom*, J. Fluid Mech. **39** (1969), 781–791.
11. C.C. Mei, *The applied dynamics of ocean surface waves*, John Wiley and Sons, New York, Chichester, Brisbane, Toronto, Singapore (1983).
12. J.W. Miles, *On the Korteweg-de Vries equation for a gradually varying channel*, J. Fluid Mech. **91** (1979), 181–190.
13. A.C. Newell *Soliton in Mathematics and Physics*, CBMS 48, SIAM 1985.
14. S.R. Pudjaprasetya and E. Fledderus, *Periodic motion of domain-walls in inhomogeneous crystals*, 1992.
15. S.R. Pudjaprasetya and E. van Groesen, *Uni-directional waves over slowly varying bottom; Part II: Quasi-static approximation of distorting waves*, 1992.
16. W. Strauss and L. Vázquez, J. Comp. Phys. **28** (1978), 271.

DEPARTMENT OF APPLIED MATHEMATICS, UNIVERSITY OF TWENTE, P.O. BOX 217, ENSCHEDE, THE NETHERLANDS
*E-mail address*: groesen@math.utwente.nl

Lectures in Applied Mathematics
Volume **29**, 1993

# Numerical Experience with Exploiting Symmetry Groups for Boundary Element Methods

JOHN WALKER

ABSTRACT. Many linear operator equations are defined on regions which are invariant under a group $\Gamma$ of symmetry transformations. If in addition, the linear operator is equivariant with respect to $\Gamma$ and if a discretization is made which respects the equivariance, the original discretized problem can be decomposed into a collection of much smaller problems, and a significant reduction of computational cost can be effected. We give a brief introduction and a detailed desription of the implementation of these ideas. As an illustration, we use the case of a boundary element method corresponding to a three-dimensional exterior Neuman problem. Finally, a two-grid method is introduced for solving the reduced problems.

## 1. Introduction

The aim of this paper is to illustrate how symmetry can be exploited to significantly facilitate the numerical treatment of boundary element methods as suggested in [**4, 5**], see also [**2**]. Furthermore, we also describe in some detail the implementation of symmetry exploiting methods discussed in the above papers.

The following example is used for the illustration. Consider the exterior boundary value problem with Neumann boundary conditions in $\mathbf{R}^3$:

(1)
$$\begin{aligned} \Delta u &= 0 \quad \text{in} \quad \mathcal{D}, \\ \tfrac{\partial u}{\partial \nu} &= g \quad \text{on} \quad \mathcal{B}, \end{aligned}$$

where $\mathcal{D}$ is the region exterior to the unit sphere centered at the origin, $\mathcal{B} = \partial \mathcal{D}$ and $\nu(y)$ denotes the outer normal to $\partial \mathcal{D}$ at $\mathrm{y} \in \partial \mathcal{D}$. We will employ a collocation method for boundary integral equations with constant elements.

1991 *Mathematics Subject Classification.* 65R20, 65F05, 65N38, 45L10, 31C20 .

*Key words and phrases.* Equivariant linear operator equations, boundary element methods, symmetry groups, two-grid method .

Partially supported by the National Science Foundation via grant number DMS-9104058

## 2. Collocation Methods for Boundary Integral Equations

Let us give a brief sketch of the boundary integral method which we employ. It is well-known, see, e.g., Atkinson [3], that the solution $u$ to the Neumann problem (1) satisfies the following Fredholm integral equation of the second kind:

$$(2) \qquad 2\pi u(x) + \int_{\mathcal{B}} K(x,y)u(y)d\mu(y) = \int_{\mathcal{B}} f(x)\frac{1}{|x-y|}d\mu(y),$$

where $d\mu$ is the standard surface element of integration and

$$K(x,y) = \frac{\partial}{\partial \nu(y)}\left(\frac{1}{|x-y|}\right).$$

Here $\nu$ denotes the outer normal on the surface. We view the integral equation as an operator equation

$$(2\pi + \mathcal{K})u = g.$$

A typical (and widely used) discretization method for solving integral equations like (2) is a collocation method. By using a suitable index set $J = \{1, 2, \ldots, n\}$, we introduce basis functions $\{\phi_j\}_{j \in J}$ on $\partial D$ and corresponding collocation points $\{x_i\}_{i \in J}$ in $\partial D$ to obtain an approximation of $u$:

$$u(x_i) \approx \sum_{j \in J} u(j)\phi_j(x_i), \qquad i \in J.$$

The following collocation equation determines the coefficients $u(j)$ of this approximation:

$$
\begin{aligned}
Au &= v, \\
\text{(3)} \qquad \text{where} \quad A(i,j) &= (2\pi + \mathcal{K})(\phi_j)(x_i) \\
\text{and} \qquad v(i) &= g(x_i) \quad \text{for } i, j \in J.
\end{aligned}
$$

## 3. Exploiting Symmetry

Frequently a surface $\mathcal{S}$ involved in a boundary element method has a certain symmetries with respect to a finite group $\Gamma$ of orthogonal transformations in $\mathbf{R}^3$, i.e., $\mathcal{S} = \gamma\mathcal{S}$ for $\gamma \in \Gamma$. In our example (1), the sphere $\mathcal{B}$ in fact allows the full orthogonal group $\mathbf{O}(3)$. In order to exploit symmetries in a boundary element method, we also need a corresponding invariance property for the kernel: we call $K$ *equivariant* with respect to $\Gamma$ if

$$K(\gamma x, y) = K(x, \gamma^{-1}y)$$

holds for $\gamma \in \Gamma$ and $x, y \in \mathcal{B}$. It is straightforward to see that the kernel in (2) is equivariant with respect to the group $\mathbf{O}(3)$. We further need the assumption that the basis functions and collocation points respect this symmetry structure:

$$\gamma x_i = x_{\bar{\gamma}i}, \quad \gamma\phi_i = \phi_{\bar{\gamma}i},$$

where the group $\Gamma$ induces the permutation group $\bar{\Gamma}$ and $\gamma \in \Gamma \mapsto \bar{\gamma} \in \bar{\Gamma}$ is a group isomorphism. Since there will be no confusion between these two groups, we will drop the distinction.

With the above properties, it can be seen, that the discretization matrix $A$ is equivariant with respect to a certain permutation group $\Gamma$, i.e.,

$$(4) \qquad\qquad A(\gamma i, j) = A(i, \gamma^{-1} j)$$

holds for $i, j = 1, ..., n$ and $\gamma \in \Gamma$.

In order to keep the exposition as simple as possible, we make one last assumption on the discretization:

ASSUMPTION 5. *The permutation group $\Gamma$ is fixed point free, i.e., if $\gamma \in \Gamma$, and $i \in \{1, \ldots, n\}$, $\gamma i = i$ holds for any $i$, then $\gamma$ is the identity element of $\Gamma$.*

It is possible to avoid this assumption, see [1].

## 4. The Reduction Method

Let us give a very brief account of the symmetry reduction method as described in [4, 5]. We introduce the projector

$$(6) \qquad\qquad P_R := \frac{1}{|\Gamma|} \sum_{\gamma \in \Gamma} R_\gamma \otimes \Pi_\gamma \in \mathcal{L}(d \cdot n)$$

where $\gamma \in \Gamma \mapsto R_\gamma \in \mathbf{U}(d)$ is a unitary irreducible representation of $\Gamma$ of dimension $d = \dim R$, and $\Pi_\gamma$ is the permutation matrix induced by the permutation $\gamma \in \Gamma$, i.e., $(\Pi_\gamma u)(i) = u(\gamma^{-1} i)$. It can be shown, see [5], that $P_R$ is an orthogonal projection. Let us remark that the above notation indicates that we are going to view a system of linear equations in block structure where $d$ is the size of the blocks.

It can be seen that

$$(7) \qquad \text{range}(P_R) = \{v : R_\gamma v(j) = v(\gamma j) \text{ for } \gamma \in \Gamma, \ j \in J\}.$$

Let $I_d$ be the $(d, d)$ identity matrix. Then the equation

$$(8) \qquad\qquad Au = g$$

is equivalent to the equation

$$(9) \qquad\qquad (I_d \otimes A)(I_d \otimes u) = (I_d \otimes g).$$

From the equivariance of $A$ it can be seen that the orthogonal projector $P_R$ and the inflated matrix $(I_d \otimes A)$ commute, and hence we obtain the inflated equation

$$(I_d \otimes A) \underbrace{P_R (I_d \otimes u)}_{=: \ u_R} = \underbrace{P_R (I_d \otimes g)}_{=: \ g_R}.$$

Let $\mathcal{O}(i) = \{\gamma i : \gamma \in \Gamma\}$ be the orbit of $i \in J$. Since the action of $\Gamma$ on the index set $J$ is fixed-point free, all orbits have the cardinality of $\Gamma$, and the index set $J$ is the disjoint union of its orbits. By a selection of indices we mean a subset $S \subset J$ which has exactly one index in each orbit. Note that by (4), the equivariant matrix $A$ is completely determined by a selection of columns (or rows).

We split the running index $j$ using our selection $S$. Furthermore, we make use of the fact that $u_R$ is in the range of the projector $P_R$. From (7) we obtain

$$\sum_{j \in S} \sum_{\gamma \in \Gamma} A(i, \gamma j) u_R(\gamma j) = \sum_{j \in S} (\sum_{\gamma \in \Gamma} A(i, \gamma j) R_\gamma) u_R(j) = g_R(i),$$

for $i \in S$. Thus, the numerical work has been reduced to the following subproblem:

$$\sum_{j \in S} \mathcal{A}_R(i, j) u_R(j) = g_R(i), \ i \in S,$$

where

$$\mathcal{A}_R(i, j) = \sum_{\gamma \in \Gamma} A(i, \gamma j) R_\gamma.$$

Once $u_R$ is caluated, we extend $u_R$ via the symmetry condition (7). We find $u(k)$, $k \in J$ by summing up over all possible irreducible representations of $\Gamma$ (one per equivalence class):

$$u(k) = \sum_R \dim R \cdot \text{trace } u_R(k).$$

## 5. The Algorithm

Let us give a brief sketch of the algorithm. First, the program will read information about the group from a file. This includes the order (order) of the group, the number (reps) of irreducible representations, the dimesion (repdim) of each representation, whether the representation is complex or real (complexity), the representations (Rep), and finally the inverse (inverse) of each group element. This information is placed in the following global structure written in C.

```
struct group_str {
    char name[50];
    int order;
    int reps;
    int *repdim;
    int *complexity;
    REAL *****Rep;
    int *inverse;
};
struct group_str group;
```

We introduce the following typedefs. The last one, namely COMPSTR, is used since we need a uniform structure which handles either the real or the complex case. (Typically, some irreducible representations of the group are real and some are complex).

```
typedef double REAL;
/* typedef float REAL; */
typedef struct{REAL r, i;} COMPLEX;
typedef struct{REAL *real; COMPLEX *complex;} COMPSTR;
```

The representations are stored in the five-dimensional array

```
group.Rep[i][j][k][l][m]
```

where i is the representation number, j is the element number, k is 0 for the real part and 1 for the imaginary part of the representation, l indicates the row and m the column index of the representation.

Once this information is entered, the reduced matrices $\mathcal{A}_R$ are constructed for each irreducible representation $R$. Below is the header of the function that constructs $\mathcal{A}_R$.

```
build_lsub(N, selection, lsub, L)
int N;
int *selection;
COMPSTR *lsub;
REAL **L;
```

The user-given array L contains a selection of columns of $A$ (note that this completely determines the equivariant matrix $A$, see (4)). The array selection contains the selected column numbers. The matrices $\mathcal{A}_R$ are stored in a long real column lsub[l].real[] if the l-th representation $R$ is real. Otherwise, they are stored in the long complex column lsub[l].complex[]. Note that this way of storing facilitates the importing of a FORTRAN subroutine for solving linear equations. The integer N denotes the size of the index set $J$. The function build_lsub uses the following user supplied function (action) which returns the action of a group element g on an index i:

```
int action(int g, int i, int N)
```

Writing the action also in dependence on the size N of the index set $J$ facilitates the reusing of the method for mesh refinements.

Next, we construct the right hand side of $\mathcal{A}_R u_R = v_R$. As before, we list the header of the function.

```
build_rsub(N, selection, rhs, rsub)
int N;
int *selection;
COMPSTR *rsub;
REAL *rhs;
```

The column rhs is the user-given right hand side of the equation $Au = v$. The array selection contains the selected column numbers. The vectors $v_R$ are stored in rsub[l].real[] if the l-th representation $R$ is real. Otherwise, they are stored in `rsub[1].complex[]`.

At this point, the subproblems are solved. The user needs to supply a linear solver. Below, we will give results for a direct solver and for a two-grid method.

Finally, we need to reconstruct the solution $u$. Below is a listing of the function call.

```
recon(N,selection, u, rsub)
int N;
int *selection;
REAL *u;
COMPSTR *rsub;
```

The column u contains the desired solution to $Au = v$. The structure rsub contains the solutions to the subproblems, and the rest of the arguments are as described above.

## 6. A Numerical Example using a Direct Solver

We will employ the collocation method for boundary integral equations as described in Section 2 for the case of constant elements. The function $g$ in equation (1) is chosen to be

$$g(x_1, x_2, x_3) := x_1 + x_2 + x_3 e^{x_2} .$$

The full symmetry group $\Gamma$ of the 3-cube will be imposed on the sphere. We view $\Gamma$ as a group of orthogonal (3,3)-matrices. A subdivison of the sphere is obtained by first subdividing the cube, then radially projecting the cube onto the sphere. To illustrate, let $\mathcal{C}$ be the surface of the cube and $E : \mathcal{C} \to \mathcal{B}$ be the radial projection of $\mathcal{C}$ onto $\mathcal{B}$. A face of the cube is subdivided into 8 congruent triangles via the diagonals and via lines joining the midpoints of opposite edges. Thus, the surface of the cube is triangulated into 48 congruent triangles. If $T$ is one of these triangles, then $\{\gamma T\}_{\gamma \in \Gamma}$ reproduces the triangulation of the surface $\mathcal{C}$ and $\{\gamma(ET)\}_{\gamma \in \Gamma}$ is a subdivision of the sphere $\mathcal{B}$. Further subdivisions of $T$ lead to refined subdivisions of $\mathcal{B}$ into $N$ pieces.

The group $\Gamma$ has 48 elements and 10 irreducible representations. Of the 10 representations, four are of dimension 1, two of dimension 2, and four of dimension 3.

Table 1 lists the time (in seconds) for solving the full system and the time for solving via the reduction method for various values of $N$. The same direct solver (an F2C conversion of LINPACK codes from the NETLIB) on an IBM 6000/550 workstation was used in both cases.

It has been shown (see [4, 5]) that the reduction factor for a direct solver with $Cn^3$ operations is:

$$(10) \qquad\qquad \rho = \frac{1}{|\Gamma|^3} \sum_R (\dim R)^3 .$$

TABLE 1.

| $N$ | symm | w/o symm | reduction factor |
|---|---|---|---|
| 768 | 2.8 | 63.0 | .044 |
| 1536 | 9.8 | 495.8 | .0198 |
| 3072 | 37.65 | 3910.3 | .0095 |
| 6144 | 165.34 | ** | .0052* |
| 12288 | 822.63 | ** | .0032* |

*Note: * indicates an extrapolated factor,   ** ran out of memory.*

For the full symmetry group of the 3-cube we have $\rho = \frac{128}{48^3} \approx 0.00116$. As can be seen, this factor is only gradually approached for large $N$. This is due to the significant role of the overhead which was neglected in (10).

## 7. Two-Grid Method

We now propose to solve the subproblems in section 6 by a two-grid method in the spirit of Atkinson [3]. Let us give a brief sketch of the method for the full system.

Let

$$(2\pi + K^{(N)})u^{(N)} = g^{(N)}$$

and

$$(2\pi + K^{(M)})u^{(M)} = g^{(M)}$$

be the linear systems ascociated with the collocation method on a coarse and fine grid, respectively. The fine grid is a refinement of the coarse grid. Let us be more specific.

DEFINITION 11. *Let $\{\mathcal{B}_i^M\}$ and $\{\mathcal{B}_i^N\}$ be the subdivision of the surface $\mathcal{B}$ associated with the fine and coarse grids, respectively. $\{\mathcal{B}_i^M\}$ is a **refinement** of $\{\mathcal{B}_i^N\}$ if for each $\mathcal{B}_m^M \in \{\mathcal{B}_i^M\}$ there exists a $\mathcal{B}_n^N \in \{\mathcal{B}_i^N\}$ such that $\mathcal{B}_m^M \subseteq \mathcal{B}_n^N$.*

The two-grid method for solving the fine grid system is given by:

$$(12) \quad \begin{aligned} r^{(l)} &= g^{(M)} - (2\pi + K^{(M)})u^{(M)(l)}, \\ p^{(l)} &= \mathcal{R}K^{(M)}r^{(l)}, \\ d^{(l)} &= \mathcal{P}(2\pi + K^{(N)})^{-1}p^{(l)}, \\ u^{(M)(l+1)} &= u^{(M)(l)} + \frac{1}{2\pi}[r^{(l)} + d^{(l)}]. \end{aligned}$$

Here $\mathcal{R} : \mathbf{R}^M \to \mathbf{R}^N$ is a restriction operator and $\mathcal{P} : \mathbf{R}^N \to \mathbf{R}^M$ is a prolongation operator. They are defined as follows:

$$(\mathcal{R}v^{(M)})(n) = v^{(M)}(m)$$

for some $m$ such that $\mathcal{B}_m^M \subseteq \mathcal{B}_n^N$, and

$$(\mathcal{P}v^{(N)})(m) = v^{(N)}(n)$$

for all $m$ such that $\mathcal{B}_m^M \subseteq \mathcal{B}_n^N$. Convergence theorems may be found in Atkinson [**3**].

The two-grid method for the subproblems is obtained by inflating the formulas in (12) as was done with (8), then applying the projector (6) to these inflated two-grid formulas. It can be shown that the restriction and prolongation operators commute with the projector (6).

Table 2 lists the time (in seconds) for solving via the reduction method using the direct solver (an F2C conversion of LINPACK codes from the NETLIB) and the two-grid method for various values of $M$ and $N$. The programs were run on an IBM 6000/550 workstation with the two-grid method being iterated from the starting value $u^{(M)(0)} = 0$ until

$$\|u^{(M)(l+1)} - u^{(M)(l)}\|_\infty \leq 10^{-7}.$$

TABLE 2.

| $M$ | direct | $N$ | two-grid |
|-----|--------|-----|----------|
| 768 | 2.8 | 48 | 2.49 |
| 1536 | 9.8 | 96 | 8.94 |
| 3072 | 37.65 | 48 | 34.18 |
| 6144 | 165.34 | 96 | 134.29 |
| 12288 | 822.63 | 48 | 604.27 |

REFERENCES

1. E. L. Allgower, K. Georg, and R. Miranda, editors. *Exploiting Symmetry in Applied and Numerical Analysis*, volume 28 of *Lectures in Applied Mathematics*, Providence, RI, 1993. American Mathematical Society.
2. E. L. Allgower, K. Georg, and J. Walker. Exploiting symmetry in 3D boundary element methods. In R. P. Agarwal, editor, *Contributions in Numerical Mathematics*, volume 2 of *World Scientific Series in Applicable Analysis*. World Scientific Publ. Comp., Singapore, 1992. to appear.
3. K. E. Atkinson. Two-grid iteration method for linear integral equations of the second kind on piecewise smooth surfaces in $R^3$. Report 14, Univ. of Iowa, 1991.
4. K. Georg and R. Miranda. Symmetry aspects in numerical linear algebra with applications to boundary element methods. Preprint, Colorado State University, 1990.
5. K. Georg and R. Miranda. Exploiting symmetry in solving linear equations. In E. L. Allgower, K. Böhmer, and M. Golubitsky, editors, *Bifurcation and Symmetry*, volume 104 of *ISNM*, pages 157–168, Basel, Switzerland, 1992. Birkhäuser Verlag.

DEPARTMENT OF MATHEMATICS, COLORADO STATE UNIVERSITY, FT. COLLINS, CO 80523, USA

*E-mail address*: walker@Math.ColoState.Edu

Lectures in Applied Mathematics
Volume **29**, 1993

# On the Shape of Solutions
# for a Nonlinear Neumann
# Problem in Symmetric Domains

ZHI-QIANG WANG

ABSTRACT. For a semilinear Neumann problem with critical exponent, we study the shape of solutions and show that if the domain possesses some symmetries the equation admits solutions being multi-peaked on the boundary, i.e. each solution attains its maximum over $\overline{\Omega}$ at finite number of points on $\partial\Omega$. The number of peaks of the solutions depends upon the symmetry the domain carries.

## 1. Introduction

Consider equation

$$(I)_\lambda \begin{cases} -\Delta u + \lambda u = u^p, \ u > 0 \quad \text{in } \Omega \\ \dfrac{\partial u}{\partial \nu} = 0 \quad \text{on } \partial\Omega \end{cases}$$

where $\Omega$ is a bounded domain in $R^N (N \geq 3)$ with smooth boundary, $\lambda > 0$ is a constant, and $\nu$ is the unit outer normal to $\partial\Omega$. Problem $(I)_\lambda$ may be viewed as a prototype of pattern formation in biology and is related to the steady state problem for a chemotoctic aggregation model by Keller and Segel([8]). In this note, we consider the so-called critical exponent problem, namely, $p = \frac{N+2}{N-2}$.

In the recent years, there have been many results concerning problem $(I)_\lambda$ and related problems([1]-[5], [10]-[14], [17]-[22]). Especially, least-energy solutions, i.e. the solutions which minimize the following energy functional

$$(1.1) \qquad E_\lambda(u) = \frac{\int_\Omega (|\nabla u|^2 + \lambda|u|^2)dx}{\{\int_\Omega |u|^{p+1}dx\}^{\frac{2}{p+1}}}$$

1991 *Mathematics Subject Classification.* Primary 35B30, 35B40; Secondary 35J20, 58F14.

The author was supported in part by an NSF grant (DMS-9201283) and a faculty research grant at Utah State University (SM-11122).

This paper is in final form and no version of it will be submitted for publication elsewhere

in $W^{1,2}(\Omega) \setminus \{0\}$, have been studied extensively (see [1] [5] [12] [17]). $E_\lambda(u)$ is called the *energy* of $u$. From the result in [20] (also see [2], [5] ), the *least energy*, which is defined by

$$(1.2) \qquad m_\lambda = \min_{u \in W^{1,2}(\Omega) \setminus \{0\}} E_\lambda(u),$$

satisfies

$$(1.3) \qquad m_\lambda \to 2^{-\frac{2}{N}} S, \quad \text{as } \lambda \to \infty.$$

A solution is called a *low energy solution (resp., high energy solution)* if its energy is less than (resp., greater than) $2^{-\frac{2}{N}} S$. In [20], we have obtained multiplicity results for low energy solutions (see also [2] [5] for multiplicity results). Furthermore, it has been shown that all these solutions share the following property: each solution attains its maximum over $\overline{\Omega}$ at exactly one point on the boundary $\partial\Omega$. We call these solutions single-peaked or single spike solutions. Contrary to Dirichlet problems, a surprising feature here is that the solutions do not preserve the rotation symmetry when $\Omega$ is invariant under some rotation transformations. A natural question which arises along the line of our study is to find out whether $(I)_\lambda$ admits other type of solutions. Especially, we want to know whether $(I)_\lambda$ possesses solutions having high energies and being multi-peaked. In [21] we have attacked this problem and have shown that $(I)_\lambda$ has solutions with high energies when $\Omega$ possesses some symmetries. Precisely, let $T(Z_k)$ be an isometric representation of the cyclic group $Z_k$ in $R^N$ and $T$ be the generator of this $Z_k$ action, i.e. $T^k = Id$. We assume that $\Omega$ has the following properties

$(\Omega_1)$. $\Omega$ is invariant with respect to $T(Z_k)$;

$(\Omega_2)$. For any $x \in \overline{\Omega}$, the orbit of $x$ under the $Z_k$ action, defined by $O(x) = \{y \in R^N \mid y = T^i x, \text{for some integer } i \}$, contains exactly $k$ ponits.

We have the following theorem in [21].

THEOREM A. *Let $(\Omega_1)$ and $(\Omega_2)$ be satisfied. For $\lambda$ large, $(I)_\lambda$ possesses a solution $u_\lambda$ satisfying*

$$(1.4) \qquad E_\lambda(u_\lambda) = k^{\frac{2}{N}} 2^{-\frac{2}{N}} S + o(1) \quad as \ \lambda \to \infty,$$

*where $S$ is the best Sobolev constant, i.e.*

$$(1.5) \qquad S = \min_{u \in W^{1,2}(R^N) \setminus \{0\}} \frac{\int_{R^N} |\nabla u|^2 dx}{\{\int_{R^N} |u|^{p+1} dx\}^{\frac{2}{p+1}}}.$$

Note that from (1.4) the solution $u_\lambda$ in Theorem A has high energy when $k \geq 2$.

The purpose of this paper is to show that the solutions given in Theorem A are k-peaked on the boundary, i.e. $u_\lambda$ attains its maximum over $\overline{\Omega}$ at exactly $k$ points on $\partial\Omega$(one $Z_k$ orbit). This will, on one hand, further reveal the relation between the energy of solutions and the shape of solutions. On the other hand, this will show that although the nonlinear Neumann problem tends to turn down the symmetry of the domain there still exist solutions which preserve certain symmetry. More precisely, we have the following theorem.

THEOREM B. *Let* $(\Omega_1)$ *and* $(\Omega_2)$ *be satisfied. Then for* $\lambda$ *large, the solution* $u_\lambda$ *in Theorem A is* $k$-*peaked on the boundary, i.e.* $u_\lambda$ *attains its maximum over* $\overline{\Omega}$ *at exactly one* $Z_k$-*orbit* $O(P_\lambda)$ *for some* $P_\lambda \in \partial\Omega$. *Moreover, letting* $\epsilon_\lambda = [u_\lambda(P_\lambda)]^{-\frac{p-1}{2}}$, *we have*

(1.6)
$$\lim_{\lambda \to \infty} \|u_\lambda - V_{\epsilon_\lambda, P_\lambda}\|_{W^{1,2}(\Omega)} = 0,$$

*where* $V_{\epsilon,y}(x)$ *is defined in (2.4) below.*

REMARK 1.1. Among the domains with properties $(\Omega_1)$ and $(\Omega_2)$, are annular domains in even dimensional space $R^N$. For instance, $\Omega = \{x \in R^N \mid R_1 \leq |x| \leq R_2\}$ with $N$ even. A solid torus in $R^3$ also satisfies $(\Omega_1)$ and $(\Omega_2)$. But standard ball domains do not possess the property $(\Omega_2)$. We shall come back to this case in a forthcoming paper([22]).

## 2. The proof of Theorem B

Define

(2.1)
$$U(x) = [\frac{N(N-2)}{N(N-2) + |x|^2}]^{\frac{N-2}{2}}$$

which is a positive solution of

(2.2)
$$-\Delta U = U^p \quad \text{in } R^N,$$

with $U(0) = 1$. Define for $\epsilon > 0$, $y \in R^N$

(2.3)
$$U_{\epsilon,y}(x) = \epsilon^{-\frac{N-2}{2}} U(\frac{x-y}{\epsilon}),$$

then $U_{\epsilon,y}$ are also solutions of (2.2). For simplicity, we shall write for $1 \leq q < \infty$,

$$\|u\|_q = \|u\|_{L^q(\Omega)}, \qquad \|u\| = \|u\|_{W^{1,2}(\Omega)}.$$

Let $T$ be the generator for the $Z_k$ action in $R^N$, i.e. $T^k = Id$. For any $y \in \overline{\Omega}$, define

(2.4)
$$V_{\epsilon,y}(x) = \sum_{i=1}^{k} U_{\epsilon,T^iy}(x) = \sum_{i=1}^{k} U_{\epsilon,y}(T^ix).$$

Because the proof of Theorem B relies upon the approach used in the proof of Theorem A, we shall firstly sketch the proof of Theorem A from [21].
Define

(2.5) $$W_k^{1,2}(\Omega) = \{u \in W^{1,2}(\Omega) \mid u(Tx) = u(x), \quad a.e. \text{ in } \Omega\}.$$

Then $W_k^{1,2}(\Omega)$ is a closed subspace of $W^{1,2}(\Omega)$ and from the principle of symmetric criticality (see [15]) any critical point of $E_\lambda(u)$ in $W_k^{1,2}(\Omega)$ gives rise to a solution of $(I)_\lambda$. We define

$$(2.6) \qquad m_{\lambda,k} = \min_{W_k^{1,2}(\Omega)\setminus\{0\}} E_\lambda(u).$$

In [21] we have proved the following result.

PROPOSITION 2.1. *For $m_{\lambda,k}$ defined in (2.6), we have*
(a). $m_{\lambda,k} \to k^{\frac{2}{N}} 2^{-\frac{2}{N}} S$, *as $\lambda \to \infty$;*
(b). $m_{\lambda,k} < k^{\frac{2}{N}} 2^{-\frac{2}{N}} S$;
(c). $m_{\lambda,k}$ *is achieved in $W_k^{1,2}(\Omega)\setminus\{0\}$ by some $u_\lambda$ which is a solution of $(I)_\lambda$.*

From this proposition ans a simple computation of energy, we see that for $\lambda$ large $(I)_\lambda$ has a nonconstant solution $u_\lambda$, which minimizes the energy functional $E_\lambda(u)$ in $W_k^{1,2}(\Omega)$ and has energy very close to $k^{\frac{2}{N}} 2^{-\frac{2}{N}} S$.

We also need the following result in [21] (Proposition 1 in [21]).

PROPOSITION 2.2. *Let $\lambda_n > 0$ and $u_n \in W_k^{1,2}(\Omega)$ be such that as $n \to \infty$, $\lambda_n \to \infty$ and*

$$E_{\lambda_n}(u_n) \to k^{\frac{2}{N}} 2^{-\frac{2}{N}} S,$$

*then there exists a subsequence (still denoted by $u_n$) of $u_n$, and $y_n \in \partial\Omega$ such that for any $\epsilon > 0$ there exists $R > 0$*

$$\frac{\int_{B_{\frac{R}{\sqrt{\lambda_n}}}(y_n)\cap\Omega} |u_n|^{p+1} dx}{\|u_n\|_{p+1}^{p+1}} \geq \frac{1-\epsilon}{k},$$

*where $B_{\frac{R}{\sqrt{\lambda_n}}}(y_n) = \{x \in R^N \mid |x - y_n| \leq \frac{R}{\sqrt{\lambda_n}}\}$.*

Now,

THE PROOF OF THEOREM B. It suffices to consider a sequence $u_{\lambda_n}$ with $\lambda_n \to \infty$, and we write $u_{\lambda_n}$ as $u_n$. Firstly, let us choose an $r > 0$ such that for any $y \in \partial\Omega$, $B_r(T^i y)$ $(i = 1, 2, \cdots, k)$ are $k$ disjoint subsets. From Proposition 2.2 and the fact that $u_n$ are solutions of $(I)_{\lambda_n}$, there exists a sequence $y_n \in \partial\Omega$ such that

$$\int_{\Omega\setminus\cup_{i=1}^k B_r(T^i y_n)} |u_n|^{p+1} dx \to 0, \quad \text{as } n \to \infty.$$

Using the equation $(I)_\lambda$ again, we get

$$(2.7) \qquad \|u_n\|_{W^{1,2}(\Omega\setminus\cup_{i=1}^k B_r(T^i y_n))} \to 0, \quad \text{as } n \to \infty.$$

From elliptic theory,

$$(2.8) \qquad \|u_n\|_{C^2(\Omega\setminus\cup_{i=1}^k B_r(T^i y_n))} \to 0, \quad \text{as } n \to \infty.$$

By maximum principle, $\max_{x\in\Omega} u_n(x) > \lambda_n^{\frac{1}{p-1}}$. Therefore, $u_n$ must assume its maximum over $\overline{\Omega}$ in $\cup_{i=1}^k B_r(T^i y_n)$.

Define

$$(2.9) \qquad w_n(x) = u_n(x)\eta\left(\frac{|x - y_n|^2}{r^2}\right)$$

where $\eta$ is a smooth nonincreasing function in $[0, +\infty)$ satisfying $\eta(t) = 1, 0 \le t \le 1$ and $\eta(t) = 0, 2 \le t < \infty$. Then from (2.7), (2.8) and the assertion (a) in Proposition 2.1, we can get

$$
\begin{aligned}
(2.10) \qquad E_{\lambda_n}(w_n) &= \frac{\|w_n\|^2 + \lambda_n\|w_n\|_2^2}{\|w_n\|_{p+1}^2} \\
&= \frac{\frac{1}{k}E_{\lambda_n}(u_n) + o(1)}{\frac{1}{k}^{\frac{2}{p+1}}(\|u_n\|_{p+1}^2 + o(1))} \\
&= k^{\frac{-2}{N}}E_{\lambda_n}(u_n) + o(1) \\
&= 2^{-\frac{2}{N}}S + o(1) \quad \text{as} \quad n \to \infty.
\end{aligned}
$$

CLAIM:.

(i). *For $n$ large, $w_n$ attains its unique maximum over $\overline{\Omega}$ at a unique point $P_n \in \partial\Omega \cap B_r(y_n)$ and $|y_n - P_n| \to 0$;*

(ii).

$$(2.11) \qquad \lim_{n \to \infty}\|w_n - U_{\epsilon_n, P_n}\| \to 0, \text{ as } n \to \infty,$$

*where $\epsilon_n = [w_n(P_n)]^{-\frac{p-1}{2}} \to 0$ as $n \to \infty$.*

Postponing the proof of the claim, we shall first use the claim to finish the proof of Theorem B. Since $u_n$ attains its maximum over $\overline{\Omega}$ in $\cup_{i=1}^k B_r(T^i y_n)$ and $w_n = u_n$ in $B_r(y_n)$, by claim (i), $u_n$ attains its maximum over $\overline{\Omega}$ at $P_n$ and $T^i P_n$ for $i = 1, 2, \cdots, k-1$. Therefore, $u_n$ attains its maximum over $\overline{\Omega}$ at exactly $k$ points, $T^i P_n, i = 0, 1, \cdots, k-1$.

From (2.7) and (2.8) we have

$$(2.12) \qquad \|u_n(x) - \sum_{i=1}^k w_n(T^i x)\| \to 0, \quad \text{as} \quad n \to \infty.$$

And from (2.11)

$$\|w_n(T^i x) - U_{\epsilon_n, T^i P_n}\| \to 0, \quad i = 1, 2, \cdots, k, \quad \text{as} \quad n \to \infty.$$

Therefore, from (2.12) we get

$$\|u_n - V_{\epsilon_n, P_n}\| \to 0, \quad \text{as} \quad n \to \infty.$$

This completes the proof of Theorem B.

Now let us go back to the proof of the claim. If $w_n$ were solutions of $(I)_{\lambda_n}$, the proof of theorem 1.2 in [5] would have prevailed. However, $w_n$ are not solutions of $(I)_{\lambda_n}$. Nevertheless, $w_n$ are approximate solutions of $(I)_{\lambda_n}$. By modifying the arguments in [5], we can prove the claim as follows.

Firstly, calculating directly, we can see that $w_n$ satisfies the following equation

$$-\Delta w_n + \lambda_n w_n = w_n^p + (\eta - \eta^p)u_n^p - \frac{4}{r^2}\langle \nabla u_n, x - y_n\rangle\eta' - \frac{2}{r^2}u_n\eta' - \frac{4}{r^2}|x - y_n|^2 u_n\eta'',$$

where all terms of $\eta, \eta'$, and $\eta''$ are evaluated at $\frac{|x-y_n|^2}{r^2}$. We can write this equation in a short form

(2.13) $$-\Delta w_n + \lambda_n w_n = w_n^p + A_n(x, u(x), \nabla u_n(x)), \text{ in } \Omega$$

with

$$A_n(x, u_n(x), \nabla u_n(x)) = 0, \text{ if } |x - y_n|^2 \le r^2 \text{ or } |x - y_n|^2 \ge 2r^2$$

and

$$|A_n(x, u_n(x), \nabla u_n(x))| \le C(|u(x)|^p + |\nabla u_n(x)|), \text{if } r^2 \le |x - y_n|^2 \le 2r^2,$$

where $C$ is independent of $n$.

Now we follow the blow up technique used in [5] (see also [13] [14]). Let $P_n$ be a point in $B_r(y_n)$ such that

(2.14) $$w_n(P_n) = \max_{x \in \bar\Omega} w_n(x).$$

Since $u_n$ and $w_n$ have the same maximum in $B_r(y_n)$, from the maximum principle, $w_n(x) > \lambda_n^{\frac{1}{p-1}}$. From the same argument as we prove (2.8), we get

(2.15) $$|P_n - y_n| \to 0, \quad n \to \infty.$$

Define

(2.16) $$\tilde w_n(x) = \epsilon_n^{\frac{N-2}{2}} w_n(\epsilon_n x + P_n)$$

where $\epsilon_n = [w_n(P_n)]^{-\frac{p-1}{2}}$. Then we have $\lambda_n \epsilon_n^2 < 1$.

Calculating directly, we get from (2.13)

$$-\Delta \tilde w_n + \lambda_n \epsilon_n^2 \tilde w_n = \tilde w_n^p + \epsilon_n^{\frac{N+2}{2}} A_n(\epsilon_n x + P_n, u_n(\epsilon_n x + P_n), \nabla u_n(\epsilon_n x + P_n))$$

or

(2.17) $$-\Delta \tilde w_n + \lambda_n \epsilon_n^2 \tilde w_n = \tilde w_n^p + B_n, \text{ in } \Omega_n = \frac{\Omega - \{P_n\}}{\epsilon_n}$$

where $B_n$ satisfies

$$B_n = 0, \quad \text{if } |\epsilon_n x - (y_n - P_n)|^2 \le r^2, \text{ or } |\epsilon_n x - (y_n - P_n)|^2 \ge 2r^2;$$

$$|B_n| \le C\epsilon_n^{\frac{N+2}{2}}(|u_n(\epsilon_n x + P_n)|^p + |\nabla u_n(\epsilon_n x + P_n)|), \text{ if } r^2 \le |\epsilon_n x - (y_n - P_n)|^2 \le 2r^2.$$

We also have $0 \le \tilde w_n \le 1$, $\tilde w_n(0) = 1$.

From (2.17) and elliptic theory,

(2.18) $$\tilde w_n \to w \text{ in } C^2_{loc}(\Omega_\infty)$$

where $\Omega_\infty = \lim_{n\to\infty} \Omega_n$. And from (2.10),

$$(2.19) \qquad \int_{\Omega_\infty} |\nabla w|^2 \leq \liminf_{n\to\infty} \int_{\Omega_n} |\nabla \tilde{w}_n|^2 = \limsup_{n\to\infty} \int_{\Omega} |\nabla w_n|^2 < \infty,$$

$$(2.20) \qquad \begin{cases} -\Delta w + aw = w^p, \ \ w(0) = 1 \ \text{ in } \ \Omega_\infty \\ \dfrac{\partial w}{\partial \nu} = 0 \ \text{ on } \ \partial\Omega_\infty. \end{cases}$$

If $\lim_{n\to\infty} \frac{\text{dist}(P_n, \partial\Omega)}{\epsilon_n} = \infty$, $\Omega_\infty = R^N$. From (2.19), (2.20) and Pohozaev's identity we have $a = 0$ and $w = U$. From (2.10), $\lim_{n\to\infty} ||\tilde{w}_n||^2 = \frac{S^{\frac{N}{2}}}{2}$. However, from (2.10) and (2.19)

$$S^{\frac{N}{2}} = \int_{R^N} |\nabla U|^2 dx \leq \lim_{n\to\infty} ||w_n||^2 = \frac{S^{\frac{N}{2}}}{2},$$

which is a contradiction. So we have $\lim_{n\to\infty} \frac{\text{dist}(P_n, \partial\Omega)}{\epsilon_n} = \alpha < +\infty$. By translation and rotation we can assume that, without loss of generality, $P_n \to 0 \in R^N$ (the origin) and $x^N$ is the inner normal direction to $\partial\Omega$ at 0. Now we introduce a diffeomorphism(See [5] [13] [20]) which straightens a boundary portion near $0 \in \partial\Omega$. Firstly, there exists a smooth function $\psi(x'), x' = (x^1, \cdots, x^{N-1})$, defined for $|x'| \leq a$ such that (i) $\psi(0) = 0$ and $\nabla\psi(0) = 0$; (ii) $\partial\Omega \cap M = \{(x', x^N) \mid x^N = \psi(x')\}$ and $\Omega \cap M = \{(x', x^N) \mid x^N > \psi(x')\}$, where $M$ is a neighborhood of 0. Now define a mapping $x = \Phi(y)$ with $\Phi(y) = ((\Phi)_1(y), \cdots, (\Phi)_N(y))$ by

$$(2.21) \qquad (\Phi)_j(y) = \begin{cases} y^j - y^N \dfrac{\partial\psi}{\partial x^j}(y') \ \text{ for } \ j = 1, \ldots, N-1 \\ y^N + \psi(y') \ \text{ for } \ j = N. \end{cases}$$

Then $\Phi$ is a diffeomorphism in the neighborhood of 0 with a smooth inverse mapping $\Psi$. It maps some neighborhood $O$ of 0 onto $B_R(0)$ with $\Phi(O \cap \{y|y^N \geq 0\}) = B_R(0) \cap \{(x', x^N) \mid x^N \geq \psi(x')\} = B_R(0) \cap \overline{\Omega}$. We can choose $R \leq r$. Consider functions $v_n(y) = w_n(\Phi(y))$ for $y \in O^+ = \{y \in O \mid y^N \geq 0\}$. Since in $B_r(0)$, $w_n(x) = u_n(x)$, then $v_n(y) = w_n(\Phi(y)) = u_n(\Phi(y))$. Following the same calculation as in [5] and [13](see also [20]), one has

$$(2.22) \qquad \begin{cases} -\sum a_{i,j}(y) \dfrac{\partial^2 v_n}{\partial y^i \partial y^j} + \sum b_j(y) \dfrac{\partial v_n}{\partial y^j} + \lambda_n v_n = v_n^p \ \text{ in } \ O^+ \\ \dfrac{\partial v_n}{\partial y^N} = 0 \ \text{ on } \ O^+ \cap \{y^N = 0\} \end{cases}$$

where $a_{i,j}, b_j$ are smooth functions with

$$a_{i,j}(y) = \delta_{i,j} + O(|y|).$$

Let $q_n$ be such that $P_n = \Phi(q_n)$. Then $v_n(q_n) = u_n(P_n)$. If we write $q_n = (q_n^1, \cdots, q_n^N)$, we have $\lim_{n\to\infty} \frac{q_n^N}{\epsilon_n} = \alpha$. Define $O_n = \frac{O^+ - \{q_n\}}{\epsilon_n}$ and for $y \in O_n$

$$\tilde{v}_n(y) = \epsilon_n^{\frac{N-2}{2}} v_n(\epsilon_n y + q_n).$$

Then

$$
\begin{cases}
-\sum(\delta_{i,j} + O(\epsilon_n y + q_n))\dfrac{\partial^2 \tilde{v}_n}{\partial y^i \partial y^j} + \sum b_j(\epsilon_n y + q_n)\dfrac{\partial \tilde{v}_n}{\partial y^j} + \lambda_n \tilde{v}_n = \tilde{v}_n^p \text{ in } O_n \\
\dfrac{\partial \tilde{v}_n}{\partial y^N} = 0 \quad \text{on} \quad O_n \cap \{y^N = -\dfrac{q_n^N}{\epsilon_n}\}
\end{cases}
$$

and

$$
\tilde{v}_n(0) = 1, \quad 0 \le \tilde{v}_n \le 1.
$$

From elliptic theory again, we get $\tilde{v}_n \to w_1$ in $C^2_{loc}(O_\infty)$, where $O_\infty = \{y \mid y^N \ge -\alpha\}$, and $w_1 \in W^{1,2}(O_\infty)$

$$
(2.23) \qquad
\begin{cases}
-\Delta w_1 + a w_1 = w_1^p, \quad \text{in } O_\infty \\
\dfrac{\partial w_1}{\partial y^N} = 0 \quad \text{on} \quad \partial O_\infty.
\end{cases}
$$

We also have $0 \le w_1 \le 1, w_1(0) = 1$. By Pohozaev's identity again, $a = 0$. Since $w_1(0) = 1$, $w_1 = U$. By $\frac{\partial w_1}{\partial y^N} = 0$ on $\partial O_\infty$, $\alpha = 0$.

Now if $P_n \notin \partial\Omega$, then $q_n \notin \{y^N = 0\}$. Since $\frac{\partial \tilde{v}_n}{\partial y^N} = 0$ on $\{y^N = -\frac{q_n^N}{\epsilon_n}\}$, and $\frac{\partial \tilde{v}_n}{\partial y^N}(0) = 0$, we get $z_n = (0, 0, \cdots, z_n^N)$ with $-\frac{q_n^N}{\epsilon_n} \le z_n^N \le 0$ such that $\frac{\partial^2 \tilde{v}_n}{\partial(y^N)^2}(z_n) = 0$. Letting $n \to \infty$, we have $z_n \to 0$ and

$$
(2.24) \qquad 0 = \lim_{n\to\infty} \frac{\partial^2 \tilde{v}_n}{\partial(y^N)^2}(z_n) = \frac{\partial^2 U}{\partial(y^N)^2}(0) < 0,
$$

which is a contradiction. Therefore $P_n \in \partial\Omega$, and $\Omega_\infty = R^N_+$.

Next, from $\tilde{w}_n \to U$ in $C^2_{loc}(R^N_+)$ and the fact that $\int_{\Omega_n} |\nabla \tilde{w}_n|^2 dx$ is bounded, we have

$$
(2.25) \qquad \|\nabla w_n - \nabla U_{\epsilon_n, P_n}\|_2^2 = \|\nabla \tilde{w}_n - \nabla U\|_{L^2(\Omega_n)}^2 \to 0, \quad n \to \infty.
$$

Hence, we get claim (ii).

Finally, to prove that $P_n$ is unique, we assume there exist $P_n^1, P_n^2 \in \partial\Omega$ such that $w_n(P_n^1) = w_n(P_n^2) = \max_{\overline{\Omega}} w_n(x)$. Then from (2.15) and (2.24), $|P_n^i - y_n| \to 0$ and $\|\nabla w_n - \nabla U_{\epsilon_n, P_n^i}\|_2 \to 0$ as $n \to \infty$ for $i = 1, 2$. This implies $\lim_{n\to\infty} \frac{|P_n^1 - P_n^2|}{\epsilon_n} = \alpha < \infty$. If $P_n^1 \ne P_n^2$, we can get a contradiction as follows. In the proof above, if we use $P_n^1$ as center to do blow up, for $\tilde{v}_n(y)$ we can make a reflection with respect to $y^N = 0$. Then $\tilde{v}_n(y)$ attains its maximum at 0 and $x_n = \frac{\Phi^{-1}(P_n^2) - \Phi^{-1}(P_n^1)}{\epsilon_n} \in \{y^N = 0\}$. For simplicity, we assume $x_n = (x_n^1, 0, \cdots, 0)$. Then there exists $z_n = \xi_n x_n$ for some $0 < \xi_n < 1$ where $\tilde{v}_n(y)$ achieves its local minimum on $\{y^2 = \cdots = y^N = 0\}$. Then

$$
\frac{\partial^2 \tilde{v}_n}{\partial(y^1)^2}(z_n) \ge 0, \quad \frac{\partial \tilde{v}_n}{\partial y^1}(z_n) = 0 \ .
$$

Since $x_n$ is bounded we can assume $z_n \to z_0$. Passing to a limit, we get $\lim_{n\to\infty} \frac{\partial \tilde{v}_n}{\partial y^1}(z_0) = \frac{\partial U}{\partial y^1}(z_0) = 0$ and we must have $z_0 = 0$. Also we have

$$(2.26) \qquad 0 \le \lim_{n\to\infty} \frac{\partial^2 \tilde{v}_n}{\partial (y^1)^2}(z_n) = \frac{\partial^2 U}{\partial (y^1)^2}(0) < 0,$$

which is a contradiction. Thus, $P_n$ is unique.

## REFERENCES

1. Adimurthi and G. Mancini, *The Neumann problem for elliptic equations with critical nonlinearity*, Estratto da Nonlinear Analysis, Scuola Normale Superiore, Pisa (1991), 9–25.
2. Adimurthi and G. Mancini, *Effect of geometry and topology of the boundary in the critical Neumann problem*, preprint.
3. Adimurthi and S.L. Yadava, *Critical Sobolev exponent problem in $R^N$ ($n \ge 4$) with Neumann boundary condition*, Proc. Indian Acad. Sci. (Math. Sci.) **100** (1990), 275–284.
4. Adimurthi and S.L. Yadava, *Existence and nonexistence of positive radial solutions of Neumann problems with Critical Sobolev exponents*, Arch. Rational Mech. Anal. **115** (1991), 275–296.
5. Adimurthi, F. Pacella, and S.L. Yadava, *Interaction between the geometry of the boundary and positive solutions of a semilinear Neumann problem with critical nonlinearity*, preprint.
6. H. Brezis, *Nonlinear elliptic equations involving the critical Sobolev exponent*, Survey and Perspectives in Directions in P.D.E. (Ed. by Grandall, Rabinowitz and Turner) (1987), 17.-36.
7. H. Brezis and L. Nirenberg, *Positive solutions of nonlinear elliptic equations involving critical exponents*, Comm. Pure Appl. Math. **36** (1983), 437–477.
8. E.F. Keller and L.A. Segel, *Initiation of slime model aggregation viewed as an instability*, J. Theor. Biol. **26** (1970), 399–415.
9. P.L. Lions, *The concentration-compactness principle in the calculus of variations*, The locally compact case, Part 1 and Part 2, Ann. Inst. H. Poincaré Anal. Nonlinéaire **1** (1984), 109–145, 223–283.
10. P.L. Lions, F. Pacella, and M. Tricarico, *Best constants in Sobolev inequalities for functions vanishing on some parts of the boundary and related questions*, Indiana Univ. Math. J. **37** (1988), 301–324.
11. C.-S. Lin, W.-M. Ni, and I. Takagi, *Large amplitude stationary solutions to a chemotaxis system*, J. Diff. Euqa. **72** (1988), 1–27.
12. W.-M. Ni, X.-B Pan, and I. Takagi, *Singular behavior of least energy solutions of a semilinear Neumann problem involving critical Sobolev exponents*, Duke Math. J. **67** (1992), 1–20.
13. W.-M. Ni and I. Takagi, *On the shape of least-energy solutions to a semilinear Neumann problem*, Comm. Pure Appl. Math. **45** (1991), 819–851.
14. W.-M. Ni and I. Takagi, *On the existence and shape of solutions to a semilinear Neumann problem*, Progress in Nonlinear Diff. Equa.(Ed. by Lloyd, Ni, Peletier and Serrin) (1992), 425–436.
15. R. Palais, *The principle of symmetric criticality*, Comm. Math. Phys. **69** (1979), 19–30.
16. M. Struwe, *A global compactness result for elliptic boundary value problems involving limiting nonlinearities*, Math. Z. **187** (1984), 511–517.
17. X.-J. Wang, *Neumann problems of semilinear elliptic equations involving critical Sobolev exponents*, J. Diff. Equ. **93** (1991), 283–310.
18. Z.-Q. Wang, *On the existence of multiple, single-peaked solutions of a semilinear Neumann problem*, Arch. Rational Mech. Anal. **120** (1992), 375–399.
19. Z.-Q. Wang, *On the existence of positive solutions for semilinear Neumann problems in exterior domains*, Comm. in P.D.E. **17** (1992), 1309–1325.

20. Z.-Q. Wang, *The effect of the domain geometry on the number of positive solutions of Neumann problems with critical exponents*, preprint.

21. Z.-Q. Wang, *Nonlinear Neumann problems with critical exponent in symmetrical domains*, To appear in Proceedings of the conference on variational methods in nonlinear analysis, Erice-Sicily, Italy (May 1992).

22. Z.-Q. Wang, *High energy and multi-peaked solutions for a semilinear Neumann problem with critical exponent*, preprint.

DEPARTMENT OF MATHEMATICS, UTAH STATE UNIVERSITY, LOGAN, UTAH 84322

*E-mail address*: wang@sunfs.math.usu.edu

Lectures in Applied Mathematics
Volume **29**, 1993

# The Numerical Analysis of Bifurcation Problems with Symmetries Based on Bordered Jacobians

BODO WERNER

ABSTRACT. The paper addresses the problem of detecting, computing and pathfollowing of symmetry breaking steady state bifurcations in parameter dependent equivariant ODE's. The approach relies upon monitoring scalar functions which are given by the solution of linear systems with certain symmetry–adapted bordered Jacobian matrices.

## 1. Introduction

We will be concerned with (symmetry breaking) steady state bifurcations of dynamical systems

(1.1) $$\dot{x} = g(x, \lambda), \quad \lambda \in \mathbb{R}, \quad x \in \mathbb{R}^N,$$

with the symmetry

(1.2) $$g(\gamma x, \lambda) = \gamma g(x, \lambda), \quad x \in \mathbb{R}^N, \lambda \in \mathbb{R}, \gamma \in \Gamma,$$

where $\Gamma$ is a compact Lie group of orthogonal $N \times N$-matrices acting on $\mathbb{R}^N$.

Following a path of $\Gamma$–symmetric equilibria

$$C^\Gamma = \{(x(s), \lambda(s)) : |s| < \delta\}, \quad (\gamma x(s) = x(s) \text{ for all } \gamma \in \Gamma),$$

we are interested in the numerical *detection, computation* and — after releasing a second parameter — *pathfollowing* of (symmetry breaking) bifurcation points. The numerical concept we advocate is that of *test functions* [19], [15], also called *bifurcation functions* [13]. This concept is of special use for path following of bifurcation points of higher codimension.

Test functions are essentially real functions defined on a neighborhood of $C^\Gamma$, being monitored during continuation and strictly changing sign at bifurcation points $(x_0, \lambda_0) = (x(s_0), \lambda(s_0))$ of interest (see Def.2.1).

1991 *Mathematics Subject Classification.* Primary 65H10, 20C30, 58F14

This paper is in final form and no version of it will be submitted for publication elsewhere.

An obvious choice for a test function for steady state bifurcation (without symmetries) is the *determinant* of the Jacobian matrices. But – besides of scaling properties – there are some shortcomings with respect to numerical differentiation which increases the effort for pathfollowing of bifurcation points considerably. Instead we will prefer a test function based on *bordered Jacobian matrices* ([9]),

$$(1.3) \qquad B(A, r, l) := \begin{pmatrix} A & r \\ l^T & 0 \end{pmatrix}, \quad r, l \in \mathbb{R}^N, \quad A := D_x g(x, \lambda),$$

see Section  which contains some new results concerning the generic behavior of these test functions along a branch of equilibria.

To find a test function for Hopf bifurcation is more difficult. We refer the interested reader to the recent paper [10], to [13] and to [19], where also bordered Jacobians are used.

For problems with symmetries (1.2), the block diagonalization of the Jacobians with respect to a symmetry–adapted basis is a powerful tool, see [19], [4] and the references therein. Testing the individual blocks for singularity is a possible choice for defining test functions for symmetry breaking bifurcation.

To obtain the block diagonal form of the Jacobians, a group theoretical machinery has to be used. We follow here another idea based on (1.3) with *symmetry–adapted bordering* $r, l$ and leading to test functions for symmetry breaking bifurcation— with a special bordering for each symmetry type of the bifurcation point. Similar ideas are contained in [12] in the framework of numerical Lyapunov–Schmidt reductions.

These test functions can be easily added to existing numerical packages based on the concept of test functions as LOCBIF [13]. The information necessary for branch switching is contained in the solution of the bordered system. Less group theoretical knowledge is needed than for the block diagonalization. An incomplete (even no) information about the underlying symmetry can lead to successful results (Section 3.3). Moreover, *block elimination methods* for the numerical solution of bordered systems using a black box solver for $A$ ([7],[8] and the references therein) can be systematically used for evaluation of test functions, for numerical continuation of equilibria and bifurcation points of higher codimension and for branch switching as well.

## 2. The concept of test function

Roughly speaking, test functions are real functions being evaluated during numerical continuation of a path $\mathcal{C}$ and vanishing at certain bifurcation points $y_0$ of interest. $\mathcal{C}$ may be a path of codimension-$k$ bifurcation points and $y_0$ a bifurcation point of codimension $k + 1$, $k \in \mathbb{N}$.

The precise definition for $k = 0$ ($\mathcal{C}$ is a path of equilibria) is as follows:

**Definition 2.1** *Let $g(x, \lambda)$ be a smooth vector field, let $\mathcal{C} = \{(x(s), \lambda(s)) : s \in J\}$ be a path of equilibria and let $\mathcal{B}$ be a certain set of bifurcation points on $\mathcal{C}$. A real $C^1$-function $t$ defined on a neighborhood $U$ of $\mathcal{B}$ is called* **test function** *for $\mathcal{B}$ if each $(x_0, \lambda_0) \in \mathcal{B}$ is a regular zero of*

(2.1) $$G : U \to X \times \mathbb{R}, \quad G(x, \lambda) := (g(x, \lambda), t(x, \lambda))$$

*e.g., if $DG(x_0, \lambda_0)$ is regular.*

Numerically, it is important to take into account the costs for the numerical evaluation of derivatives $Dt(x, \lambda)$, see Section 3.1.

The regularity of the *minimally augmented system* (2.1) implies that $\tau(s) := t(x(s), \lambda(s))$ strictly changes sign at bifurcation points $y \in \mathcal{B}$:

(2.2) $$\tau(s_0) = 0, \quad \tau'(s_0) \neq 0 \quad \text{for } y_0 = (x(s_0), \lambda(s_0)) \in \mathcal{B}.$$

In [19] we used the notation *regular t-point* for regular zeros of $G$. This regularity can be interpreted as a transversal crossing of the manifold defined by $t = 0$ by the path $\mathcal{C}$.

The neighborhood $U$ in Def.2.1 may be chosen only relatively open in some subspace $X$ of $\mathbb{R}^N$ being invariant under $g(\cdot, \lambda)$. We have certain fixed point spaces $X^\Sigma$ in mind.

Since bifurcation often depends on (the eigenvalues of) the Jacobians $A(s) := D_x g(x(s), \lambda(s))$, test functions $t$ we are going to consider here, will have the form

$$t(x, \lambda) = T(D_x g(x, \lambda)),$$

where the Jacobian derivative of $g$ with respect to the state variables $x$ is denoted by $D_x g$ and where $T(A)$ is defined on some open set of real $N \times N$–matrices $A$.

Again, $T(A) = 0$ in general defines a manifold of codimension 1 in some matrix space and (2.2) describes the transversal crossing of this manifold by the path $A(s)$ of the Jacobians. Usually, certain *eigenvalue crossing conditions* are reflected by (2.2).

We will use the name *test function* for $t$ and for $T$ as well.

The generalization of Def. 2.1 to the case of bifurcation points of arbitrary codimension is obvious: Now $g(x, \lambda)$ may depend on $k$ parameters $\lambda \in \mathbb{R}^k$ and $\mathcal{C}$ is a path of codimension-$(k-1)$ bifurcation points defined implicitly by

$$g(x, \lambda) = 0, \quad t_j(x, \lambda) = 0, \quad j = 1, ..., k-1,$$

where $t_1, ..., t_{k-1}$ are test functions used for the location of codimension-j bifurcation points $(j = 1, ..., k-1)$. Now the real function $t_k$ is a test function for certain codimension-k bifurcation points $y_0$ if the $N + k$ dimensional system

$$g(x, \lambda) = 0, \quad t_j(x, \lambda) = 0, \quad j = 1, ..., k$$

has a regular root at $y_0$.

$T(A) := \det(A)$ seems to be the most obvious choice for a test function ($X = \mathbb{R}^N$) for steady state bifurcation points. But the evaluation of the analytical formula for the derivative of $det(A)$,

$$(2.3) \qquad D_A \det(A)H = \text{trace}(Adj(A)H), \quad H \in \mathbb{R}^{N,N},$$

or the numerical differentiation by finite differences is too expensive. Here $Adj(A)$ denotes the matrix of the adjoints of $A$ appearing in Cramer's rule

$$(2.4) \qquad (\det A)I = Adj(A)A.$$

## 3. Bordering and test functions

Given an $N \times N$-matrix $A$ and certain nonvanishing *bordering vectors* $r, l \in \mathbb{R}^N$, we consider the bordered $(N+1) \times (N+1)$-matrix (1.3) and the linear system

$$(3.1) \qquad \begin{pmatrix} A & r \\ l^T & 0 \end{pmatrix} \begin{pmatrix} u(A) \\ T(A) \end{pmatrix} = \begin{pmatrix} 0 \\ 1 \end{pmatrix}.$$

The following Theorem ([9]) essentially recommends $T(A)$ as a competitor of the determinant $det(A)$ as a test for singularity of $A$ and hence for steady state bifurcations.

**Theorem 3.1** *With $B(A, r, l)$ defined as in (1.3) let*

$$(3.2) \qquad R_{r,l} := \{A \in \mathbb{R}^{N,N} : B(A, r, l) \text{ is regular}\}.$$

*By solving the bordered system (3.1), one obtains a $C^\infty$-function $T : R_{r,l} \to \mathbb{R}$ satisfying $T(A) = 0$ if $A \in R_{r,l}$ is singular, and $T(A) = -(l^T A^{-1} r)^{-1}$ if $A \in R_{r,l}$ is regular.*

*The (Fréchet) derivative of $T$ is given by*

$$(3.3) \qquad D_A T(A)H = -v(A)^T H u(A), \quad H \in \mathbb{R}^{n,n},$$

*where $u(A) \in \mathbb{R}^N$ is specified by (3.1) while $v(A) \in \mathbb{R}^N$ is defined as the solution of the transposed bordered system*

$$(3.4) \qquad \begin{pmatrix} A^T & l \\ r^T & 0 \end{pmatrix} \begin{pmatrix} v(A) \\ T(A) \end{pmatrix} = \begin{pmatrix} 0 \\ 1 \end{pmatrix}.$$

**3.1  Numerical differentiation.** As a consequence of Th. 3.1, $T(A)$ is a test for singularity of $A$ as long as $B(A, r, l)$ is regular ($A \in R_{r,l}$). Before we investigate $R_{r,l}$ and its complement $S_{r,l}$ we like to comment on formula (3.3):

In numerical computations we need the derivative $Dt(y)$ of the test function $t(y) := T(D_x g(y))$, $y := (x, \lambda)$. Setting $A(y) := D_x g(y)$ we obtain the following formulas from (3.3) ([9])

$$(3.5) \quad D_x t(y) \quad \approx \quad -v(A(y))^T \frac{A(x + h \cdot u(A(y)), \lambda) - A(y)}{h},$$

$$(3.6) \quad D_\lambda t(y) \quad \approx \quad -v(A(y))^T \frac{A(x, \lambda + h) - A(y)}{h} u(A(y)) \quad \text{with small } h.$$

The effort for the evaluation of these formulas is considerably less than for the determinant $det(A)$, compare also with Prop. 3.5.

**3.2 Generic behavior of $\tau(s) = T(A(s))$.** We wish to get more insight into the sets $R_{r,l}$ and its complement $S_{r,l}$ to understand the behavior of $\tau(s)$ along a branch $\mathcal{C}$.

We start with the following well known *Keller's Lemma*:

**Lemma 3.2** *If $A$ is regular, then the bordered matrix $B(A, r, l)$ is regular if and only if the Schur complement $l^T A^{-1} r$ is non-zero.*

*If $A$ is singular, then the borderd matrix $B(A, r, l)$ is regular if and only if*

$$rank(A) = N - 1, \quad l \notin \mathcal{N}(A)^\perp, \quad r \notin \mathcal{N}(A^T)^\perp.$$

**Proof:** In contrast to other proofs in the literature, we make use of the matrix of adjoints of $A$, $Adj(A)$, which will be also of help in the proof of the next Theorem.

First note that

$$(3.7) \qquad \det B(A, r, l) = -l^T Adj(A) r.$$

The first claim follows from (3.7) and Cramer's rule, $Adj(A) = \det(A) A^{-1}$.

If rank(A)$< N - 1$, then $Adj(A) = 0$ and $B(A, r, l)$ is singular by (3.7).

Let rank(A)$=N - 1$. Because of (2.4), there exist nonzero $\varphi$ and $\psi$ spanning the kernel $\mathcal{N}(A)$ and cokernel $\mathcal{N}(A^T)$ respectively with

$$(3.8) \qquad Adj(A) = \varphi \cdot \psi^T.$$

Therefore the second claim follows from (3.7). ∎

Recall the definition (3.2) of $R_{r,l}$. Let $S_{r,l}$ be the complement of $R_{r,l}$ in $\mathbb{R}^{N,N}$. Then by (3.7), $A \in R_{r,l}$ if and only if $l^T Adj(A) r \neq 0$. Thus $R_{r,l}$ is an open set.

**Theorem 3.3** *The codimension-1 manifold*

$$S_{r,l}^0 := \{A : A \text{ is regular with } l^T A^{-1} r = 0\}$$

*is an open and dense subset of $S_{r,l}$.*

**Proof:** By Lemma 3.2, $S_{r,l}^0$ is an (open) subset of $S_{r,l}$. Let $A \in S_{r,l}$, but $\notin S_{r,l}^0$. Then $l^T Adj(A) r = 0$. We have to show that in every neighborhood of $A$ there is a regular $\tilde{A}$ with $l^T \tilde{A}^{-1} r = 0$. To show this, we can assume without loss of generality that $r = e_j, l = e_i$ are standard cartesian basis vectors. Then $l^T Adj(A) r = A_{ij} = 0$. Now small perturbations of $A$ can be found which do not change the element $A_{ij}$ of $Adj(A)$, but result into a regular matrix: first perturb the submatrix after deleting the $i^{th}$ row and the $j^{th}$ column to a rank=$N-2$-matrix. Then a small perturbation of the $i^{th}$ row and the $j^{th}$ column, not changing $A_{ij}$, can produce a regular $N \times N$–matrix $\tilde{A}$.                         ∎

During path following of equilibria we want to evaluate $\tau(s) := T(A(s))$ whith $A(s) := D_x g(x(s), \lambda(s))$ defining an 1-dimensional manifold $\mathcal{J}$ of matrices. The Theorem 3.3 tells us that generically $\tau(s)$ is not defined if and only if $A(s) \in S^0_{r,l}$. Since $S^0_{r,l}$ is a codimension-1 manifold, the crossing of $S^0_{r,l}$ by the path $\mathcal{J}$ of the Jacobians $A(s)$ will be generically transverse. At the crossing parameters $s$ we will have simple poles of $\tau(s)$.

For $A(s) \in R_{r,l}$, our candidate $\tau(s)$ for a test function is well defined with $\tau(s) = 0$ if and only if $A(s)$ is singular. Denote the singular matrices in $R_{r,l}$ by $R^0_{r,l}$. Then $R^0_{r,l}$ is a codimension-1 manifold implicitly defined by $T(A) = 0$. Again, generically the crossing of $R^0_{r,l}$ by $\mathcal{J}$ will be transverse which results in simple zeros of $\tau(s)$ which is just the condition (2.2) for a test function.

**Definition 3.4** *We call a path $\mathcal{J} := \{A(s) | s \in J\}$ of $N \times N$-matrices ($J$ being an interval) a* **generic path** *with respect to the bordering vectors $r, l$ if $A(s) \notin R_{r,l}$ if only if $A(s) \in S^0_{r,l}$ and if $\mathcal{J}$ crosses $S^0_{r,l}$ and $R^0_{r,l}$ transversally.*

Along a generic path, $\tau(s)$ will have only simple zeros and simple poles as in Figure 1 (left). We have just shown that the term *generic* is justified. Observe that there is always a sign change of $\tau(s)$ not only at the desired zeros but also at the simple poles. But the latter can be easily distinguished numerically from the zeros.

By investigating the regularity condition for the minimally augmented system (2.1), it can be shown that in the sense of Def. 2.1, $t(x, \lambda) := T(D_x g(x, \lambda))$ defines a test function for quadratic turning points and simple subspace breaking bifurcation points as $\mathbb{Z}_2$–pitchforks (for these notions see [19]).

### 3.3 Comparison with the determinant.

The test function $T(A)$ is superior to the determinant $det(A)$ with respect to numerical differentiation. But for singular matrices $A$ the derivatives of $T(A)$ and of $det(A)$ essentially coincide. This follows from

**Proposition 3.5** *If $A \in R_{r,l}$ is singular, then $D_A T(A) = c \cdot D_A \, det(A)$, $c \neq 0$.*

**Proof:**     The derivative of $det(A)$ is given by (2.3), and (3.8) yields $Adj(A) = c\, u(A) v(A)^T$ with $c \neq 0$. Then the claim follows directly from (3.3).                         ∎

The disadvantage of $T(A)$ is that the bordering vectors $r, l$ must be chosen suitably. Our test function will fail if the path $A(s)$ of matrices meets a singular matrix which is not in $R_{r,l}$. Though this cannot happen for a generic choice of $r$ and $l$, their might be numerical difficulties if a simple zero and a simple pole of $\tau(s)$ are very close.

Problems can also occur at pathfollowing of codimension-1 bifurcation points defined by $t(x, \lambda) = 0$ if we do not update $r, l$. Good choices are $r = \psi \in \mathcal{N}(A^T)$ and $l = \varphi \in \mathcal{N}(A)$ which can easily be approximated by previous solutions $u(A)$ and $v(A)$ of the bordered systems (3.1) and (3.4).

## 4. Test functions for symmetry breaking bifurcation

We assume now that there is a linear representation $D$ of a compact Lie group $G$ on $\mathbb{R}^N$ such that $\Gamma := D(G)$ satisfies (1.2). Then $A := D_x g(x, \lambda)$ for $\Gamma$-symmetric $x$ is $\Gamma$-*symmetric* in the sense that

$$\gamma A = A \gamma \quad \text{for all } \gamma \in \Gamma.$$

Call $\mathcal{M}^\Gamma$ the space of all $\Gamma$–symmetric, real $N \times N$-matrices.

Hence along the path $\mathcal{C}^\Gamma$ of $\Gamma$-symmetric equilibria, the 1-dimensional manifold $\mathcal{J}$ of Jacobians $A(s)$ stays in $\mathcal{M}^\Gamma$.

We are interested in detecting singular $A(s)$. Because of the group symmetry, there are different possibilities of $A(s)$ getting singular depending on how $G$ is acting on the kernel of $A(s)$.

Generically we may assume that $G$ is acting absolutely irreducible on the kernel ([6]). Each *symmetry type* of singularity and the corresponding (symmetry breaking) bifurcation is characterized by an (absolutely) irreducible representation $d$ of $G$. Hence in the sequel, $d$ always denotes an absolutely irreducible representation.

The issue is to find individual test functions for each absolutely irreducible representation $d$. One possibility — which has been shortly described in [19] — is based on a blockdiagonal form of all $A \in \mathcal{M}^\Gamma$ with respect to a symmetry–adapted basis. Roughly speaking, for each irreducible representation $d$ of dimension $n_d$ and multiplicity $c_d$ in $\mathbb{R}^N$, there is a subspace $Y_d$ of dimension $c_d$, invariant under all $A \in \mathcal{M}^\Gamma$, such that the eigenvalues of the restriction $A_d$ of $A$ to $Y_d$ occur in $n_d$ copies (see [2], [18], [3]). Hence only $A_d$ has to be tested for singularity — for instance by the bordering method described above.

But instead of performing the group theoretical computation of $A_d$, we will investigate $T(A)$ in (3.1) and $\tau(s) = T(A(s))$ for the "big" matrix $A \in \mathcal{M}^\Gamma$. The main idea is to choose the bordering vectors $r, l$ somehow symmetry–adapted, but we will also discuss the choice of arbitrary chosen bordering vectors.

An obvious problem is that the 1-dimensional manifold $\mathcal{J}$ is not a generic path in the sense of Def. 3.4 if—due to multidimensional irreducible representations—multidimensional kernels of $A_0 := A(s_0)$ occur. Then $A_0$ is not in $R_{r,l} \cup S^0_{r,l}$: the bordered matrix $B(A_0, r, l)$ and $A$ itself are simultaneously singular, so that $T(A_0)$ seems not to be defined.

**4.1 Symmetry–adapted bordering.** To make the basic principle as clear as possible, we replace the symmetry–adapted subspace $Y_d$ by some subspace

$Y$ of $\mathbb{R}^N$. Let $\mathcal{M}^Y$ denote all real $N \times N$-matrices $A$ for which $Y$ and $Y^\perp$ are invariant subspaces. (Observe that for $Y := Y_d$, $\mathcal{M}^\Gamma$ is a subset of $\mathcal{M}^Y$.) Call $A_Y$ the restriction of $A \in \mathcal{M}^Y$ to $Y$.

In this context, symmetry–adapted bordering means that we choose $r, l \in Y$, but consider the "big" bordered matrix $B(A, r, l)$ instead of the "small" one $B(A_Y, r, l)$. Note that the latter is just the restriction of the former to $Y \times \mathbb{R}$ and that $B(A, r, l)$ might be singular without $B(A_Y, r, l)$ being so. Nevertheless we will see that $T(A)$ (but not necessary $u(A)$ in (3.1)) is well defined:

**Theorem 4.1** Let $B(A_Y, r, l)$ be regular and $T_Y(A_Y)$ given by the unique solution of the "small" bordered system

$$(4.1) \qquad \begin{pmatrix} A_Y & r \\ l^T & 0 \end{pmatrix} \begin{pmatrix} u_Y(A_Y) \\ T_Y(A_Y) \end{pmatrix} = \begin{pmatrix} 0 \\ 1 \end{pmatrix}.$$

Then the "big" bordered system

$$(4.2) \qquad \begin{pmatrix} A & r \\ l^T & 0 \end{pmatrix} \begin{pmatrix} u \\ \tau \end{pmatrix} = \begin{pmatrix} 0 \\ 1 \end{pmatrix}.$$

has at least one solution $(u, \tau)$ whith the same $\tau = T_Y(A_Y)$ for all solutions.

**Proof:**   $(u, \tau) = (u(A_Y), T_Y(A_Y))$ solves (4.2). For any other solution $(u, \tau)$ we decompose $u = u_Y + u_Z$ with $u_Y \in Y$ and $u_Z \in Z := Y^\perp$. Since $r, l \in Y$, it follows that $A_Y u_Y + \tau r = 0$, $\quad A u_Z = 0$ and $l^T u_Y = 1$. Hence $u_Y = u_Y(A_Y)$ and $\tau = T_Y(A_Y)$ (while $u$ may not be unique due to $u_Z \neq 0$).   ∎

Coming back to the situation of the group symmetry, Theorem 4.1 tells us that testing the $d$–block $A_d$ for singularity by bordering, is equivalent to the bordering test of $A$ provided that the bordering vectors $r, l$ are chosen in some symmetry–adapted subspace $Y_d$. It is not necessary to compute $A_d$ itself! In the next Section we discuss what that means for $r, l$.

**4.2   Source vectors.** Given an irreducible representation $d$, it remains to examine how vectors $r, l \in Y_d$ can be chosen. To this end we define

**Definition 4.2** Every nonzero vector $v$ of an irreducible subspace $V$ of type $d$ is called a **d-source vector**.

Here an irreducible subspace $V$ of type $d$ is a $\Gamma$–irreducible subspace where $\Gamma$ is acting *equivalently* (in the sense of representation theory) to $d$.

**Definition 4.3** Let the irreducible representation $d$ have multiplicity $c_d$. A **symmetry–adapted subspace** $Y_d$ of type $d$ is a $c_d$-dimensional subspace of $\mathbb{R}^N$, invariant under all $A \in \mathcal{M}^\Gamma$, such that the restriction $A_d$ to $Y_d$ is similar to the $d$-block of the block diagonal form of $A$ (see [2], p. 40).

A symmetry–adapted subspace $Y_d$ of type $d$ can be constructed by projection, $Y_d = P_d \mathbb{R}^N$, where the projection $P_d$ is specified by

$$(4.3) \qquad P_d := \frac{\dim d}{|G|} \sum_{g \in G} \tilde{d}(g^{-1})_{ii} D(g)$$

for some irreducible representation $\tilde{d}$ equivalent to $d$ and any $i = 1, ..., N$ (see [2], p. 146).

It is not a difficult task (use the idea of construction in [2], p. 40) to show what follows

**Proposition 4.4** *A vector $v$ belongs to a symmetry–adapted subspace of type $d$ if and only if $v$ is a d-source vector.*

Hence to construct a test function $T_d(A)$ for $A \in \mathcal{M}^\Gamma$ in order to find symmetry breaking bifurcation points of type $d$, according to Th. 3.1, we can use a bordering with $d$-source vectors $r, l$.

It remains to discuss how to find $d$-source vectors.

For 1-dimensional irreducible representations $d$, source vectors coincide with vectors from the isotypic components $V_d$ of type $d$ (see [2], p. 36), These vectors can be found easily, often naively. For instance in the most common case of a $\mathbb{Z}_2$-symmetry with reflection $S$, source vectors $v$ are either symmetric ($Sv = v$) or anti–symmetric ($Sv = -v$). The latter source vectors lead to test functions for $\mathbb{Z}_2$-pitchforks.

For higher-dimensional irreducible representations $d$, source vectors can be constructed via projection by $P_d$ in (4.3) or by some orthogonality considerations.

If no information about irreducible representations or even about the group $G$ itself is available, we suggest another method: solve a complete eigenvalue problem for a Jacobian $A$ on the path $\mathcal{C}^\Gamma$. Since eigenspaces corresponding to real eigenvalues are generically absolutely irreducible, every such eigenvector is expected to be a $d$-source vector for some irreducible representation $d$.

For example take a coupled 6-box network, where the six boxes are arranged in a ring. Call a state $x = (x_1, ..., x_6)$, where $x_i$ is the state of the $i^{th}$ box. The group $G$ is the dihedral group $D_6$.

Source vectors $r$ are given by $r = (a, a, a, a, a, a)$ for the trivial irreducible representation, $r = (a, -a, a, -a, a, -a)$ for another one-dimensional irreducible representation, $r = (2a, -a, -a, -2a, a, a)$ and $r = (2a, a, -a, -2a, -a, a)$ (or $r = (0, -a - a, 0, a, a)$) for the two two-dimensional irreducible representations. Here all $d$-source vectors have the symmetry of a bifurcation subgroup for $d$, see below.

**4.3  Branch switching.** For the evaluation of the test function $T_d(A)$ specified by (3.1) or (4.2) with symmetry–adapted $d$–source vectors $r_d, l_d$ as bordering, we need not to care about component $u \in \mathbb{R}^N$ of the solution. But $u$ is important for an efficient differentiation (see (3.5) and (3.6)) and for branch switching, since $u$ contains valuable directional informations, because of $u \in \mathcal{N}(A)$ if $T_d(A) = 0$.

If $d$ is nontrivial, tangential directions of bifurcating branches from a bifurcation point of type $d$ are essentially given by those vectors in the kernel of $D_x g^0$ which belong to the fixed point space $X^\Sigma$ for some *bifurcation subgroup* $\Sigma$ of $\Gamma$ ([1]).

Hence we suggest to choose the bordering $r, l$ to have as an additional property the symmetry of a bifurcation subgroup $\Sigma$.

If $d$ is a 1-dimensional irreducible representation, $\Sigma$ is unique and $d$-source vectors $r, l$ automatically have the symmetry of $\Sigma$.

If $d$ is multidimensional, we suggest to choose for each bifurcation symmetry $\Sigma$ of interest a specific bordering.

All source vectors given above for a 6-box network have the symmetry of a bifurcation subgroup.

The aim is to force the bordered system solution $u \in \mathcal{N}(A)$ to belong at least approximately to $X^\Sigma$ and to be a good predictor candidate for numerical continuation of the bifurcating branch.

Problems may arise if the bordered matrix $B(A, r_d, l_d)$ is singular due to the symmetry (see case 4 below).

**4.4   Computational considerations.** Assume that we have chosen a fixed symmetry–adapted bordering $r_d, l_d$ for each irreducible representation $d$ and that we start continuation of the $\Gamma$–symmetric path $\mathcal{C}^\Gamma$ by computing the manifold $\mathcal{J}$ of $\Gamma$–symmetric Jacobians.

The first observation is that to evaluate each test function $\tau_d(s)$, we have to solve several bordered systems with the *same* upper block $A(s)$. It is a well–established fact that a black–box solver for $A(s)$ can be used for the solution of all bordered systems even if $A(s)$ is nearly singular, see [7].

Let us discuss the behavior of $\tau_d(s)$ and numerical problems in connection with the computation of $\tau_d(s)$. To this end, the "big" bordered matrices— $A(s)$ bordered by $r_d, l_d$—will be called $B(s)$, the diagonal $d$-subblocks of $A(s)$ are denoted by $A_d(s)$ while the "small" bordered matrices—$A_d(s)$ bordered by $r_d, l_d$—are called $B_d(s)$.

**Case 1:** $A(s)$ and $B(s)$ are regular.
Then $\tau_d(s) \neq 0$ and there are no problems.

**Case 2:** $A(s)$ is singular and $B(s)$ is regular.
Then $A_d(s)$ is singular, the dimension of $d$ is 1, all other diagonal blocks $A_{\tilde{d}}(s)$ are regular, $\tau_d(s)$ will have a simple zero. Theoretically there can be problems with using a black–box solver for the solution of the bordered system by block elimination methods ([7]). But this was not the case in my practice.

**Case 3:** $A(s)$ regular, $B(s)$ singular.
Then $B_d(s)$ is singular, $\tau_d(s)$ will have a simple pole. There are no numerical problems.

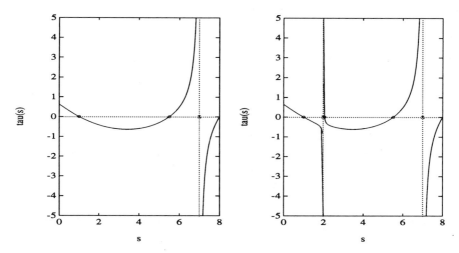

Figure 1: The graph of $\tau_d(s)$ without (left) and with perturbation (right)

So far we have described the generic situations described in Def. 3.4 without symmetries. But with symmetries, our path $A(s)$ is not generic in the sense of Def. 3.4, since

**Case 4:** $A(s), B(s)$ singular

cannot be avoided. It is caused by a bifurcation point of type $\tilde{d} \neq d$ (**case 4a:** singular $A_{\tilde{d}}(s)$, regular $A_d(s)$) or by a bifurcation point of type $d$ with dimension $d \geq 2$ (**case 4b**).

The value $\tau_d(s)$ is not affected by case 4. Hence, generically, $\tau_d(s)$ will have only simple zeros and simple poles as in the generic situation. But the computation of $\tau_d(s)$ (and the component $u_d(s)$ of the solution) may suffer from the singularity of $B(s)$.

If we solve the singular (but solvable!, see Th. 4.1,) bordered systems by using the expensive pseudo–inverse of $B(s)$ there will be no numerical problems. Morerover, the solution component $u_d \in \mathbb{R}^N$ will be in the symmetry–adapted subspace $Y_d$ which contains $r_d, l_d$. This is important for the numerical differentiation of $T_d$ and for branch switching.

But even the use of a conventional black–box solver for $A(s)$ has turned out to be successful. For an explanation we examine $\tau_d(s)$ slightly perturbing $A(s)$ to a generic path (Def. 3.4). Then case 4 does not occur, case 4a is perturbed into a closely neighbored pair of a simple zero and a simple pole of $\tau_d(s)$, see Fig.1. If dim $(d) = 2$, we expect that case 4b is perturbed into a closely packed triple of two simple zeros and one simple pole.

If the step size of continuation is not too small in comparison with the perturbation, these closely neighbored pairs of poles and zeros will not be detected by a sign change - a desirable thing.

**Remark 4.5** *Case 4b can be avoided by using a wider bordering, replacing $r, l$*

*by (N × m)-matrices with m=dim (d), the columns of which span an irreducible subspace of type d, see [12].*

**Remark 4.6** *If we follow codimension-1 bifurcation points of type $\tilde{d} \neq d$, the evaluation of $\tau_d(s)$ along such a path will be always concerned with (at least) case 4a, hence we have to deal with singular bordered matrices $B(s)$ along the the whole path.*

**4.5   Arbitrary bordering.** We close this section by studying the case that the $\Gamma$–symmetric $A(s)$ are bordered by arbitrary $r, l$.

Denote the diagonal blocks in the diagonal normal form of $A(s)$ as $A_k(s)$ $k = 1, ..., m$ (including multiplicities for multidimensional $d$). Then $\mathbb{R}^N$ is the direct sum of the corresponding symmetry–adapted subspaces $Y_k$, $k = 1, .., m$.

**Theorem 4.7** *Assume that*

$$r = \oplus_{k=1}^{m} r_k, \quad l = \oplus_{k=1}^{m} l_k$$

*are such that $B(A_{d_k}, r_k, l_k)$ is regular, $k = 1, ..., m$. Denote the test function, the existence of which is guaranteed by Th. 4.1, by $T_k(A)$.*

*If at least one $A_k$ is singular, the bordered system (3.1) is solvable with $T(A) = 0$ for each solution.*

*If all $A_k$ are regular, then $B(A, r, l)$ is regular if and only if $\sum_k \frac{1}{T_k(A)} \neq 0$. In that case, (3.1) is uniquely solvable with*

(4.4) 
$$\frac{1}{T(A)} = \sum_k \frac{1}{T_k(A)}.$$

The proof is simple linear algebra and will be omitted.

The formula (4.4) reminds us of a formula for the total resistance $T(A)$ in a circuit with parallel arrangement of resistances $T_k(A)$, where also negative resistances are allowed.

Concerning the behavior of $\tau(s) = T(A(s))$ near a singular matrix $A(s)$, the claim (which we do not prove here) is, that generically we will have a strict sign change of $\tau(s)$ even if several singular blocks $A_k(s)$ are involved (this is the case if dim $d_k \geq 2$). This is a big difference with the competitor $d(s) := \det A(s)$!

Hence, at least theoretically, every bifurcation point of arbitrary symmetry type could be detected by a generic choice of bordering. In practice this is not true. For example, if one pair $r_k, l_k$ is underrepresented in comparison with another pair, $\tau(s)$ may look like in Fig. 1, right, where the closely neighbored pair of zero and pole indicates a singular $A_k(s)$ which will be difficult to detect.

**4.6    Example.** The following equivariant dynamical system arises in population dynamics ([14], [11]). The state variables $x_i, i = 1, 2, .., 6$ are the frequencies of genotypes $AA, AB, BB$,
$BC, CC, CA$ in a diploid population with three alleles $A, B, C$. Assuming that the three alleles cannot be distinguished as regards to fertility, the following system models selection:

$$\dot{x}_1 = ax_1^2 + \tfrac{b}{4}(x_2^2 + x_6^2) + cx_1(x_2 + x_6) + \tfrac{f}{2}x_2x_6 - x_1\varphi(x)$$
$$\dot{x}_2 = \tfrac{1}{2}bx_2^2 + cx_2(x_1 + x_3) + 2dx_1x_3 + e(x_1x_4 + x_3x_6)$$
$$\qquad + \tfrac{f}{2}(x_2x_4 + x_2x_6 + x_4x_6) - x_2\varphi(x)$$
$$\dot{x}_3 = ax_3^2 + \tfrac{b}{4}(x_2^2 + x_4^2) + cx_3(x_2 + x_4) + \tfrac{f}{2}x_2x_4 - x_3\varphi(x)$$
$$\dot{x}_4 = \tfrac{1}{2}bx_4^2 + cx_4(x_3 + x_5) + 2dx_3x_5 + e(x_2x_5 + x_3x_6)$$
$$\qquad + \tfrac{f}{2}(x_2x_4 + x_2x_6 + x_4x_6) - x_4\varphi(x)$$
$$\dot{x}_5 = ax_5^2 + \tfrac{b}{4}(x_4^2 + x_6^2) + cx_5(x_4 + x_6) + \tfrac{f}{2}x_4x_6 - x_5\varphi(x)$$
$$\dot{x}_6 = \tfrac{1}{2}bx_6^2 + cx_6(x_1 + x_5) + 2dx_1x_5 + e(x_1x_4 + x_2x_5)$$
$$\qquad + \tfrac{f}{2}(x_2x_4 + x_2x_6 + x_4x_6) - x_6\varphi(x)$$

where $a, b, c, d, e, f$ are fertility coefficients and

$$\varphi(x) = a(x_1^2 + x_3^2 + x_5^2) + b(x_2^2 + x_4^2 + x_6^2)$$
$$\qquad + 2c\left(x_2(x_1 + x_3) + x_4(x_3 + x_5) + x_6(x_1 + x_5)\right)$$
$$\qquad + 2d(x_1x_3 + x_1x_5 + x_3x_5) + 2e(x_1x_4 + x_2x_5 + x_3x_6)$$
$$\qquad + 2f(x_2x_4 + x_2x_6 + x_4x_6)$$

denotes the mean fertility.

This system is equivariant with respect to the permutation group $S_3$ of three elements (the three alleles $A, B, C$).

We choose $a = 0.01, b = 0.14, c = 0.9, d = 0.5, e = 0.2$ and vary $\lambda := f$ starting with a symmetric stable equilibrium at $f = 0.9$ ($x_1 = x_3 = x_5, x_2 = x_4 = x_6$). Since $S_3$ has a two dimensional irreducible representation $d$ we expect bifurcation points of this type. To detect them, we choose $d$–source vectors $r = l = (1, 0, 1, 0, -2, 0)$ for bordering. We detected and computed a $d$–bifurcation point at $f = 0.80005$ from which a branch of equilibria with the symmetry $x_1 = x_3 \neq x_5, x_4 = x_6 \neq x_2$ and two further conjugate branches emerge. This $\mathbb{Z}_2$–symmetry can be broken at $\mathbb{Z}_2$–pitchforks which we were able to detect and compute using the bordering vectors $r = l = (1, 0, -1, 0, 0, 0)$. We found two of these $\mathbb{Z}_2$–pitchforks at $f = 0.70567$ and $f = 0.63712$. Fig. 2 shows a bifurcation diagram where (un)stable solutions are represented by solid (dashed) lines. Note that the bifurcation point on the symmetric branch is transcritical with a pitchfork degeneration.

**4.7    Concluding remarks.** We have also successfully applied the preceding concept to steady state bifurcations of symmetric truss structures (for example

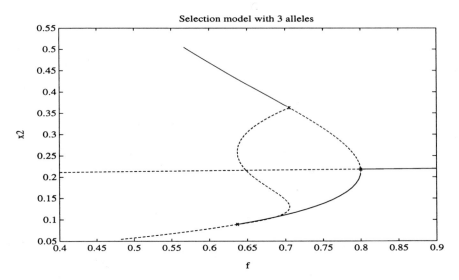

Figure 2: Bifurcation diagram of the selection model

the hexagonal lattice dome in [16]) and to coupled Brusselators (c. [1], [5]). Clearly, this method is inferior to the more sophisticated method based on the group theoretical block diagonalization ([16], [3], [4]) especially if secondary and even tertiary and higher bifurcations are taken into account. But this method provides a simple tool for performing quick calculations with a minimum of group theoretical knowledge using standard techniques.

Another advantage is the possibility of the use of special block elimination methods for the solution of bordered systems based on black–box solvers for the "big" Jacobian matrices ([7], [8]). Hence all computations in connection with numerical continuation, detection of bifurcation points, branch switching, and continuation of bifurcation points of higher codimension are based on (wider) bordered matrices with the same Jacobian matrices—a conceptually simple, but in case of symmetries not a highly optimized procedure.

We have applied the test function concept also to the continuation of bifurcation points of type $\tilde{d}$ with respect to a second parameter using the defining equation $T_{\tilde{d}}(A) = 0$. Then, taking a second test function $T_d(A)$ for another type $d$, it is possible to detect and to compute codimension-2 bifurcation points where two bifurcation points of different types $d$ and $\tilde{d}$ coalesce (mode interaction). Moreover, under some group theoretical assumptions, from this codimension-2 bifurcation point bifurcates a branch of codimension-1 bifurcation points of $d$-like type, see [5], which are secondary symmetry breaking bifurcation points on a branch emerging from the $d$-bifurcation point. The numerical branch switching procedure is the same as for codimension-1 bifurcation points of type $d$.

### Acknowledgement:
Lijun Yang has pointed out to me the usefulness of $Adj(A)$ in Section 3. I also

should like to thank Vladimir Janovsky for helpful discussions.

## REFERENCES

1. M. Dellnitz, B. Werner, Computational methods for bifurcation problems with symmetries - with special attention to steady state and Hopf bifurcation points, J. of Comp. and Appl. Math. **26**, 97-123, 1989.
2. A. Fässler, E. Stiefel, *Group Theoretical Methods and their Applications,* Birkhaüser, Boston-Basel-Berlin, 1992.
3. K. Gatermann, A. Hohmann, Symbolic Exploitation of symmetry in Numerical Pathfollowing, Impact of Computing in Science and Engeneering **3**, 330-365, 1991.
4. K. Gatermann, Computation of Bifurcation Graphs, Konrad-Zuse-Institut, Berlin, Report, 1992.
5. K. Gatermann, B. Werner, Group Theoretical Mode Interactions with Different Symmetries, In preparation, 1992.
6. M. Golubitsky, I. Stewart, D. Schaeffer, *Singularities and Groups in Bifurcation Theory, Vol. II,* Springer, New York, 1988.
7. W. Govaerts, J.D. Pryce, Mixed block elimination for linear systems with wider borders, To appear in IMA Journ. of Numer. Anal., 1992.
8. W. Govaerts, Solution of bordered singular systems in numerical continuation and bifurcation, Preprint, University of Gent, Belgium, 1992.
9. A. Griewank, G.W. Reddien, Characterization and computation of generalized turning points, SIAM J. Numer. Anal. **21**, 176-185, 1984.
10. J. Guckenheimer, M. Myers, B. Sturmfels, Computing Hopf Bifurcations, Preliminary version of a manuscript, Cornell University, 1992.
11. K.P. Hadeler, U. Libermann, Selection models with fertility differences, Journ. of Math. Biol. 2, 19-32, 1975.
12. V. Janovsky, P. Plechac, Numerical Applications of equivariant reduction techniques, in: *Bifurcation and Symmetry* (E. Allgower, K. Böhmer, M. Golubitsky (eds)), ISNM **104**, 203-213, Birkhäuser, Basel, 1992.
13. A. Khibnik, Y. Kuznetsov, V. Levetin, E. Nicolaev, Continuation techniques and interactive software for bifurcation analysis of ODEs and Iterated maps, Physica D (special issue, NATO Workshop on Homoclinic Chaos), 1991.
14. D. Otte, unpublished notes, 1992.
15. R. Seydel, On detecting stationary bifurcations, Int. J. of Bifurcation and Chaos 1, 1-5, 1991.
16. P. Stork, B. Werner, Symmetry adapted block diagonalization in equivariant steady state bifurcation problems and its numerical applications, Advances in Mathematics (China), Vol. **20,4**, 455-487, 1991.
17. B. Werner, V. Janovsky, Computation of Hopf branches bifurcating from Takens-Bogdanov points for problems with symmetries, in: *Bifurcation and Chaos* (R. Seydel, F.W. Schneider, T. Küpper, H. Troger, eds.), ISNM **97**, 377-388, Birkhäuser, Basel, 1991.
18. B. Werner, Eigenvalue problems with the symmetry of a group and bifurcations, in: *Continuation and Bifurcations: Numerical Techniques and Applications,* D. Roose, B. de Dier, A. Spence (eds.), NATO ASI Series C, Vol. **313**, 71-88, Kluwer Academic Publishers, Dordrecht, 1990.
19. B. Werner, Test Functions for Bifurcation Points and Hopf Points in Problems with Symmetries, in : *Bifurcation and Symmetry* (E. Allgower, K. Böhmer, M. Golubitsky (eds)), ISNM **104**, 317-327, Birkhäuser, Basel, 1992.

*Current address:* Institut für Angewandte Mathematik der Universität Hamburg, Bundesstr.55, D-2000 Hamburg 13

*E-mail address:* am90080@dhhuni4.bitnet